New Technologies for Constructing Complex Agricultural and Environmental Systems

Petraq Papajorgji
European University of Tirana, Albania

François Pinet
Cemagref - Clermont Ferrand, France

Information Science
REFERENCE

Managing Director: Lindsay Johnston
Senior Editorial Director: Heather Probst
Book Production Manager: Sean Woznicki
Development Manager: Joel Gamon
Development Editor: Development Editor
Acquisitions Editor: Erika Gallagher
Typesetter: Jennifer Romanchak
Cover Design: Nick Newcomer, Lisandro Gonzalez

Published in the United States of America by
Information Science Reference (an imprint of IGI Global)
701 E. Chocolate Avenue
Hershey PA 17033
Tel: 717-533-8845
Fax: 717-533-8661
E-mail: cust@igi-global.com
Web site: http://www.igi-global.com

Library of Congress Cataloging-in-Publication Data

New technologies for constructing complex agricultural and environment systems / Petraq Papajorgji and Francois Pinet, editors.
p. cm.
Includes bibliographical references and index.
ISBN 978-1-4666-0333-2 (hardcover) -- ISBN 978-1-4666-0334-9 (ebook) -- ISBN 978-1-4666-0335-6 (print & perpetual access) 1. Agriculture--Information services. 2. Environmental sciences--Information services. 3. Information storage and retrieval systems. 4. Data warehousing. I. Papajorgji, Petraq J. II. Pinet, Francois, 1973-
S494.5.I47N49 2012
635.028'5--dc23
2011045660

British Cataloguing in Publication Data
A Cataloguing in Publication record for this book is available from the British Library.

All work contributed to this book is new, previously-unpublished material. The views expressed in this book are those of the authors, but not necessarily of the publisher.

Table of Contents

Section 2
Ontologies

Section 3
GIS-Based Systems and Spatial Analyses

Detailed Table of Contents

This paper examines the multidimensional modeling of a data warehouse for simulation results. Environmental dynamics modeling is used to study complex scenarios like urbanization, climate change and deforestation while allowing decision makers to understand and predict the evolution of the environment in response to potential value changes in a large number of influence variables. In this context, exploring simulation models produces a huge volume of data, which must often be studied extensively at different levels of aggregation due to there being a great need to define tools and methodologies specifically adapted for the storage and analysis of such complex data. Data warehousing systems provide technologies for managing simulation results from different sources. Moreover, OLAP technologies allow one to analyze and compare these results and their corresponding models. In this paper, the authors propose a generic multidimensional schema to analyze the results of a simulation model, which can guide modelers in designing specific data warehouses, and an adaptation of an OLAP client tool to provide an adequate visualization of data. As an example, a data warehouse for the analysis of results produced from a savanna simulation model is implemented using a Relational OLAP architecture.

This article describes Simexplorer, a computer framework for managing simulation experiments and, to some extent, the scientific quality of the modelling process. An information system, included in the framework, insures the traceability of the experiments and their reproducibility and thus contributes to the modelling process quality management. Moreover, this information system provides facilities for sharing and exchanging components of experiment scenarios. The authors illustrate the use of the framework on a simple example of modelling process.

Spatial OLAP (SOLAP) integrates spatial data into OLAP systems, and SOLAP models define spatial dimensions while measuring spatio-multidimensional operators. In this paper, the author presents the concepts of geographic and complex measures that allow integrating geographic and complex information as subjects of analysis in spatial data warehouses. The concept of geographic measure extends the concept of spatial measure to the semantic component of geographic information. The concept of complex measure allows introducing complex data as subjects of multidimensional analysis. To reduce the gap in flexibility between spatial and multidimensional analysis, this paper proposes a symmetrical representation of measures and dimensions. Additionally, the author presents a Web-based SOLAP prototype, GeWOlap, that enriches existing SOLAP tools by effectively and easily supporting symmetrical geographic/complex measures and dimensions for modeling and visualization. To validate this approach, the simulated environmental data concerning the pollution of the Venice lagoon is used.

Geospatial datacubes are the database backend of novel types of spatiotemporal decision-support systems employed in large organizations. These datacubes extend the datacube concept underlying the field of Business Intelligence (BI) into the realm of geospatial decision-support and geographic knowledge discovery. The interoperability between geospatial datacubes facilitates the reuse of their content. Such interoperability, however, faces risks of data misinterpretation related to the heterogeneity of geospatial datacubes. Although the interoperability of transactional databases has been the subject of several research works, no research dealing with the interoperability of geospatial datacubes exists. In this paper, the authors support the semantic interoperability between geospatial datacubes and propose a categorization of semantic heterogeneity problems that may occur in geospatial datacubes. Additionally, the authors propose an approach to deal with the related risks of data misinterpretation, which consists of evaluating the fitness-for-use of datacubes models, and a general framework that facilitates making appropriate decisions about such risks. The framework is based on a hierarchical top-down structure going from the most general level to the most detailed level, showing the usefulness of the proposed approach in environmental applications.

Section 2
Ontologies

The Health-e-Waterways Project is a multi-disciplinary collaboration between the University of Queensland, Microsoft Research and the South East Queensland Healthy Waterways Partnership (SEQ-HWP). This project develops the underlying technological framework and set of services to enable streamlined access to the expanding collection of real-time, near-real-time and static datasets related to water resource management in South East Queensland. More specifically, the system enables water resource managers to access the datasets being captured by the various agencies participating in the SEQ HWP Ecosystem Health Monitoring Program (EHMP). It also provides online access to the statistical data processing tools that enable users to analyse the data and generate online ecosystem report cards dynamically via a Web mapping interface. The authors examine the development of ontologies and semantic querying tools to integrate disparate datasets and relate management actions to water quality indicators for specific regions and periods. This semantic data integration approach enables scientists and resource managers to identify which actions are having an impact on which parameters and adapt the management strategies accordingly. This paper provides an overview of the semantic technologies developed to underpin the adaptive management framework that is the central philosophy behind the SEQ HWP.

This paper presents a research experiment that started in 2007 on the French "blue and green infrastructures". At the prompting of a French local administration, this experiment takes into account ecological stakes and involves actors in territorial planning at a metropolitan scale. Founded on previous works, the goal of the approach builds a participative workshop based on GIS and MCA coupling in order to think collectively and better understand the initial foundation of the method, the problems and proposed solutions, and the theoretical and the practical prerequisites. The final goal suggests methodological solutions to facilitate its generalization and its diffusion. The objectives are: (1) to analyse as objectively as possible the initial approach, methodological and technical choices, constraints, limits, and potential improvements; (2) to take into account questioning about the participative workshop conception and functioning; (3) to test new methodological proposals.

The protection and the maintenance of the exceptional environment of New Caledonia are major goals for this territory. Among environmental problems, erosion has a strong impact on terrestrial and coastal ecosystems. However, due to the volume of data and its complexity, assessment of hazard at a regional scale is time-consuming, costly and rarely updated. Therefore, understanding and predicting environmental phenomenons need advanced techniques of analysis and modelization. In order to improve the understanding of the erosion phenomenon, this paper proposes a spatial approach based on co-location mining and GIS. Considering a set of Boolean spatial features, the goal of co-location mining is to find subsets of features often located together. This system provides useful and interpretable knowledge based on a new interestingness measure for co-locations and a new visualization of the discovered knowledge. The interestingness measure better reflects the importance of a co-location for the experts, and is completely integrated in the mining process. The visualization approach is a simple, concise and intuitive representation of the co-locations that takes into consideration the spatial nature of the underlying objects and the experts practice.

This paper is composed of two parts dealing with the modeling of environmental phenomena. The first part presents the traditional ER and OO formalisms dedicated to geographic information modeling. These languages focus mainly on representing the spatial and temporal properties of this type of information. Many of these languages express these properties visually by using pictograms. After a quick historical presentation of the languages, the authors show the various types of spatiality and temporality usually encountered in these languages. Often qualified as primitive, some of these spatialities and temporalities are simple. Others, which are more complex, result from combinations of simple spatialities and simple temporalities. Still others are used in very specific situations encountered during the development of geographical information systems. These different spatialities and temporalities are presented via examples provided in the field of environmental dynamics.

This paper presents a theoretical and methodological framework to take into consideration the social dimension in a sustainable development project. To do this, the authors have developed the SocLab software environment, which implements a formalization of a well-established sociological theory, and enables the modeling of social organizations, to analyze their properties and to simulate social actors' behaviors. SocLab was used to assess the social acceptability of new agricultural practices more in line with the preservation of water resources and natural environments, in a well defined context. The paper shows how it was used and presents the main results.

Medium term management of territorial risk increase requires decision support tools able to represent and simulate spatial dynamics. For forest fire risk, simulators of spatial dynamics of both fuel zones and vulnerable zones are produced by specialised disciplines, like ecology and geography. In order to integrate simulators of spatial dynamics of different themes at different scale levels, this paper proposes a spatial agent and GIS based software infrastructure called Pyroxene. This infrastructure is designed to implement "models for models integration", specified by domain expert users. To do so, end users have to use a specification framework, specifically designed for spatio-dynamic models integration. The purpose of the implemented platform is to execute and synchronise integrated simulations. To operate the semantic and syntax requirements of such integration, the platform is structured in an HLA-like architecture, and implemented as a multi-agents system for models integration, compliant with the FIPA specification. It is also organised around a GIS. First steps of validation confirm the validity of the system at the functional level and precise the limits of using the approach on the operational level.

Preface

Agricultural and environmental related sciences face the daunting challenge of dealing with problems of an increasing number of factors of different nature that requires to be approached with a multidisciplinary and multi-institutional team. Addressing these complex scientific issues demands the refinement of existing, as well as the creation of new, analytical theories with potential uses in agriculture and the environment. Therefore, the scientific approach needs to combine the newest achievements in many scientific domains such as agronomy, mathematics, economics, and computer science to name a few.

The objective of this book is to present the novel work on the design and implementation of information systems in the field of agriculture and environment. The main goal is to present how the new technologies can improve the construction of these complex systems. The work presented in this book is selected from the International Journal of Agricultural and Environmental Information Systems (IJAEIS) vol.1 & 2. In this preface, we introduce the topics of these articles.

DATA WAREHOUSES

Recently, we have seen a large increase in sources of geo-referenced agricultural and environmental data:

- Certain data are obtained through sensors and remote detection systems – e.g. through systems integrated within spreaders to monitor work carried out, using networks of sensors and satellite images,
- Data can be produced from environmental simulation models,
- Other data are entered using specialized computer programs (e.g. using a computer application which records practices).

All this information requires effective storage methods, as well as effective integration and analysis methods. Data warehouses (DW) (Berson & Smith, 1997; Martin R., 2008; Rizzi, Abello, Lechtenborger, & Trujillo) are the most appropriate modern systems for applying such methods. This new type of database is used to integrate, accumulate and analyze data from various sources (Calì, Lembo, Lenzerini, & Rosati, 2003). Various information from different databases are integrated into a data warehouse for combined analysis (List, Bruckner, Machaczek, & Schiefer, 2002). Depending on requirements, this can be loaded every week, every month, every year or even less frequently. These data are usually organized in a form referred to as being multidimensional in order to facilitate the calculation of indicators by combining different criteria. The indicators are made up of aggregated information obtained by functions such as sum, average, variance etc. Data from a data warehouse can be combined to provide hitherto unknown causal links. For this purpose, users can view data from the data warehouse using OLAP (On-line Analytical Processing) type tools (Thomsen, 1997). Causal links can also be obtained automatically by using data-mining algorithms (Breault J., Goodall C., & P., 2002).

As indicated in (Pinet, 2010), the use of data warehouses is therefore important within a decision-making context. For example, a data warehouse containing economic, urban and environmental information can be used to help find the best place for establishing a new infrastructure. Data warehouses can supply data about algorithms for optimization or simulation (e.g. for optimizing the way agricultural work is organized). The concept of data warehouses is relatively recent and has great potential for assessing the impact of actions, practices, scenarios and programs both from a socio-economic as well as an environmental point of view (Nilakanta, Scheibe, & Rai, 2008; Pinet, et al., 2010; Schneider, 2008; Schulze C., Spilke J., & W., 2007)(Burmann & Gómez, 2007).

The papers of IJAEIS (Bimonte, 2010; Mahboubi, Bimonte, Faure, & Pinet, 2010; Sboui, Salehi, & Bédard, 2010) presented in this book deal with DW technologies.

The authors of (Mahboubi, et al., 2010) show that DW provides interesting technologies for managing and visualizing simulation results. This technology allows modelers to analyze and compare these results and their corresponding simulation models. (Mahboubi, et al., 2010) proposes a generic schema and an OLAP tool to analyze the results. The proposed schema can guide modelers in designing specific data warehouses and analyzing data. In the paper, the authors implement a data warehouse for the analysis of the savanna evolution. These simulations have been performed using SimExplorer, a tool also presented in IJAEIS (Chuffart, Dumoulin, Faure, & Deffuant, 2010). This tool is dedicated to facilitate the design of computer experiments on any simulation model. The joint use of DW technologies and SimExplorer allow modelers to obtain a traceability of simulations.

In (Bimonte, 2010), the author presents the implementation of a spatial DW-based tool called GeWOlap. GeWOlap allows users to model and visualize of complex and geographic indicators in a very flexible way. The author describes the use of the tool on simulated environmental data concerning the pollution of the Venice lagoon. He presents how the tool could be used to analyze water pollution phenomena in the Venice Lagoon in terms of several factors, such as time, pollutant, and location.

(Sboui, et al., 2010) deals with the issue related to spatial DW interoperability. The interoperability between spatial DW aims at facilitating the reuse of their content. In many situations, such as a simultaneous and rapid intervention in environmental emergencies using different spatial DW, decision-makers need to interoperate several heterogeneous data to support environmental applications.

ONTOLOGIES

As mentioned in (Papajorgji, Pinet, Miralles, Jallas, & Pardalos, 2010; Pinet, Roussey, Brun, & Vigier, 2009), Gruber defined an ontology as "a formal explicit specification of a shared conceptualization" (Gruber, 1993a). According to Gruber, conceptualization refers to a model of phenomena in the world after identifying the relevant concepts of these phenomena. Explicit means that the type of concepts used, and the constraints on their use are explicitly defined. Formal refers to the fact that the ontology should be machine readable. Shared reflects the idea that ontology should capture consensual knowledge accepted by the communities (Gruber, 1993b). Hendler defines also ontology as “a set of knowledge terms, including the vocabulary, the semantic interconnections, and some simple rules of inference and logic for some particular topic” (Hendler, 2001). Thus, an ontology specifies the semantics of relations between the modeled concepts and should enable some sort of reasoning.

Usually, one considers that an ontology represents a domain of knowledge. It can be used to formalize complex knowledge of various domains. Ontologies can be represented in various forms (diagrams, formal models, thesauruses, etc.). They should contain:

- A vocabulary of terms,
- A set of term definitions which identify concepts and fix the term interpretation (in the domain of knowledge considered),
- A representation of relationships between concepts,
- An agreement of a community of ontology users about term definitions and the ontology structure.

The paper of IJAEIS (Hunter, Becker, Alabri, Ingen, & Abal, 2011) focuses on the use of ontologies to relate resource management actions to environmental monitoring data in South East Queensland. The authors propose ontologies and tools to integrate different datasets and relate management actions to water quality indicators. This paper describes the Health-e-Waterways Project, which is a cross-disciplinary collaboration between information scientists, water resource managers and stakeholders. It provides an overview of the ontologies that have been developed, the system architecture and the query, reporting and visualization interfaces - that are enabling scientists and policy makers to identify trends in ecosystem health indicators both geographically and temporally and to understand the impact that water resource management actions are having on water quality. This data integration approach enables scientists and resource managers to identify which actions are having an impact on which parameters and to adapt the management strategies accordingly.

In the paper “A Conceptual Model of Grassland-Based Beef Systems” (Martin R., 2008), authors address the challenging issue of designing management strategies with regard to cattle demand. The article describes the application of an ontology of agricultural production systems to the generic conceptual model SEDIVER that supports the representation and dynamic farm-scale simulation of specific grassland-based beef systems. The most salient and novel aspects of SEDIVER concern the explicit modeling of (a) the diversity in plant, grassland, animal and farmland, (b) management strategies that deal with the planning and coordination of activities whereby the farmer controls the biophysical processes. By using the SEDIVER conceptual framework, one can expect to capture part of the subjective and context-specific knowledge used in farm management and, in this way, enable scientific investigation of management practices.

(Karetsos, Haralampopoulos, & Kotis, 2011) presents an ontology-based method for the production of learning designs. These works are applied on the domain of sustainable energy education. The proposed framework includes an ontology of the sustainable energy development domain and an educational model designed in compliance with widespread standards. The authors envisage this framework both as a means to support the authoring of learning scenarios and as a provisioning of a field for conversation about which should be the appropriate form of an authoring tool in this area.

GIS-BASED SYSTEMS AND SPATIAL ANALYSES

Nowadays, numerous environmental and agricultural data are geo-referenced (Miralles, Pinet, & Bedard, 2010). They can be generated by remote sensing systems or other computer applications. In order to take benefits of these valuable data, some new solutions propose to integrate them into Geographical Information Systems (GIS) (Goodchild, et al., 1996; Laurini & Thompson, 1992). These systems allow taking into account characteristics of these complex data and phenomena (such as semantics, spatial components, etc.).

The papers published in IJAEIS (Bimonte, 2010) and (Sboui, et al., 2010), cited above, provide two examples of GIS-based technologies and spatial analyzes applied to environment.

The paper "Application of Support Vector Machines to Melissopalynological Data for Honey Classification" (Aronne, De Micco, & Guarracino, 2010) addresses the problem of the discrimination of geographical origin and the selection of marker species of honeys using Support Vector Machines and z-scores (Louveaux et al. 1978). The methodology is based on the elaboration of palynological data with statistical learning methodologies. This innovative solution provides a simple yet powerful tool to detect the origin of honey samples. In case of honeys from Sorrento Peninsula, the discrimination from other Italian honeys is obtained with high accuracy.

The authors of the IJAEIS paper "Urban versus Rural: The Decrease of Agricultural Areas and the Development of Urban Zones Analyzed with Spatial Statistics" (Murgante & Danese, 2011) use spatial statistics and point pattern analysis to characterize the "periurban areas". They apply their method to the Potenza Municipality in Italy.

The paper entitled "Coupling Geographic Information System (GIS) and Multi-Criteria Analysis (MCA) for Modelling the Ecological Continuum in Participative Territorial Planning" (Batton Hubert, Bonnevialle, Joliveau, Mazagol, & Paran, 2011) concerns participative territorial planning at a metropolitan scale taking into account ecological stakes. GIS and MCA techniques are proposed to take into account participative issues and to test new methodological proposals.

The paper "Spatial Pattern Mining for Soil Erosion Characterization" (Selmaoui-Folcher, Flouvat, Gay, & Rouet, 2011) describes the protection and the maintenance of the environment of New Caledonia. Authors state that among environmental problems, erosion has a strong impact on terrestrial and coastal ecosystems. However, due to the volume of data and their complexity, assessment of hazard at a regional scale is time-consuming, costly and rarely updated. Therefore, understanding and predicting environmental phenomenons need advanced techniques of analysis and modelization. In order to improve the understanding of the erosion phenomenon, they propose in this paper a spatial approach based on co-location mining and GIS.

Entity-Relationship (ER) (Chen, 1976) and Object-Oriented (OO) formalisms (Din & Idris, 2009) are widely used to describe environmental information of databases or computer programs. Numerous recent examples are available - see for example: (Almeida, Martins Ferreira, Eiras, Obermayr, & Geier, 2010; Anselme, et al., 2010; Asseng, Dray, Perez, & Su, 2010; Bimonte, 2010; Campo, Bousquet, & Villanueva, 2010; Farolfi, Müller, & Bonté, 2010; Goodall, Fay, & Bollinger Jr, 2010; Kraft, Vaché, Frede, & Breuer, 2011; Lagabrielle, et al., 2010; Lenz-Wiedemann, Klar, & Schneider, 2010; Merot & Bergez, 2010; Miralles, Pinet, & Bedard, 2010; Moglia, Perez, & Burn, 2010; Papajorgji & Pardalos, 2006; Papajorgji, Pinet, Miralles, Jallas, & Pardalos, 2010; C. Parent, S. Spaccapietra, & E. Zimanyi, 2006; C. Parent, S. Spaccapietra, & E. Zimányi, 2006; Perez & Dragicevic, 2010; Pinet, 2010; Pinet, et al., 2009; Pinet, Duboisset, & Soulignac, 2007; Pinet, et al., 2010; Raffaetà, et al., 2008; Saqalli, Gérard, Bielders,

& Defourny, 2010; Simon & Etienne, 2010; Spaccapietra, Parent, & Zimányi, 2007; Stempliuc, Lisboa Filho, Andrade, & Borges, 2009). For 20 years, several research teams adapted ER and OO to facilitate the modeling of spatio-temporal information (Bédard, Larrivee, Proulx, & Nadeau, 2004; Bédard & Paquette, 1989; Miralles, Libourel, Papajorgji, & Pardalos, 2009a, 2009b; Pinet, Kang, & Vigier, 2005). Because this type of information is complex, one goal of researchers is to propose specific notations to clarify its representation in models. According to the experiments presented in (Parent, et al., 1998), using a formalism specifically dedicated to spatio-temporal information allows for a 22% reduction in the number of entities and relationships in an ER diagram (without losing semantics), compared to a traditional ER model. The UML-based formalism called Plug-in for Visual Languages (PVL) is a well-known method to model spatio-temporal information (Bédard, 1999, 2009). The IJAEIS paper (Miralles, et al., 2010) provides a PVL tutorial. The authors illustrate PVL on different environmental examples.

DECISION SUPPORT SYSTEMS

The paper "Using Soclab for a Rigorous Assessment of the Social Feasibility of Agricultural Policies" (Adreit, Roggero, Sibertin-Blanc, & Vautier, 2011) presents a theoretical and methodological framework to take into consideration the social dimension in a sustainable development project. To do this, the authors have developed the SocLab software environment, which implements a formalization of a well-established sociological theory, and enables to model social organizations, to analyze their properties and to simulate social actors' behaviors. The authors used SocLab to assess the social acceptability of new agricultural practices more in line with the preservation of water resources and natural environments, in a well-defined context. The paper shows how they used it and presents the main results.

The paper "Pyroxene: A Territorial Decision Support System Based on Spatial Simulators Integration for Forest Fire Risk Management" (Maillé & Espinasse, 2011) introduces an architecture, called Pyroxene, dedicated to the integration, the execution and the synchronization of simulation models. The final goal of Pyroxene is to help decision makers to manage forest fire risks. Pyroxene is a multi-agents system. It takes into account models for the ecosystems dynamics, the urban dynamics and the forest fire risks.

In the paper "On the Use of Abduction as an Alternative to Decision Trees in Environmental Decision Support Systems " (Wotawa, 2011), the author discusses the use of rule-based systems and decision trees for representing the knowledge used in decision support systems. According to the author, neither decision trees nor rule-based systems are very well suited for knowledge representation for decision support systems and in particular diagnostic systems. As an alternative he proposes the use of abductive diagnosis, which allows for deriving root causes from effects where the underlying knowledge base represents cause effect relationships directly. Such models are usually available in the natural sciences. In order to show the use of abductive diagnosis in the environmental field, the author uses a model that was used for diagnosing a wastewater treatment plant.

SURVEY ON NEW COMPUTER-BASED TECHNIQUES APPLIED TO AGRICULTURE AND ENVIRONMENT

Three papers present surveys on computer-based technologies related to object-oriented and UML-based modeling, data mining and crop systems simulation.

The paper "Modeling: A Central Activity for Flexible Information Systems Development in Agriculture and Environment" (Papajorgji, et al., 2010) aims to show author's vision on using model-based approaches to design complex and flexible agricultural and environmental information systems. At the center of this modeling approach is the Unified Modeling Language that facilitates expressing visually concepts of a problem domain and their relationships (Papajorgji, 2005; Papajorgji, 2009; Papajorgji & Clark, 2005; Papajorgji & Pardalos, 2006; Papajorgji & Shatar, 2004; Pinet, et al., 2007). UML has a core of notations that are generic and that can be used to model problems in any domain (OMG, 2009). Furthermore, UML can be extended to create profiles in order to take into consideration modeling concerns in a particular problem domain. UML profiles are created to use UML in designing spatial systems, ontologies and model driven architecture-based systems. A UML profile is created to design web-based systems and a recent profile makes it possible to use UML for business modeling purposes. The paper presents the state of the art in modeling agricultural and environmental systems and provides discussions for future directions.

In the paper "Data Mining Techniques in Agricultural and Environmental Sciences" (Chinchuluun, Xanthopoulos, Tomaino, & Pardalos, 2011) authors show their vision on the importance of knowing and efficiently using data mining and machine learning-related techniques for knowledge discovery in the field of agriculture and environment. Data mining is the process of extracting important and useful information from large sets of data. This information can be converted into useful knowledge that could help to better understand the problem in study and to better predict future developments.

The paper, "The Role of Crop Systems Simulation in Agriculture and Environment" (Boote, Jones, Hoogenboom, & White, 2011) describes the evolution of the simulation of crop systems in the last 30-40 years. The goal of this paper is to give author's vision on how crop systems simulation can serve important future roles in agriculture and environment and to prioritize research to better meet supporting these roles. The most important roles and uses of crop systems simulation are seen in five primary areas: 1) Basic research synthesis and integration, where simulation is used to integrate and synthesize our understanding of physiology, genetics, soil characteristics, management, and weather effects, 2) Strategic tools for research planning/policy to evaluate strategies and consequences of genetic improvement or management of resources or decisions to produce biofuel crops rather than food crops, 3) Applications for management purposes, where crop systems simulations are used to evaluate impacts of climate variability on production, consequences of weather and nutrient management on water use and nutrient use, consequences on economics, water use, and nutrient leaching, 4) Real time decision support to assist in management decisions (irrigation, N fertilization, sowing date, projected harvest, yield forecast, pest management), and 5) Education both in class rooms and extension contexts, to explain how crop systems function and are managed. Authors see a good potential to link crop models to molecular genetics, in effect, modeling from knowledge of genes of different cultivars to phenotypic performance in different environments. They see a continuing need to improve the crop models for simulating root growth and nutrient uptake, coupling of diseases and pests, fully-coupled energy balance, and response to climate change.

Petraq Papajorgji
European University of Tirana, Albania

François Pinet
Cemagref - Clermont Ferrand, France

REFERENCES

Adreit, F., Roggero, P., Sibertin-Blanc, C., & Vautier, C. (2011). Using Soclab for a Rigorous Assessment of the Social Feasibility of Agricultural Policies. *International Journal of Agricultural and Environmental Information Systems*, *2*(2). doi:10.4018/jaeis.2011070101

Almeida, S. J. d., Martins Ferreira, R. P., Eiras, Á. E., Obermayr, R. P., & Geier, M. (2010). Multi-agent modeling and simulation of an Aedes aegypti mosquito population. *Environmental modelling & software*, *25*(12), 1490-1507. doi:10.1016/j.envsoft.2010.04.021

Anselme, B., Bousquet, F., Lyet, A., Etienne, M., Fady, B., & Le Page, C. (2010). Modelling of spatial dynamics and biodiversity conservation on Lure mountain (France). *Environmental modelling & software, 25*(11), 1385-1398. doi:10.1016/j.envsoft.2009.09.001

Aronne, G., De Micco, V., & Guarracino, M. (2010). Application of Support Vector Machines to Melissopalynological Data for Honey Classification. *International Journal of Agricultural and Environmental Information Systems*, *1*(2), 85–94. doi:10.4018/jaeis.2010070105

Asseng, S., Dray, A., Perez, P., & Su, X. (2010). Rainfall-human-spatial interactions in a salinity-prone agricultural region of the Western Australian wheat-belt. *Ecological Modelling, 221*(5), 812-824. doi: 10.1016/j.ecolmodel.2009.12.001

Batton Hubert, M., Bonnevialle, M., Joliveau, T., Mazagol, P.-O., & Paran, F. (2011). Coupling Geographic Information System (GIS) and Multi-Criteria Analysis (MCA) for Modelling the Ecological Continuum in Participative Territorial Planning. *International Journal of Agricultural and Environmental Information Systems*, *2*(2), 29–51. doi:10.4018/jaeis.2011070103

Bédard, Y. (1999). Visual Modelling of Spatial Database towards Spatial PVL and UML. *Geomatica*, *53*(2), 169–185.

Bédard, Y. (2009). Perceptory Web site. <http://sirs.scg.ulaval.ca/Perceptory/>

Bédard, Y., Larrivee, S., Proulx, M. J., & Nadeau, M. (2004). Modeling geospatial databases with plug-ins for visual languages: A pragmatic approach and the impacts of 16 years of research and experimentations on perceptory. *Lecture Notes in Computer Science*, *3289*, 17–30. doi:10.1007/978-3-540-30466-1_3

Bédard, Y., & Paquette, F. (1989). *Extending entity/relationship formalism for spatial information systems.* Paper presented at the AUTO-CARTO 9, Baltimore.

Berson, A., & Smith, S. (1997). *Data Warehousing, Data Mining, and OLAP (Data Warehousing/Data Management).* Computing Mcgraw-Hill.

Bimonte, S. (2010). A Web-Based Tool for Spatio-Multidimensional Analysis of Geographic and Complex Data. *International Journal of Agricultural and Environmental Information Systems, 1*(2), 42–67. doi:10.4018/jaeis.2010070103

Bimonte, S. (2010). A Web-Based Tool for Spatio-Multidimensional Analysis of Geographic and Complex Data. *International Journal of Agricultural and Environmental Information Systems, 1*(2), 42–67. doi:10.4018/jaeis.2010070103

Boote, K. J., Jones, J. W., Hoogenboom, G., & White, J. W. (2011). The Role of Crop Systems Simulation in Agriculture and Environment. *International Journal of Agricultural and Environmental Information Systems, 1*(1), 41–54. doi:10.4018/jaeis.2010101303

Breault, J., & Goodall, C., & P., F. (2002). Data mining a diabetic data warehouse. *Artificial Intelligence in Medicine, 26*(1), 37–54. doi:10.1016/S0933-3657(02)00051-9

Burmann, A., & Gómez, J. (2007). *Data Warehousing with Environmental Data* (pp. 153–160). Information Technologies in Environmental Engineering.

Calì, A., Lembo, D., Lenzerini, M., & Rosati, R. (2003). Source Integration for Data Warehousing *Multidimensional Databases: Problems and Solutions* (pp. 361-392).

Campo, P. C., Bousquet, F., & Villanueva, T. R. (2010). Modelling with stakeholders within a development project. *Environmental modelling & software, 25*(11), 1302-1321. doi: 10.1016/j.envsoft.2010.01.005

Chen, P. (1976). The Entity-Relationship Model: Toward a Unified View of Data. *ACM Transactions on Database Systems, 1*(1), 9–35. doi:10.1145/320434.320440

Chinchuluun, A., Xanthopoulos, P., Tomaino, V., & Pardalos, P. M. (2011). Data Mining Techniques in Agricultural and Environmental Sciences. *International Journal of Agricultural and Environmental Information Systems, 1*(1), 26–40. doi:10.4018/jaeis.2010101302

Chuffart, F., Dumoulin, N., Faure, T., & Deffuant, G. (2010). SimExplorer: Programming Experimental Designs on Models and Managing Quality of Modelling Process. *International Journal of Agricultural and Environmental Information Systems, 1*(1), 55–68. doi:10.4018/jaeis.2010101304

Din, J., & Idris, S. (2009). Object-Oriented Design Process Model . *International Journal of Computer Science and Network Security, 9*(10), 71–79.

Farolfi, S., Müller, J.-P., & Bonté, B. (2010). An iterative construction of multi-agent models to represent water supply and demand dynamics at the catchment level. [doi: doi: 10.1016/j.envsoft.2010.03.018]. *Environmental modelling & software, 25*(10), 1130-1148.

Goodall, J. L., Fay, J. P., & Bollinger Jr, D. L. (2010). A software library for quantifying regional-scale nitrogen transport within river basin systems. *Environmental modelling & software, 25*(12), 1713-1721. doi: 10.1016/j.envsoft.2010.04.007

Goodchild, M. F., Steyaert, L. T., Parks, B. O., Johnston, C., Maidment, D., Crane, M., et al. (1996). *GIS and environmental modeling: progress and research issues.*

Gruber, R. (1993a). *Towards Principles for the Design of Ontologies Used for Knowledge Sharing. In International Workshop on Formal Ontology, Padova, Italy, 1993. Available as technical report KSL-93-04.*

Gruber, R. (1993b). A Translation Approach to Portable Ontology Specifications. *Knowledge Acquisition, 5*(2), 199–220. doi:10.1006/knac.1993.1008

Hendler, J. (2001). Agents and the semantic web. *IEEE Intelligent Systems and Their Applications, 16*(2), 30–37. doi:10.1109/5254.920597

Hunter, J., Becker, P., Alabri, A., Ingen, C., & Abal, E. (2011). Using Ontologies to Relate Resource Management Actions to Environmental Monitoring Data in South East Queensland. *International Journal of Agricultural and Environmental Information Systems, 2*(1), 1–19. doi:10.4018/jaeis.2011010101

Karetsos, S., Haralampopoulos, D., & Kotis, K. (2011). An Ontology-Based Framework for Authoring Tools in the Domain of Sustainable Energy Education . *International Journal of Agricultural and Environmental Information Systems, 2*(1), 40–62. doi:10.4018/jaeis.2011010103

Kraft, P., Vaché, K. B., Frede, H.-G., & Breuer, L. (2011). CMF: A Hydrological Programming Language Extension For Integrated Catchment Models. *Environmental modelling & software, 26*(6), 828-830. doi: 10.1016/j.envsoft.2010.12.009

Lagabrielle, E., Botta, A., Daré, W., David, D., Aubert, S., & Fabricius, C. (2010). Modelling with stakeholders to integrate biodiversity into land-use planning - Lessons learned in Réunion Island (Western Indian Ocean). *Environmental modelling & software, 25*(11), 1413-1427. doi: 10.1016/j.envsoft.2010.01.011

Laurini, R., & Thompson, D. (1992). Fundamentals of spatial information systems. *Fundamentals of spatial information systems.*

Lenz-Wiedemann, V. I. S., Klar, C. W., & Schneider, K. (2010). Development and test of a crop growth model for application within a Global Change decision support system. *Ecological Modelling, 221*(2), 314-329. doi: 10.1016/j.ecolmodel.2009.10.014

List, B., Bruckner, R. M., Machaczek, K., & Schiefer, J. (2002). A Comparison of Data Warehouse Development Methodologies Case Study of the Process Warehouse. *Lecture Notes in Computer Science, 2453*, 203–215. doi:10.1007/3-540-46146-9_21

Louveaux, J., Maurizio, A., & Vorwohl, G. (1978). Methods of Melissopalynology . *Bee World, 59*, 139–157.

Mahboubi, H., Bimonte, S., Faure, T., & Pinet, F. (2010). Data warehouse and OLAP for Environmental Simulation Data. *International Journal of Agricultural and Environmental Systems, 1*(2), 1–19. doi:10.4018/jaeis.2010070101

Maillé, E., & Espinasse, B. (2011). Pyroxene: a Territorial Decision Support System Based on Spatial Simulators Integration for Forest Fire Risk Management. *International Journal of Agricultural and Environmental Information Systems*, *2*(2), 73–92. doi:10.4018/jaeis.2011070104

Martin, R. (2008). *Data Warehouse 100 Success Secrets - 100 most Asked questions on Data Warehouse Design, Projects, Business Intelligence, Architecture, Software and Models*. Emereo Pty Ltd.

Merot, A., & Bergez, J. E. (2010). IRRIGATE: A dynamic integrated model combining a knowledge-based model and mechanistic biophysical models for border irrigation management. *Environmental modelling & software, 25*(4), 421-432. doi: 10.1016/j.envsoft.2009.11.003

Miralles, A., Libourel, T., Papajorgji, P. J., & Pardalos, P. M. (2009a). *Application of a Model Transformation Paradigm in Agriculture: A simple environmental system case study Advances in Modelling Agricultural Systems* (pp. 37–54). Springer.

Miralles, A., Libourel, T., Papajorgji, P. J., & Pardalos, P. M. (2009b). *A new methodology to automate the transformation of GIS models in an iterative development process Advances in Modelling Agricultural Systems* (*Vol. 25*, pp. 19–36). Springer.

Miralles, A., Pinet, F., & Bedard, Y. (2010). Describing Spatio-Temporal Phenomena for Environmental System Development: An Overview of Today's Needs and Solutions. *International Journal of Agricultural and Environmental Systems*, *1*(2), 68–84. doi:10.4018/jaeis.2010070104

Miralles, A., Pinet, F., & Bedard, Y. (2010). Describing Spatio-Temporal Phenomena for Environmental System Development: An Overview of Today's Needs and Solutions. *International Journal of Agricultural and Environmental Systems*, *1*(2), 68–84. doi:10.4018/jaeis.2010070104

Moglia, M., Perez, P., & Burn, S. (2010). Modelling an urban water system on the edge of chaos. *Environmental modelling & software, 25*(12), 1528-1538. doi: 10.1016/j.envsoft.2010.05.002

Murgante, B., & Danese, M. (2011). Urban versus Rural: the decrease of agricultural areas and the development of urban zones analyzed with spatial statistics. *International Journal of Agricultural and Environmental Information Systems*, *2*(2), 16–28. doi:10.4018/jaeis.2011070102

Nilakanta, S., Scheibe, K., & Rai, A. (2008). Dimensional issues in agricultural data warehouse designs. *Computers and Electronics in Agriculture*, *60*(2), 263–278. doi:10.1016/j.compag.2007.09.009

OMG. (2009). *Unified Modeling Language (OMG UML), Infrastructure Version 2.2.*

Papajorgji, P. (2005). A plug and play approach for developing environmental models. *Environmental Modelling & Software*, *20*(10), 1353–1357. doi:10.1016/j.envsoft.2004.11.016

Papajorgji, P. (2009). State of the Art in Modeling Agricultural Systems *Encyclopedia of Optimization* (pp. 3693-3704).

Papajorgji, P., & Clark, R. (2005). *A Model Driven Approach to Agricultural Systems (slides), EFITA 2005, Portugal.*

Papajorgji, P., & Pardalos, P. M. (2006). *Software Engineering Techniques Applied to Agricultural Systems An Object-Oriented and UML Approach*. Springer.

Papajorgji, P., Pinet, F., Miralles, A., Jallas, E., & Pardalos, P. M. (2010). Modeling: a central activity for flexible information systems development in agriculture and environment. *International Journal of Agricultural and Environmental Information Systems*, *1*(1), 1–25. doi:10.4018/jaeis.2010101301

Papajorgji, P., Pinet, F., Miralles, A., Jallas, E., & Pardalos, P. M. (2010). Modeling: a central activity for flexible information systems development in agriculture and environment. *International Journal of Agricultural and Environmental Information Systems*, *1*(1), 1–25. doi:10.4018/jaeis.2010101301

Papajorgji, P., & Shatar, T. M. (2004). Using the Unified Modeling Language to develop soil water-balance and irrigation-scheduling models. *Environmental Modelling & Software*, *19*(5), 451–459. doi:10.1016/S1364-8152(03)00160-9

Parent, C., Spaccapietra, S., & Zimanyi, E. (2006). *Conceptual Modeling for Traditional and Spatio-temporal Applications*. Springer.

Parent, C., Spaccapietra, S., & Zimányi, E. (2006). The MurMur project: Modeling and querying multi-representation spatio-temporal databases. *Information Systems*, *31*(8), 733–769. doi:10.1016/j.is.2005.01.004

Parent, C., Spaccapietra, S., Zimanyi, E., Donini, P., Plazanet, C., & Vangenot, C. (1998). *Modeling Spatial Data in the MADS Conceptual Model.* Paper presented at the the International Symposium on Spatial Data Handling, SDH 98, Vancouver, Canada, July 11-15,1998.

Perez, L., & Dragicevic, S. (2010). Modeling mountain pine beetle infestation with an agent-based approach at two spatial scales. *Environmental modelling & software, 25*(2), 223-236. doi: 10.1016/j.envsoft.2009.08.004

Pinet, F. (2010). Precise Design of Environmental Data Warehouses. *Operations Research*, *10*(3), 349–369.

Pinet, F., Duboisset, M., Demuth, B., Schneider, M., Soulignac, V., & Barnabe, F. (2009). *Constraints modeling in Agricultural Databases Advances in Modeling Agricultural Systems*. Springer.

Pinet, F., Duboisset, M., & Soulignac, V. (2007). Using UML and OCL to maintain the consistency of spatial data in environmental information systems. *Environmental Modelling & Software*, *22*(8), 1217–1220. doi:10.1016/j.envsoft.2006.10.003

Pinet, F., Kang, M. A., & Vigier, F. (2005). Spatial constraint modelling with a GIS extension of UML and OCL: application to agricultural information systems. *Lecture Notes in Computer Science*, *3511*, 160–178. doi:10.1007/11518358_13

Pinet, F., Miralles, A., Bimonte, S., Vernier, F., Carluer, N., Gouy, V., et al. (2010). *The use of UML to design agricultural data warehouses.* Paper presented at the International Conference on Agricultural Engineering, AGENG 2010.

Pinet, F., Miralles, A., Bimonte, S., Vernier, F., Carluer, N., Gouy, V., et al. (2010). *The use of UML to design agricultural data warehouses.* Paper presented at the International Conference on Agricultural Engineering, AGENG 2010.

Pinet, F., Roussey, C., Brun, T., & Vigier, F. (2009). *The Use of UML as a Tool for the Formalisation of Standards and the Design of Ontologies in Agriculture Advances in Modeling Agricultural Systems*. Springer.

Raffaetà, A., Ceccarelli, T., Centeno, D., Giannotti, F., Massolo, A., & Parent, C. (2008). An application of advanced spatio-temporal formalisms to behavioural ecology. *GeoInformatica, 12*(1), 37–72. doi:10.1007/s10707-006-0016-6

Rizzi, S., Abello, A., Lechtenborger, J., & Trujillo, J. *Research in data warehouse modeling and design: dead or alive?* Paper presented at the Proceedings of the 9th ACM international workshop on Data warehousing and OLAP.

Saqalli, M., Gérard, B., Bielders, C., & Defourny, P. (2010). Testing the impact of social forces on the evolution of Sahelian farming systems: A combined agent-based modeling and anthropological approach. *Ecological Modelling, 221*(22), 2714-2727. doi: 10.1016/j.ecolmodel.2010.08.004

Sboui, T., Salehi, M., & Bédard, Y. (2010). A systematic approach for managing the risk related to semantic interoperability between geospatial datacubes. *International Journal of Agricultural and Environmental Information Systems, 1*(2), 20–41. doi:10.4018/jaeis.2010070102

Schneider, M. (2008). A general model for the design of data warehouses. *International Journal of Production Economics, 112*(1), 309–325. doi:10.1016/j.ijpe.2006.11.027

Schulze, C., & Spilke, J., & W., L. (2007). Data modeling for Precision Dairy Farming within the competitive field of operational and analytical tasks. *Computers and Electronics in Agriculture, 59*(1), 39–55. doi:10.1016/j.compag.2007.05.001

Selmaoui-Folcher, N., Flouvat, F., Gay, D., & Rouet, I. (2011). Spatial Pattern Mining for Soil Erosion Characterization. *International Journal of Agricultural and Environmental Information Systems, 2*(2), 73–92. doi:10.4018/jaeis.2011070105

Simon, C., & Etienne, M. (2010). A companion modelling approach applied to forest management planning. *Environmental modelling & software, 25*(11), 1371-1384. doi: 10.1016/j.envsoft.2009.09.004

Spaccapietra, S., Parent, C., & Zimányi, E. (2007). *Spatio-temporal and Multi-representation Modeling: A Contribution to Active Conceptual Modeling* (pp. 194–205). Active Conceptual Modeling of Learning.

Stempliuc, S. M., Lisboa Filho, J., Andrade, V., & Borges, K. (2009). *Extending the UML-GeoFrame data model for conceptual modeling of network applications.* Paper presented at the International Conference on Enterprise Information Systems, Milão, Italy.

Thomsen, E. (1997). *OLAP Solutions: Building Multidimensional Information Systems*. John Wiley & Sons.

Wotawa, F. (2011). On the use of abduction as an alternative to decision trees in environmental decision support systems. *International Journal of Agricultural and Environmental Information Systems, 2*(1), 63–82. doi:10.4018/jaeis.2011010104

Section 1
Data Warehouses and Simulation Results

Chapter 1
A Multidimensional Model for Data Warehouses of Simulation Results

Hadj Mahboubi
Cemagref, France

Thierry Faure
Cemagref, France

Sandro Bimonte
Cemagref, France

Guillaume Deffuant
Cemagref, France

Jean-Pierre Chanet
Cemagref, France

François Pinet
Cemagref, France

ABSTRACT

This paper examines the multidimensional modeling of a data warehouse for simulation results. Environmental dynamics modeling is used to study complex scenarios like urbanization, climate change and deforestation while allowing decision makers to understand and predict the evolution of the environment in response to potential value changes in a large number of influence variables. In this context, exploring simulation models produces a huge volume of data, which must often be studied extensively at different levels of aggregation due to there being a great need to define tools and methodologies specifically adapted for the storage and analysis of such complex data. Data warehousing systems provide technologies for managing simulation results from different sources. Moreover, OLAP technologies allow one to analyze and compare these results and their corresponding models. In this paper, the authors propose a generic multidimensional schema to analyze the results of a simulation model, which can guide modelers in designing specific data warehouses, and an adaptation of an OLAP client tool to provide an adequate visualization of data. As an example, a data warehouse for the analysis of results produced from a savanna simulation model is implemented using a Relational OLAP architecture.

DOI: 10.4018/978-1-4666-0333-2.ch001

1. INTRODUCTION

Data warehousing, combined with OLAP (Online Analytical Processing) technologies, provides an innovative support for business intelligence (Codd et al., 1994). It has recently become a topic of great interest in the business world as well as in the research community. Typical OLAP applications concern marketing and organization monitoring (Kimball & Ross, 2002). However, new fields of application have been studied such as medical diagnosis (Bentayeb et al., 2010; Arigon et al., 2007) and web data and environmental data monitoring (Boussaid et al., 2008; Bimonte et al., 2005). The success of these technologies is due to several factors. First, they provide versioning and centralization of huge amounts of heterogeneous historical data. Second, decision-makers can easily explore, analyze and compare these data with user-friendly tools. These systems provide OLAP operators (which aggregate and select data) by manipulating pivot tables and graphic displays.

On the other hand, environmental dynamics modeling is extensively used to study complex phenomena and scenarios, such as urbanization, climate change and deforestation. This modeling allows researchers and stakeholders to understand and predict the evolution of the environment in response to changes in a large number of influence indicators (i.e., input data).

These models are complex because they contain several sub-models in interaction. They often involve coupled models (biological, meteorological and/or hydrological models) or individual-based models, which explicitly represent individuals of a given population (plants, animals or human). They can output a huge volume of data because they include such a high number of objects and variables, especially when simulations run for long periods of time. Moreover, many of these models include stochastic processes, which require making several replications of each simulation to get representative results. This again increases the quantity of result data for storage and analysis. One way to deal with this situation is to use statistical methods of experimental design to define the most efficient settings for simulation experiments (Saltelli, 2000 or Kleijinen, 2008 for review). Another possibility is to exploit large computing architectures (grid architectures or distributed computation infrastructures) (Chuffart et al., 2008).

To manage huge amounts of simulation result data, scientists or "model users" need to extract certain condensed data, such as typical regularities or synthetic indicators. In addition, researchers and stakeholders need to perform comparative analyses of results issued from different models and their inputs. For that purpose we need tools to extract and construct regularities and indicators as well as to analyze and compare models.

The modelers often need to compare results issued from different models or results produced by different input sets. These needs are important in the context of environmental data and models. These models can be very different, as they are created by various scientists. They can also be based on different modeling paradigms (e.g., partial differential equation, discrete event, cellular automata, and other complex modeling frameworks). Data used or produced by these models can thus be very diverse. For example, some are continuous, whereas others are discrete; thus, a storage system taking into account of all these heterogeneities should be highly flexible.

Motivated by these needs, in this paper we propose a methodology for warehousing and analyzing simulation results. This methodology aims at including important facilities for modelers, such as the ability to estimate parameter values, to redo explorations and to produce analysis reports. We also define and provide a general multidimensional data schema (i.e., a warehouse data model) that is derived from a general-purpose conceptual data warehouse schema. Our work takes into account the specificities of simulation results (data type heterogeneity) and can guide modelers in defining their warehouse schemas according to their analytic needs. We also propose an extension of an OLAP client to allow decision-makers to analyze, explore, validate and compare results produced from simulation models.

To highlight the benefits of our proposal, we have applied our methodology to a savanna simulation model (Jeltsch et al., 1996, 1998, 1999). This model includes the dynamics of various types of vegetation in savannas (trees, grass, bushes), coupled with dynamics of fire and scenarios for precipitation. The model is spatially explicit, and output results consist of numerical and raster data.

The remainder of this paper is organized as follows. First, we introduce the main concepts of the data warehouse, OLAP systems and simulation models. Then, we present our methodology for warehousing simulation results. We describe its steps and define our simulation result warehouse meta-model. After that, we illustrate the practical use of this methodology with an environmental case study. In the last section, we present conclusions and future work.

2. BACKGROUND

2.1. Data Warehousing and OLAP Technologies

Data warehouses are subject-oriented, integrated, time-variant and non-volatile collections of data in support of management decision processes (Inmon, 2005). They constitute effective supports for integrating and managing a huge volume of data from heterogeneous sources, and for providing multidimensional analysis (Inmon, 2005). As shown in Figure 1, a typical data warehousing architecture consists of four levels:

- **Back-end level:** The back-end level is composed of operational databases and other sources that are integrated into the data warehouse through the *Extraction, Transformation and Loading* (ETL) process. The extraction step gathers data from multiple and heterogeneous sources. The

Figure 1. Data warehousing and OLAP architecture

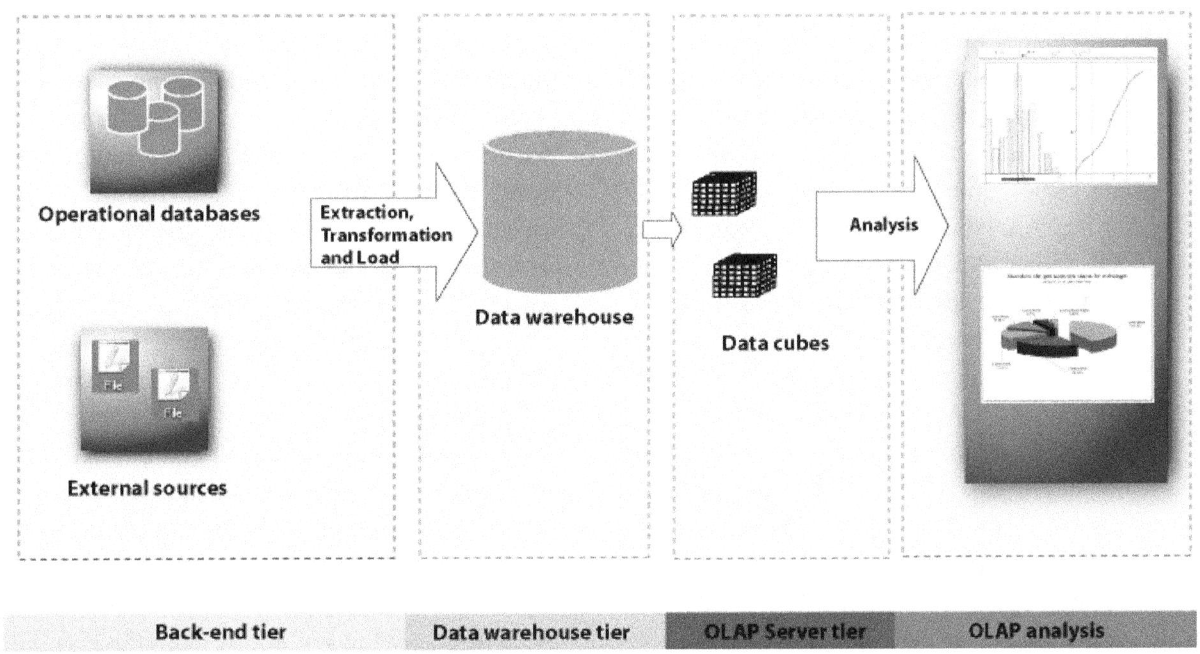

transformation step modifies the format of the source data. Finally, the loading step feeds the data warehouse with the transformed data. More details of such a process can be found in (Kimball & Caserta, 2004).

- **Data warehouse level:** This level manages the data integrated by the ETL process in a multidimensional manner. Several proposals aim at designing and building data warehouses. Some store symbolic and numeric data (classical data warehouses) (Inmon, 2005; Kimball & Ross, 2002), others integrate spatial data (spatial data warehouses) (Bimonte et al., 2005) and finally some manage semi-structured and XML data (XML data warehouses) (Mahboubi et al., 2009).
- **OLAP Server level:** This level optimizes data access, pre-calculates and aggregates data, and implements OLAP operators. Several implementations can be envisaged for the OLAP level (Kimball & Ross, 2002): (1) ROLAP (Relational Online Analytical Processing), which accesses the data stored in a relational database via SQL queries; and (2) MOLAP (Multidimensional Online Analytical Processing), which stores the data in an optimized multi-dimensional array. This alternative requires the pre-computation and storage of information into data cubes. Finally, (3) HOLAP (Hybrid Online Analytical Processing) combines ROLAP and MOLAP technologies.
- **OLAP Client level:** This level offers a user interface with reporting, interactive analysis and/or data mining capabilities. Users can trigger OLAP operators by interacting with a pivot table (Stolte et al., 2003). At this stage, data mining techniques can be performed for knowledge discovery and data mining (KDD) purposes. Many data mining tools are designed to operate on data stored in a warehouse or fully integrated in data warehouse systems.

The schema of a data warehouse is called a "multidimensional model". It is based on the concepts of facts and dimensions. A *fact* represents a business measurement (Kimball, 2002), usually described by a set of numeric values and *measures* (such as sale amounts, for instance). These measures are observed according to *dimensions,* which are textual descriptors (*members*) of the business being studied (Kimball & Ross, 2002).

A common logical design for multidimensional schema for ROLAP architecture (at the data warehouse level) is the star schema (Kimball & Ross, 2002). Shaped like a star, it consists of a single central *fact table,* representing facts, which is linked to peripheral *dimensions tables*. Star-like schemata including hierarchies in different dimensions (e.g., city, region and country granularity levels in a geographical dimension) are termed *snowflake schemas*. Star-modeled warehouse data with several fact tables are termed *constellation schemas*.

Figure 2 shows a UML[1] (Unified Modeling Language) class diagram representing a snowflake schema. It represents the classical sale facts case study. These facts are characterized by the *quantity* (of purchased products) measure. Four dimensions (products, customers, suppliers and dates) describe the criteria of analysis. Three dimensions contain hierarchies: the dimension "Time" (Day<Month<Year), the dimension "Customers" (Customer<Region<Nation) and the dimension (Supplier<Region<Nation). These three dimensions allow us to aggregate fact measures at different levels. Several queries can be expressed in the schema in Figure 2. The schema allows combining two or more dimensional levels to aggregate measure values (e.g., per day, per day and per customers, per month and per nation). Several aggregation functions can be used (e.g., sum, min, max, avg and so forth). Some queries that can be expressed with the schema in Figure 2 are presented in Table 1.

These kinds of queries are performed using OLAP tools (at the server and client levels). The OLAP server implements OLAP operators: drill

Figure 2. Warehouse snowflake schema

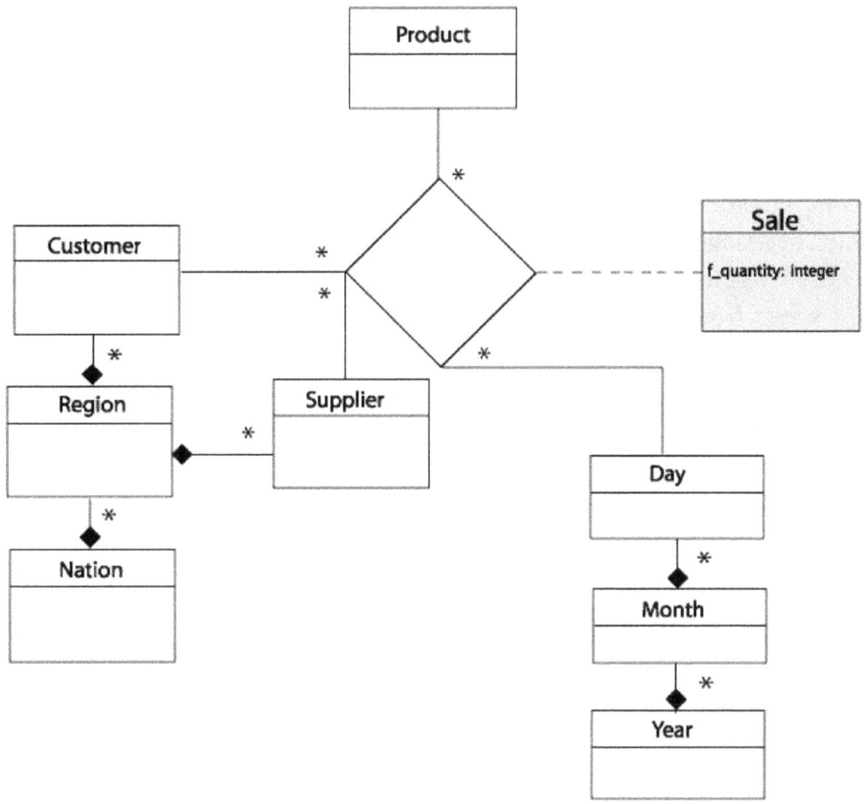

Table 1. Examples of analysis

	Query examples
1	What is the **average** quantity of purchased products per **year** and **customer?**
2	What is the **maximum** quantity of purchased products per **month** and **supplier?**
3	What are the **maximum** and the **minimum** quantities of purchased products per **year** and **supplier nation?**
4	What is the **average** quantity of purchased products per **month** and **product?**
5	What is the **average** quantity of products purchased per **month** and **customer?**
6	What is the **minimum** quantity of purchased products per **day** and **supplier region?**
7	What is the **maximum** quantity of purchased products per **year** and **customer nation?**
8	What is the **total** quantity of purchased products per **year** and **supplier?**

operators that can navigate dimensions and aggregate measures, and slicing operators that can select a part of the warehouse data (Inmon, 2005). For this purpose, the schema exploits a specific data structure, called data cubes (Kimball & Ross, 2002).

In this context, the design of a data warehouse is a very challenging issue. Currently, there are no well-established strategies for the development process (Niedrite et al., 2009). Existing data warehouse development methods fall within three basic categories: *data driven*, *user-driven* and *goal driven* (List et al., 2002).

1. The data-driven method explores data sources and derives possible data warehouse conceptual schemas (Golfarelli et al., 1998; Inmon, 2002). The conceptual schemas are defined via transformations of source sche-

mas. The analytic needs of an organization are not identified at all, or are only partially identified. They are entirely elucidated after the implementation of the data warehouse (Inmon, 2002).
2. The user-driven method first elicits user requirements, in order to understand the way in which decisions are made (Kimball & Ross, 1998). This elicitation is performed in different ways: from users and experts and through group interviews. Afterward, it derives the warehouse schema using the identification of facts and dimensions (Poe, 1996).
3. Finally, the goal-driven method consists of several steps (Kimball, 1996). It starts by choosing the business process, determines the grain of the process, and chooses dimensions and facts. This approach considers a business process (Niedrite et al., 2009) as a major operational process.

Unlike the user-driven and the goal-driven methods, the data-driven method is the fastest way to define a data warehouse schema (Niedrite et al., 2009). In our context, this method can help simulation experimenters to discover needs by analyzing the simulation results' sources. In consequence, it helps to derive the data warehouse schema from the simulation results' sources. We have adopted this data-driven method to design our simulation result data warehouse schema and to conceive the entire data warehousing and analysis process.

2.2. Exploring Simulation Models

Increasing computing power makes it possible to develop more and more complex models, coupling sub-models of different types representing industrial, natural or social processes, or representing explicitly all the individuals of populations (individual or agent based models). Understanding the behavior of such models through an analytical analysis of the equations has become unfeasible because of their complexity. For example, meteorological models have benefitted from computing power increases by a factor of millions during the last 20 or 30 years (with the Moore laws) but the prediction period has only increased by a few days. This is due to the nonlinearity and instability in the physical equations used to model the atmosphere. In general, a lot of simulations are necessary to analyze such models (models with hundreds of parameters are common), especially to make sensitivity analyses or to explore critical regimes. To perform these sets of simulations, scientists have applied experimental designs initially developed in agriculture; see for example (Sacks et al., 1989; Santner et al., 2003; Helton et al., 2006).

To help researchers in this task, this situation has led to the development of specific tools that aim at:

- A programming environment devoted to the development of simulation experiments via a dedicated user interface that is compatible with the largest set of possible models,
- Easy access to well-established libraries for experiment designs and data treatments (R, Scilab integration, etc),
- An information system in a client/server configuration that provides the possibility of sharing or exchanging components of experiments and tracing them.

Our aim to develop data warehousing methodologies is particularly relevant if they can be coupled with such a tool.

3. SIMULATION RESULTS: WAREHOUSING AND ANALYSIS METHODOLOGY

In this section, we propose a methodology for warehousing/analyzing simulation results and describe our general data warehouse multidimensional schema.

3.1. The Four Steps of the Methodology

Our proposed methodology for simulation results warehousing is depicted in Figure 3. It consists of three steps:

1. The "experiment design" step, in which simulations are designed and performed.
2. The "data warehouse design" step, which provides a multidimensional schema for the results warehouse.
3. The "simulation results integration" step, which helps to load simulation results into the data warehouse.
4. The "results analysis" step, which provides tools to analyze the simulation results.

The "Experiment Design" Step

In the initialization phase, all the parameters used by the model are provided for a given scenario, which creates a correspondence between a set of inputs and exploration factors. Then, the initialized model is launched. After the model run reaches its final step, the delivered outputs (often over time) are generally aggregated (see Faure & Deffuant, 2006, for a more detailed description of these steps).

The "Data Warehouse Design" Step

In the literature, several works propose multidimensional schemas for data warehouses. Surveys can be found in (Pedersen, 2001; Abello et al., 2006; Malinowsky & Zimanyi, 2008). They focus on particular aspects of facts and dimension concepts, such as complex hierarchies and measures. In addition, extensions to new types of data have also been proposed: spatial data (Bédard et al., 2001), temporal data (Eder et al., 2002), medical data (Midouni et al., 2009), and multimedia data (Arigon et al., 2007). However, to the best of our knowledge, no work proposes a conceptual multidimensional schema that takes into account all the characteristics of simulation results: heterogeneous simulation result types (e.g., numeric, raster, vector) and relevant dimensions (e.g., models, inputs).

Figure 3. Warehousing of simulation results and methodology

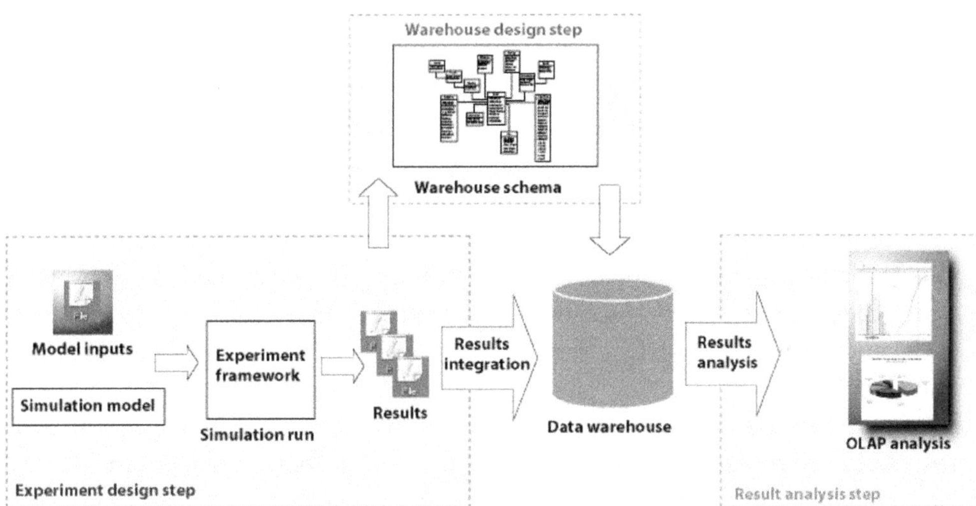

The "Simulation Result Integration" Step

In this step, simulation results are loaded into the data warehouse. For this purpose, we use ETL tools that provide periodic extraction, transformation, and integration of simulation results into a data warehouse. These tools should be adapted to the structural specificities of simulation results, e.g., raster images of environmental simulations. Taking into account images in the ETL process involves the use of specific functionalities in order to process vector and raster data.

The "Simulation Result Analysis" Step

This step requires an OLAP tool that provides efficient analysis and visualization capabilities of data (simulation results in our case). Its goals are:

- **Exploration of simulation results:** this task is performed using specific, customized decision-support queries. Such queries allow explorations through two or more dimensions. For example, they allow the exploration and the visualization of phenomena evolution over time, as well as of simulation model inputs. In this step, indicators (e.g., *distance or ratios*) can be also defined.
- **Comparison of models and their versions**: this task helps the experimenter to analyze and to compare a large number of observed scenarios using different model versions. It aims at evaluating the discrepancy and the synergies between models. Analysis can be also performed for validating the model. This analysis consists of comparing models' results from measured data. Such validations help modelers to improve the model.
- **Production of reports and charts:** experimenters usually investigate and report the general trends produced by the model runs. Charts and graphic displays can help reveal hidden information. With OLAP tools, such tasks can be easily and dynamically performed. However, these tools should also provide facilities to visualize and explore spatial data (vector and raster) that characterize simulation results.

3.2. The Meta-Model of a Simulation Experiments Warehouse

We present a generic multidimensional schema dedicated to simulation results by describing its corresponding meta-model. Our meta-model is inspired from those proposed in the literature (Abello, 2002; Luján-Mora et al., 2006). Our meta-model specifies the concepts of measures and dimensions (and levels) in order to represent simulation result heterogeneity and dimensionality for analysis. Figure 4 shows the schema for the meta-model. We use UML to describe the main classes of this meta-model, which includes the following main objects.

Facts

Simulation results are usually the common analysis subject. Modelers tend to observe the evolution of phenomena by analyzing results obtained from different models. In consequence, simulation results are the warehouse facts. These facts (*Fact class*) consist of measures (*Measure class*) that represent the simulation results (*Simulation result class*). These results are obtained at several points of granularity (by years, by models and by inputs—*Dimension classes*) e.g., estimating the appearance of dead trees in a savanna over the next 10 years and by considering specific scenarios for rain.

Each measure is an attribute (*Attribute class*). It corresponds to a simulation result. This attribute can be simple, (*Simple class*, which represents a numeric value, such as an average number), complex (*Complex class*, which is a composite

Figure 4. Meta-model of simulation experiments warehouse

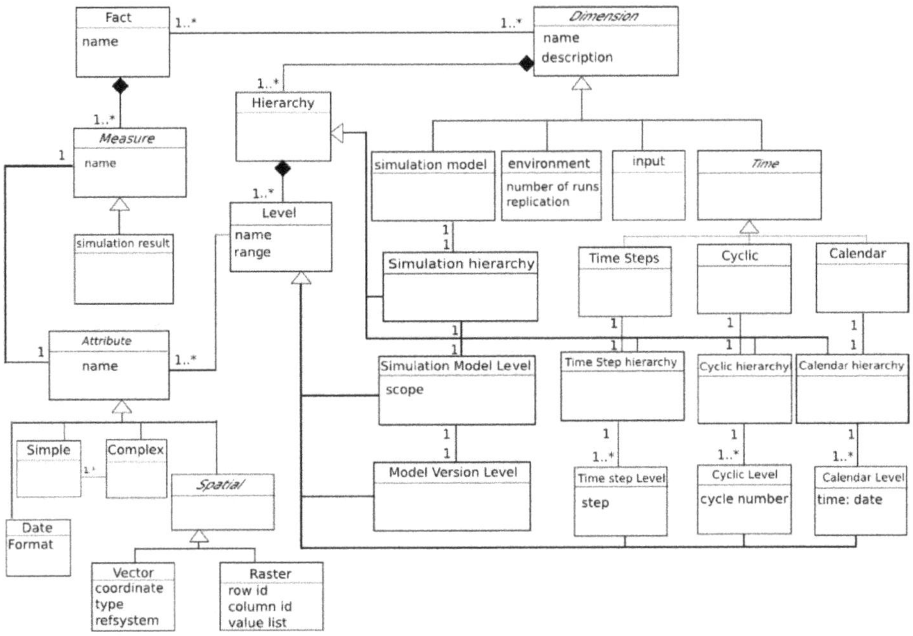

or a sequence of simple attributes) or spatial (*Spatial class*). Spatial information can be based on two models: discrete (vector) and continuous field (Longley et al., 2001). The former associates each geographic object with a spatial component (geometry), represented as a point, line, or surface; and a descriptive component, represented as attributes. The most important characteristic of this model is the explicit representation of geometry, which allows topological, directional and distance analysis. The latter model represents space as a continuous field. Fields can be discretized (Longley et al., 2001). For example, the raster representation is a discretization of the phenomenon using a complete grid (*Raster class*). Raster representation is fit for modeling pure spatial phenomena such as pollution or temperature. It allows a point-by-point analysis through the map algebra operators (Tomlin, 1990).

This heterogeneity of attributes is due to the fact that different types of models and different exploration scopes can be used. For example, automaton models, which consist of a regular grid (raster) of cells, output a set of raster data representing the system states at distinct time steps, while event-discrete models can only output a set of statistics at each event occurrence.

Dimensions

A dimension is characterized by a name and a description. It is composed of one or several hierarchies (*Hierarchy class*). Each hierarchy belongs to only one dimension and contains one or more levels (*Level class*). A level is characterized by a name and a range (its position in the hierarchy) and some attributes (*Attribute class*). In the context of simulation modeling, several specific types of dimensions can be identified:

- **The Simulation Model dimension (*Simulation model class*):** this dimension can have several model versions or scenarios. This dimension is important for the comparison of model versions or for the evaluation of different models. It is organized in a hierarchy (*Simulation hierarchy class*) composed of two levels: the root is the model version level (*Model version level class*), and the second level is the model level (*Simulation model level class*). It groups all model versions and defines the simulation model's scope.

- **The Environment dimension (*Environment class*):** this dimension provides information about simulation runs. It concerns the number of runs and replications (how many times the simulation should repeat itself). This class also defines the parameters of the initialization state (simple attributes).
- **The Input dimension (*Input class*):** this dimension represents the variables of the simulation model. These variables, their number and their associated values depend on the model and the scope of the simulation.
- **The Time dimension (*Time class*):** this dimension indicates the possible types of temporal exploration of the models. Different types of time can be considered:
  1. Calendar time (*Calendar class*), which includes dates and organizes them by days, months and years (*Time hierarchy class*). This hierarchy is composed of levels (*Calendar level class*) characterized by a date (*Date class*).
  2. Cyclic time (*Cyclic class*) defines time in a cyclic manner, such as seasons (e.g., for vegetation model simulations) or school and holiday periods (e.g., to observe traffic evolution). It can be organized in a hierarchy composed of levels (*Cyclic level class*). These levels are characterized by an attribute called cycle number (simple attribute).
  3. Time steps (*Time steps class*), which conceptualize time points where changes occur. It can be organized in hierarchy composed of levels (*Time steps levels class*) characterized by a step attribute (a simple attribute).

3.3. Example of Instantiation

Figure 5 shows a class diagram representing an instance example of our meta-model. This instance represents a generic multidimensional schema for a data warehouse for simulation results. It has a result fact class composed of two measures: measure 1 (a simple measure) and measure 2 (a raster measure).

These facts are associated with the input, environment, simulation and time dimensions. All these dimensions have one hierarchy. Each hierarchy is composed of one or more levels. In addition, each level is characterized by a name and a description. In this example, the time dimension is a calendar time specialization that is composed of date, month and year levels. All of these levels contain a level name and a description.

4. CASE STUDY

4.1. Savanna Simulation Model

To illustrate our methodology, we use the savanna model presented in (Jeltsch et al., 1999). This model was proposed to study tree spacing and coexistence in semiarid savannas. It explores the long-term and spatial consequences of alterations in the events impacting the savanna vegetation such as rain, fire, disturbance, etc. Variants of this model have also been used to better understand the partial contributions/interactions of rain and grazing to shrub encroachment on the savanna's area (Jeltsch et al., 1996, 1998, 1999).

The model is based on a cellular automaton (Jeltsch et al., 1999). It consists of a regular grid of cells (raster cells); for a specific period, each one has a value corresponding to a state. Over the time, each cell has a finite number of states (shrubs, perennial grasses and herbs, and trees). In the model, the grid comprises 200*100 equal-sized cells. Each cell corresponds to a geographical location.

Figure 5. Meta-model instance example

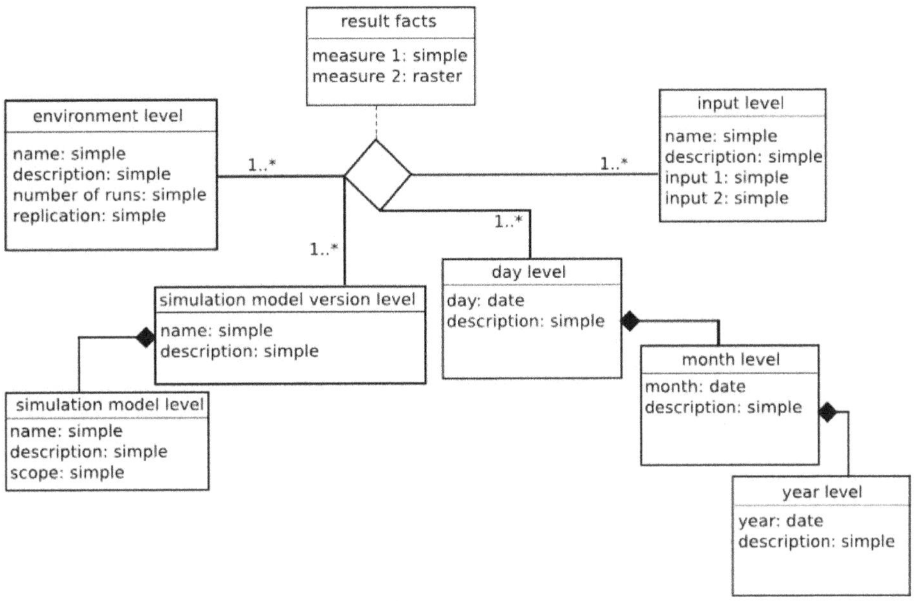

The model inputs are a set of parameter values. These parameters belong to the following categories: (1) *Environment*: defines the number of exploration replications, time steps and the model version. (2) *Local disturbances*: defines the number of cells annually disturbed. This type of parameter is needed for the model. (3) *Grazing*: defines the grazing intensity which corresponds to the impacts of animals on the savanna. Grazing causes deterioration in the condition of the herbaceous vegetation, possibly leading to extinction. (4) *Fire*: this parameter corresponds to fire occurrence and intensity. (5) *Rain*: there are different rain scenarios based on measured precipitation or on some mathematical law.

A simulation yields three distinct outputs. (1) exploration characteristics: the name, the run and the run's outcome, which represent the dominant vegetation types (grass, shrub or tree); (2) summarizing statistics for every time step, like the number of trees, shrubs, perennials, annuals, biomass, occurrence of fire, rain values and grass/tree ratio; and (3) complete grids. Each grid is associated with a time period.

4.2. Savanna Data Warehouse Schema

The data warehouse schema for the results analysis of the savanna simulation model (a star schema) is depicted in Figure 6. This schema consists in simulation result facts (*Results* class) described by the following measures: number of trees, shrubs, perennials, annuals, biomass, occurrence of fire, rain values, grass/tree ratio and raster data. These facts can be analyzed according to the time, environment, model and input dimensions. The input dimensions are: rain scenario, disturbance, grazing, fire and vegetation.

The *environment* dimension defines the initialization parameters and provides information about simulation runs (such as name, replication and the number of time steps). The *time-step* dimension defines the different time steps in which system changes occur. The *model version* dimension defines the savanna model variants; the different model versions presented in (Jeltsch et al., 1996, 1997a, 1997b) are considered. The *fire* dimension represents fire occurrence and the *rain*

Figure 6. Savanna warehouse schema

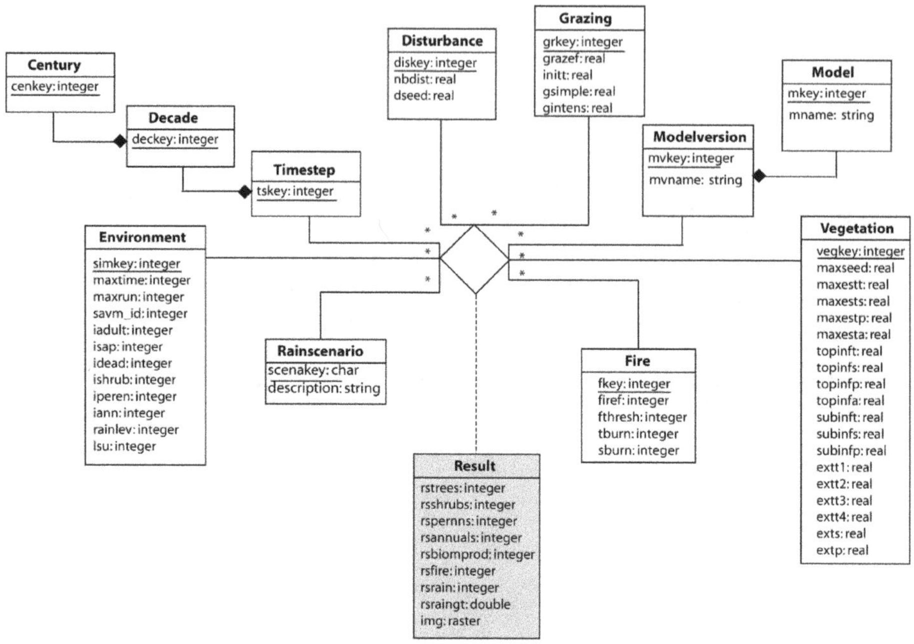

scenario dimension provides all possible rain scenarios (e.g., defined by a value or a random rain around a value). The *disturbance* dimension describes added seeds and the number of disturbed cells. The *grazing* dimension describes the type of grazing and defines its density. Finally, the *vegetation* dimension defines vegetation rules.

Figure 6 also shows that the model version and time-step dimensions contain hierarchies. Such hierarchies allow results analysis on several levels of granularity. For example, statistics from results can be observed over every ten years.

4.3. Savanna Model Results Warehousing and Analysis

Implementation

In order to build our test simulation result warehouse (called savanna warehouse), we first performed four simulations of the savanna model using SimExplorer.

SimExplorer is a tool dedicated to facilitate the design and the analysis of computer experiments on any simulation model (Chuffart et al., 2010, 2008). It allows the factorizations of simulation model exploration components. It also promotes and provides quality insurance by using traceability features with an information system.

The results obtained are then loaded into the savanna warehouse (according to the schema described in Figure 6). Load tasks have been performed by specific Java classes that we have developed. The warehouse has been implemented using a ROLAP (Relational OLAP) with the Postgres DBMS (DataBase Management System). We have also used the OLAP server Mondrian. Mondrian is an open source Java OLAP server, which provides an easily customizable, full-featured OLAP server based on a relational database backend.

Examples of Analysis

In order to analyze simulation results, we used the JRubik[2] (Java Rubik) OLAP tool. JRubik is an OLAP client developed in Java/Swing and based on JPivot[3] project components. The main components of JRubik are: OLAP navigator, MDX query editor, table viewer, chart viewer, map viewer, menus and bookmarks and statistical data.

The pivot tables are well-adapted to the analysis of data warehouses because their structure visually represents the main concepts of the multidimensional schema. A pivot table is a 2D spreadsheet that displays detailed and aggregated measure values. We have extended the OLAP functionalities of the JRubik client to allow raster visualization in the cells of the pivot table. Also, we have made the cells of the pivot table clickable. Users can see a general vision of raster measures in the pivot table. Then, when they want more details, they can click on the cell, and the corresponding raster is visualized at a more detailed resolution in another window.

Figure 7 shows a screenshot of the JRubik interface. This figure shows the temporal evolution of the numbers of trees and shrubs calculated by the savanna model. More precisely, the interface enables users to visualize the average numbers of trees and shrubs per decade time steps and model versions (versions 1.0 and 2.0); see Figure 7a. Ten decades are presented in the table. For each decade, the average numbers (of trees/shrubs) of all the simulation runs of a model version (1.0 or 2.0) are calculated. This table is produced by combining the levels *Decade* and *Model Version* of the multidimensional schema (Figure 6). The average function aggregates the measures *rsshrubs* and *rstrees* per decade and model version. The corresponding chart is presented in Figure 7b.

Figure 7 shows that the obtained values are different and depend on the model version. For example, using model version 1.0, we obtained a larger number of trees than when we used model version 2.0. Such observations can help users choose a model.

Figure 7. Analysis example 1

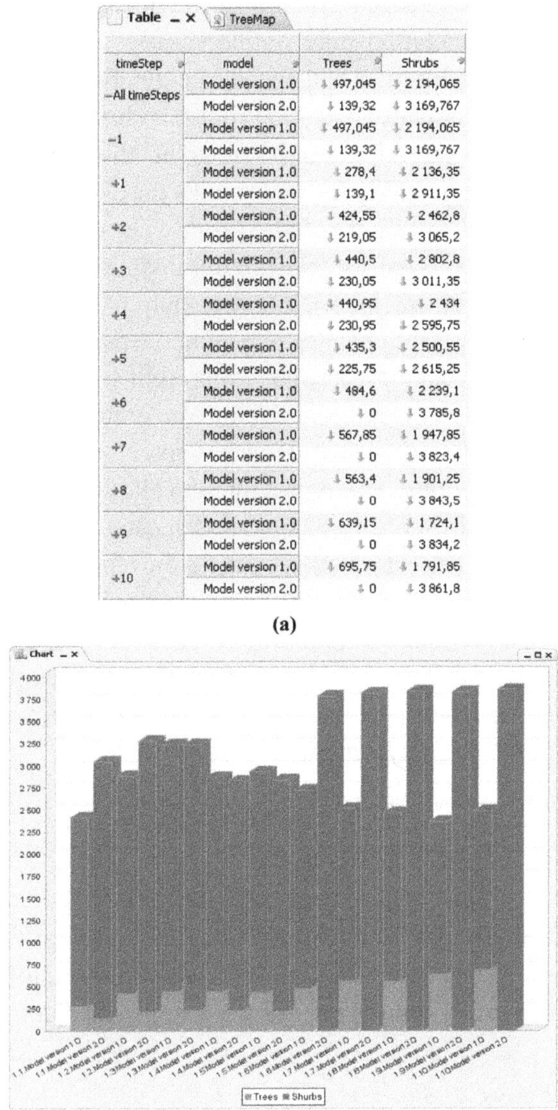

timeStep	model	Trees	Shrubs
All timeSteps	Model version 1.0	497,045	2 194,065
	Model version 2.0	139,32	3 169,767
1	Model version 1.0	497,045	2 194,065
	Model version 2.0	139,32	3 169,767
1	Model version 1.0	278,4	2 136,35
	Model version 2.0	139,1	2 911,35
2	Model version 1.0	424,55	2 462,8
	Model version 2.0	219,05	3 065,2
3	Model version 1.0	440,5	2 802,8
	Model version 2.0	230,05	3 011,35
4	Model version 1.0	440,95	2 434
	Model version 2.0	230,95	2 595,75
5	Model version 1.0	435,3	2 500,55
	Model version 2.0	225,75	2 615,25
6	Model version 1.0	484,6	2 239,1
	Model version 2.0	0	3 785,8
7	Model version 1.0	567,85	1 947,85
	Model version 2.0	0	3 823,4
8	Model version 1.0	563,4	1 901,25
	Model version 2.0	0	3 843,5
9	Model version 1.0	639,15	1 724,1
	Model version 2.0	0	3 834,2
10	Model version 1.0	695,75	1 791,85
	Model version 2.0	0	3 861,8

(a)

(b)

Figure 8 shows the average numbers of trees and shrubs per decade time-steps. Unlike the analysis presented in Figure 7, the two versions of the savanna model are not distinguished. This table is produced by combining the levels *Decade* and *Model* of the multidimensional schema (Figure 6).

Figure 9 depicts raster images obtained for the first twelve time steps and for the first vegetation rule. Vegetation rules define values such as establishment rates and speed of extinction of different kinds of vegetation (trees or shrubs). Displayed raster images represent the studied savanna area. These images show the distribution of vegetation in the studied area. They consist of set of pixels representing the dominant vegetation at each point (pixel), and they help to visualize the area state changes over time steps. In these images, dark green pixels represent trees, light green pixels represent dead trees, orange pixels represent shrubs and white pixels represent empty points. This representation allows a visual comparison of results obtained at each time step and can be performed by considering other dimensions (model versions, for example).

Other analyses (e.g., queries) can be considered by combining two or more dimension levels. Diverse aggregation functions can be used. Some possible queries are presented in Table 2.

5. CONCLUSION

In this paper, we have proposed a methodology for designing data warehouses for simulation results. Our methodology involves four steps: simulation, data warehouse design, results integration and analysis. We have introduced a general multidimensional schema that is derived from traditional conceptual multidimensional schemas and have presented a specific guide for modelers to design data warehouse schemas.

Figure 8. Analysis example 2

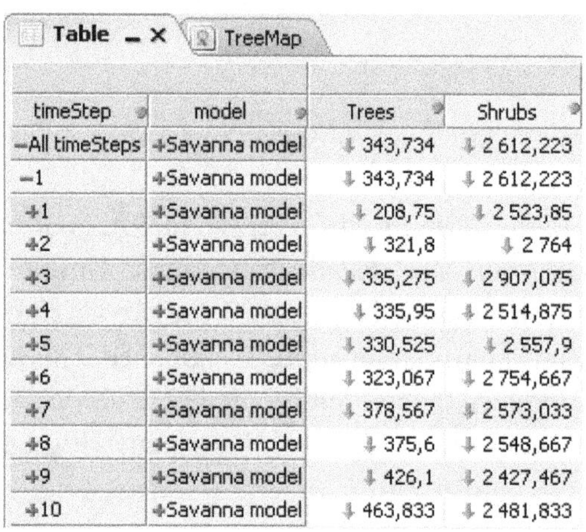

timeStep	model	Trees	Shrubs
-All timeSteps	+Savanna model	343,734	2 612,223
-1	+Savanna model	343,734	2 612,223
+1	+Savanna model	208,75	2 523,85
+2	+Savanna model	321,8	2 764
+3	+Savanna model	335,275	2 907,075
+4	+Savanna model	335,95	2 514,875
+5	+Savanna model	330,525	2 557,9
+6	+Savanna model	323,067	2 754,667
+7	+Savanna model	378,567	2 573,033
+8	+Savanna model	375,6	2 548,667
+9	+Savanna model	426,1	2 427,467
+10	+Savanna model	463,833	2 481,833

We have applied our methodology to a savanna simulation model that studies vegetation in savannas and outputs results in numerical and raster formats. We have also described the implementation of our methodology using ROLAP architecture with Postgres and an OLAP client, and we have extended the JRubik client OLAP to allow raster image visualization. Finally, we have presented examples of results analysis and comparisons of model versions.

Our research perspectives concern the definition of a conceptual schema for simulation results using UML profiles and the Model-Driven Approach (MDA) (Trujillo et al., 2001). We also plan to extend the dimension class of our meta-model in order to consider complex spatial data such as TIN (Triangulated Irregular Network). This extension allows displaying measures on digital terrain models.

Moreover, we aim to integrate the data warehouse system into the functionalities of the SimExplorer simulation framework (Chuffart et al., 2008). Ideally, the framework would automatically derive the warehouse schema from the definition of the simulation experiments.

Finally, OLAP tools can be used for a data mining purposes. In this context, several studies have addressed the issue of performing data mining tasks on data warehouses (Tjioe & Taniar, 2005). Some of them focus on mining patterns and association rules in data cubes (Imieliński et al., 2002). They propose to use methods such as agglomerative hierarchical clustering or frequent item-sets mining. Such operators could be very valuable in the context of simulation results warehousing and analysis.

Figure 9. Analysis example 3

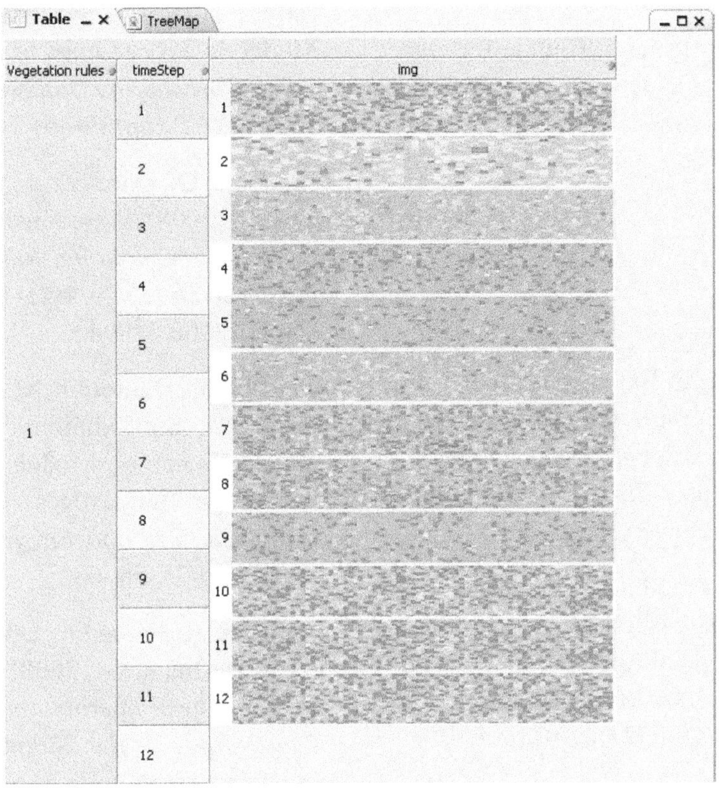

Table 2. Analysis example

	Query examples
1	What is the **average** number of **trees** per **year** and **rain scenario?**
2	What is the **average** number of **trees** and **shrubs** per **year** and **rain scenario?**
3	What is the **average** number of occurred **fire** per **decade** and **environment?**
4	What is the **total** number of occurred **fire** per **year, rain scenario** and **model version?**
5	What are the **minimum** and the **maximum** numbers of **trees** per **year, rain scenario, model version** and **fire scenario?**
6	Draw the **ratio variance** histogram per **time step** and **disturbance** configuration
7	Display the **raster images** per **time step, disturbance** configuration, **rain scenario** and model version
8	Display the **raster images** and draw **tree variance** histogram per **time step, disturbance** configuration, **rain scenario** and **model version**
9	What is the **maximum** number of **trees** per **century** and **rain scenario?**
10	What is the **average** number of **trees** and **shrubs** per **decade** and **model version?**
11	What is the **average** number of **trees** per **year** and **rain scenario?**
12	Display the **raster images** and draw **tree variance** histogram for the last ten years
13	Display the **raster images** and draw total number of **trees** histogram for the first ten years
14	What is the **minimum** number of **biomass production** and **shrubs** per **year** and **model version?**
15	What is the **average** number of **shrubs** per **model version** and **rain scenario?**

ACKNOWLEDGMENT

The authors would like to thank Romain Tertiaux, who participated in coding and extending the JRubik OLAP tool in order to allow the display of raster data.

REFERENCES

Abelló, A. (2002). YAM2 (Yet Another Multidimensional Model): An Extension of UML. In *Proceedings of the International Database Engineering & Applications Symposium (IDEAS 02)*, Edmonton, Canada (pp. 172-181). Washington, DC: IEEE Computer Society.

Arigon, A. M., Miquel, M., & Tchounikine, A. (2007). Multimedia Data Warehouses: a Multiversion Model and a Medical Application. *Journal of Multimedia Tools and Applications, 35*(1), 91–108. doi:10.1007/s11042-007-0118-7

Bédard, Y., Merrett, T., & Han, J. (2001). Fundamentals of Spatial Data Warehousing for Geographic Knowledge Discovery. In *Geographic Data Mining and Knowledge Discovery* (pp. 53–73). New York: Taylor & Francis. doi:10.4324/9780203468029_chapter_3

Ben Messaoud, R., Boussaid, O., & Loudcher Rabaséda, S. (2006). Efficient Multidimensional Data Representation Based on Multiple Correspondence Analysis. In *Proceedings of the 2th ACM SIGKDD International Conference on Knowledge Discovery and Data Mining (KDD'2006)*, Philadelphia (pp. 662-667). New York: ACM Press.

Bentayeb, F., Boussaid, O., Darmont, J., Harbi, N., & Loudcher, S. (Eds.). (2010). Warehousing and Mining Complex Data: Applications to Biology, Medicine, Behavior, Health and Environment. *International Journal of Biomedical Engineering and Technology, 3*(1-2).

Bimonte, S., Tchounikine, A., & Miquel, M. (2005). Towards a Spatial Multidimensional Model. In *Proceedings of the 8th ACM International Workshop on Data warehousing and OLAP (DOLAP 05)* (pp. 39-46). New York: ACM Press.

Boussaid, O., Darmont, J., Bentayeb, F., & Loudcher, S. (2008). Warehousing complex data from the Web. *International Journal of Web Engineering and Technology, 4*(4), 408–433. doi:10.1504/IJWET.2008.019942

Chuffart, F., Dumoulin, N., Faure, T., & Deffuant, G. (2010). SimExplorer: Programming Experimental Design on Models and Managing Quality of Modelling Process. *International Journal of Agricultural and Environmental Information Systems, 1*(1), 55–68.

Chuffart, F., Faure, T., Dumoulin, N., Reuillon, R., & Deffuant, G. (2008). Design and Analysis of Computer Experiments with SimExplorer. In *Proceedings of the 2008 European Simulation and Modelling Conference (ESM 08)*, Le Havre, France (pp. 575-577).

Codd, E., Codd, S., & Salley, C. (1994). *Providing OLAP (On-line Analytical Processing) to User-Analysts: An IT Mandate*. E. F. Codd Associates.

Darmont, J., Boussaid, O., Ralaivao, J. C., & Aouiche, K. (2005). An Architecture Framework for Complex Data Warehouses. In *Proceedings of the 7th International Conference on Enterprise Information Systems (ICEIS 05)*, Miami, FL (pp. 370-373).

Eder, J., Koncilia, C., & Morzy, T. (2002). The COMET Metamodel for Temporal Data Warehouses. In *Proceedings of the 14th International conference on Advanced Information Systems Engineering (CAiSE 02)*, Toronto, Canada (LNCS 2348, pp. 83-89). New York: Springer.

Elzbieta, M., & Zimányi, E. (2008). *Advanced Data Warehouse Design*. New York: Springer.

Faure, T., & Deffuant, G. (2006). *SimExplorer: A Software Tool for Programming and Executing Designs on Complex Models*. Paper presented at the SCS Conference on Modeling and Simulation Methodology, Tools, Software Applications, Calgary, Canada.

Helton, J. C., Johnson, J. D., Salaberry, C. J., & Storlie, C. B. (2006). Survey of sampling based methods for uncertainty and sensitivity analysis. *Reliability Engineering & System Safety, 91*, 1175–1209. doi:10.1016/j.ress.2005.11.017

Imieliński, T., Khachiyan, L., & Abdulghani, A. (2002). Cubegrades: Generalizing Association Rules. *Data Mining and Knowledge Discovery, 6*(3), 219–257. doi:10.1023/A:1015417610840

Inmon, W. (2005). *Building the Data Warehouse* (4th ed.). New York: John Wiley & Sons.

Jeltsch, F., Milton, S. J., Dean, W. R. J., & van Rooyen, N. (1996). Tree spacing and coexistence in semiarid savannas. *Journal of Ecology, 84*, 583–595. doi:10.2307/2261480

Jeltsch, F., Milton, S. J., Dean, W. R. J., van Rooyen, N., & Moloney, K. A. (1998). Modelling the impact of small-scale heterogeneities on tree-grass coexistence in semi-arid savannas. *Journal of Ecology, 86*, 780–794. doi:10.1046/j.1365-2745.1998.8650780.x

Jeltsch, F., Moloney, K. A., & Milton, S. J. (1999). Detecting process from snap-shot pattern: lessons from tree spacing in the southern Kalahari. *Oikos, 85*, 451–467. doi:10.2307/3546695

Kimball, R., & Caserta, J. (2004). *The Data Warehouse ETL Toolkit: Practical Techniques for Extracting, Cleaning, Conforming, and Delivering Data*. New York: John Wiley & Sons.

Kimball, R., & Ross, M. (2002). *The Data Warehouse Toolkit* (2nd ed.). New York: John Wiley & Sons.

Langley, P., Goodchild, M., Maguire, D., & Rhind, D. (Eds.). (2001). *Geographic Information Systems and Science* (p. 517). New York: John Wiley & Sons.

List, B., Bruckner, R. M., Machaczek, K., & Schiefer, J. (2002). A Comparison of Data Warehouse Development Methodologies Case Study of the Process Warehouse. In *Proceedings of the 13th International Conference on Database and Expert Systems Applications (DEXA 02)*, Aix-en-Provence, France (LNCS 2453, pp. 203-215). New York: Springer.

Luján-Mora, S., Trujillo, J., & Song, I. Y. (2006). A UML Profile for Multidimensional Modeling in Data Warehouses. *Data & Knowledge Engineering, 59*(3), 725–769. doi:10.1016/j.datak.2005.11.004

Mahboubi, H., Ralaivao, J. C., Loudcher, S., Boussaid, O., Bentayeb, F., & Darmont, J. (2009). X-WACoDa: An XML-based approach for Warehousing and Analyzing Complex Data. Advances in Data Warehousing and Mining. In Bellatreche, L. (Ed.), *Data Warehousing Design and Advanced Engineering Applications: Methods for Complex Construction* (pp. 38–54). Hershey, PA: IGI Global.

Niedrite, L., Treimanis, M., Solodovnikova, D., & Grundmane, L. (2009). Development of Data Warehouse Conceptual Models - Method Engineering Approach. In Taniar, D. (Ed.), *Progressive Methods in Data Warehousing and Business Intelligence: Concepts and Competitive Analytics* (pp. 1–23). Hershey, PA: IGI Global.

Poe, V. (1996). *Building a Data Warehouse for Decision Support*. Upper Saddle River, NJ: Prentice-Hall.

Ravat, F., Teste, O., Tournier, R., & Zurfluh, G. (2007). Integrating Complex Data into a Data Warehouse. In *Proceedings of the 9th International Conference on Software Engineering Knowledge Engineering (SEKE 07)*, Boston, MA (p. 483).

Sacks, J., Welch, W. J., Mitchell, T. J., & Wynn, H. P. (1989). Design and analysis of computer experiments. *Statistical Science*, *4*, 409–435. doi:10.1214/ss/1177012413

Saltelli, A., Ratto, M., Andres, T., Campolongo, F., Cariboni, J., & Gatelli, D. (2008). *Global Sensitivity Analysis*. New York: John Wiley & Sons.

Santner, T. J., Williams, B. J., & Notz, W. I. (2003). *Design and Analysis of Computer Experiments*. New York: Springer Verlag.

Stolte, C., Tang, D., & Hanrahan, P. (2003). Multiscale Visualization Using Data Cubes. *IEEE Transactions on Visualization and Computer Graphics*, *9*(2), 176–187. doi:10.1109/TVCG.2003.1196005

Tjioe, H. C., & Taniar, D. (2005). Mining Association Rules in Data Dare Warehouses. *International Journal of Data Warehousing and Mining*, *1*(3), 28–62.

Trujillo, J., Palomar, M., Gomez, J., & Song, I. S. (2001). Designing Data Warehouses with OO Conceptual Models. *Computer*, *34*(12), 66–75. doi:10.1109/2.970579

Vaisman, A. A., & Zimànyi, E. (2009). A multidimensional model representing continuous fields in spatial data warehouses. In *Proceedings of the 17th ACM SIGSPATIAL international Symposium on Advances in Geographic Information Systems (ACM SIG 09)*, Seattle, WA (pp. 168-177). New York: ACM Press.

ENDNOTES

1. http://www.uml.org/
2. http://rubik.sourceforge.net/jrubik/intro.html
3. http://jpivot.sourceforge.net/

This work was previously published in International Journal of Agricultural and Environmental Information Systems, Volume 1, Issue 2, edited by Petraq Papajorgji and François Pinet, pp. 1-19, copyright 2010 by IGI Publishing (an imprint of IGI Global).

Chapter 2
SimExplorer:
Programming Experimental Designs on Models and Managing Quality of Modelling Process

Florent Chuffart
Cemagref- Laboratoire d'Ingénierie des Systèmes Complexes, France

Nicolas Dumoulin
Cemagref- Laboratoire d'Ingénierie des Systèmes Complexes, France

Thierry Faure
Cemagref- Laboratoire d'Ingénierie des Systèmes Complexes, France

Guillaume Deffuant
Cemagref- Laboratoire d'Ingénierie des Systèmes Complexes, France

ABSTRACT

This article describes Simexplorer, a computer framework for managing simulation experiments and, to some extent, the scientific quality of the modelling process. An information system, included in the framework, insures the traceability of the experiments and their reproducibility and thus contributes to the modelling process quality management. Moreover, this information system provides facilities for sharing and exchanging components of experiment scenarios. The authors illustrate the use of the framework on a simple example of modelling process.

INTRODUCTION

The theory of design of experiments (DoEs) has been initially developed for real, non-simulated experiments in agriculture in the 1920s (Chen (2003) refers to 1926 publication by Fisher), and in engineering, psychology, and so forth, since the 1950s. In real experiments it is impractical to investigate many factors; ten factors seems a maximum, because otherwise the number of experiments to perform becomes too high. Moreover, it is hard to experiment with factors that have more than a few values.

DOI: 10.4018/978-1-4666-0333-2.ch002

These restrictions do not apply when experimenting on models (simulations). Indeed, computer codes may have hundreds of inputs and parameters each with many values. The experiment is a simulation, which generally has a very low cost, compared with real experiments. Consequently, a multitude of scenarios may be simulated. Therefore a change of mindset of simulation experimenters took place, to adapt theory and practice to this new situation (Kleijnen et al., 2005).

In this context, it is particularly important to easily access the advances of the research on simulation experiments. The goals of such a numerical experiment can be (Oden, 2006; Bayarri et al., 2007)

- Calibration and validation,
- Sensitivity Analysis—either global or local—or "What If" analysis,
- Optimization,
- Risk Analysis.

Each of these types of experiments requires particular methodological approaches, and specific experimental designs. Using the adequate method is also a strong requirement to insure a good scientific quality of modelling.

The practise of experiments on models is already widespread, and will most probably keep increasing in the future. Indeed, we notice the following evolutions of modelling processes:

- The increasing computing power makes it possible to develop more and more complex models, coupling sub-models of different types representing industrial, natural or social processes, or representing explicitly all the individuals of populations (individual or agent based models).
- Such complex models generally lack an analytical theoretical framework that fits them. Therefore, it is necessary to explore experimentally their behaviours and confront them with observation data or expertise.
- In this context, the modelling process involves loops where simulation experiments on the model suggest new hypotheses and possible changes, leading again to experiments.
- These experiments are performed through a specific application, developed for the particular model and generally poorly reusable in other modelling processes. Moreover, keeping the trace of all the modelling steps becomes rapidly difficult without a specific tool. Yet, this trace is often necessary to justify the modelling choices, and to insure the scientific quality of the process.
- There is a growing requirement of Reproducible research (RR) (Buckheit & Donoho, 1995), stating that all the results of a paper can be reproduced. This means that the whole simulation process is reproducible.

Last, the increasing complexity of the models implies frequently a team development, which enhances the necessity to take into account a multi-user context. This multi-user perspective is also important for sharing methods and libraries which are connected to different types of experiments.

SimExplorer aims at addressing the needs appearing in this context, in particular, to provide:

- A programming environment devoted to the development of simulation experiments, that can be used with any executable model,
- An easy access to well established libraries for experiment designs and corresponding data treatment,
- An information system, in a client server configuration, providing the possibility for sharing or exchanging components of experiments, and for tracing them.

We believe that gathering these functionalities in a single tool matches a growing need which is not satisfied by existing tools. Indeed, one can distinguish three categories in the tools related to experimental designs:

- General statistical software with experimental design library: open source (R), and commercial (SAS, Matlab, …)
- Modelling platforms with batch system and limited experimental design functions; for instance: Repast (North et al., 2007), Capsis (De Coligny, 2007), Isis-Fish (Mahévas, 2004), VLE (Quesnel, Duboz, & Ramat, 2009). The experimental designs are usable only for models developed in these platforms.
- Dedicated experimental design tools. A lot of them are commercial and devoted to online data analysis (à la Excel and not for batch treatment), for instance Moode, Design Expert, doe++. Some are specialized on specific problems: OpenTURNS (Dutfoy et al., 2009) focuses on liability testing, SimLab (Saltelli et al., 2004), from the JRC Saltelli group is dedicated mainly to sensitivity analysis. Dakota (Eldred et al., 2008) from Sandia University is a general purpose library and offers also optimisation functions.

None of these tools offers an environment for designing the exploration workflow, together with an information system insuring the traceability of experiments, as SimExplorer does.

In the next section, we present an overview of SimExplorer's architecture. Then, we describe its use for designing simulation experiments and result treatments, on a particular example. Then we elaborate on this example to illustrate the role of the information system in the modelling process.

SimExplorer Architecture Overview

SimExplorer's architecture evolved significantly since its first versions (Faure & Deffuant, 2006). SimExplorer is now based on a client / server architecture. A rich client, SimExplorer IDE (Integrated Development Environment) is used to develop exploration applications, to design and execute explorations, to update and synchronize the different versions between the client and the archived versions on server information system. The information system (IS) is also accessible through a web interface. An overview of the architecture is shown on Figure 1.

The IS has been designed to store all data and codes defining an exploration application. These include the codes for: generating DoE, generating initial conditions, treating model outputs, specifying values of parameters involved, in particular the paths of different data files (input and output model, treatment of outputs). The information system (IS) archives different versions of code and data of an exploration application. Code and data are exchanged between the server system and clients. The IS can be seen as a storage engine with versioning features.

In the next versions of SimExplorer, the IS server will be able to launch and drive simulations through the API (*application programming interface*) server. To do this, the necessary data must be sent to this interface, and the simulation results must be retrieved from the server. The server should be able to provide information on the status of simulations launched from the API. This part of SimExplorer is not the main focus of this article and will be described in a forthcoming article.

The development environment (a screenshot is shown on Figure 3), allows the user to program exploration applications, with code editing features for the components of such applications:

Figure 1. Global architecture

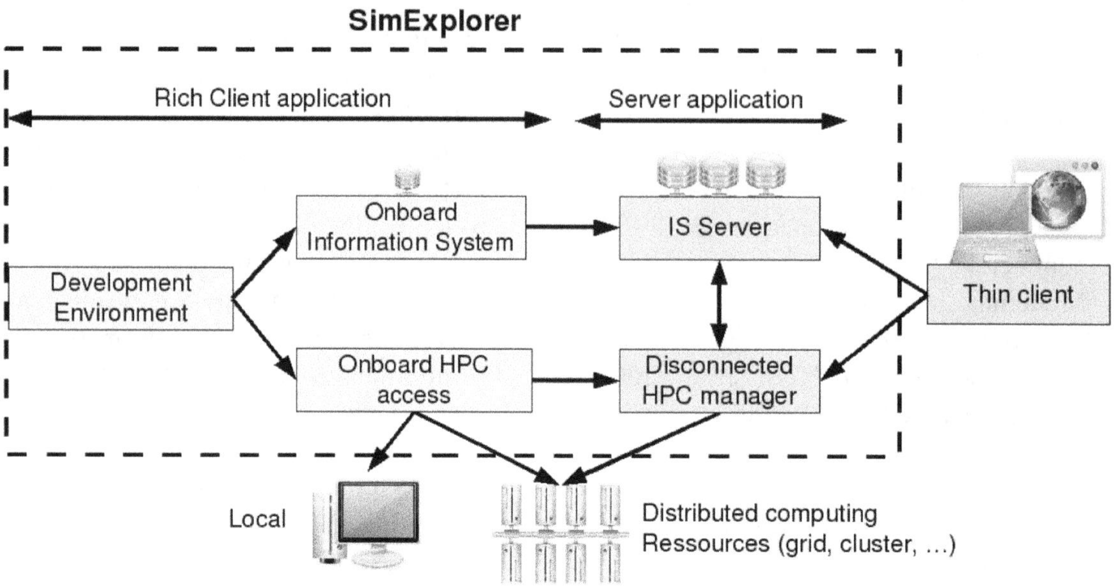

initialization, DoE, generation of model inputs (initial conditions), execution of simulation, treatment of model outputs. A dedicated programming interface allows the user to visualize data and code from the IS interface. Various libraries, such as DoE, random number generators, diverse statistical treatments (R), help programming. When the exploration application has been developed, the rich and light clients can choose the resources on which the computations are launched. The light client allows the administrator, from a web browser, to manage user rights.

SimExplorer information system is based on a distributed architecture that is usable on a computer disconnected from the network. It means that each client can be connected to a central IS server, and can also have its own standalone IS. The centralized approach is essential to bring collaborative and backup facilities. On the other hand, the standalone approach is useful for interacting with the IS without an access to a centralized server. Thus, a user can keep the access to his archives and traceability features directly on his own PC.

Example of Model Exploration

In order to better understand the interest of SimExplorer, we now describe its use on a particular modelling process. The aim is to design a model which reproduces qualitatively the general behaviour of a differential equation model, but using explicit individuals moving and interacting in an explicit space (also called individual based model or IBM) instead of variables representing densities. Indeed, working in an explicit space with explicit individuals implies to introduce new hypotheses, which should be experimentally adjusted to get the desired qualitative properties. Our main focus is on the use of SimExplorer in this modelling process, but we need to give some information about the problem and the models to illustrate this use more precisely.

The differential equation model rules the evolution of the densities of preys and predators, where predators can adopt a cooperative (dove) or aggressive tactic (hawk) (Auger et al., 2006). It is supposed that predators encounter each other by pairs. When two doves meet, they share their preys, when a hawk meets a dove it takes the whole prey, and when two hawks meet, they fight to keep their prey. The predators update their tactic periodically by considering the success of each tactic in the whole population. Introducing these tactics changes significantly the qualitative evolution of the predator and prey densities compared with standard predator-prey models. In particular, for some parameter values, the model shows two stable attractors (see Figure 2), while it is never the case in the standard predator-prey models. In the first attractor, there is a high density of preys and low density of predators, and all predators are hawks. The second attractor, on the contrary, there is a low density of preys, a high density of predators, and a mixture of cooperative and aggressive tactics, because the cooperative tactic (dove) becomes more rewarding when the density of predators is high.

We defined a first version of an IBM, derived from the density model. We assumed that predators and preys are uniformly distributed on 2D grid. For technical reason of no relevance here, we had to make an important change in the hypotheses of the model: we considered that predators share or fight with all other predators located within a given neighbourhood around them when they catch a prey. We extended the rule for rewarding the predators according to the proportion of hawks and doves in the neighbourhood. Periodically the predators observe the gain of the other predators in a neighbourhood, and tend to adopt the most rewarding tactic.

We want to check if the new model we elaborated can show two attractors and a saddle point for some parameters, as does the initial differential equation model.

The parameters which are relevant for this exploration are *C*, the cost of fights between hawks

Figure 2. Auger et al. (2006) model of predator prey dynamic with two stable attractors and a saddle point

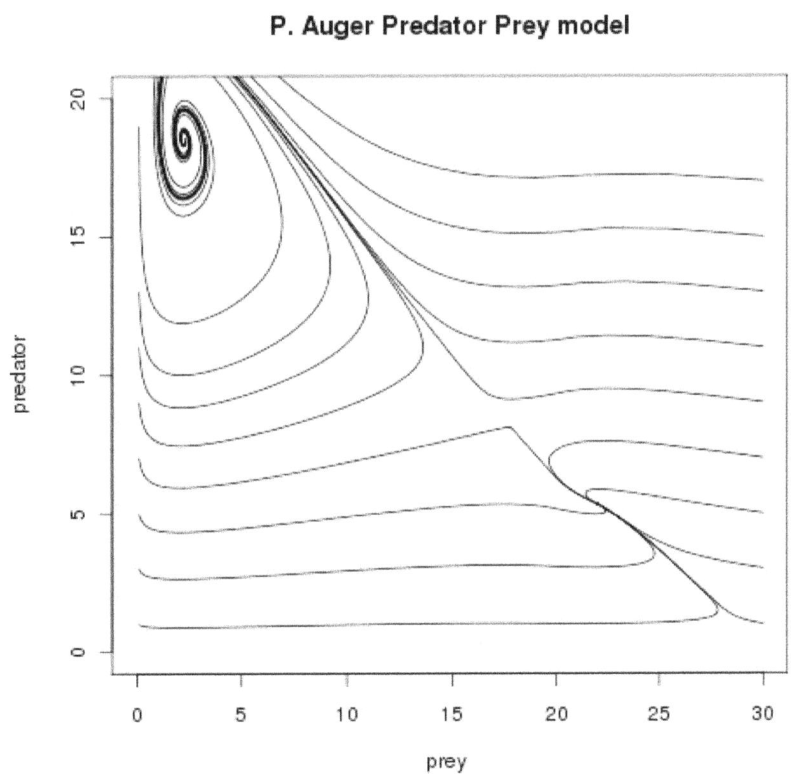

and v, the size of the neighbourhood. For each couple (C,v), we want to get the phase portrait in space (n, p), with n the number of preys and p the number of predators, similarly to Figure 2. Therefore, for each couple (C,v), we need to compute several simulations starting with different initial conditions (n_0, p_0). We characterise this set of trajectories by computing the number of attractors and their type (only one tactics, or mixture of tactics).

Figure 3 shows a part of SimExplorer development environment for building our exploration application. The panel on the left shows the structure of the application. Each node of the tree is called exploration component and can contain itself nested exploration components. The rest of the interface is used to configure each component, to define the exploration factors and to manipulate the variables used by the components.

Hence we define two distinct designs:

- Initialisations (n_0, p_0) to get the phase portrait (like in Figure 3) for a given (C,v). We chose initial numbers of predators p_0 taken in {128, 256, 512, 1024, 2048}, and initial number of preys n_0 taken in {100, 1600, 3100}.
- Values of C and v. We choose to C-values taken in {0.0, 0.2, 0.4, 0.6, 0.8} and v-values in {11, 21, 31, 41, 51}.

For each scenario (given values of v, C, n_0, p_0), we generate the right input file for the model and we launch the model. This job is done in the *Final processing* component (Figure 4).

There is no configuration with two attractors found in the experiments (see Figure 6). The typical result with only one pure hawk attractor is shown on Figure 5a, and those with mixed attractor is shown on Figure 5b. This suggests that the model does not reproduce qualitatively the

Figure 3. IDE, Exploration Application and components

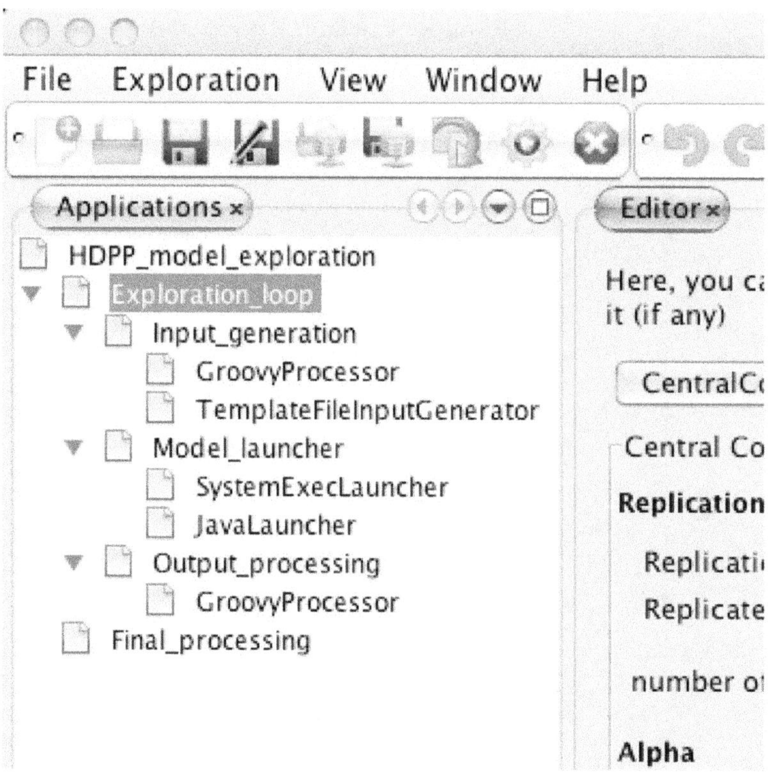

Figure 4. Aggregating results using Groovy script

```
import fr.cemagref.HDPP.*
def results = []          //new array of results
output.C.unique().each{C->          //for all C
  output.v.unique().each{v->        //for all v
    def phase = output.find{exp->   //filter
      exp.C==C && exp.v==v}
    conf = HDPP.analysePhase(phase)
    results << ['C':C, 'v':v, 'conf':conf]
  }     //collecting hierarchy of results
}
HDPP.plot2DMap(results);    //plotting 2D MAP
```

results of the density model, and thus fails to match our objective. Of course, we could have missed a configuration of two attractors because the exploration which is too gross or in a too restricted part of the parameter space. Actually, a more detailed analysis that we omit here suggests that the configurations with two attractors cannot be reach with this version of the model. Therefore, we need to modify this model, and we start a classical modelling process, where a model is iteratively modified.

Using the Information System in a Modelling Process

Now we shall illustrate how the information system is used in a modelling process. We design and explore a new version of the model. The modification of the first version is driven by an analysis of the model dynamics that is omitted here. It leads to modify the reward rules.

In the information system, the exploration application is stored in two main nodes (Figure 7): the exploration components structure that contains components organization and the exploration data that contains factors, constants values. The exploration data node also contains the constant values used in all the application components.

In the new version of the model, the doves can have some food even if they have hawks around. We suppose that doves have a probability to hide a prey to the hawks, and this probability is higher when they are numerous. This rule depends on a new parameter α.

Now we want to experiment the new version of our IBM. Thus, we have to define a new protocol to explore model parameters. With SimExplorer, we can easily use the components developed for the first version of the model, and adapt them to the second version.

We derive *SystemExecLauncher* (a dedicated component for launching executable) component by duplicating it and modifying the copy. We keep *Model_launcher* component unchanged, it still contains the structure. The second protocol reuses many components of the first protocol.

Similarly, we modify the input structure in order to make it consistent with new model signature. As shown on Figure 8, we also add a new factor (*e*) to explore parameter α, taking values in {0, 0.06, 0.1, 0.2, 0.3, 0.4, 0.5}. We affect the *input* and the new *factor*. It impacts *Input_generation* component. Finally, we modify *Final_processing* component in order to visualize the results on a 3D map. As shown on the Figure 8, it impacts *Final_processing* component.

We reuse a large part of the first exploration components, slightly modifying them to adapt them to the second exploration. This approach shows several advantages:

Figure 5. (a) Example of predator prey dynamics for the first version of the IBM with only one pure hawk attractor (C=0.2, v=10); (b) Example of predator prey dynamics for the first version of the IBM with one single mixed hawk and dove attractor (C=0.6, v=10)

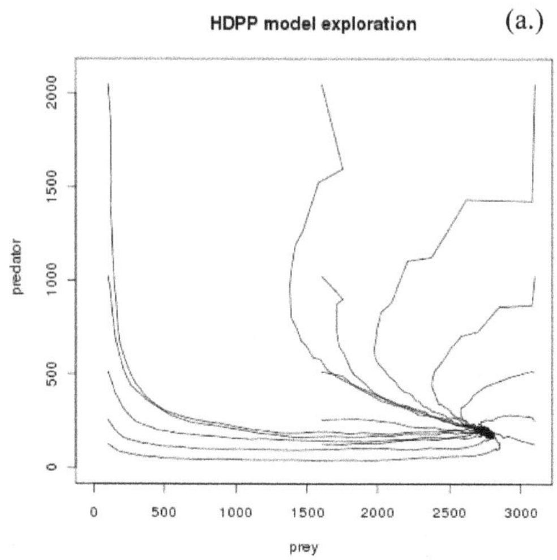

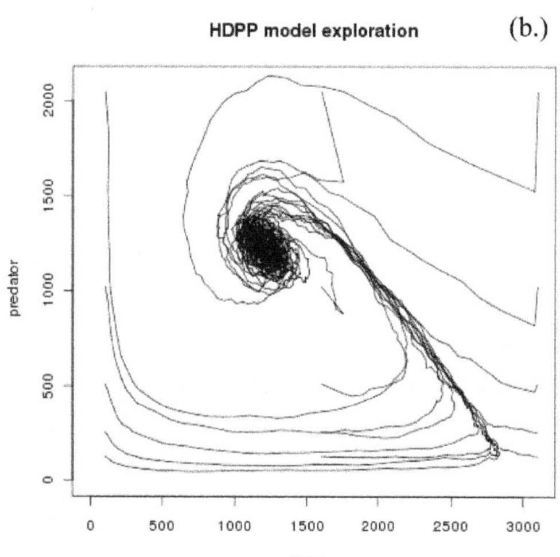

- Time saving by reusing existing components,
- Providing traceability of experiments by storing and versioning all tasks and data structures in SimExplorer information system.

As expected, for some values of α, v, C, the exploration of the second model shows two attractors with one containing only hawks and the other one mixing hawks and doves (see Figures 9 and 10).

The information system stores all the data and components used in both versions of the model. It gives an easy access to these components, in order to build a new exploration of one of these versions, or of a possible third version.

Working with Multi-Users Features

We illustrated the use of the IS by one single user on a standalone computer. Now, we suppose that at least two users collaborate in the modelling process. Actually collaborative work is often needed for designing and studying complex models, especially when specialists of different fields are required.

Figure 11 shows how the centralized information system allows several users to work on the same exploration application. A first user can create the exploration application and submit it to the server with a dedicated synchronizing tool integrated into SimExplorer. A second user can retrieve the application, and, after some modifications, this second user can synchronize his work and send his new version on the server. It is then available to the first user for potential changes.

This collaborative facility allows for instance a first user to build the main design of experiments regarding the considered model, and a second user to study more deeply some aspects depending on his interest or skills. Of course, this can be extended to more than two users, and can also be useful when the same user is working from several computers.

Furthermore, each user can set access rights to an exploration application or an exploration component that he submitted to the server.

Figure 6. 2D map of the IBM dynamic (red squares for pure hawk attractor, green triangles for mixed one)

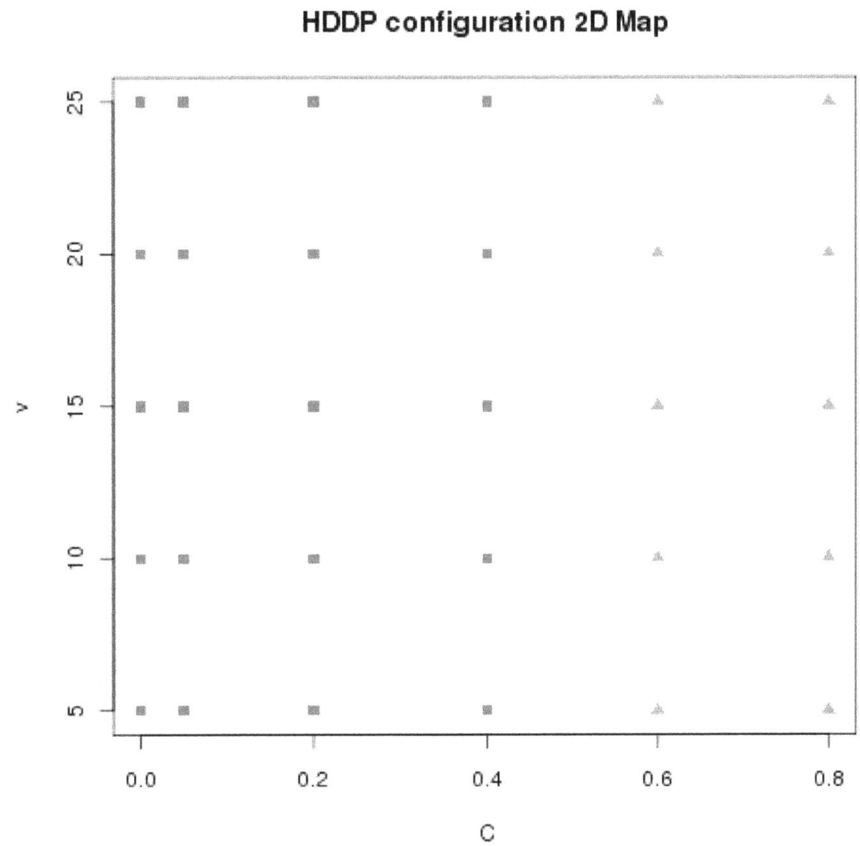

CONCLUSION AND DISCUSSION

This article illustrates how SimExplorer can help to design and execute experiments on models and insure the traceability and reproducibility of the modelling process.

SimExplorer is more specifically designed for middle to large models which are developed in a medium to long term perspective. In these cases the traceability and multi user features provided by the IS are particularly useful.

The necessary investment for mastering the structure of SimExplorer and its script language is certainly too high for the single exploration of a simple model. Nevertheless, once this investment is made, our experience shows that one tends to use it even with simple models and simple explorations. The main reason is that it provides efficiency and time gains when the exploration must be done again with a small change. And this happens frequently, even with simple explorations.

The current version of SimExplorer is being tested internally in our team and by a few partner teams. Most of these tests are related to agricultural and environmental applications:

- In PATRES European project (NEST 043268), SimExplorer is used to perform sensitivity analyses and dynamical behaviour studies of several models, in particular a model of Savana dynamics, in collaboration with UFZ (Centre for Environmental Research of Leipzig, Germany),
- In PRIMA European project (ENV 212345), SimExplorer will support the modeling process of demographic, economic and

Figure 7. IS view of the first exploration (first version of the model)

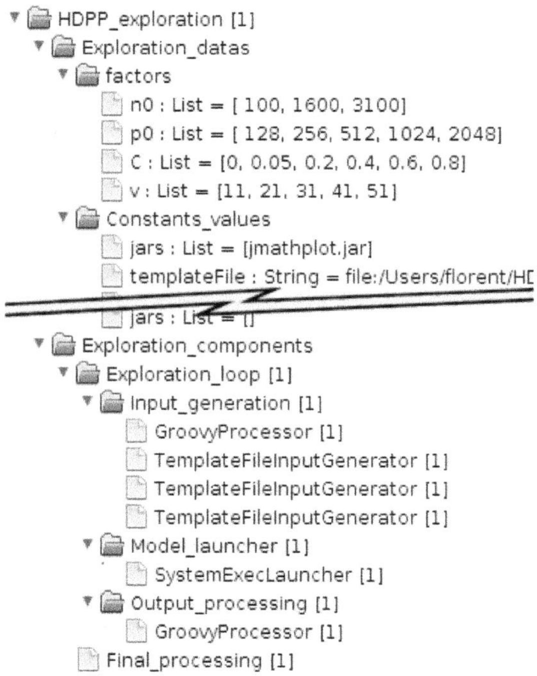

Figure 8. IS view of the second exploration (second version of the model)

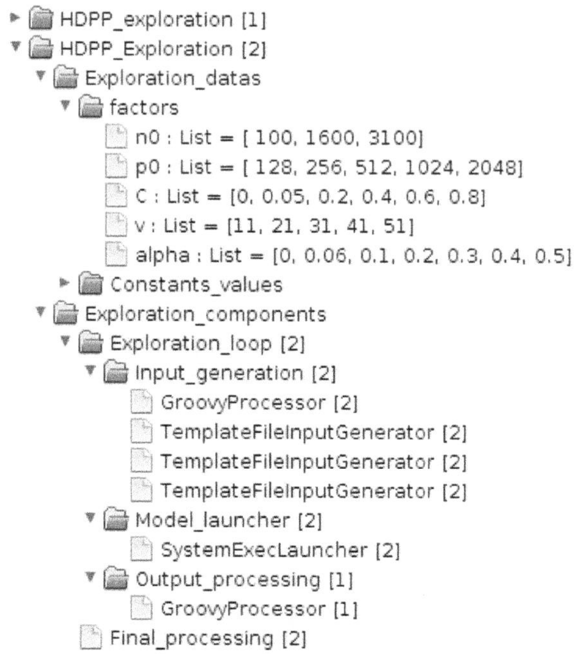

Figure 9. 2D map of the IBM dynamic (red squares for pure hawk attractor, green triangles for mixed one, blue circles for both)

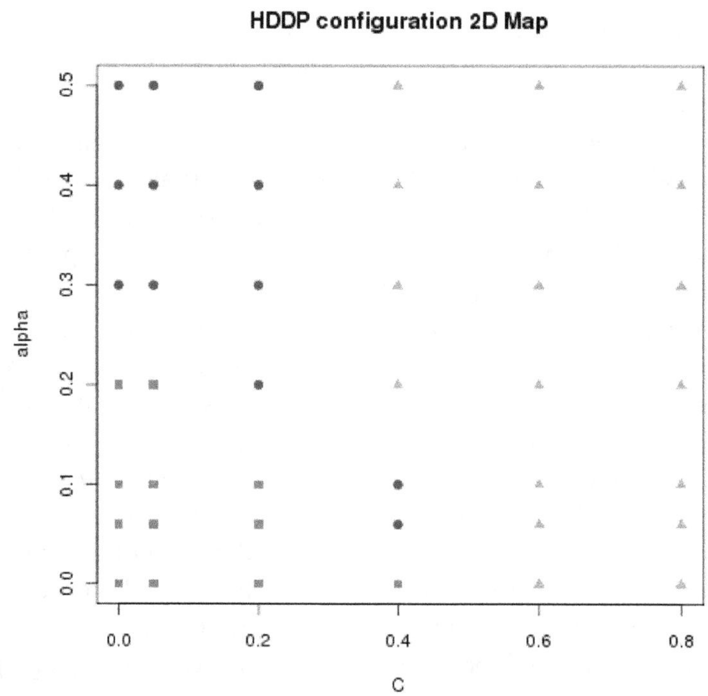

Figure 10. Predator prey dynamics for the second version of the IBM with α =0.2 and C=0.2

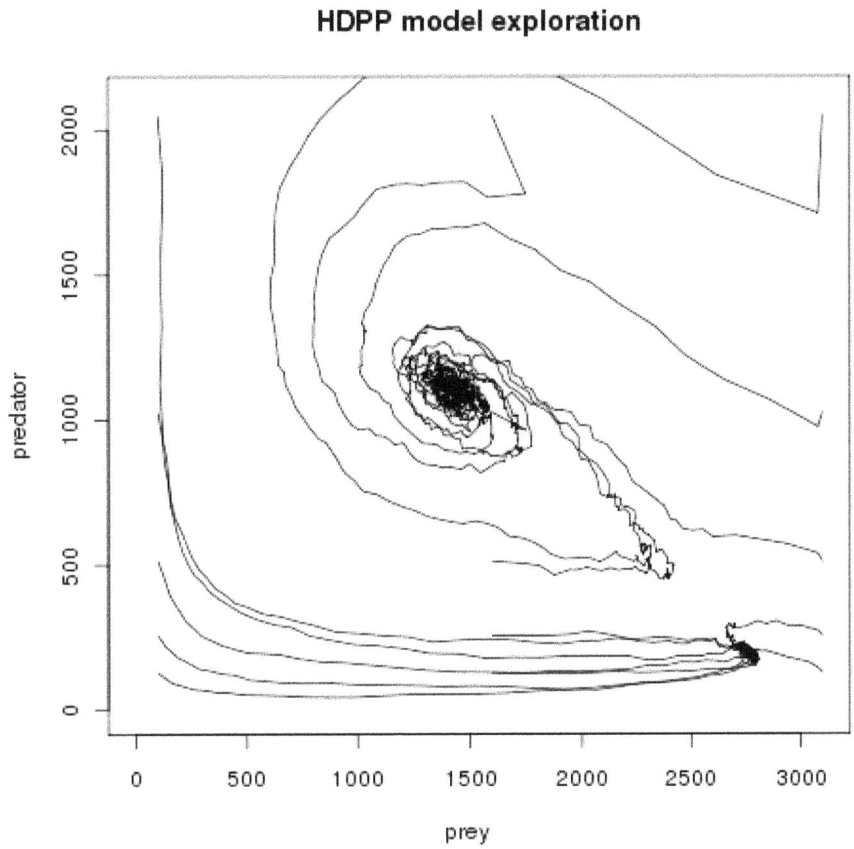

Figure 11. Sequence diagram of an example of collaborative work

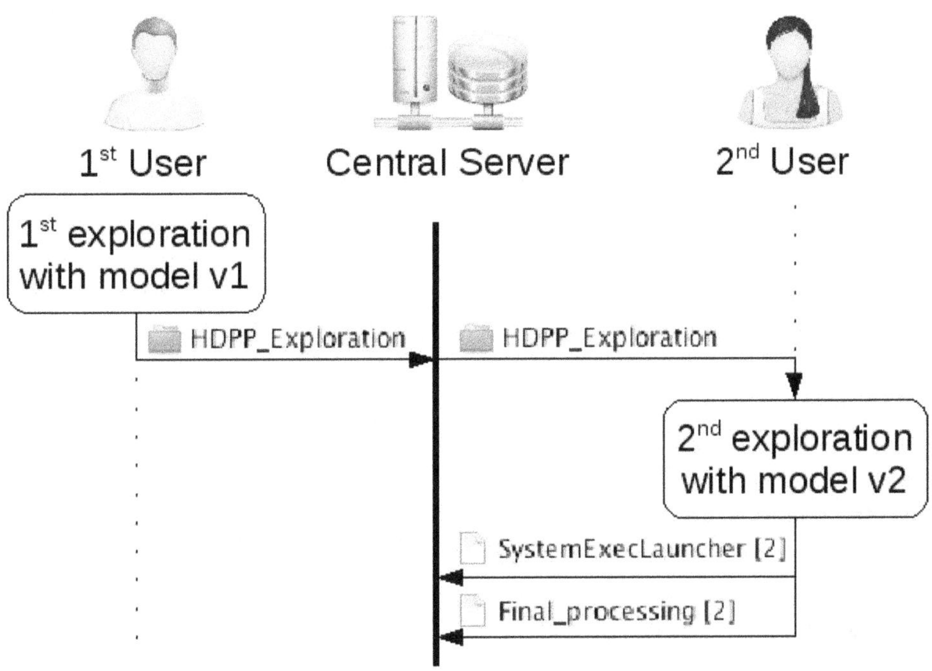

ecological evolution of rural municipalities in Europe. The study will take into account the uncertainties on parameters and scenario variations. This work is carried out in collaboration with 11 partner teams, and more particularly Groningen University, Netherlands, and Leibniz Institute of Agricultural Development in Central and Eastern Europe, in Halle, Germany.

- In DISCO project funded by the French research agency (SYSCOM call), SimExplorer will support the modelling process of coupling between biodiversity and structure in bacteria biofilms. This study will be performed in collaboration with the French Research Institute in Agronomy (INRA) and the French Research Institute in Informatics and Automatics (INRIA).

These tests and improvement phase should last still a few months, before we get an easily usable version by external users. The current version is nevertheless already accessible from http://www.simexplorer.org.

In addition to our first objective to make the current version more robust and user friendly, in the future, we envisage several directions for SimExplorer development:

- SimExplorer can already be used for model coupling exploration. The information system keeps track of the different versions of the global model. But some specific features could be added, like for instance a specific workflow to represent and perform model coupling.
- Another improvement we consider is related to result storage and access, in particular when the results are spatially distributed with different levels of aggregation. Then the use of data warehouse techniques could be interesting to facilitate frequent computations.

ACKNOWLEDGMENT

This research is partially funded by European Project PATRES (NEST 043268), and has also been supported by LifeGrid regional program (Auvergne – France).

REFERENCES

Auger, P., Kooi, B., Bravo de la Parra, R., & Poggiale, J. C. (2006). Bifurcation analysis of a predator-prey model with predators using hawk and dove tactics. *Journal of Theoretical Biology, 238*, 597–607. PubMed doi:10.1016/j.jtbi.2005.06.012

Bayarri, M. J. (2007). A framework for validation of computer models. *Technometrics, 49*(2), 138–154. doi:10.1198/004017007000000092

Buckheit, J., & Donoho, D. L. (1995). *Wavelab and reproducible research.* http://www-stat.stanford.edu/~donoho/Reports/1995/

Chen, V. C. P. (2003). A review of design and modeling in computer experiments. In Khattree, R., & Rao, C. R. (Eds.), *Handbook of Statistics* (*Vol. 22*). Elsevier.

De Coligny, F. (2007). Efficient building of forestry modelling software with the capsis methodology. In Fourcaud, T., & Zhang, X. P. (Eds.), *Proceedings of PMA06: Plant Growth Modeling and Applications* (pp. 216–222). Los Alamitos, CA: IEEE Computer Society.

Dutfoy, A., Dutka-Malen, I., Pasanisi, A., Lebrun, R., Mangeant, F., Sen Gupta, J., et al. (2009). OpenTURNS, an open source initiative to treat uncertainties, risks'N statistics in a structured industrial approach. In *41èmes Journées de Statistique*. Bordeaux: SFdS.

Eldred, M. S., Bichon, B. J., Adams, B. M., & Mahadevan, S. (2008). Overview of reliability analysis and design capabilities in DAKOTA with application to shape optimization of MEMS. In Tsompanakis, Lagaros & Papadrakakis (Ed.), *Structural Design Optimization Considering Uncertainties. Structures and Infrastructures series, 1*, 401-432.

Faure, T., & Deffuant, G. (2006). SimExplorer: A software tool for programming and executing designs on complex models. In *SCS International Conference on Modeling and Simulation - Methodology, Tools, Software Applications*. Calgary, Canada.

Kleijnen, J. P. C., Sanchez, S. M., Lucas, T. W., & Cioppa, T. M. (2005). State-of-the-art review: a user's guide to the brave new world of designing simulation experiments. *INFORMS Journal on Computing, 17*(3), 263–289. doi:10.1287/ijoc.1050.0136

Mahévas, S., & Pelletier, D. (2004). ISIS-Fish, a generic and spatially-explicit simulation tool for evaluating the impact of management measures on fisheries dynamics. *Ecological Modelling, 171*, 65–84. doi:10.1016/j.ecolmodel.2003.04.001

North, M. J., Tatara, E., Collier, N. T., & Ozik, J. (2007). Visual agent-based model development with repast simphony. In *Proceedings of the Agent 2007 Conference on Complex Interaction and Social Emergence*. Argonne, USA: Argonne National Laboratory.

Oden, J. T. (2006). *Revolutionizing engineering science through simulation. Blue Ribbon Panel on Simulation-Based Engineering Science. National Science Foundation*. NSF.

Quesnel, G., Duboz, R., & Ramat, E. (2009). The virtual laboratory environment -- an operational framework for multi-modelling, simulation and analysis of complex dynamical systems. *Simulation Modelling Practice and Theory, 17*, 641–653. doi:10.1016/j.simpat.2008.11.003

Saltelli, A., Tarantola, S., Campolongo, S., & Ratto, M. (2004). *Sensivity analysis in practice - A guide to assessing scientific models*. Willey.

This work was previously published in International Journal of Agricultural and Environmental Information Systems, Volume 1, Issue 1, edited by Petraq Papajorgji and François Pinet, pp. 55-68, copyright 2010 by IGI Publishing (an imprint of IGI Global).

Chapter 3
A Web-Based Tool for Spatio-Multidimensional Analysis of Geographic and Complex Data

Sandro Bimonte
Cemagref, France

ABSTRACT

Spatial OLAP (SOLAP) integrates spatial data into OLAP systems, and SOLAP models define spatial dimensions while measuring spatio-multidimensional operators. In this paper, the author presents the concepts of geographic and complex measures that allow integrating geographic and complex information as subjects of analysis in spatial data warehouses. The concept of geographic measure extends the concept of spatial measure to the semantic component of geographic information. The concept of complex measure allows introducing complex data as subjects of multidimensional analysis. To reduce the gap in flexibility between spatial and multidimensional analysis, this paper proposes a symmetrical representation of measures and dimensions. Additionally, the author presents a Web-based SOLAP prototype, GeWOlap, that enriches existing SOLAP tools by effectively and easily supporting symmetrical geographic/complex measures and dimensions for modeling and visualization. To validate this approach, the simulated environmental data concerning the pollution of the Venice lagoon is used.

1. INTRODUCTION

Data warehousing and On-Line Analytical Processing (OLAP) systems are technologies intended to support business intelligence. OLAP models are based on the concepts of dimensions and facts (Inmon, 1996). A fact is a concept that is relevant for the decision-making process, and it is described by a set of numerical indicators (measures). Dimensions, composed of hierarchies, allow for the analysis of facts along different analysis axes and at different levels of details.

Nowadays, new information and communication technologies allow the collection of huge amounts of geo-referenced agricultural and environmental data. These data come from remote sensing systems or other computer applications that record agricultural practices. OLAP systems

DOI: 10.4018/978-1-4666-0333-2.ch003

were recently successfully used to analyze agricultural and environmental data (Bédard et al., 2008). Therefore, a new kind of Decision Support System, called Spatial OLAP (SOLAP), has been developed in order to introduce Geographic Information Systems (GIS) functionalities (memorizing, analyzing and visualizing) into OLAP and data warehousing systems. SOLAP tools extend the tiers of the classical OLAP system architecture in order to manage spatial data, provide a cartographic representation of facts on maps, and perform SOLAP operations through simple user interactions with the maps. Spatial OLAP integrates spatial data into multidimensional models by defining the concepts of spatial dimension, spatial measure, and spatio-multidimensional operators.

In our opinion, SOLAP models only partially exploit the complex nature of geographic information in the multidimensional decision-making process, as they only partially use the semantics of geographic information and the flexibility of spatial analysis. The subjects of analysis SOLAP models (spatial measures) are generally spatial objects or numerical values derived from geometry (i.e., surface area, distance, etc.). Thus, in (Bimonte et al., 2006) we define the concept of geographic measure, introducing the descriptive attributes and spatial/thematic/ "map generalization" relationships of geographic information in multidimensional analysis. We also provide a symmetrical representation of measures and dimensions in spatial data warehouses: each level of a dimension can be used also as a subject of analysis. This increases the multidimensional analysis capabilities since it allows the system to analyze and discover possible trends and relationships among all the elements (levels and measures) of the data warehouse (Pinet & Schneider, 2009). The need for symmetry of measures and dimensions is very important for multidimensional geographic applications, since the process of spatial analysis is flexible, in the sense that the analyst can iteratively change and/ or reformulate its goals (subjects of analysis) throughout the spatial analysis process. Therefore, in this paper we extend the concept of geographic measure to that of complex measure. This enables the multidimensional analysis of complex data in functions of geographic information.

In this paper, we describe the implementation of these concepts in our prototype, called GeWOlap. GeWOlap enriches the functionalities of existing SOLAP tools in order to support the modeling and visualization of complex/geographic measures and the symmetry of dimensions/measures more easily and fully. For this purpose, we have selected a particular logical multidimensional model, redefined the aggregation process of the OLAP server aggregation process, and applied some geovisualization techniques. GeWOlap extends our previous work presented in (Bimonte et al., 2006; Bimonte et al., 2007) and is based on the spatio-multidimensional model GeoCube (Bimonte et al., 2006b). To validate the method, we use simulated environmental data concerning the pollution of the Venice lagoon. We present some SOLAP applications to analyze water pollution phenomena in the Venice Lagoon in terms of several factors, such as time, pollutant, and location. However, it is important to note that we have used simulated pollution data, and these applications have not been used by the Consortium for Coordination of Research Activities concerning the Venice Lagoon System (CORILA) experts to draw any conclusions on the phenomenon of pollution.

The paper is structured as follows. Section 2 presents the functionalities of GIS and the main concepts of data warehousing, OLAP, and Spatial OLAP. Section 3 describes our new concepts of geographic and complex measures. Section 4 describes the three tiers of our prototype. Section 5 presents the environmental case study, and Section 6 discusses related work.

2. BACKGROUND

2.1. Geographic Information Systems

Geographic information represents objects or phenomena of the real world described by a *spatial* and a *semantic component* (Laurini & Thompson, 2001). The spatial component corresponds to the geometry of the object and its position on the earth's surface. The semantic component is: (i) a set of descriptive properties and (ii) spatial, thematic and "map generalization" relationships that the object shares with other (geographic) objects. A spatial relationship is defined between geographic objects using spatial predicates (such as inside, cover, intersect, etc.), whereas a thematic relationship uses non-spatial predicates (e.g., belonging to, managed by, etc). A "map generalization" relationship links geographic objects of less detailed scales and/or secondary themes (for example, buildings at two different topographic scales) (Weibel & Dutton, 2001). As an illustration, let us consider the geographic object "Unit" corresponding to an environmental region of the Venice lagoon in Italy. Figure 1 shows its spatial component. A "Unit" is characterized by a semantic component, represented by a set of descriptive properties such as the name, an index of salinity, a list of plants, an area and a socio-economic type. A spatial relationship represents the inclusion of a "Unit" in a "Zone" (a set of units) (see Figure 1).

Geographic Information Systems (GIS) are the main tools used to handle geographic information (Longley et al., 2001). GIS allow for storing, visualizing and analyzing geographic data. The spatial analysis process is a complex, iterative and flexible process. Indeed, Mitchell (2005) defines the spatial analysis process as an iterative process, which consists of the following steps: identifying the problem and the goal, identifying tools to use, identifying data and spatial operations to use, creating and analyzing a plan, executing the plan, and visualizing the results. A plan corresponds to a set of spatial operations that can be applied sequentially; the result of one operation can be used as input for other operations. Moreover, the tools and data have to be flexible. Users cannot define data a priori, but they can change and transform them (for example, using transformation methods) at each iteration of the process, taking into account the intermediary results obtained up to then.

2.2. Data Warehouse and OLAP

A data warehouse is "a subject-oriented, integrated, non-volatile and time-variant collection of data stored in a single site repository and collected from multiple sources" (Kimball, 1996). Data warehouses are represented by multidimensional models. These models define dimensions, facts and measures (Inmmon, 1996). A dimension is composed of hierarchies, which in turn are composed of levels. Instances of the levels are called members. Facts are the subjects of analysis, and are described by numerical indicators called measures. Dimensions are the axes of analysis, and the levels of the hierarchies define the granularity with which the facts are analyzed. At coarser levels, the values of the measures are aggregated using SQL aggregation functions.

The following is an example of a multidimensional model to analyze of the pollution values of the Venice Lagoon. The dimensions are Time, Pollutants and Location, with the following hierarchies

- "Time" has a hierarchy composed of three levels, "day < month < year";
- "Pollutants" has a hierarchy composed of two levels, "pollutant < type of pollutants";
- "Location" has a hierarchy composed of two levels, "unit < zone".

The measure is the pollution value, which is aggregated using the averaging operation.

An example data offset in this multidimensional model is shown on Table 1. The values

Figure 1. Spatial components of Venice Lagoon's units and zones

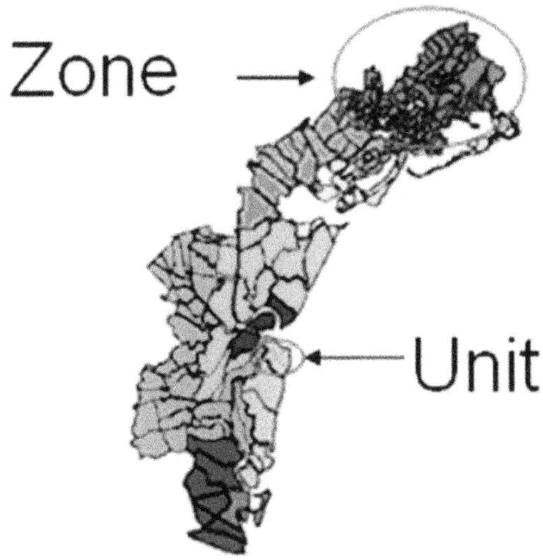

Table 1. Example data in the multidimensional model for the analysis of Venice Lagoon pollution

pollutant	day	unit	value
Zinc	1-1-01	Murano	10
Zinc	2-1-01	Murano	46
Sulfur Trioxide	26-9-05	Coa di Latte	25

Table 2. Aggregation of pollution values on the temporal dimension using the average

pollutant	Month	unit	value
Zinc	1-01	Murano	28
Sulfur Trioxide	9-05	Coa di Latte	25

of the columns "pollutant", "day" and "unit" are members of the most detailed levels of the dimensions "Pollutants", "Time", and "Location," respectively; the column "value" contains the values of the measures.

This multidimensional model allows us to answer queries like "*what is the average pollution value per month, pollutant and unit*?" (Table 2) or "*what is the average pollution value per day, pollutant and zone*?"

The data warehouse is explored using OLAP operators (Inmon, 1996). The main OLAP operators are drill operators (Roll-up and Drill-Down) that make it possible to navigate in the hierarchies by aggregating measures, and cutting operators (Slice and Dice) that select and project pieces of the data warehouse. Some other operators (i.e., split, nest, pull, push, etc.) have been also implemented (a survey can be found in Rafanelli, 2003). The multidimensional decision process consists in exploring the data warehouse through OLAP operators. The user navigates through the multidimensional data along the different analysis axes, looking for useful information in an interactive process. The paths of the multidimensional analysis paths are not defined a priori (Tang et al., 2003). For example, the user can visualize pollution values per day, pollutant and unit. Then, by means of a Roll-up operation, the user can climb into the time dimension in order to analyze pollution values at a less detailed level of temporal granularity in order to look for temporal trends.

The classical architecture of OLAP systems is a three-tier architecture (Kimball, 1996). The first tier is the data warehouse, which contains data extracted from transactional databases before being cleaned, homogenized and summarized by ETL tools (Extraction, Transformation, and Loading). This critical and important step aims at consolidating heterogeneous schemata and reducing the data in order to comply with the data warehouse model. The second tier is the OLAP server. It implements the OLAP operators; it pre-calculates some aggregated measures, which are stored in data structures called materialized views, and optimizes the forthcoming accesses to the data. The main approaches to the implementation of OLAP servers are Multidimensional OLAP (MOLAP), Relational OLAP (ROLAP) and Hybrid OLAP (HOLAP). MOLAP servers stock and retrieve multidimensional data using particular data structures (multidimensional arrays) and indexing and hashing techniques. ROLAP servers use relational databases. In the case of ROLAP, data warehouses are modeled at a logical level using the star or the snowflake schema (Kimball, 1996). The star schema represents dimensions as denormalized tables (dimension tables) where attributes define the levels of the hierarchies. A table (called the facts table) represents facts with the foreign keys to the dimension tables, and a set of numerical attributes representing the measures. The snowflake schema normalizes some of the dimension tables. Finally, the HOLAP approach is the combination of the ROLAP and the MOLAP techniques.

The third level is an OLAP client that offers a user interface with reporting, interactive analysis and/or data mining tools. The most commonly used visualization and interaction technique is the pivot table. A pivot table is a 2D spreadsheet that displays detailed and aggregated measures values by nesting members of the different dimensions on the x- and y-axes. Pivot tables enable users to visualize the results of OLAP operations. The data are displayed on multiple pages. Pivot tables are well adapted to the analysis of data warehouses because their structure visually represents the main concepts of the multidimensional model. It generates a set of multiple displays that make it easy to compare the data.

2.3. Spatial OLAP

The integration of valuable spatial data into data warehouses has become a challenging research area, and has led to the definition of new kind of Spatial Decision Support System: Spatial OLAP (SOLAP). SOLAP has been defined by Y. Bédard as "a visual platform built especially to support rapid and easy spatio-temporal analysis and exploration of data following a multidimensional approach comprised of aggregation levels available in cartographic displays as well as in tabular and diagram displays" (Bédard et al., 2001). Spatial information could be used to enhance the decision-making process by improving the user interface through interactive maps.

2.3.1. SOLAP Main Concepts

The multidimensional models of SOLAP introduce geographic information as analysis axes thanks to spatial dimensions (Bédard et al., 2001; Malinowski & Zimányi, 2004; Silva et al., 2008). A spatial dimension is described by spatial hierarchies that are characterized by the presence of the geometric attribute in the different levels (called spatial levels). A spatial hierarchy has at least one spatial level, and it can also have non-spatial levels. Members of different spatial levels are linked by the topological relationships of intersection or inclusion (Malinowski & Zimányi, 2007). An example is shown in Figure 2a. It reconsiders the multidimensional application previously described in Section 2.2.1 adding a spatial attribute to the levels "Unit" and "Zone", thus transforming them into spatial levels.

Spatial information can be used in the decision process as axes of analysis, but also as subjects of analysis (spatial measures). Spatial measures can be analyzed through non-spatial and/or spatial dimensions. A spatial measure is sometimes represented as a collection of spatial objects (Silva et al., 2008; Malinowski & Zimányi, 2004; Marchand et al., 2003; Rivest et al., 2005; Stefanovic et al., 2000, Sampaio et al., 2006, Glorio & Trujillo,

Figure 2. SOLAP model with a) spatial dimension, b) spatial measure

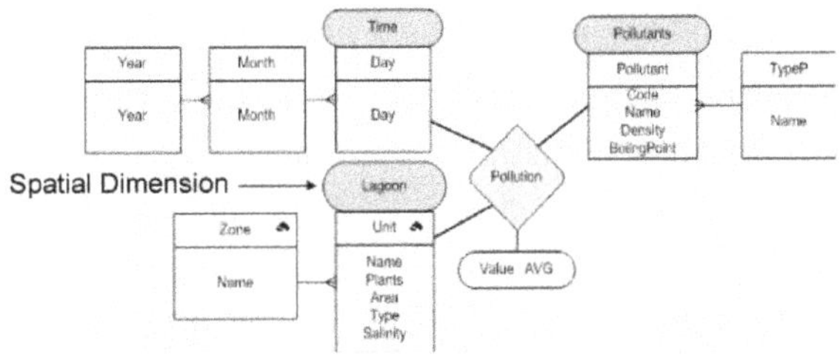

(a)

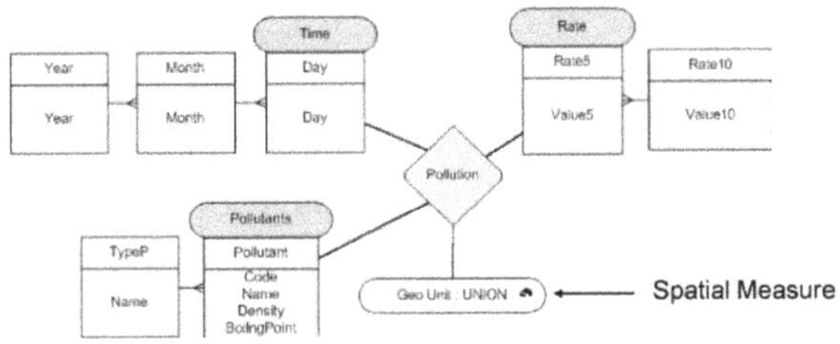

(b)

2008) and/or results of spatial operations (Malinowski & Zimányi, 2004; Rivest et al., 2005). Therefore, spatial objects have to be aggregated using spatial functions (i.e., intersection, union, etc.) (Shekar et al., 2001). The conceptual model is provided in Figure 2b, using the formalism proposed by Malinowski and Zimányi (2004). In this model, the geometric attributes are represented using pictograms. In this example, the geometry of units is the measure. The dimensions are Time, Pollutants, and the Pollution values. The pollution values are represented by the dimension "Rate". The level "Rate5" represents groups of pollution values close to 5 mg/l, for example 0-5 mg/l, 6-10 mg/l, etc. The level "Rate10" represent groups of pollution values close to 10 mg/l. The spatial measure is aggregated using the geometric union. Thus, a possible query is "*What is the geometric union of units polluted by Zinc with a pollution rate value of '40-45 mg/l' per day?*"

The spatial data warehouse is explored using specific spatio-multidimensional operators. Hernandez et al. (2005), Matias and Moura-Pires (2007), Rivest et al. (2005), Sampaio et al. (2006), and Scotch and Parmanto (2006) define "spatial-drill down" and "spatial roll-up" drill operators that permit the navigation within spatial dimensions and the aggregation of measures (i.e., climbing from the "Unit" to the "Zone" level in the spatio-multidimensional model of Figure 2a). Rivest et al. (2005) call "spatial drill-across" a slice operation on the members of the spatial dimension selected directly by the user through the user-interface of the SOLAP client. These operators are classical OLAP operators applied to spatial dimensions. Bimonte et al. (2007) define the "spatial slice" operator as a slice operation on a spatial data warehouse. The spatial slice operator uses functions to select a part of the spatial data warehouse. For instance, using the spatio-multidimensional model of Figure 2a, a spatial slice operator could be applied to select spatial members of the "Unit" level located at 50 Km from Venice city. Scotch and Parmanto (2006) define two new slice operators: "buffer" and "spatial drill-out". The operation "buffer" uses a function that creates a buffer around a spatial object to cut the spatial data warehouse. Only the spatial members that are topologically included in the buffer region are used to slice the spatial data warehouse. The operation "spatial drill-out" slices the spatial data warehouse on members that are topologically adjacent to a particular spatial member selected by the user.

2.3.2. SOLAP Tools

Hernandez et al. (2005) show that GIS functionalities are necessary and complementary to OLAP functionalities for an effective spatio-multidimensional analysis. This is the reason why some Spatial OLAP tools have been developed in the last years that integrate GIS functionalities. These solutions are called OLAP-GIS integrated tools (Rivest et al., 2005), and are based on a three-tier architecture.

The first tier of SOLAP is the spatial data warehouse that has to handle the complex data structures used to represent spatial measures and spatial dimensions (Malinowski & Zimányi, 2007). It is usually implemented using a Spatial DBMS.

The second tier is the SOLAP server, which adds spatial functions to the traditional OLAP functions (Kimball, 1996):

- Spatio-multidimensional operators,
- Spatial measures and spatial dimensions (Rivest et al., 2005)
- Avoiding multiple counting (i.e., considering the same fact twice) during the aggregation process (Bimonte et al., 2006).
- Providing a mechanism to calculate materialized spatial views (Colonnese et al., 2005).

The third tier is the SOLAP client. The functionalities of the OLAP client are adapted in order to permit spatial and multidimensional analysis at the same time. The SOLAP client has to combine and synchronize classical OLAP displays with interactive maps. Maps show members of spatial dimensions and/or spatial measures. Numerical measures are visualized on maps using visual variables and graphic displays (i.e., bars, pies, etc.). Interactive maps are the main instruments for the analysis of spatio-temporal data (Andrienko et al., 2003). Indeed, SOLAP maps have to allow for exploring and analyzing spatial data warehouses without using any ad-hoc query language or knowing its structure. The simple interaction with the cartographic component has to trigger spatio-multidimensional operators so that the visualization reflects the user's actions. Tabular, graphic and cartographic displays are synchronized (Rivest et al., 2005); in other words, an action (for example a drill-down or a roll-up) on a component (for instance a map) is propagated to the other components (pivot table and graphic displays). In fact, this visualization and exploration technique, called multiple coordinated windows (Shneiderman & Plaisant, 2004), is necessary for an effective spatio-temporal exploration process (Andrienko et al., 2003). Moreover, interactive maps can sometimes be coupled with advanced geo-visualization techniques. Geo-visualization is the integration of scientific visualization, cartography, image analysis, and data exploration. Geo-visualization defines a theory, analysis methods and tools to represent and discover spatial knowledge (MacEachren & Kraak, 1997). In this context, cartographic procedures pay more attention to the usability and interaction issues of user interfaces. Geo-visualization techniques are characterized by their high interactivity and flexibility.

Finally, in addition to navigation operators, a SOLAP tool needs to provide traditional GIS functionalities to analyze spatial dimensions and/or measures. Indeed, the user interface has to allow adding raster and vector layers, personalizing cartographic displays, using metric operators (i.e., calculating distance, surface area, etc.), querying the spatial database, using spatial analysis operators and data mining methods (Scotch & Parmanto, 2006), etc.

3. GEOGRAPHIC AND COMPLEX MEASURES

In this section, we introduce the concepts of complex and geographic measures. They enrich the SOLAP paradigm by introducing the semantic component of geographic information and the flexibility of the spatial analysis. They enable the analysis of geographic and complex data in a symmetrical way. On the other hand, they raise several modeling and implementation issues.

3.1 Definitions

The semantic component of geographic information modeled by the descriptive attributes is mandatory for an effective spatial analysis process (Longley et al., 2001). However, as described in the previous section, spatial measures represent only the spatial component of geographic information, i.e., the geometry and its position on the earth's surface. To overcome this limitation, we introduce the concept of geographic measure.

Definition:

- **Geographic measure:** We define a geographic object as a geo-referenced entity of the real world described by spatial and alphanumeric attributes, and associated to other geographic/objects through spatial, thematic and "map generalization" relationships. A Geographic measure is a geographic object (Bimonte et al., 2006b). Its formalization is provided in Section 3.3.1. Our definition is similar to that of (Damiani & Spaccapietra, 2006).

Geographic measures enable the extension of spatio-multidimensional analysis to the semantic component of geographic information (alphanumeric attributes and spatial/thematic/"map generalization" relationships). An example of a multidimensional application with a geographic measure is shown in Figure 3a.

The dimensions are the time, the pollutants and the pollution values ("Rate"). The measure is the geographic object "Unit," which represents a unit of the Venice lagoon. This geographic object is composed of a spatial description (attribute "Geom") and several descriptive attributes. This model allows us to analyze the units of the lagoon as functions of time, pollutants and pollution value. It permits us to answer the query "*what is the geometric union of all units polluted by zinc per month, their salinity, their type and their plants?*" An aggregation of units corresponds to the aggregation of its descriptive and spatial attributes. In this application, we use the geometric union for the geometry, a concatenation operator that returns a list of plants (called "list") for the plants, an area-weighted average (called "ratio") for the type, and the average for the salinity. The surface area is not aggregated because it can be calculated using the geometry of the union. These descriptive attributes can be useful in the decisional process, for example, a high salinity value could be associated with a high iron pollution value. Note that some descriptive attributes may be irrelevant to the analyses (e.g., the name of a unit in Figure 3a).

Moreover, a geographic object can also be involved in some relationships. These relationships open new analysis multidimensional analysis scenarios because they can be used to change the subject of the multidimensional analysis dynamically (Damiani & Spaccapietra, 2006). For example, since a unit belongs to a zone, it is possible to perform the following query: "*which zones of the lagoon are polluted by zinc, per day and the pollution value?*". Therefore, exploiting the relationships between geographic objects al-

Figure 3. SOLAP model with a) Geographic measure "Unit", b) Complex Measure "Pollutant"

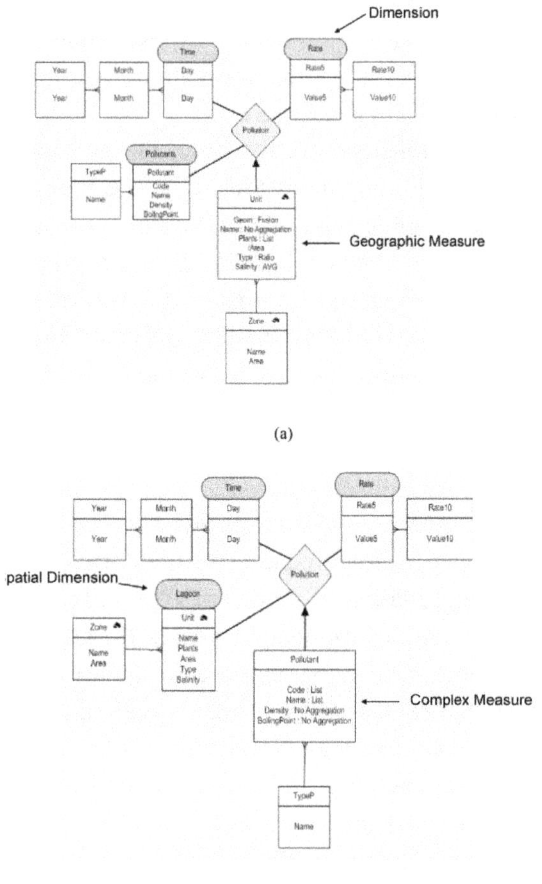

lows for multigranular spatial analysis (Camossi et al., 2008). Here, the data are represented at different levels of details (granularities) i.e., units and zones.

Existing SOLAP models use different data structures to represent spatial members and spatial measures. Users of SOLAP cannot permute measures with dimensions at will. This asymmetry obliges users to design two different spatio-multidimensional models: one with a spatial measure and another with a spatial dimension. Thus, since it is important to try to include geographic information as dimensions as well as measures, a symmetric vision of measures and dimensions, which should permit changing the multidimensional model in real time, will improve the multidimensional capabilities of traditional SOLAP models. In the context of OLAP models, the need for symmetry of measures and dimensions has been already noted in (Pinet & Schneider, 2009). In the case of SOLAP, spatial analysis processes cannot always define a priori the subjects of analysis; a certain level of flexibility is needed (Mitchel, 2005). For example, it would be very interesting to allow users of SOLAP to transform the model of Figure 3a into the model of Figure 3b. The data of the two models are the same, but the subjects of analyses are different. The new SOLAP application permits us to answer queries like "*what are the pollutants in each zone of the lagoon per month?*" In this way, the spatial analyst could be able to visualize polluted units (Figure 3a) as functions of pollutants and time. Then, she/he can decide to exchange the geographic measure "Unit" with the "Pollutants" dimension (Figure 3b) to analyze the spatial evolution of pollutants in the lagoon. This could allow the discovery of possible trends and relationships between pollutants and their geo-localization.

Extending the work of (Bimonte et al., 2006), we now define the concept of complex measure. A complex measure is equivalent to a geographic measure. The only difference is that a complex measure does not present a geometric attribute.

Definition:

- **Complex measure:** A complex measure is an object described by alphanumeric attributes and associated to other geographic objects through thematic relationships (its formalization is provided in Section 3.3.1).

3.2. Requirements

Supporting geographic/complex measures and dimensions in a symmetrical way raises some modeling and visualization problems.

First, in order to support the symmetry of complex/geographic measures and dimensions, the logical model implemented into the spatial data warehouse must not distinguish between dimensions and measures. Thus, spatio-multidimensional logical models using the star schema have to force measures to be dimensional keys and the aggregation of complex and geographic measures (set of geometric, numerical and non-numerical values) cannot be done using the traditional SQL functions provided by existing OLAP servers. Indeed, the (S)OLAP server has to provide methods to define, directly or indirectly (calculated measures, user-defined functions, etc.), user-defined aggregation functions taking as inputs the values of several spatial, numerical and non-numerical attributes. These functions must return a result for each attribute (e.g., area-weighted average for the socio-economic type of units). In addition, the logical model and the SOLAP server have to support two advanced multidimensional features: many to many relationships and multi-valued measures, i.e., a unit can be polluted by several pollutants, and a pollutant can pollute several units.

Finally, simple textual labels are not adequate to analyze geographic and complex measures visually. It is essential to combine adequate semiology rules (Bertin & Bonin, 1992) and advanced geo-visualization techniques (such as space-time cube, multi-maps, etc.) for an effective analysis of spatio-multidimensional data. Then, the SOLAP thematic maps have to be reformulated, adopting adequate visual variables to display alphanumeric attributes of geographic/complex measures, and allowing users to compare them visually along the various dimensions of the spatio-multidimensional model.

3.3. An Overview of the GeoCube Model

Bimonte et al. (2006) present a multidimensional model called GeoCube that supports all the requirements related to geographic/complex objects and the symmetrical representation of measures and dimensions. Our SOLAP prototype is based on that model. In the following sections, we present the main concepts of the model.

3.3.1. Data Model

The following are the main concepts of the multidimensional data model: Entity, Hierarchy and Base Cube. Measures and dimension members are represented as Entities. An Entity is a set of attributes and functions used to represent derived attributes (data calculated using other data). For example, derived attributes are necessary to store the calculated surface area, the perimeter, etc. of geographic objects.

Definitions:

- **Entity Schema and Entity Instance:** An Entity Schema S_e is a tuple $\langle a_1, \ldots a_n, [F] \rangle$ where
    - a_i *is an attribute defined on a domain* $dom(a_i)$
    - *F, if it exists, is a tuple* $\langle f_1, \ldots f_m \rangle$*, where* f_i *is a function defined on a subset of attributes* $a_i, \ldots, a_k$.
    - An Instance of an Entity Schema S_e is a tuple t_i such that,
    - *If F exists then* $t_i = \langle val(a_1), val(a_n), val(b_1), \ldots, val(b_m) \rangle$ *where* $val(a_i) \in dom(a_i)$ *and* $val(b_j) = f_j(val(a_i), \ldots, val(a_k))$
    - *If F does not exist, then* $t_i = \langle val(a_1), \ldots, val(a_n) \rangle$ *where* $val(a_i) \in dom(a_i)$
    - *F* is used to model derived measures and/or dimension attributes.

Definition:

- **Geographic Entity Schema:** An Entity Schema S_e is a Geographic Entity Schema if the domain of one attribute is a set of spatial objects. The Geographic Entity Schema representing lagoon units is S_{unit} = ⟨geometry, plants, type, salinity, f_{area}⟩, where f_{area}: dom(geometry) → N is a function to calculate the area of a unit (see Figure 3). An instance (the real geographic objects) of S_{unit} is ⟨pt01, Sparta Marittima, Commercial, 21, 110⟩. Its cartographic representation is shown in Figure 1.

We call $I(S_e)$ the set of all instances of n (Geographic) Entity Schema S_e. Entities are organized in hierarchies thanks to the concepts of Hierarchy Schema and Hierarchy Instance. A level of a hierarchy is an Entity Schema, and a member is an Entity Instance. The Hierarchy Schema organizes levels into a lattice. The levels of members (i.e., Entity Instances) are organized in a tree structure. The root is an instance of the Entity Schema that corresponds to the top level of the lattice. Leaves are the instances of the Entity Schema that represent the bottom level of the lattice.

Definition:

- **Hierarchy:** A Hierarchy Schema is a tuple $H_h = \langle L_h, \lfloor_h, \lceil_h, ⇞_h \rangle$, where
  - *L_h is a set of Entity Schemata,*
  - *$\lfloor_h$ and $\lceil_h$ are two Entity Schemata and $\lceil_h$ contains one instance ('all'),*
  - *$⇞_h$ is a partial order defined on the levels of the hierarchy $(L_h \cup \lfloor_h \cup \lceil_h)$ and $⇞_h$ is a lattice where $\lfloor_h$ and $\lceil_h$ are the bottom and the top levels of the order, respectively.*

Let $L(H_h)$ be the set of all levels of H_h: $L(H_h) = \{ L_h \cup \lfloor_h \cup \lceil_h \}$.

An Instance of a Hierarchy H_h is a partial order $\uparrow_h$ defined on L (H_h), such that:

- *if $t_i \uparrow_h t_j$ then $S_i ⇞_h S_j$, where $t_i \in I(S_i)$ and $t_j \in I(S_j)$*
- *$\forall\, t_i$ not belonging to the top level, then $\exists\, t_j$ such that $t_i \uparrow_h t_j$*
- *$\forall\, t_i$ not belonging to the bottom level, $\exists\, t_j$ such that $t_j \uparrow_h t_i$*

An example of a Hierarchy Schema representing the spatial hierarchy of the Venice lagoon, "Unit"<"Zone," is $H_{lagoon\_spatial}$ = ⟨{S_{zone}}, S_{unit}, S_{all_unit}, $⇞_{lagoon\_spatial}$⟩ where (S_{unit} $⇞_{lagoon\_spatial}$ S_{zone}) and (S_{zone} $⇞_{lagoon\_spatial}$ S_{all_unit}) (see Figure 2a).

The *Base Cube* denoted by BC represents the facts table.

Definition:

- **Base Cube:** A Base Cube Schema BC_{bc} is a tuple $\langle H_1, \ldots H_n, \delta \rangle$ where:
  - *$\forall\, i \in [1,\ldots n]$, H_i is a Hierarchy Schema and all the hierarchies H_i are disjoint*
  - *δ is a Boolean function defined on $I(\lfloor_1) \times \ldots \times I(\lfloor_n)$* (δ is defined on the bottom levels of $H_1, \ldots H_{m+1}$ and it returns a Boolean value: true if the set of Entity Instances exist in the spatial complex data warehouse, false otherwise).

An Instance of a Base Cube Schema BC_{bc}, denoted $I(BC_{bc})$, is a set of tuples $I(BC_{bc}) = \{\langle t_1^j, \ldots t_n^j \rangle, j=1,\ldots p\}$ where $\forall\, i \in [1,\ldots n]$, $t_i^j \in I(\lfloor_i)$ et $\delta(t_1^j, \ldots t_n^j)$ is true

The Base Cube does not explicitly define whether the bottom level of hierarchies uses geographic or complex object as measures. Therefore, the representation of dimensions and measures is symmetric.

The Base Cube Schema of our case study is $BC_{corila} = \langle H_{pollutants}, H_{time}, H_{lagoon\_spatial}, H_{rate}, \delta \rangle$, where δ:

$I(S_{pollutant}) \times I(S_{day}) \times I(S_{unit}) \times I(S_{rate5}) \rightarrow$ Boolean, i.e., the instances of the bottom levels of the hierarchies $H_{pollutants}$, H_{time}, $H_{lagoon\_spatial}$ and H_{rate}. An example of an instance of BC_{corila} is shown in Table 3.

In this example, each bottom level of a hierarchy (object and/or geographic object), with all its attributes, can be used as a measure.

3.3.2. Multidimensional Query Model

Since the Base Cube does not define what entity is used as a measure, in order to define a spatio-multidimensional query we introduce the concept of a View. A View defines the geographic or complex measures, aggregation functions, and dimensions levels used by a SOLAP query. We define an "Aggregation Mode" as a set of aggregation functions applied to the attributes of the geographic/complex object used as the measure. For example, considering the spatio-multidimensional model of Figure 3a, the Aggregation Mode for the measure "Unit" is composed of the aggregation functions "fusion", "ratio", "list" and "average". A "View" is composed of a Base Cube, a set of dimensions levels, an Aggregation Mode and a Boolean function that plays the role of the Base cube Boolean function. Consider the following SOLAP query: "*what are the regions of the lagoon (aggregation of units using the geometric union), their salinities, their types and their plants polluted per pollutant, month and pollution value*?" The View associated to this query is: $V_{corila\text{-}unit\text{-}month} = \langle BC_{corila}, \langle S_{pollutant}, S_{month}, S_{rate5} \rangle, \Theta_{fusion\text{-}unit}, \gamma \rangle$, where BC_{corila} is the Base Cube that represents

Table 3. Instance of BC_{corila}

$S_{pollutant}$.name_pollutant	S_{day}.day	S_{unit}.name_unit	S_{rate5}.value5
Zinc	1-1-01	Murano	10-14
Methane	2-1-01	Mazzorbo	45-49
…	…	…	…
Sulfur Trioxide	26-9-05	Coa di Latte	20-24

the SOLAP application, $S_{pollutant}$, S_{month}, and S_{rate5} are the dimensions levels used by the query, and $\Theta_{fusion\text{-}unit}$ is a set of aggregation functions for the attributes of the geographic measure "Unit" as shown in Figure 3a.

Another example of View is $V_{corila\text{-}unit\text{-}day} = \langle BC_{corila}, \langle S_{pollutant}, S_{day}, S_{rate5} \rangle, \Theta_{fusion\text{-}unit}, \gamma \rangle$, which represents the polluted units per day, the pollutant and the pollution value.

The symmetrical representation of measures and dimension provided by the Base Cube BC_{corila} allows us to define queries with pollutants as measures, always using the same Base Cube (facts table data): $V_{corila\text{-}pollutant} = \langle BC_{corila}, \langle S_{unit}, S_{day}, S_{rate5} \rangle, \Theta_{list\text{-}pollutant}, \gamma \rangle$ i.e., the pollutants present in units per day and the pollution value.

4. GEWOLAP: A TOOL FOR THE ANALYSIS OF COMPLEX SPATIAL DATA WAREHOUSES

In this section, we present our OLAP-GIS integrated solution. GeWOlap supports spatial dimensions, traditional measures, and geographic and complex measures. GeWOlap also supports a symmetrical representation of measures and dimensions. It extends drill operators to aggregate geographic and/or complex objects, and slice operators to use spatial and alphanumeric operations. It provides three operators that create on the fly new spatial members, resulting from the spatial transformation methods buffer, overlay and dissolve. GeWOlap is based on a three-tier architecture composed of a Spatial DBMS, a ROLAP Server and a front-end client integrating a Web OLAP client and a Web GIS client.

The relational model of the Spatial Data warehouse is based on a logical model that supports symmetrical geographic/complex measures and dimensions. The OLAP Server extends the aggregation process, and the user interface integrates ad-hoc geo-visualisation techniques for the effective visualization, exploration and analysis of geographic and complex multidimensional data.

4.1. Spatial Data Warehouse

The data warehouse tier is implemented using Oracle (ORACLE, 2009). A main advantage of Oracle is that it provides native support for spatial data. Spatial data are stored using the vector format, according to the strongly coupled approach. The strongly coupled approach integrates spatial and alphanumeric data into a single DBMS (Spatial DBMS). Spatial DBMSs allow GIS applications to take advantage of the main properties of DBMS: security, concurrency, flexibility and scalability.

GeWOlap supports numerical and spatial measures (as existing SOLAP tools). In addition, it supports symmetric geographic/complex measures and dimensions through a particular logical model and aggregation process. GeWOlap uses a facts table that makes no difference between measures and dimensions. It contains only the identifiers of the complex and/or geographic objects representing the most detailed levels of the dimensions. In other words, this is a facts table without measures (i.e., a fact-less table) (Kimball, 1996). The symmetry between dimensions and measures is broken only in the XML schema representing the multidimensional application of the OLAP Server. In our logical model, measures are defined as the most detailed level of dimensions. This logical model allows the modeling of many to many relationships between dimensions and measures. Aggregated measures are represented using another table, because they may be different from measures. Aggregated measures could have more or less of attributes of detailed measures (i.e., aggregated pollutants are identified only by names, but do not have a code or density). An example is shown in Section 5.1. In the proposed logical model, the facts table corresponds to the BaseCube concept, and the dimensions correspond to the Hierarchy concept of the GeoCube model.

4.2. OLAP Server

The second tier of the architecture is implemented using the OLAP server Mondrian (PENTAHO, 2009). Mondrian is an open source Java OLAP server, which provides an easily customizable, full featured OLAP server, based on a relational database backend. Multidimensional models (facts, measures, dimensions, etc.) are specified in an XML file. This file represents the logical model.

We propose an extension of Mondrian. In our system, a Mondrian XML schema is defined for each multidimensional model with a complex and/or geographic measure. On the other hand, since at the logical level there is symmetry between measures and dimensions, these schemata all present the same facts table and the same dimensions. Therefore, in order to support geographic and complex measures, we have modified Mondrian to handle user defined aggregation functions in addition to classical SQL functions. This type of aggregation function is called "UserDefinedAggregation". It is defined in Oracle through a user-defined aggregation function implemented in PL/SQL.

In particular, the XML schema file references the aggregate function "UserDefinedAggregation" defined on the attribute representing the identifier of the measure. Mondrian translates a multidimensional query into a SQL query using this aggregation function instead of the classical SQL aggregation functions. Figure 4-1 shows the structure of the SQL queries produced. Note that the SQL statement "distinct" is used in the aggregation function in order to avoid the multiple counting of measures. This SQL query is processed by Oracle. The values of attributes of the detailed measures are aggregated using PL/SQL user-defined aggregation functions. As a result, a new aggregated measure is created (Figure 4-2). Its identifier is returned to the SOLAP client (Figure 4-3).

Therefore, GeWOlap easily supports the symmetry of dimensions and geographic/complex measures (set of geometric and alphanumeric values) using a simple OLAP Server model, because the aggregations of spatial and alphanumeric values are handled transparently by the spatial data warehouse tier. The user-defined aggregation functions implement the concept of Aggregation Mode of the GeoCube model.

Geographic multidimensional applications are usually characterized by multi-valued measures. For example, in the model of Figure 3a, several units can be polluted by zinc. In order to support multi-valued measures, GeWOlap uses materialized views, which we also call aggregate tables. A materialized view is a relational table containing aggregated measures values to improve the query performance. Then, we define an aggregated table collapsing the dimension containing the measure, and having all other dimensions at the most detailed levels. Therefore, the queries that concern detailed measures do not aggregate the measures using the aggregation function "UserDefinedAggregation", but they use the values of the measure from the aggregated table. Unlike the facts table, which contains references to geographic/complex objects, the aggregated table contains a reference to a list of geographic/complex measures, allowing GeWOlap to support multi-valued measures.

4.3. User Interface

The client tier is implemented using JPivot (JPIVOT, 2009) to provide JSP pages and a MapXtreme Java (MAPINFO, 2009) applet for map visualization and interaction. The client provides an interactive user interface that synchronizes the tabular component, the graphic displays and a cartographic component. Multidimensional and GIS operators are available through simple interactions with the pivot table and the interactive map.

The user interface is composed of two main panels: the Data Panel and Control Panel (Figure 5).

Figure 4. Aggregation process for geographic/complex measures

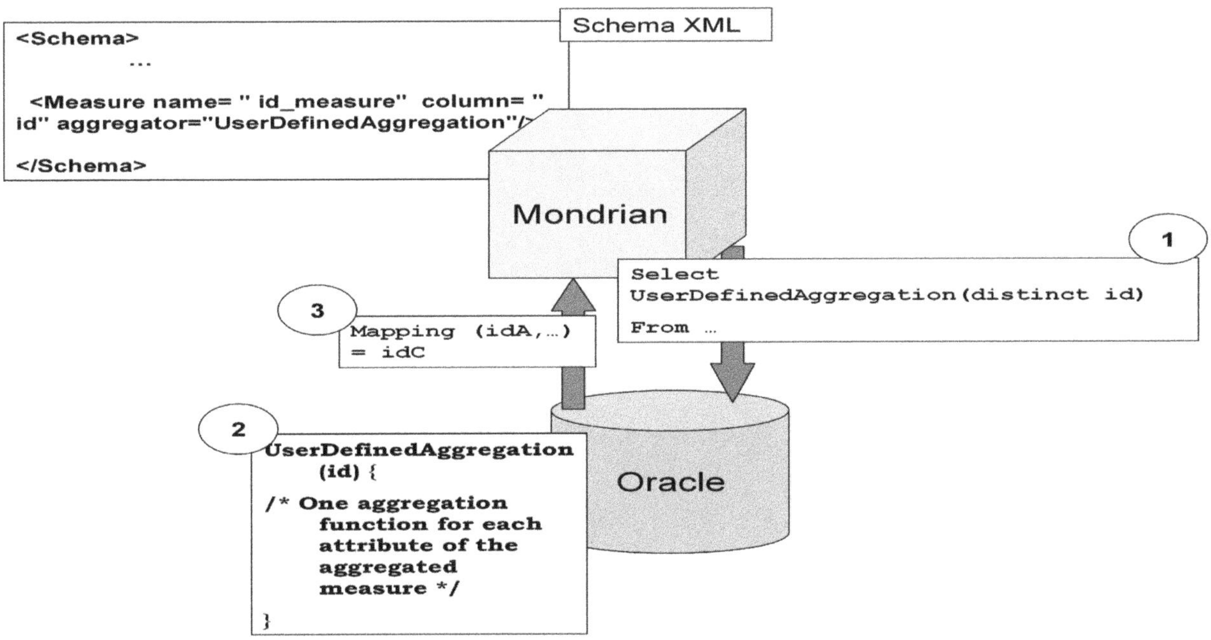

The Control Panel contains three parts:

1. Navigation lists all available cubes, sorted by project.
2. Display/Export permits the user to select the visible pivot table, graphic display and/or map and to export them to various formats.
3. Analysis provides the Cube Browser tool. The Cube Browser allows the user to map the hierarchies on table axes and to select members to slice or dice the cube.

The Data Panel is composed of three frames: the pivot table, the graphic display and the interactive map. Multidimensional data are visualized using the pivot table, the graphic displays and the map. Members of the spatial dimension represent the geographic objects of the map; measures are visualized using colors, bars, or pie charts. Geographic measures can be displayed on the map. All the multidimensional operators are accessible through the interaction with the pivot table and the map. These visual interactive components are associated with the GIS Toolbar, the Geographic Multidimensional Toolbar and the Multidimensional Toolbar. The GIS toolbar offers the following GIS functionalities: Zoom in/out, Pan, Retrieve, Map Print, Map Export, Map Refresh, Rule Tool and Control Layer. The Geographic Multidimensional Toolbar provides all spatio-multidimensional operators through interaction with the map. The drill operators are Drill-down Replace, Drill-down Position, Expand-all and Roll-up. The Drill-down Replace operator enables drilling from a selected spatial member to its child members in the dimension hierarchy,

Figure 5. User interface of GeWOlap: Control Panel and Data Panel

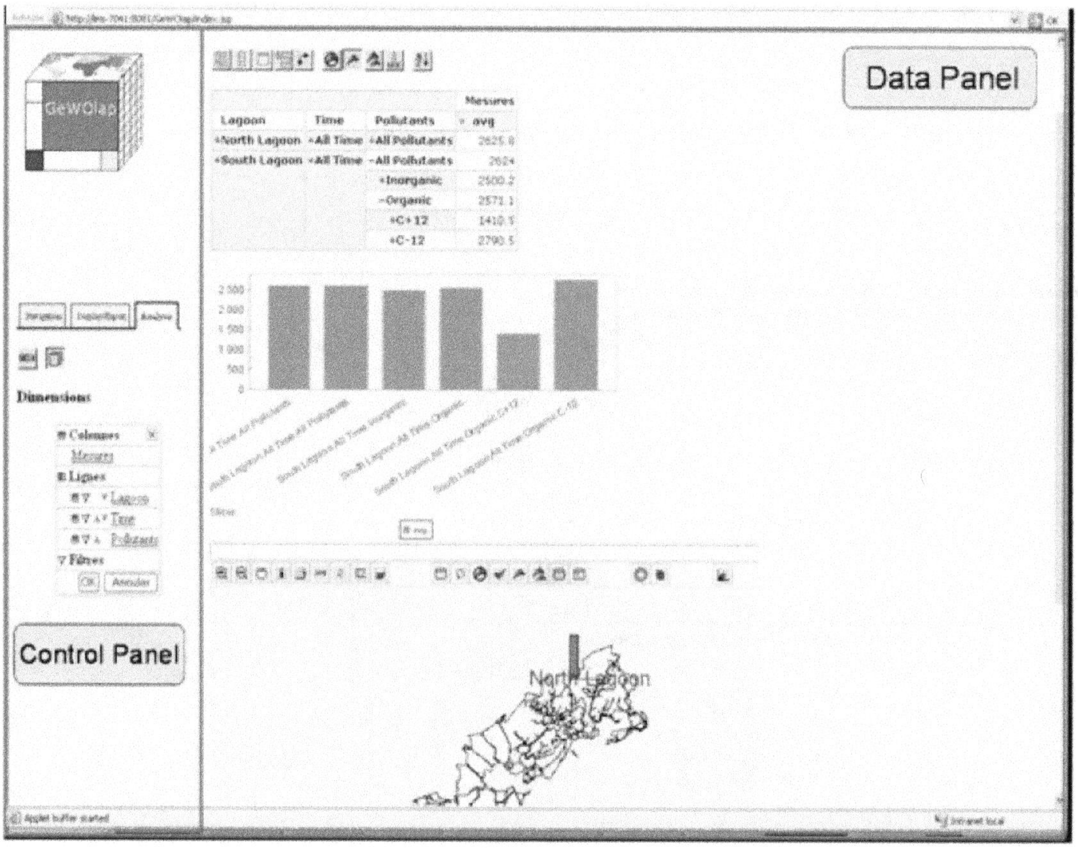

hiding the parents, whereas the Drill-down Position shows its parents. The Expand-all operator enables drilling from all visible spatial members to their child spatial members. The Roll-up permits climbing up in a spatial dimension. In addition to the Cube Browser of the Control Panel, GeWOlap offers two operators to select spatial members: Slice Position and Slice Predicate. The Slice position selects members using mouse pointing, whereas Slice Predicate uses spatial and/or alphanumeric operations defined as Spatial SQL queries. Three operators that modify dynamically spatial dimensions are also implemented: Overlay, Buffer and Dissolve. The Multidimensional Toolbar provides the same drill operators of the Geographic Multidimensional Toolbar on spatial and classical dimensions through the interaction with the pivot table.

GeoWOlap also implements spatio-multidimensional drill operations on spatial dimensions; the details can be found in (Bimonte et al., 2007).

The introduction of geographic information as a subject of analysis suggests a reformulation of the visualization techniques used to represent numerical measures. In particular, the values of the cells of the pivot table do not contain numerical values, but rather contain the identifiers of the geographic measures. This feature motivates an extension of Mondrian and JPivot to handle alphanumeric measures. Textual labels are insufficient for an effective spatio-multidimensional analysis (Rivest et al., 2005). Therefore in order to visualize geographic measures, the cartographic component of GeWOlap displays two maps. One map represents the set of all geographic measures and the other map shows the geographic measures chosen dynamically by the user. In the following, we discuss how GeWOlap allows the visualization and the exploration of complex measures. The visualization of the measures in the pivot table is similar to that of geographic measures - the identifiers of the (aggregated) measures are shown in the cells of the pivot table. Traditional visualization techniques used for numerical measures are not adequate to visualize complex measures because they are not numerical values but sets of alphanumeric attributes. Thus, in order to visualize complex measures on a spatial dimension, we use a set of choropleth maps (i.e., multimaps) where each complex measure or set of complex measures is represented by a color, as this visual variable is well adapted to the representation of nominal values (Bertin & Bonin, 1992).

In conclusion, GeWOlap allows the interactive exploration and analysis of multidimensional complex and geographic data through different geo-visualization techniques (multiple coordinate windows, multi-maps and interactive maps) coupled with pivot tables whose cells are textual labels.

5. AN ENVIRONMENTAL CASE STUDY

In this section we present the application of our prototype with simulated environmental data for the SOLAP models of Figure 3.

5.1. Logical Model

As described in the previous section, the spatial data warehouse provides a symmetrical representation of measures and dimensions. Figure 6 shows the logical model including units of Venice as measures (table "Unit"). Aggregated units are represented by the table "Unitagg". This logical model also allows using pollutants as measures. A table representing the aggregated pollutants is defined, and denoted "Pollagg". Note that in this type of models, all of the most detailed levels of dimensions can be used as measures but in practice, not all levels are interesting. For example, in our case study, the time could not be used as a subject of analysis. This logical model corresponds to *Base Cube* $BC_{corila} = \langle H_{pollutants}, H_{time}, H_{lagoon\_spatial}, H_{rate}, \delta \rangle$.

5.2. OLAP Model and Aggregation Process

The representation of measures and dimensions is not symmetric in the OLAP Server. The common part of schemata of the OLAP Server using the units and the pollutants as measures is shown in Figure 7. There are four dimensions, "Time", "Lagoon", "Rate" and "Pollutants," and the facts table "Corilafacts". They do not depend on the measure used. Only the measure and the definition of the aggregate table change as functions of the measure. The XML tags for the measures "Unit" and "Pollutant" are shown, respectively, in Figure 8 (a and b).

In order to model multi-valued units and pollutants, two materialized views are defined, as described in the previous section. Figure 9 (b and c) show Aggregated Table tags for units and pollutants. We can notice that the spatial dimension is collapsed when units are used as measures (Figure 9b), whereas in the case when the measures are pollutants, then the pollutant dimension is collapsed (Figure 9c). Figure 9a shows the relational representation of the materialized view for units. Here, the aggregated measure "12" in table "AGGFACTSUNITS", which represents the materialized view, is a list of units ("UNITSLIST" table). Thus, queries using most detailed dimension levels will use the materialized view that contains lists of units. The same relational representation is used for pollutants. This allows for *multi-valued measures*.

We now describe the aggregation process for the spatio-multidimensional model, using pollutants as measures. Suppose we wish to aggregate pollutants with identifiers "15", "17" and "24". The first step of the aggregation process (see Figure 4-1) is the generation of the SQL aggregation statement by Mondrian "Select UserDefinedAggregation(distinct POLLUTANT_CODE)...". The second step (Figure 4-2) corresponds to the "real" aggregation of these complex objects. The result of the aggregation is the identifier ("15-17-24"). For the aggregation of the name attribute, we have defined a user-defined aggregation function, called LIST; this function uses the "distinct" statement of SQL. Finally, the identifier "15-17-24" is returned to the SOLAP client (Figure 4-3).

5.3. Navigation and Visualization

In this section, we describe the visualization of the geographic measure "Unit" and the complex measure "Pollutant" in GeWOlap's user interface.

Suppose that the user wants to answer the query *"what are the polluted units, per day, pollutant and pollution value?"* ($V_{corila\text{-}unit\text{-}day} = \langle BC_{corila}, \langle S_{pollutant}, S_{day}, S_{rate5}\rangle, \Theta_{fusion\text{-}unit}, \gamma\rangle$)

The pivot table representing the result of this query is shown in Figure 10. This query uses the aggregated table; all the most detailed levels of the dimensions are used. The query returns the identifiers of the sets of geographic measures.

The cells of the pivot table display the names of the units. For instance, on 19-9-05, the set of units with the identifier "5" were polluted by methane with a pollution value of 30-34 mg/l. The user selects a cell of the pivot table using a text field, and GeWOlap displays the map representing all the units in white and the polluted units in blue - for example, "Murano", "Coa di Latte" and "Mazzorbo". A geographic measure is characterized by a set of alphanumeric attributes. For instance, by clicking on "Mazzorbo", the user can visualize the values of its alphanumeric attributes. Suppose that the analyst decides to aggregate units on the month level in the temporal dimension; he selects the member "September 2005" with the Cube Navigator. This action is a Roll-Up operation and implies the aggregation of the geographic measures ($V_{corila\text{-}unit\text{-}month} = \langle BC_{corila}, \langle S_{pollutant}, S_{month} S_{rate5}\rangle, \Theta_{fusion\text{-}unit}, \gamma\rangle$). The "UserDefinedAggregation" function is applied, and the associated ad-hoc functions for the spatial and alphanumeric attributes of the units are performed. The result of the Roll-Up is shown in Figure 11. Now

Figure 6. Logical model with geographic measure "Unit" and complex measure "Pollutant"

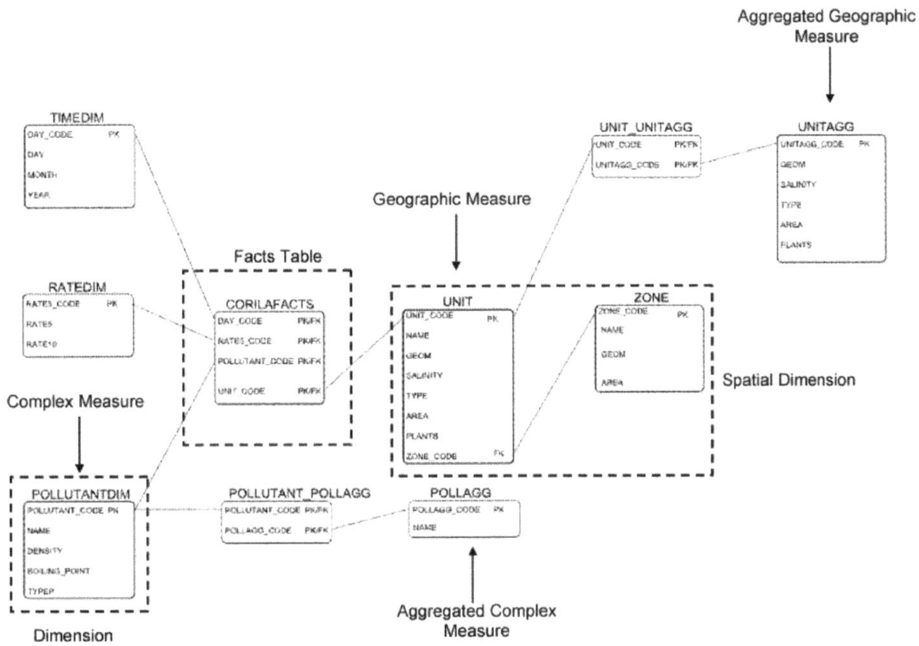

the pivot table contains the month "September 2005," and the cells are the identifiers of the aggregated geographic measures. For example, inn 9-05, the aggregated zone with the identifier "231418" is polluted by methane with a pollution value of 30-34 mg/l. The geometry of this geographic measure is the geometric union of the geometries of the detailed measures, "Coa di Latte", "Mazzorbo", "Murano" and "Tessera". Note that this geographic object has no name and that its salinity is calculated using the average (25.5). The list of plants contains all the plants of all the units, and the type is "commercial".

In the following, we discuss how GeWOlap enables the analysis (visualization and exploration) of complex measures.

Let us consider the spatio-multidimensional model with pollutants as measures (see Figure 3b). This application presents a spatial dimension and a complex measure. As described in the previous section, in order to effectively visualize complex measures, GeWOlap uses a set of choropleth maps (multi-maps). In this type of representation, each complex measure or set of complex measures is represented by a color. To clarify our solution, we describe how GeWOlap allows for visualizing the result of this query "*what pollutants are present in each unit, for each day and each pollution value?*" ($V_{corila\text{-}pollutant} = \langle BC_{corila}, \langle S_{unit}, S_{day}, S_{rate5}\rangle, \Theta_{list\text{-}pollutant}, \gamma\rangle$). The pivot table is shown in Figure 12a.

For this pivot table, GeWOlap generates a choropleth map for each combination of non-spatial dimension members (45-49; 19-9-05), (45-49; 26-9-05), (30-34; 19-9-05), (30-34; 26-9-05), (20-24; 19-9-05) and (20-24; 26-9-05). The choropleth map for the combination of members "30-34" mg/l and "19-9-05" is shown on Figure 12b. This map shows that "Coa di latte" and "Murano" are polluted by the same pollutants and that "Tessera" is not polluted. The user can visualize the alphanumeric attributes of the complex measures, i.e., the alphanumeric attributes (the code, the name and the density) of the pollutants present in "Mazzorbo" (Figure 12b). The choropleth map for the combination of members "30-34" mg/l and "26-9-05" is shown on Figure 12c. These

Figure 7. Common XML Mondrian schema of the applications of Figure 3

```xml
<xml version="1.0" ?>
    <Schema name="CORILACUBE">
        <Cube name="POLLUTION_UNIT">
            <Table name="CORILAFACTS">
            Aggregate Table
      </Table>
            <Dimension name="Lagoon" foreignKey="UNIT_CODE">
            <Hierarchy hasAll="true"
            primaryKey="UNIT_CODE"primaryKeyTable="UNIT_CODE">
              ...
              <Level name="Zone" table="ZONE" column="NAME" uniqueMembers="false"
            />
             <Level name="Unit" table="UNIT" column="NAME" uniqueMembers="false" />
              </Hierarchy>
      </Dimension>
            <Dimension name="Rate" foreignKey="RATE5_CODE">
            <Hierarchy hasAll="true" allMemberName="All rate"
            primaryKey="RATE5_CODE">
              <Table name="RATEDIM" />
              <Level name="Rate10" column="RATE10" uniqueMembers="false" />
              <Level name="Rate5" column="RATE5" uniqueMembers="false" />
              </Hierarchy>
      </Dimension>
            <Dimension name="Time" foreignKey="DAY_CODE">
            <Hierarchy hasAll="true" allMemberName="All Time"
            primaryKey="DAY_CODE">
              <Table name="TIMEDIM" />
              <Level name="Year" column="YEAR" uniqueMembers="false"/>
              <Level name="Month" column="MONTH" uniqueMembers="false" />
              <Level name="Day" column="DAY" uniqueMembers="false" />
              </Hierarchy>
      </Dimension>
            <Dimension name="Pollutant" foreignKey="POLLUTANT_CODE">
            <Hierarchy hasAll="true" allMemberName="All Pollutant"
            primaryKey="POLLUTANT_CODE">
              <Table name="POLLUTANTDIM" />
              <Level name="Type" column="TYPEP" uniqueMembers="false" />
             <Level name="Pollutant" column="POLLUTANT" uniqueMembers="false" />
              </Hierarchy >
      </Dimension>
             Measure
     </Cube>
   </Schema>
```

multiple cartographic representations (multi-maps) permit the user to visualize the temporal evolution of the pollution in the Venice lagoon and to see that over the course of a week a week the pollution values changed in Tessera.

6. COMPARISON WITH EXISTING SYSTEMS

Some SOLAP tools have been developed to integrate OLAP and GIS functionalities (Bédard et al., 2006; Sampaio et al., 2006; Matias & Moura-Pires, 2007; Silva et al., 2008; Escribano et al., 2007). In Bédard et al. (2006), the authors present an OLAP-GIS integrated tool, which provides several types of cartographic, tabular and graphic displays that are accessible by different coordinated windows. The visual components are synchronized to form a uniquely flexible and interactive user interface. The tool supports numerical measures and spatial measures, defined as the set of all geometries of spatial objects representing a particular combination of spatial members. The system allows also the association of hypermedia documents (i.e., photos, text documents, etc.) to the multidimensional elements. In this system, all possible queries are pre-calculated.

SOVAT (Scotch & Parmanto, 2006) is an OLAP-GIS integrated tool that combines OLAP, GIS and (spatial) data mining functionalities. It provides statistic and spatial analysis methods, such as the buffer operator to slice the spatial data warehouse. The tool synchronizes interac-

Figure 8. XML Mondrian Tag for a) Geographic Measure b) Complex Measure

```
<Measure name="UNIT" column="UNIT_CODE"
aggregator="UserDefinedAggregation" />
```

(a)

```
<Measure name="POLLUTANT" column="POLLUTANT_CODE"
aggregator="UserDefinedAggregation" />
```

(b)

tive maps, tabular and graphic displays. SOVAT supports numerical and spatial measures as collections of spatial objects.

The prototype presented in (Matias & Moura-Pires, 2007) integrates basic GIS and OLAP functionalities. The interface combines and synchronizes cartographic, tabular and graphic displays. The cartographic component is used to visualize spatial measures and members of the spatial dimension and numerical measures through simple graphical displays. Moreover, the prototype supports spatial measures, and it enables the user to cut the spatial data warehouse using spatial relationships, and to superimpose one spatial data warehouse on another using their spatial dimensions.

In Silva et al. (2008), the authors present a model and a SOLAP tool. The model introduces a set of aggregation functions that combine numerical and spatial functions. These aggregation functions are classified according to whether the numerical aggregation is scalar, distributive or holistic, and whether spatial aggregation is a unary or n-ary function. The system extends the OLAP Server Mondrian with spatial operators and its user interface couple the OLAP tool JRubik and the GIS OpenJUMP. The tool supports spatial and numerical measures.

Escribano et al. (2007) presents an OLAP-GIS integrated solution. It provides a mechanism to improve the aggregation on spatial dimensions, called overlay precomputation. The system is based on a model that defines measures as numerical values mapped onto spatial dimensions. It does not support spatial measures. OLAP operations and aggregations are limited to the alphanumeric part of the geographic objects.

Since some of these tools are based on advanced OLAP Servers (such as Microsoft Analysis Services), they support multi-valued and many to many relationships between dimensions and facts.

However, they do not support complex/geographic measures because it is not possible to define, aggregate and visualize non-numerical measures.

On the contrary, GeWOlap provides well adapted techniques to model, aggregate and visualize geographic/complex measures. Indeed, it enriches thematic maps using diagrams such as pie charts and bars to show numerical measures, with textual labels into the pivot table, extending the functionalities of Mondrian and JPivot, and interactive multi-maps for complex/geographic measures.

Moreover, as described in the previous section, implementing complex and geographic measures and dimensions symmetrically requires the use of a particular logical model. The authors of (Glorio & Trujillo, 2008; Malinowsky & Zimanyi, 2007; Rivest et al., 2005; Silva et al., 2008; Sampaio et al., 2006) use the classical star schema defining (spatial) measures as a simple geometric attribute of the fact. Therefore, in order to support symmetry between dimensions and measures, we have had

Figure 9. XML Mondrian tags for the aggregated table for a) Geographic Measure "Unit", b) Complex Measure "Pollutant"

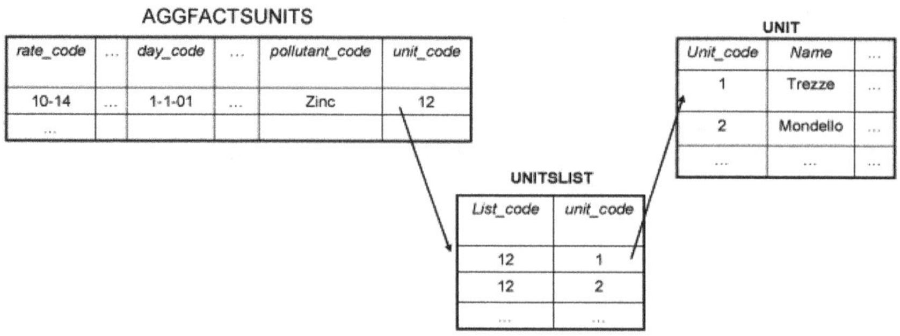

(a)

```
AggName name="AGGFACTSUNIT">
  <AggFactCount column="FACT_COUNT" />
  <AggMeasure name="[Measures].[UNIT]" column="UNIT_CODE" />
  <AggLevel name="[Pollutant].[Type]" column="TYPE_CODE" />
...
<AggLevel name="[Pollutant].[Pollutant]" column="POLLUTANT_CODE" />
<AggLevel name="[Rate].[Rate10]" column="RATE10_CODE" />
  <AggLevel name="[Rate].[Rate5]" column="RATE5_CODE" />
  <AggLevel name="[Time].[Year]" column="YEAR_CODE" />
   <AggLevel name="[Time].[Month]" column="MONHT_CODE" />
  <AggLevel name="[Time].[Day]" column="DAY_CODE" />
</AggName>
```

(b)

```
<AggName name="AGGFACTSPOLLUTANT">
  <AggFactCount column="FACT_COUNT" />
  <AggMeasure name="[Measures].[POLLUTANT]"column="POLLUTANT_CODE" />
  <AggLevel name="[Location].[Zone]" column="ZONE_CODE" />
  <AggLevel name="[Location].[Unit]" column="UNIT_CODE" />
  <AggLevel name="[Rate].[Rate10]" column="RATE10_CODE" />
  <AggLevel name="[Rate].[Rate5]" column="RATE5_CODE" />
<AggLevel name="[Time].[Year]" column="YEAR_CODE" />
   <AggLevel name="[Time].[Month]" column="MONHT_CODE" />
  <AggLevel name="[Time].[Day]" column="DAY_CODE" />
</AggName>
```

(c)

to extend this approach by forcing measures to be dimensional keys. Indeed, our spatio-multidimensional logical model is equivalent to an approach using a fact-less table (Kimball, 1996). However, our system does not restrict the aggregation functions to simple count functions or use complex OLAP queries and models (such as calculated members, auxiliary dimensions, user-defined functions, etc.) to define them. Indeed, GeWOlap uses a very simple OLAP Server model that uses the user-defined aggregation function provided by our Mondrian extension (in a quite similar way the BIOLAP system introduces in Mondrian numerical aggregation functions for genomic data), and enabling the easy implementation of aggregation functions for spatial, numerical and non-numerical

Figure 10. Pivot Table and map for the geographic measure "Unit"

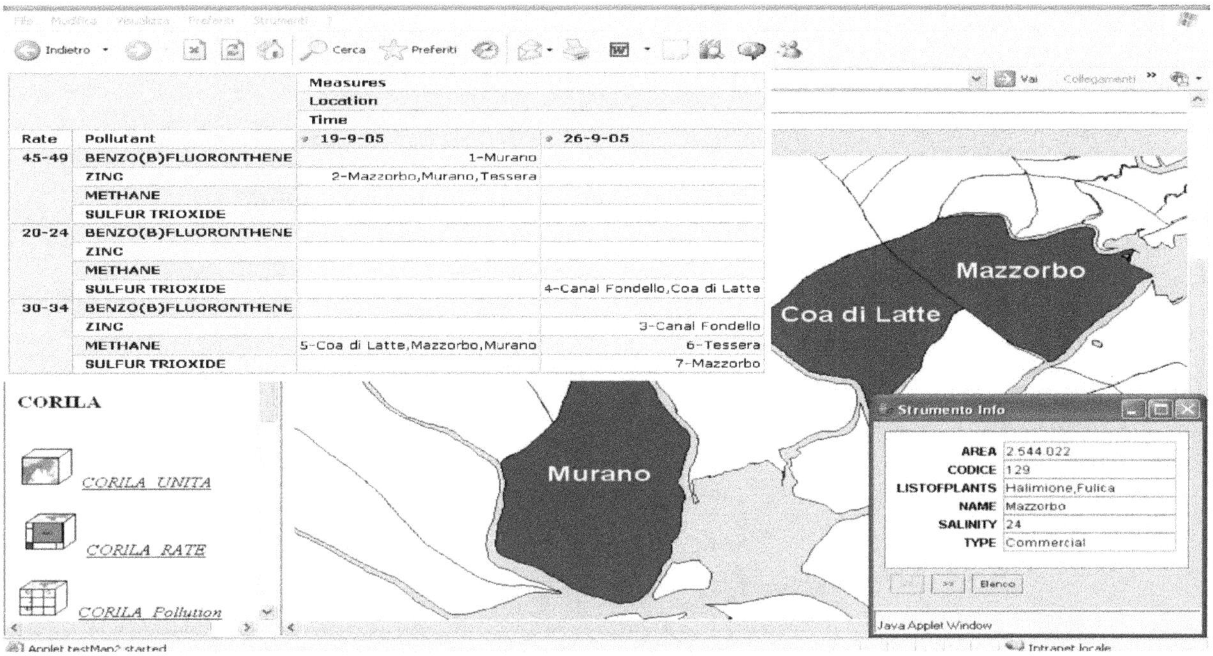

Figure 11. Pivot Table and map for aggregated geographic measures "Unit"

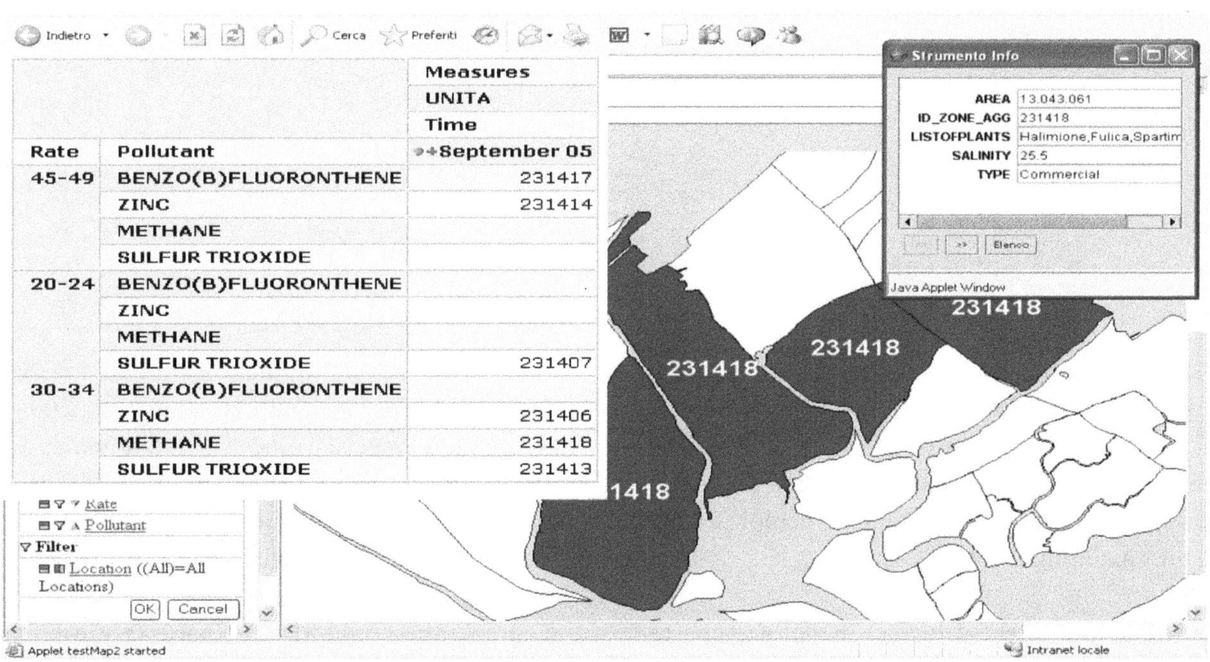

Figure 12. Visualization of complex measure "Pollutant"

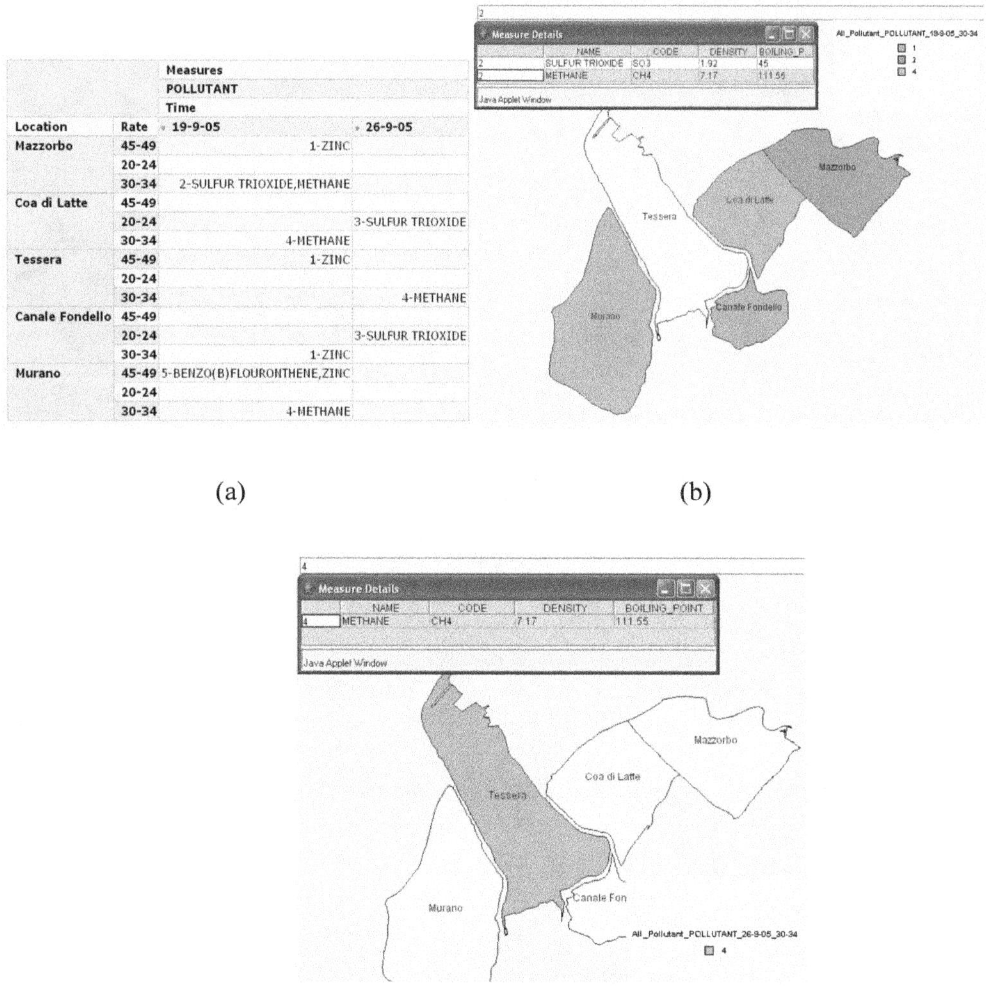

Location	Rate	Measures POLLUTANT Time 19-9-05	26-9-05
Mazzorbo	45-49	1-ZINC	
	20-24		
	30-34	2-SULFUR TRIOXIDE,METHANE	
Coa di Latte	45-49		
	20-24		3-SULFUR TRIOXIDE
	30-34	4-METHANE	
Tessera	45-49	1-ZINC	
	20-24		
	30-34		4-METHANE
Canale Fondello	45-49		
	20-24		3-SULFUR TRIOXIDE
	30-34	1-ZINC	
Murano	45-49	5-BENZO(B)FLOURONTHENE,ZINC	
	20-24		
	30-34	4-METHANE	

(a) (b)

(c)

attributes in the spatial data warehouse tier using the DBMS functionalities.

In conclusion, compared with other existing SOLAP tools, GeWOlap effectively and easily supports symmetric geographic/complex measures and dimensions by using

- A well adapted logical model and the redefinition of the aggregation process in the OLAP Server,
- A variety of geo-visualisation techniques.

However, several limitations remain to overcome. First, indexing methods (Rao et al., 2006), spatial view materializations (Stefanovic et al., 2000) and computational techniques (Escribano et al., 2007) have to be integrated to speed up the operations that dynamically change the structure of the spatial hypercube and the aggregation of geographic measures. It is also necessary to define semiology rules to synchronize the pivot table and the interactive map correctly and to introduce spatial data mining methods.

7. CONCLUSION AND FUTURE WORK

Spatial OLAP refers to the integration of spatial data into OLAP and data warehouse systems.

This work presents a SOLAP Web-based system, GeWOlap, supporting the concepts of geographic and complex measures. Geographic and complex measures (Bimonte et al., 2006b) are defined as geographic and complex objects, respectively, that belong to a hierarchy schema. These measures allow for the introduction of geographic and complex data into SOLAP models as analysis subjects. In order to reduce the in subject flexibility gap between spatial and multidimensional analysis, we propose a symmetrical representation of measures and dimensions.

GeWOlap is an OLAP-GIS integrated solution that provides a visual interactive method to explore and analyze complex spatio-multidimensional databases. The system is based on a specific logical model, an extension of the aggregation process and some geo-visualization techniques. Indeed, we propose to use a fact-less table and an aggregation process coupled with the Spatial DBMS and the OLAP Server. This scheme enables our system to support a symmetrical representation of dimensions and geographic and complex measures in a very simple way. From the perspective of visualization, we combine multi-maps, multiple coordinate windows, interactive maps and pivot tables displaying textual labels to provide for the effective visualization and exploration of complex and geographic data.

Our main work at present concerns the implementation of SOLAP operators that allow for changing the granularity of the measure using a level of its hierarchy. Indeed, we plan to integrate two operators, Classify and Specialize (Bimonte, 2007) that allow for climbing into the hierarchy of the measure. Classify and Specialize use two different rules to change the measure. Classify requires all the descendants of the new measure to be in the facts table. Specialize requires only one descendant. Using a naïve implementation, the Classify and Specialize operators are associated to different facts tables with each level of the measure hierarchy. The main limitation of this implementation is the balance of time and storage performance. Materializing all possible SOLAP queries provides good response times, but it is unfeasible with large data warehouses in terms of data storage. On the other hand, calculating all possible spatial queries on the fly with the "all" and "any" SQL predicates implies the storage of one facts table, but the time performance worsens.

Finally, we will introduce advanced geo-visualisation techniques such as Space-Time Cube and geo-browser tools in order to improve the analysis capabilities of GeWOlap. The usability studies will be conducted using a real dataset.

ACKNOWLEDGMENT

This work has been partially done by Sandro Bimonte during his Phd Thesis at INSA-Lyon, France.

REFERENCES

Andrienko, N., Andrienko, G., & Gatalsky, P. (2003). Exploratory spatio-temporal visualization: an analytical review. *Journal of Visual Languages and Computing*, *14*(6), 503–541. doi:10.1016/S1045-926X(03)00046-6

Bédard, Y., Merrett, T., & Han, J. (2001). Fundaments of Spatial Data Warehousing for Geographic Knowledge Discovery. In Miller, H., & Han, J. (Eds.), *Geographic Data Mining and Knowledge Discovery* (pp. 53–73). London: Taylor & Francis. doi:10.4324/9780203468029_chapter_3

Bédard, Y., Proulx, M., Bernier, E., & Gosselin, P. (2008). *How the new analytical geomatics technologies can help environmental health professionals and decision-makers to make further use of mapping than what is offered traditionally by geographic information systems (GIS) and web mapping*. Vancouver, Canada: The National Collaborating Centre for Environmental Health.

Bédard, Y., Proulx, M., Rivest, S., & Badard, T. (2006). Merging Hypermedia GIS with Spatial On-Line Analytical Processing: Towards Hypermedia SOLAP. In Stefanakis, E., Peterson, M., & Delis, V. (Eds.), *Geographic Hypermedia: Concepts and Systems* (pp. 167–185). Berlin: Springer. doi:10.1007/978-3-540-34238-0_10

Bertin, J., & Bonin, S. (1992). *La graphique et le traitement graphique de l'information*. Paris: Flammarion.

Bimonte, S. (2007). *Vers l'intégration de l'informtion géographique dans les entrepôts de données et l'analyse en ligne: de la modélisation à la visualisation*. Unpublished doctoral dissertation, INSA Lyon, France.

Bimonte, S., Tchounikine, A., & Miquel, M. (2006). GeoCube, a Multidimensional Model and Navigation Operators Handling Complex Measures: Application in Spatial OLAP. In T. Yaknho, M. Tatyana, & E. Neuhold (Eds.), *Proceedings of International Conference in Advances in Information Systems* (pp. 100-109). Berlin: Springer.

Bimonte, S., Tchounikine, A., Miquel, M., & Laurini, R. (2007). *Proposition pour l'intégration de l'analyse spatiale et de l'analyse multidimensionnelle*. Revue des Nouvelles Technologies de l'Information, Numéro Spécial. Cépaduès-Éditions.

Bimonte, S., Wehrle, P., Tchounikine, A., & Miquel, M. (2006). GeWOlap: A Web Based Spatial OLAP Proposal. In R. Meersman, Z. Tari, & P. Herrero (Eds.), *Proceedings of Workshop on Semantic-Based Geographical Information Systems* (pp. 1596-1605). Berlin: Springer.

Camossi, E., Bertolotto, M., & Bertino, E. (2008). Querying Multigranular Spatio-temporal Objects. In S. Bhowmick, J. Küng, & R. Wagner (Eds.), *Proceedings of International Conference on Database and Expert Systems Applications* (pp. 390-403). Berlin: Springer.

Colonnese, G., Manhães, R., Montenegro, S., Carvalho, R., & Tanaka, A. (2005). PostGeoOlap: an Open-Source Tool for Decision Support. In *Proceedings of the 2nd Simpósio Brasileiro de Sistemas de Informação*.

Damiani, M., & Spaccapietra, S. (2006). Spatial Data Warehouse Modeling. In *Processing and Managing Complex Data for Decision Support* (pp. 1–27). Hershey, PA: IGI Global. doi:10.4018/978-1-59140-655-6.ch001

Escribano, A., Gómez, L., Kuijpers, B., & Vaisman, A. (2007). Piet: a GIS-OLAP implementation. In I. Song & T. Pedersen (Eds.), *Proceedings of the International Conference on DataWarehousing and Knowledge Discovery* (pp. 73-80). New York: ACM

Glorio, O., & Trujillo, J. (2008). An MDA Approach for the Development of Spatial Data Warehouses. In I. Song, J. Eder, & T. Nguyen (Eds.), *Proceedings of International Conference of DataWarehousing and Knowledge Discovery* (pp. 23-32). New York: ACM.

Hernandez, V., Voss, A., Göhring, W., & Hopmann, C. (2005). Sustainable decision support by the use of multi-level and multi-criteria spatial analysis on the Nicaragua Development Gateway. In *Proceedings of the 8th International Conference on the Global Spatial Data Infrastructure*.

Inmon, W. H. (1996). *Building the Data Warehouse*. New York: John Wiley & Sons.

JPIVOT. (n.d.). Retrieved October 10, 2009, from http://jpivot.sourceforge.net

Kimball, R. (1996). *The Data Warehouse Toolkit: Practical Techniques for Building Dimensional Data Warehouses*. New York: John Wiley & Sons.

Laurini, R., & Thompson, D. (2001). *Fundamentals of Spatial Information Systems*. New York: Academic Press.

Longley, P., Goodchild, M., Maguire, D., & Rhind, D. (Eds.). (2001). *Geographic Information Systems and Science*. New York: John Wiley & Sons.

MacEachren, A., & Kraak, M. (1997). Exploratory cartographic visualization: advancing the agenda. *Computers & Geosciences, 23*(4), 335–343. doi:10.1016/S0098-3004(97)00018-6

Malinowski, E., & Zimányi, E. (2004). Representing spatiality in a conceptual multidimensional model. In D. Pfoser, I. Cruz, & M. Ronthaler (Eds.), *Proceedings of ACM International Workshop on Geographic Information Systems* (pp. 12-22). New York: ACM Press.

Malinowski, E., & Zimányi, E. (2007). Logical Representation of a Conceptual Model for Spatial Data Warehouses. *GeoInformatica, 11*(4), 431–457. doi:10.1007/s10707-007-0022-3

MAPINFO. *MapXtreme Java* (n.d.). Retrieved October 10, 2009, from http://extranet.mapinfo.com/products/Overview.cfm?productid=1162

Marchand, P., Brisebois, A., Bédard, Y., & Edwards, G. (2003). Implementation and evaluation of a hypercube-based method for spatio-temporal exploration and analysis. *Journal of the International Society of Photogrammetry and Remote Sensing, 59*(1), 6–20. doi:10.1016/j.isprsjprs.2003.12.002

Matias, R., & Moura-Pires, J. (2007). Revisiting the OLAP Interaction to Cope with Spatial Data and Spatial Analysis. In J. Cardoso, J. Cordeiro, & J. Filipe (Eds.), *Proceedings of 9th International Conference on Enterprise Information Systems*. Portugal: INSTICC.

Mitchell, S. (2005). The ESRI Guide to GIS Analysis: *Vol. 2. Spatial Measurements and Statistics*. Redlands, CA: ESRI Press.

ORACLE. (n.d.). *Oracle9i® Database*. Retrieved October 10, 2009, from http://www.oracle.com/technology/products/oracle9i/index.html

PENTAHO. (n.d.). Retrieved October 10, 2009, from http://mondrian.pentaho.org

Pinet, F., & Schneider, M. (2009). A unified object constraint model for designing and implementing multidimensional systems. *Journal of Data Semantics, 13*.

Rafanelli, M. (2003). Operators for Multidimensional Aggregate Data. In Rafanelli, M. (Ed.), *Multidimensional databases: problems and solutions* (pp. 116–165). Hershey, PA: IGI Global. doi:10.4018/978-1-59140-053-0.ch005

Rao, F., Zhang, L., Yu, X., Li, Y., & Chen, Y. (2003). Spatial Hierarchy and OLAP-Favored Search in Spatial Data Warehouse. In S. Rizzi & Y. Song (Eds.), *Proceedings ACM International Workshop on Data Warehousing and OLAP* (pp. 48-55). New York: ACM Press.

Rivest, S., Bédard, Y., Proulx, M., Nadeaum, M., Hubert, F., & Pastor, J. (2005). SOLAP: Merging Business Intelligence with Geospatial Technology for Interactive Spatio-Temporal Exploration and Analysis of Data. *Journal of International Society for Photogrammetry and Remote Sensing, 60*(1), 17–33. doi:10.1016/j.isprsjprs.2005.10.002

Sampaio, M., Sousa, A., & Baptista, C. (2006). Towards a logical multidimensional model for spatial data warehousing and OLAP. In Y. Song & P. Vassiliadis (Eds.), *Proceedings of ACM International Workshop on Data Warehousing and OLAP* (pp. 83-90). New York: ACM Press.

Scotch, M., & Parmanto, B. (2006). Development of SOVAT: a numerical-spatial decision support system for community health assessment research. *International Journal of Medical Informatics, 34*(10), 771–784. PubMed doi:10.1016/j.ijmedinf.2005.10.008

Shekhar, S., Lu, C., Tan, X., Schawla, S., & Vatsavai, R. (2001). Map Cube: A Visualization Tool for Spatial Data Warehouses. In Miller, H., & Han, J. (Eds.), *Geographic Data Mining and Knowledge Discovery* (pp. 74–109). London: Taylor & Francis. doi:10.4324/9780203468029_chapter_4

Shneiderman, B., & Plaisant, C. (2004). *Designing the User Interface: Strategies for Effective Human-Computer Interaction*. Reading, MA: Addison-Wesley.

Silva, J., Times, V., Salgado, A., Souza, C., Fidalgo, R., & Oliveira, A. (2008). A set of aggregation functions for spatial measures. In I. Song & A. Abelló (Eds.), *Proceedings of the ACM International Workshop on Data Warehousing and OLAP* (pp. 25-32). New York: ACM Press.

Stefanovic, N., Han, J., & Koperski, K. (2000). Object-Based Selective Materialization for Efficient Implementation of Spatial Data Cubes. *IEEE Transactions on Knowledge and Data Engineering, 12*(6), 938–958. doi:10.1109/69.895803

Tang, D., Stolte, C., & Bosch, R. (2003). Design Choices when Architecting Visualizations. In *Proceedings of IEEE Symposium on Information Visualization* (pp. 65-79). Washington, DC: IEEE Computer Society.

Weibel, R., & Dutton, G. (2001). Generalizing Spatial Data and Dealing with Multiple Representations. In Longley, P., Goodchild, M., Maguire, D., & Rhind, D. (Eds.), *Geographic Information Systems and Science* (pp. 125–15). New York: John Wiley & Sons.

This work was previously published in International Journal of Agricultural and Environmental Information Systems, Volume 1, Issue 2, edited by Petraq Papajorgji and François Pinet, pp. 42-67, copyright 2010 by IGI Publishing (an imprint of IGI Global).

Chapter 4
A Systematic Approach for Managing the Risk Related to Semantic Interoperability between Geospatial Datacubes

Tarek Sboui
Université Laval and NSERC, Canada

Mehrdad Salehi
Université Laval and NSERC, Canada

Yvan Bédard
Université Laval and NSERC, Canada

ABSTRACT

Geospatial datacubes are the database backend of novel types of spatiotemporal decision-support systems employed in large organizations. These datacubes extend the datacube concept underlying the field of Business Intelligence (BI) into the realm of geospatial decision-support and geographic knowledge discovery. The interoperability between geospatial datacubes facilitates the reuse of their content. Such interoperability, however, faces risks of data misinterpretation related to the heterogeneity of geospatial datacubes. Although the interoperability of transactional databases has been the subject of several research works, no research dealing with the interoperability of geospatial datacubes exists. In this paper, the authors support the semantic interoperability between geospatial datacubes and propose a categorization of semantic heterogeneity problems that may occur in geospatial datacubes. Additionally, the authors propose an approach to deal with the related risks of data misinterpretation, which consists of evaluating the fitness-for-use of datacubes models, and a general framework that facilitates making appropriate decisions about such risks. The framework is based on a hierarchical top-down structure going from the most general level to the most detailed level, showing the usefulness of the proposed approach in environmental applications.

DOI: 10.4018/978-1-4666-0333-2.ch004

INTRODUCTION

In order to derive the maximum profit from the power of the geospatial data and the efficiency of the datacube structure in the decision making process, geospatial datacubes have been introduced. Geospatial datacubes integrate spatial data with the datacube structure and are recognized as one of the most promising decision-support systems (Rafanelli, 2003). In some situations, we need the interoperation between geospatial datacubes. The situations where such needs arise are: (1) a simultaneous and rapid navigation through different geospatial datacubes, (2) a rapid insertion of data in a datacube from another one, and (3) an interactive and rapid analysis of phenomena changes by comparing the content of different geospatial datacubes (Sboui et al., 2007).

With the emergence of software agents, semantic interoperability has been viewed as the technical analogue to human communication (Brodeur, 2004; Kuhn, 2005; Sboui et al., 2007). According to this view, each agent tries to interpret the exchanged data as it has been originally intended by another one. However, due to the uncoordinated use of data (i.e., semantic heterogeneity), agents may faulty interpret data or be uncertain about its intended meaning. That is, there is a risk of misinterpreting the exchanged data.

The risks of data misinterpretation are even more pronounced when interoperating geospatial datacubes developed for strategic decision purposes. In fact, strategic decisions made on the basis of inappropriate interpretations of data may lead decision analysts to have inappropriate judgment and to make unwarranted inferences about some aspects of the problem to be solved, and thus to make faulty decisions.

This article aims to support the semantic interoperability between geospatial datacubes. It proposes a risk management approach that allows identifying and asses the related risks of data misinterpretation in a systematic manner based on the quality of datacubes models (i.e., metadata and schema). In the next section, we discuss the risks of data misinterpretation in the interoperability involving geospatial datacubes. Then, we propose a categorization of the semantic heterogeneity that may occur during such interoperability. Then, we propose an approach to identify and evaluate such risks. After that, we propose a method that aims at supporting an intervener (human or software agent) to respond to these risks. Then, we provide an example of application, and we show the usefulness of the proposed approach in environmental applications. Finally, we conclude this paper.

RISK OF DATA MISINTERPRETATION RELATED TO SEMANTIC INTEROPERABILITY BETWEEN GEOSPATIAL DATACUBES

This section discusses the risks of data misinterpretation related to semantic interoperability in general and the one involving geospatial datacubes in particular.

Interoperability between Geospatial Datacubes

A datacube is composed of a set of measures aggregated according to a set of dimensions with different levels of granularity. Both dimensions and measures of a geospatial datacube may contain geospatial components (Bédard et al., 2001). Interoperating geospatial datacubes may involve one or the combination of the following actions on their components: 1) integrating measures which may refer to adding a new measure to a datacube from another one based on common dimensions and members, or creating a new common measure based on existing measures of different datacubes, 2) integrating dimensions which may refer to creating a new dimension based on the dimensions of different datacubes, adding one or several dimensions of one datacube to another, or

Figure 1. Example of the spatial aggregation-generalization mismatch (adapted from Bédard et al., 2005)

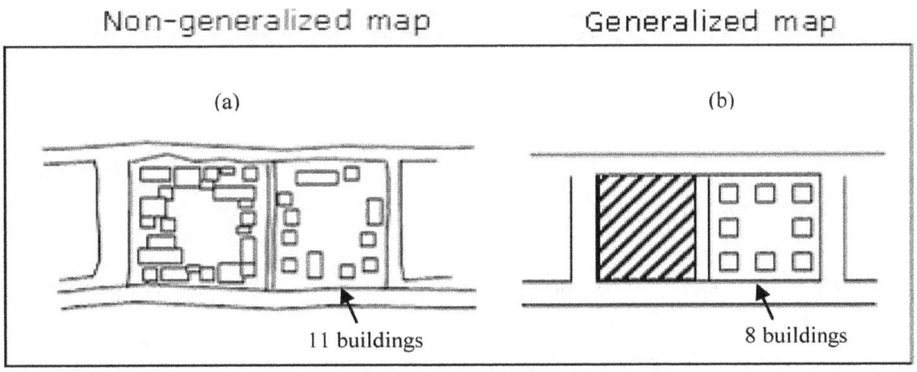

modifying a dimension of a datacube by using existing dimension's levels of another datacube, or 3) comparing a dimension or a measure against another.

Overview of the Risks of Data Misinterpretation in Geospatial Datacubes Interoperability

With the emergence of software agents, semantic interoperability has been viewed as the technical analogue to human communication (Brodeur, 2004; Kuhn, 2005; Sboui et al., 2007). According to this view, each agent tries to interpret the exchanged data by comparing them with his/her knowledgebase content (i.e., ontology). In order for the interoperability process to work properly, the receiver should interpret data as it was originally intended. However, this is not always the case, since agents (the sender and the receiver) may use different data to represent the same real-world phenomena. Also, agents may use same signs to represent different real-world phenomena, depending on the context. This uncoordinated use of data among agents may cause a risk of misinterpreting. That is, the receiver may faulty interpret data or be uncertain about its intended meaning. The uncoordinated use of data is known as the semantic heterogeneity which is know as the major barrier for the semantic interoperability (Bishr, 1998; Brodeur, 2004; Kuhn, 2005; Staub et al., 2008).

For example, as shown in Figure 1, due to type of cartographic generalization being carried out, the same real-world phenomena (i.e., a building block) may be represented differently (Figure 1 a and b), and therefore, data may be interpreted differently. Figure 1 (a) is interpreted as a block with 11 buildings, while Figure 1 (b) is interpreted as a block with 8 buildings. Consequently, there is a risk of misinterpreting the Figure 1 (b) due to the cartographic generalization process.

Data misinterpretation may lead to a poor understanding of real-world phenomena which may harm the reuse of geospatial data. The risks of data misinterpretation is even more pronounced in the context of the interoperability between geospatial datacubes as they may cause inappropriate judgments, and unwarranted inferences about some aspects of the problem to be solved, resulting in bad strategic decisions.

Furthermore, dealing with the risk of data misinterpretation in geospatial datacubes is more complicated than in transactional databases as geospatial datacubes undergo the complex ETL (Extract, Transform and Load) procedures that can impact the meaning of their content. Some interpretations may be formed, several rules may be applied to the collected data in order to fit the business needs, and different methods may be used to aggregate data. This adds another level of complexity of interpretation as users may need to understand the method or rules used in the ETL procedures.

Semantic Heterogeneity Conflicts in Geospatial Datacubes: The Main Cause of Data Misinterpretation Risk

As for transactional databases, semantic heterogeneity in geospatial datacubes occurs when there are differences in schemas and metadata. Schema heterogeneity refers to the difference in structure of datacube elements (e.g., hierarchy structure) such as the difference between the number of levels of semantically related[1] dimensions of two different geospatial datacubes. For instance, the hierarchy of dimension *Administrative Region* can have the following levels in one datacube: *Country, Province, County, Municipality*; and the following levels in the other one: *Country, Region, Municipality*. Metadata heterogeneity refers to the difference of all other elements of the model of a datacube such as the difference of the precision of the cartographic localization of the geospatial members.

In order to better illustrate our categorisation approach, we present a running example of using two geospatial datacubes (see Figure 2). To determine the risk of forest fire on population, we can interoperate two geospatial datacubes C_1 and C_2, modelled respectively in Figure 2 (a) and Figure 2 (b). The first datacube C_1 is used to determine the distribution of the population in specific areas and periods. The second geospatial datacube C_2 is used to control the forest fire extent. The example is extracted from a real case which consists in determining the risk of forest fire on the Canadian population.

The geometric geospatial levels as well as geometric geospatial measures are represented using geospatial pictograms which refer to the geometric primitives in the geospatial data modeling (Bédard & Larrivée, 2008). In our example, we use the pictograms developed in Perceptory[2] tool where the pictogram «⊡» represents a point type, the pictogram «☑» represents a line type and the pictogram "«" represents a polygon type.

In this example, the metadata of the geospatial datacubes contain information related to the geospatial referencing system, scale and precision, year of creation, stand forest method as well as the geospatial cover of some dimensions.

The problems of heterogeneity of geospatial datacube models can exist at five different levels: datacubes, dimensions, hierarchy, levels, and measures. At each level, the heterogeneity can be occurred at the schema and metadata. In order to categorise the problems of heterogeneity, we formalize the elements of datacubes as below:

1. A level $L = \{a_1, a_2, \ldots, a_n\}$ is a set of n attributes where n is called the cardinality of the set, for example, the level *City* = {city_name, city_surface}.
2. A hierarchy $H = (\{L_1, L_2, \ldots, L_h\}, <)$ is a set of h levels and an order relation between these levels. h is called the cardinality of the set. The order relation is defined as follows: $\forall L_1, L_2 \in H, L_1 < L_2$ if L_1 rolls-up to L_2, for example, the hierarchy (*City*, *Province*, *Country*) of the dimension *Administrative Region* = (*City* < *Province* < *Country*).
3. A dimension $D = \{H_1, H_2, \ldots, H_d\}$ is a set of d hierarchy. d is called the cardinality of the set and is usually equal to 1, for example, the cardinality of the dimension *Region* = {(*City* < *Province* < *Country*)} is 1.
4. A measure m is an attribute which describes the subject of analysis.
5. A datacube $C = (\{D_1, D_2, \ldots, D_c\}, \{m_1, m_2, \ldots, m_q\})$ is a set of c dimensions and q measures.

The following are the five categories of heterogeneities that may occur between geospatial datacube models:

Figure 2. Two models of datacubes (C_1 and C_2)

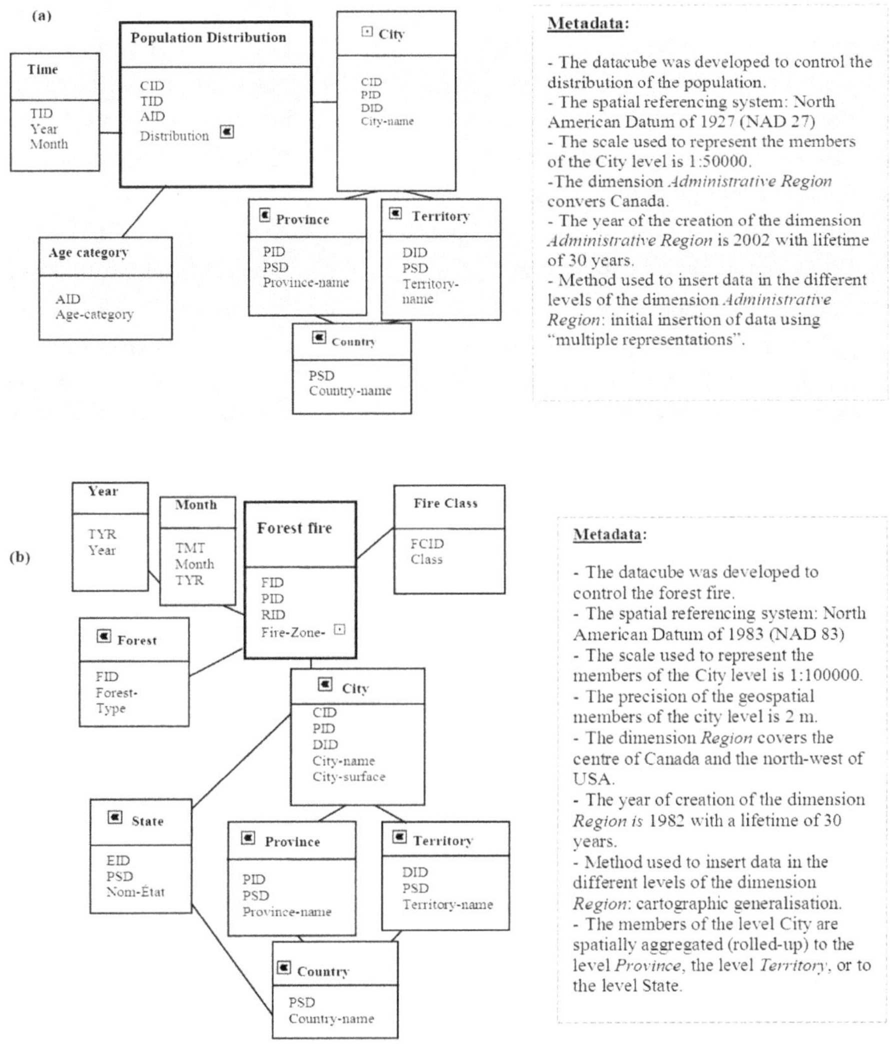

1. **Cube-to-Cube heterogeneity**
   a. **Schema Heterogeneity:** It appears when two datacubes have semantically related measures which are represented according to various numbers of dimensions. For example, the two measures *road intersection* and *points of intersection* can be analyzed according to two dimensions in one datacube, and to three dimensions in another.
   b. **Metadata Heterogeneity:** *Difference of the creation date of datacube*. It appears when two datacubes were created in two different dates.
2. **Measure-to-Measure Heterogeneity**
   a. **Schema Heterogeneity:** *Difference in geometric primitives*. It appears when two semantically related measures have different geometric primitives (e.g., line versus polygon).
   b. **Metadata Heterogeneity:** *Heterogeneity of aggregation*. Appears when different functions were used to aggregate semantically related geo-

spatial measures. For example, the functions "geometric union" and "centre of gravity" are used to aggregate the measure *density* in two different datacubes.

3. **Dimension-to-Dimension Heterogeneity**
   a. **Schema Heterogeneity:** *Inequality of the number of hierarchies.* It appears when the cardinalities of semantically related dimensions are different. Let us suppose that n_1 is the cardinality of dimension D_1, n_2 is the cardinality of dimension D_2. If $n_1 \neq n_2$ then there is a heterogeneity Dimension-to-Dimension at the schema level. In Figure 2, the dimension *Administrative Region* has only one hierarchy: (Country, Province, and City), whereas the dimension *Region* has two hierarchies: (Country, Province, Territory, and City) and (Country, State, and City).
   b. **Metadata Heterogeneity:** *Not-correspondence of the dimensions constraints*[3]. It appears when constraints of two semantically related dimensions are incoherent. For example, the constraint of dimension *Administrative Region* indicates that all the members of the level *City* roll-up to the level *Province*, while the constraint of dimension *Region* indicates that the members of the level *City* roll-up to level *Province*, to the level *Territory*, or to the level *State*.
4. **Hierarchy-to-Hierarchy Heterogeneity**
   a. **Schema Heterogeneity:** *(1) Inequality of the number of levels.* It appears when the cardinalities of the hierarchies of semantically related dimensions are unequal. Let n_1 be the cardinality of the hierarchy H_1 and n_2 the cardinality of the hierarchy H_2. If $n_1 \neq n_2$, then there is a schema heterogeneity for two hierarchies. For example, in Figure 2, the geospatial hierarchy (Country, Province, Territory and City) of the dimension *Administrative Region* contains four levels, whereas the geospatial hierarchy (Country, State, and City) of dimension *Region* contains three levels. *(2) Inequality of order of levels.* It occurs when hierarchies of semantically related dimensions have different orders of levels. More precisely, for each combination of couples of semantically related levels $((n_1, n_2), (n_1', n_2'))$, if $n_1 < n_2$ and $\neg (n_1' < n_2')$, then there is a hierarchy heterogeneity (inequality of order of levels). For example, in a given datacube, the order of the levels of a geospatial hierarchy is: *City, County,* and *Province*. Whereas, in another datacube, the same levels have the following orders: 1) *City, County,* and *Province*, and 2) *City* and *Province*.
   b. **Metadata Heterogeneity:** *Heterogeneity of geospatial coverage.* It appears when the members of the hierarchies of two semantically related dimensions have different territorial coverage. For example, the hierarchy of dimension *Administrative Region* (Country, Province, Territory and City) covers Canada, while the hierarchy of dimension *Region* (Country, State and City) covers the North-West of the United States.
5. **Level-to-Level Heterogeneity**
   a. **Schema Heterogeneity:** *Difference in geometric primitives.* It arises when, in two geospatial datacubes, two semantically related levels have different geometric primitives. For example, in datacube C_1, the level City is represented with a point, whereas in datacube C_2, the same level City is represented with a polygon.

b. **Metadata Heterogeneity:** *(1) Difference in geospatial referencing system.* It occurs when there is a difference of geospatial referencing systems used for the members of semantically related levels. In example of Figure 2, the levels of the dimension *Administrative Region* are based on the North American Datum 1927, while those of dimension *Region* are based on the North American Datum 1983. *(2) Heterogeneity of cartographic scale.* It appears when semantically related levels are represented with different cartographic scales. In our example, the cartographic scale of the level Province of datacube C_1 is 1:50000, whereas the scale of the level Province of the datacube C_2 is 1:100000.

Only some examples were mentioned in the above categorization but the aspects of heterogeneity can include many others, such as geospatial positioning methods in referencing systems, techniques of data acquisition, algorithms of data transformation, cartographic generalization, and so on.

These problems of heterogeneity may cause risk of misinterpreting data in the semantic interoperability between geospatial datacubes. This paper aims to manage such risks. In the following sections, we propose a risk management approach to manage such risks. The approach consists of evaluating the quality of the conceptual models of datacubes involved in the interoperability process to identify and evaluate the risks of data misinterpretation, and of framework to facilitate making decisions to respond to such risks. The proposed framework corresponds to the proposed categorization of semantic heterogeneity suggesting to respond to the risks at one category at a time (from the more general level to the most detailed level).

AN APPROACH TO IDENTIFY AND ASSESS THE RISK OF DATA MISINTERPRETATION: MODEL QUALITY PERSPECTIVE

Phases of Managing the Risks of Data Misinterpretation

Risk management paradigm constitutes an established framework for evaluating and managing malfunction or potential hazard that may cause bad consequences (Renn, 1998). Inspired from risk management paradigm in the project management literature, we define four iterative phases to manage the risk related to data misinterpretation (see Figure 3): identifying the risks, assessing the risks (i.e., determining their probability of occurrence and their degree of harm), and responding to the risks by making appropriate decisions related to the semantic interoperability process. The possible responses to risk include: reduction, absorption, and transfer (Agumya & Hunter, 2002; Bédard, 1986). We should note that, in the last phase, the risk may be ignored and users will assume the consequences of such choice.

Risk identification and assessment are essential steps to determine how to manage risks. In order to identify and evaluate the risks of data misinterpretation, we measure the external quality (fitness-for-use) of the conceptual models of geospatial datacubes, i.e., metadata and schema.

CONCEPTUAL MODEL QUALITY: AN ASPECT TO IDENTIFY AND EVALUATE THE RISKS OF DATA MISINTERPRETATION

Conceptual models are central to information systems design and are used as a basis for developing information systems to meet user's requirements at different levels (Moody, 2005).

While a good fitness-for-use of conceptual model (metadata and schema) indicates it is less

Figure 3. Phases for managing the risk of data misinterpretation

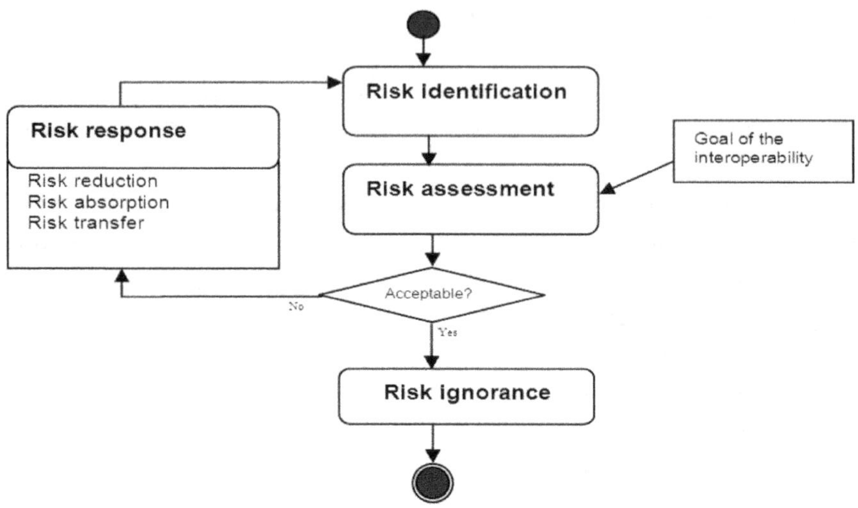

likely to have a risk of data misinterpretation, a poor quality indicates a higher risk. Consequently, the fitness-for-use of conceptual model allows identifying and evaluating the risk of data misinterpretation. Moreover, evaluating the fitness-for-use of conceptual models facilitates the response to the risks of data misinterpretation as it facilitates the comparison of exchanged data. Indeed, based on such fitness-for-use interveners (software agents or humans) involved in semantic interoperability can be advised:

a. Not to use one of the heterogeneous elements of the conceptual models (the one which has a lower quality compared to the other).
b. To consider a model element which has an excellent quality.
c. To use a model element which has a good quality to create new one.
d. Not to consider two heterogeneous elements of two conceptual models if they both have a poor quality (that is likely to produce a poor result).

Consequently, our method to identify, evaluate and respond the risks of data misinterpretation is based on evaluating the fitness-for-use of conceptual model: the fitness-for-use of metadata (e.g., coverage of geospatial data) and the one of geospatial datacube schema (e.g., geometric primitives, the number of levels).

The subject of conceptual model quality evaluation occupies a substantial part of the effort devoted to conceptual modelling (Genero et al., 2007). This subject received further emphasis with the Model Driven Development (MDD) paradigm in which development effort is focused on the design of models, rather than on coding (Genero et al., 2007). However, research works focused more on schema quality than the quality of the metadata. Moreover, while a range of quality frameworks have been proposed in the literature, none of these have been widely accepted in practice and none has emerged as a potential standard. As a result, conceptual models have been evaluated in an *ad-hoc* manner (Moody, 2005).

The Quality of Schema

The impact of conceptual schema quality is of central concern to computer scientists, as well as to end-users (Cherfi et al., 2007). However, while there is a little agreement among experts as to what makes a "good" schema, there are neither guidelines nor standards for evaluating the quality of conceptual schema. Consequently, people continue to evaluate conceptual schemas in an *ad-hoc* and subjective manner based on common sense and experience (Moody, 2005).

The Quality of Metadata

While a good metadata quality indicates it is less likely to have a risk of misinterpreting data, a poor quality indicates a higher risk, and may undermine the reuse of geospatial data (Agumya & Hunter, 2002), i.e., the main aim of semantic interoperability. Consequently, evaluating the quality of metadata would help to identify and assess the risk of data misinterpretation. While significant research efforts have been carried out to evaluate and enhance the quality of geospatial data (e.g., Agumya & Hunter, 2002; Boin, 2008; Devillers et al., 2007; Frank, 2007), there has been no work on the fitness-for-use of geospatial metadata.

INDICATORS TO IDENTIFY AND EVALUATE THE RISKS OF DATA MISINTERPRETATION

In this section, we define a restricted set of indicators and a method for evaluating the fitness-for-use of metadata and geospatial datacube schema. The definition and evaluation of these indicators takes into account data requirements of end-users which are represented in the form of a geospatial datacube model (see Figure 4).

We define two categories of indicators: a category related to the schema (*relevance of the geometric primitive*, *relevance of the structure* and *relevance of the hierarchy order*) and another category related to the metadata (*relevance of the metadata* and *freshness of metadata*). Each indicator is evaluated according to a function. The resulting quality value is within the interval (0, 1). The value 1 indicates perfect quality, and hence a low risk. The value 0 indicates completely poor quality, and hence a higher risk. This value is defined in a pragmatic and mathematical way as explained in the following paragraphs. Although quantitative values are used, the quality value represents a scale of ordinal measure. Consequently, we can apply the operators =, >, < to compare the fitness-for-use of elements of different metadata or different schemas. In other words, a quality 0.8 is passably better than a quality 0.4, but it is not precisely twice better. By using such ordinal scale of measures, we are avoiding to attribute a "too quantitative" valued to the result provided to the interveners. We should notice that the proposed indicators do not aim at being complete or precise but rather at making agents globally aware of the fitness-for-use of geospatial datacube models.

Indicators for Evaluating the Fitness-for-Use of Schema Elements

For evaluating the quality of the schema, we define the following three indicators:

1. **Relevance of the geometric primitive:** this indicator evaluates the relevance of the geometric primitive used to represent an element of the datacube model compared to the geometric primitive required by end-users. The indicator can be evaluated based on predefined correspondences between the two primitives. Table 1 represents an example for defining the degree of correspondence between these primitives. The values of this table are defined based on the rules of cartographic generalization and the common practice of cartographers. We set the values

Figure 4. Geospatial datacube model describing the needs of the end-user

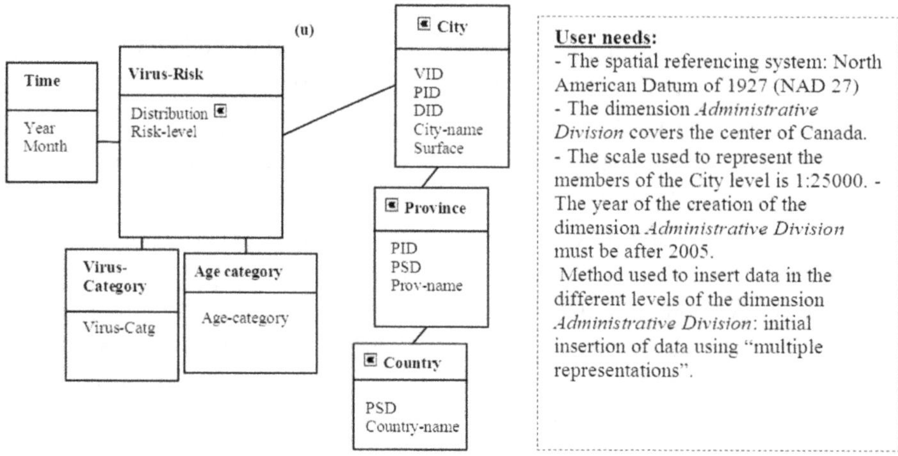

0, 0.25, 0.5, 0.75, and 1 respectively when the primitive of the model: 1) is not sufficient to generalize the required primitive (e.g., a point should not be used to generalize a line), 2) can be used to generalize the required primitives, but the result of generalization will be a bad representation of reality (e.g., a point (0D) and a line (3D)), 3) can be used to generalize the required primitives, but the result of generalization will not faithfully reflect reality (e.g., a point (0D) and a line (1D)), 4) are sufficient to generalize the required primitives, this result will faithfully reflect reality (e.g., a line (1D) and a point (0D)), and 5) is similar to those required (e.g., a line is relevant to represent a line). Geometric primitives are defined for the levels or measures of a geospatial datacube.

2. **Relevance of the structure:** this indicator evaluates the relevance of an element of the geospatial datacube schema with regard to an element of the required schema. It indicates the ratio of the schema elements which are semantically related to those of the required schema. Examples of relevance of the structure are the relevance of the number of hierarchies, relevance number of levels and relevance of the number of attributes. The relevance of the structure P_S is calculated as follows:

$$P_s = \begin{cases} \dfrac{N_{CE}}{N_{CR}} & ; \ if \quad N_{CE} < N_{CR} \\ 1 & ; \ Otherwise \end{cases}, \qquad (1)$$

where NR_{CR} is the total number of components required by the user, and NR_{CE} is the number of components which are semantically related to the required components.

This indicator is evaluated for all elements of geospatial datacubes (cube, measure, dimension, hierarchy, and level).

3. **Relevance of the hierarchy order:** it indicates the relevance of the order of each heterogeneous hierarchy with regard to the structure of the required hierarchy. Suppose that n_1 and n_2 are two levels belonging to the hierarchy H of a given datacube and n_1' and n_2' are two levels belonging to the hierarchy B of the datacube expressing the user's needs (i.e., needs for interoperability). The relevance of the hierarchy order (O_s) is the average of all the elementary relevance between each pair of levels which belong

to the hierarchy of each datacube (o_s). The elementary relevance o_s is determined using the following expression:

$$\forall n_1, n_2 \in H, \; \forall n_1', n_2' \in B : \\ \left[sem\_rel(n_1, n_1') \wedge sem\_rel(n_2, n_2') \wedge \; n_1 < n_2\right], \\ \Rightarrow n_1' < n_2' \tag{2}$$

where *sem_rel* is a function that verifies if the levels are semantically related or not. Value 1 (or 0) is set to the elementary relevance (o_s) when the expression (2) is true (or false). Thus, if for example the structure quality of the element 1 is greater than that of the element 2, we can conclude that the structure of the element 1 is more relevant than that of the element 2. As its name indicates, the indicator *relevance of the hierarchy order* is evaluated only for the hierarchies of a geospatial datacube.

Indicators for Evaluating the Fitness-for-Use of Metadata

The fitness-for-use of metadata refers to the relevance of the metadata that should be considered for a specific application. We propose a set of indicators and a quantitative approach to evaluate the fitness-for-use of the geospatial datacubes metadata. The following two indicators are proposed.

Relevance of metadata: It indicates the degree of relevance of metadata with regard to the needs of end-users. The relevance is evaluated at various levels: thematic, geospatial and temporal. It is evaluated based on the ratio of the number of metadata elements, which are semantically related to the required elements, to the total number of required elements. Relevance of the metadata *Pm* is evaluated using the following formula:

$$P_m = \begin{cases} w \times \dfrac{N_{Elt}}{N_{ReqElm}} & ; if \quad N_{Elt} < N_{ReqElm} \\ 1 & ; \; Otherwise \end{cases}, \tag{3}$$

where N_{Elt} is the number of thematic, geospatial, or temporal elements of metadata which are semantically related and required by users, N_{ReqElm} is the total number of required metadata elements, and *w* a predefined value between 0 and 1 which indicates the weight of each type (i.e., thematic, geospatial, and temporal) for end-users. If N_{Elt} is equal or larger than N_{ReqElm}, the metadata are perfectly relevant and has 1 as value.

Freshness of metadata: This indicator indicates the degree of freshness of the metadata of the geospatial datacubes. It is evaluated according to the age of metadata with regard to their lifetime. The age of metadata is the time passed since the date of metadata definition (*Tdef*) until the desired date of freshness of this metadata (*Treq*). The *lifetime* is the number of years after which the metadata will not be valid anymore. Freshness of metadata *Am* is evaluated as follow:

$$A_m = \begin{cases} 1 - \dfrac{\left|T_{req} - T_{def}\right|}{DV} & ; \; if \quad \left|T_{req} - T_{def}\right| < DV \\ 0 \quad ; & Otherwise \end{cases}, \tag{4}$$

where *DV* is the *liftime* of the metadata. A low value of the freshness decreases the fitness-for-use of metadata. This value decreases when its age increases. The *lifetime* and the date of metadata definition can be provided by the metadata producer.

The way of presenting these indicators to the user has a great importance on decision-making. Generally, decision-support systems use a restricted number of indicators (Few, 2006). Accordingly, we suggest presenting only two indicators to the user: one for schema quality and the other for metadata quality. If the user would like to get more information about the quality of schema or the quality of metadata, the user would go into each of the previously defined indicator. The two indicators are calculated as follows:

$$Q_5 = \frac{\sum (a \times 1_5)}{n} \quad (5)$$

$$Q_m = \frac{\sum (b \times 1_m)}{m}, \quad (6)$$

where a and b are predefined values between 0 and 1 that indicate the weight of to each quality indicator of schema and metadata respectively. *Is* and *Im* refer, respectively, to the value of each schema and metadata indicator. The variables n and m are the numbers of the indicators.

In the following section, we propose a general framework to support agents in responding to the risk of data misinterpretation. We will show an example of how the proposed indicators can help making appropriate decisions about the risks of data misinterpretation.

RESPONDING TO THE RISKS OF DATA MISINTERPRETATION

This section proposes an approach to help the end-user to make appropriate decisions about the risk of data misinterpretation in the context of the interoperability between geospatial datacubes. The approach consists of proposing a general framework that presents the previously defined indicators to agents in an intuitive way in order to help them making appropriate decisions about the risks of data misinterpretation. The framework is based on the well established possibilities of responding to the risks (i.e., reducing, absorbing, or transferring the risks).

General Framework to Respond to the Risks of Data Misinterpretation

The framework consists of five successive phases of analysing and responding to the risks of data misinterpretation. These phases correspond to the proposed categorization of semantic heterogeneity, from the more general level to the most detailed level: cube-to-cube, measure-to-measure, dimension-to-dimension, hierarchy-to-hierarchy, and level-to-level. At each level, the intervener analyzes the fitness-for-use of metadata and of schema based on the previously identified indicators, and makes one of the following decisions (see Figure 5):

- To suspend the interoperability process of the geospatial datacubes if it presents a high risk leading to harmful consequences. In this case, the intervener is invited not to continue with the remaining levels.
- To continue the interoperability process of the geospatial datacubes if there is no risk of data misinterpretation or if the risks are low. Consequently, two decisions can be made:
    - Solving the causes of the risks.
    - Doing nothing to solve the causes of the risks, and enduring the risks if they do not significantly affect data use.

The proposed indicators play a key role within the proposed framework. They allow to identify the risks of data misinterpretation and to draw some conclusions about them. More specifically, these indicators have three principal aims:

1. First, to help the intervener to identify the risks of data misinterpretation. In fact, at each level of the proposed framework, while a good quality of metadata and of schema indicates it is less likely to have a risk of data

Figure 5. A general framework to respond to the risks of data misinterpretation

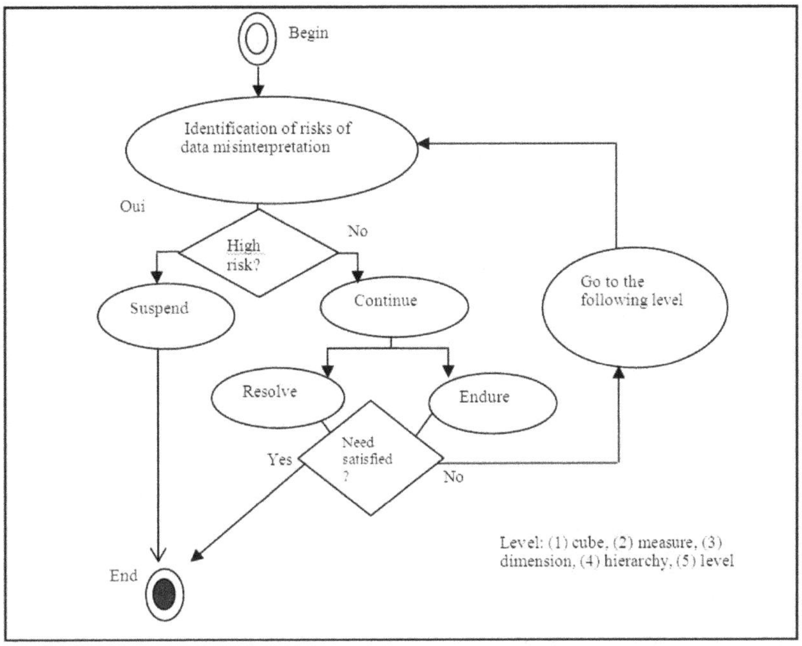

misinterpretation, a poor quality indicates a higher risk.

2. Second, to help agents to make appropriate decisions at each level (to suspend the interoperability process, to solve or endure the problems). For example, if the two heterogeneous elements of different models, which are essential for the interoperability, have a poor quality, the user can be advised not to consider both elements in any interoperability result.
3. Third, to help the intervener to solve the problems at each level of the proposed framework (cube-to-cube, measure-to-measure, dimension-to-dimension, hierarchy-to-hierarchy, level-to-level).

At the end of each level, according to the decision made, the intervener should write a report explaining the followings: the reasons of the suspension, how the problem was solved and the noted comments, or the reasons for which the agent decided to endure the problems.

The risk identification and management are carried out according to a hierarchical top-down approach which has two advantages:

1. The approach allows the intervener to make relevant decisions about the risks of data misinterpretation at an early stage of the interoperability process allowing them to put less time and effort dealing with such risks. At the end of each level, interveners can suspend the interoperability before going into details and hence reduce the costs of the interoperability. Interveners can also continue the interoperability by taking into account the observations made at the general level to better deal with the detailed levels.
2. The proposed approach is in accordance with the mental model of human (Yougworth, 1995). Indeed, this framework is based on a hierarchical structure which is one of the essential principles of human cognition. This principle stipulates that humans gather data in categories according to their own knowl-

edge (Mennis et al., 2000). These categories are organized in a hierarchical way in order to allow the maximum re-use of data with the minimum effort (Rosch, 1978).

Figure 5 illustrates the proposed framework. Based on the previously identified indicators of the fitness-for-use of metadata and schema, the agent starts by identifying the risks of data misinterpretation at the more general level (i.e., the cube level) and making decisions to solve their causes (e.g., problems of semantic heterogeneity).

We should note that the problems of heterogeneity at the measure level must be dealt with before investigating the problems at the dimension level. The reason is that measures are the subject of analysis of the datacubes while dimensions represent the context of this analysis. After analyzing the risks at the dimension level, the user identifies the risks at the hierarchy level and makes decisions to solve them. Later on, we provide a continuation of the example presented in Figure 2 to show how the proposed framework can help to deal with the risks of data misinterpretation. .

Along with the content of the resulting message at each level, in the next section, we define a set of symbols indicating the degree of the risk of data misinterpretation during the interoperability process. This allows interveners to intuitively understand the message.

Symbolic Notation of Quality Indicators

In order to help interveners to intuitively understand the degree of such a risk, we provide a set of warnings which are based on standard danger symbols proposed by ANSI Z535.4 (2006) and previously used in the geospatial domain by Levesque et al. (2007). These symbols are enriched by other symbols to suitably assist interveners in geospatial datacubes interoperability. The symbols show the degree of the risk of data misinterpretation at each level, and thereby stimulate appropriate responses to such a risk. For example, if the warning is 'Danger', it would be better to suspend the interoperability process which may lead to a considerable risk. Figure 6 shows an example of how warnings can be predefined with regards to quantitative values of the fitness-for-use of geospatial datacube model.

To the above list, we add: the symbol → which invites the user to continue the evaluation of others semantically related elements, and the symbol "!" which indicates an absence of information about the semantically related elements (e.g., the absence of metadata), and thus the impossibility of evaluating their fitness-for-use.

In the following section, we provide a continuation of our example to show how the proposed approach can be used to deal with such risks.

EXAMPLE OF APPLICATION

We continue with our example, i.e., two geospatial datacubes presented in Figure 2 and the requirements of the end-user in the form of a geospatial datacube model illustrated in Figure 4. In this example, we suppose that the goal of the interoperability is to define a common model that helps end-users to navigate simultaneously through different geospatial datacubes.

We evaluate the indicators at each level (cube-to-cube, measure-to-measure, dimension-to-dimension, hierarchy-to-hierarchy, level-to-level). To simplify the evaluation, we set the value 1 for a and b (which indicate the weight of each quality indicator of schema and metadata respectively; see Equations (6) and (7)), and that for all levels.

For the two first levels (cube-to-cube and measure-to-measure), there is no problem of heterogeneity. Therefore, the symbol "↓" will be shown for both levels.

Figure 6. A definition of the symbols according to the quality value

Quality value	Representation of the quality indicators	Significance of the symbol
Q = 0	STOP	Interveners are advised to stop the process of interoperability.
0 < Q <= 0.25	⚠ DANGER	There is a high risk if the intervener decides to continue integration.
0.25 < Q <= 0.5	⚠ WARNING	The user is informed about the existence of potential risks.
0.5 < Q <= 0.75	⚠ CAUTION	Interveners are invited to pay attention if they decide to continue the interoperability.
0.75 < Q < 1	NOTICE	Information will be shown to interveners.
Q = 1	↓	Interveners are invited to continue to a more detailed level.

Dimension-to-Dimension

There is a semantic heterogeneity between two dimensions *Administrative Region* of datacube C_1 and *Region* of datacube C_2, therefore the quality of both dimensions is evaluated for their structure and their metadata:

a. **Schema heterogeneity:**
   - *Relevance of the number of hierarchies*

The dimension *Administrative Region* of the datacube C_1 has only one hierarchy: (*Country*, *Province*, *Territory* and *City*), but the dimension *Region* of the datacube C_2 has two hierarchies: (*Country*, *Province*, *Territory*, and *City*) and (*Country*, *State*, and *City*). Since the user needs only one hierarchy: (*Division Administrative*), the quality indicator *Relevance of the number of hierarchies* is evaluated according to the Equation (1):

For C_1, *Ps* (*Administrative Region*) = 1/1 = 1

For C_2, since the number of hierarchy of the dimension *Region* is larger than that of the dimension semantically related of the required datacube C_3 (*Administrative Division*), then:

P_S *(Region)* = 1

According to the Equation (5) the quality of datacubes schemas:

Q_S *(Administrative Region)* = 1

Q_S *(Region)* = 1

Consequently, the symbol "↓" will be shown for the two dimensions *Administrative Region* and *Region*.

b. **Metadata heterogeneity:**
   - *Relevance of metadata*

Metadata associated with the dimension *Administrative Region* of C_1 have two elements which are semantically related to the elements required by the user (the spatial coverage and the year of creation of the dimension). On the other hand, metadata associated with the dimension *Region* of C_2 have only one element which is semantically related to a required element (the spatial coverage of the dimension). If we consider that the weight of spatial information is 1, then based on Equation (3):

P_m *(Administrative Region)* = 2/2 = 1

P_m *(Region)* = 1/2 = 0.5

- *Freshness of metadata*

Metadata of the dimensions *Administrative Region* of C_1 and *Region* of C_2 were created respectively in 2002 and 1982. Moreover, these two dimensions have the same lifetime: 30 years. Consequently, according to Equation (4):

A_m *(Administrative Region)* = 1 - (2005-2002/30) = 0.9

A_m *(Region)* = 1 - (2005-1982/30) = 0.23.

According to the Equation (6) the quality of metadata:

Q_m *(Administrative Region)* = (1 + 0.9) /2 = 0.95

Q_m *(Region)* = (0.5 + 0.23) /2 = 0.36.

Consequently, the symbol "↓" will be shown for the dimension *Administrative Region*. On the other hand, the symbol "! Warning" will be shown for the dimension *Region*. The user is then invited to be careful when considering the dimension *Region* in the process of interoperability, and to evaluate the quality of the detailed levels of these two dimensions (i.e., hierarchy) by taking into account the difference in quality of these dimensions.

Hierarchy-to-Hierarchy

There is a semantic heterogeneity between the hierarchies (H_1: *City*, *Province*, *Territory* and *Country*) of dimension *Administrative Region* of datacube C_1 and (H_2: *City*, *Province*, *Territory* and *Country*) of the dimension *Region* of the datacube C_2, then we evaluate the quality of schema and metadata of the two hierarchies with regards to the requirement's hierarchy (H_3: *City*, *Province*, *Territory* and *Country*).

a. **Schema heterogeneity:**
    - *Relevance number of levels*

Each hierarchy (H_1, H_2 and H_3) contains 4 levels. According to the Equation(1):

$P_S (H_1, H_3)$ = 4/4 =1

$P_S (H_1, H_3)$ = 4/4 = 1.

- *The order of levels*

The hierarchy of the dimension *Administrative Region* has the following order: (*City* < *Province*), (*City* < *Territory*), (*Province* < *Country*) and (*Territory* < *Country*). The hierarchy of dimension *Region* has the following order: (*City* < *Province*), (*City* < *Territory*), (*Province* < *Country*) and (*Territory* < *Country*). The hierarchy of dimension *Administrative Division* of the datacube C_3 has the following order: (*City* < *Province*) and (*Province* < *Country*).

For the datacube C_1, according to expression (2):

The levels *City*, *Province* (or *Territory*) and *Country* (C_1) are respectively, semantically related to the levels *City*, *Province* and *Country* (C_3). Moreover, the level orders *City* < *Province* (in C_1) and *City* < *Province* (in C_3), therefore elementary relevance of the order (*City* < *Province*) o_s = 1. Also, o_s (*City* < *Territory*) = o_s (*Province* < *Country*) = o_S (*Territory* < *Country*) = 1. Therefore, relevance of the structure is as follows:

$O_S (H_1, H_3)$ = (1+1+1+1) /4 = 1.

Similarly, the relevance of the hierarchy (*City*, *Province*, *Territory* and *Country*) of the datacube C_2 is calculated based on the expression (2):

$O_S (H_2, H_3) = (1+1+1+1) /4 = 1.$

Finally, according to the Equation (5):

$Q_S (H_1, H_3) = 1$

$Q_S (H_2, H_3) = 1.$

Thus, the quality of schema of both hierarchies is very good. Consequently, the symbol "↓" will be shown and the intervener is invited to continue with the remaining level.

b. **Metadata heterogeneity:**
 ◦ *Relevance of the metadata*

Metadata associated with the hierarchy (*City*, *Province*, *Territory* and *Country*) of the datacube C_1 contain an element which is semantically related to the required element (i.e., using the "multiple representation"). On the other hand, the metadata associated with the hierarchy (*City*, *Province*, *Territory* and *Country*) of C_2 do not contain any element that is semantically related to the required element. If we consider that the weight of spatial information is = 1, then, according to the Equation (3): " "

$P_m (H_1, H_3) = 1/1 = 1$

$P_m (H_2, H_3) = 0/1 = 0.$

Also, according to the Equation (6):

$Q_m (H_1, H_3) = 1/1 = 1$

$Q_m (H_2, H_3) = 0/1 = 0.$

Therefore, the symbol "↓" for the hierarchy (*City*, *Province*, *Territory* and *Country*) of C_1 and the symbol "! Danger" for the hierarchy (*City*, *Province*, *Territory*, and *Country*) of C_2 will be shown. Accordingly, selecting the second hierarchy creates risks to harm the interoperability between geospatial datacubes. Thus, the user is invited to continue to evaluate only the levels of the first hierarchy (*City*, *Province*, *Territory* and *Country*) of C_1.

Heterogeneity Level-to-Level

a. **Schema heterogeneity (Level City of C_1):**
 ◦ *Relevance of the number of attributes*

The number of attributes of the level *City* of the datacube C_1 is one (*City-name*). Since the user needs two attributes for this level (*City-name* and *Surface*), then according to the Equation (1):

$P_S (City, City) = 1/2 = 0.5.$

 ◦ *Relevance of the geometric primitive*

In C_1, each member of the level *City* is represented by a point (0D), whereas the users need a polygon (2D) to represent the members of the same level *City*. According to the Table 1:

$P_P (City, City) = 0.5$

Then, according to the Equation (5), the structure quality of the level *City* of C_1 is calculated as follow:

$Q_S (City, City) = (0.5+0.5) /2 = 0.5.$

Consequently, the quality is reasonably satisfactory, hence the symbol "! Warning" is shown to inform the user about potential risks that may occur when considering this level. Based on the proposed framework, the user can decide to solve the problems related to this level or to endure the potential consequences of these problems.

Table 1. Evaluation of the relevance of the geometric primitive

	Elements of the required model				
Element of the models to be integrated		0D	1D	2D	3D
	0D	1	0	0.5	0.75
	1D	0	1	0	0
	2D	0.75	0	1	0.5
	3D	0.5	0	0.75	1

b. **Metadata heterogeneity (Level City of C_1):**

- *Relevance of the metadata*

Metadata of the datacube C_1 contain two elements (the spatial referencing system and the scale of representation) which are semantically related to the required elements. If we consider that the weight of spatial information is 1, then according to the Equation (3):

P_m *(City, City)* = 2/2 = 1.

According to the Equation (6), the quality of the metadata of the level *City* of C_1 is calculated as:

Q_m *(City, City)* = (1+1) /2 = 1.

Consequently, the structure quality of the level *City* of C_1 is quite satisfactory. The symbol "↓" will be illustrated.

Similarly, the qualities of the levels *Province*, *Territory* and *Country* datacube C_1 are evaluated based on the Equations (5) and (6):

Q_S *(Province, Province)* = 1

Q_m *(Country, Country)* = 1.

We should notice that, if an element of one of the datacubes sources is not semantically related to any other element of another datacube, and that it fits the user's requirement, then this element (measure, dimension, hierarchy, level) is integrated in the common model. The dimensions *Age category* and *Forest* in our example are of this case.

Figure 7 shows an example of model which could be obtained to enable the interoperability between geospatial datacubes C_1 and C_2 according to the different levels of the general framework and using the proposed quality indicators.

USEFULNESS OF THE INTEROPERABILITY OF GEOSPATIAL DATACUBES FOR ENVIRONMENTAL APPLICATIONS

In many situations, such as a simultaneous and rapid intervention in environmental emergencies using different geospatial datacubes, a rapid acquisition of decisional data for environmental applications, and an interactive analysis of environmental factor changes, we may need to interoperate several heterogeneous datacubes to support environmental applications. In the following, we provide some typical examples of these situations.

1. **Simultaneous and rapid intervention in environmental emergencies using different geospatial datacubes:** Users intervening in environmental emergency situations may need to access and navigate simultaneously through heterogeneous geospatial datacubes. Navigating separately through each datacube would be an arduous work for users, since they likely need to make extra efforts to manually resolve the problems of heterogeneity between datacubes (e.g., comparing the meaning of concepts and establishing a mapping between them). The principal aim of interoperability is to automatically overcome such differences which can considerably facilitate the navigation task and, hence, facilitate human intervention in emergency situation. For example,

Figure 7. Example of the resulting common model

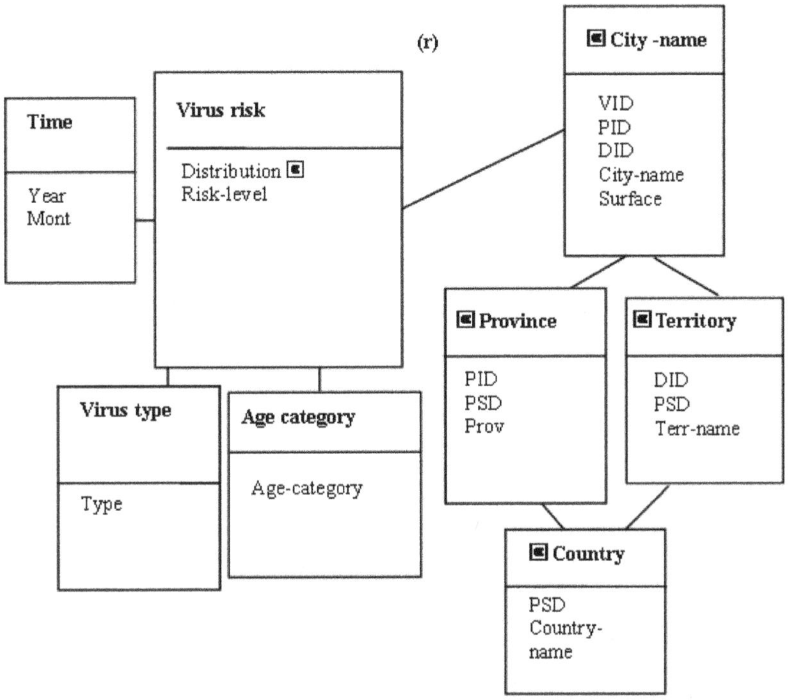

a health organization willing to analyze the risk of the West Nile virus to the population, may need to simultaneously navigate two geospatial datacubes; one containing data related to water bodies and the other containing data related to the location of dead birds reported by the population. The interoperability between these geospatial datacubes would enable, for example, to create a common multidimensional structure that allows to rapidly navigate through data stored in both geospatial datacubes. Our proposed approach can help intervener in the interoperability process to define such common structure.

2. **Rapid acquisition of decisional data for environmental applications:** While data in datacubes are usually collected from legacy systems, they can be imported from other heterogeneous datacubes (Bédard et al., 2001). We may need to rapidly acquire data from different heterogeneous geospatial datacubes to be used for environmental applications. For example, we may need to rapidly insert data in a geospatial datacube which contains data about forest stand from another geospatial datacubes that contains soil type of the same geographic region. The proposed approach in this paper can make intervener aware of potential heterogeneity problems that may occur during data insertion, and can help him/her to make appropriate decisions to overcome such problems.
3. **An interactive analysis of environmental factor changes:** In order to analyze the changes of environmental factors (e.g., air pollution evolution), we may need to compare data describing these factors at different epochs, and stored in geospatial datacubes build also at different epochs. Interoperating such datacubes would enable to interactively compare data and analyze

these factors' changes. For example, we may need to rapidly compare air pollution changes following a volcanic eruption. For that, we interoperate two existing geospatial datacubes for analyzing air pollution one based on data before the volcanic eruption and another one using data after eruption. Based on this study, we can realize the impact of the volcanic eruption on the air pollution. In our approach, we proposed a set of quality indicators that can help intervener in the interoperability process to compare the content of both geospatial datacubes.

Our proposed approach can be used in any of the above mentioned or similar situations to solve the problems a risen during the process of interoperability between geospatial datacubes. As the above examples illustrate, many of these situations can happen in environmental applications.

CONCLUSION

The interoperability between geospatial datacubes faces risks of data misinterpretation that may hinder the aim of strategic decision-making. These risks are typically caused by the semantic heterogeneity of the content of geospatial datacubes. In this paper, we proposed an approach to deal with such risks. We introduced a set of indicators of the fitness-for-use of conceptual model of datacube (i.e., quality of schema and of metadata with regard to a specific use), and quantitatively evaluated them. These indicators facilitate the identification and the evaluation of the risks, and hence, making users aware of the risks and their severity. In fact, while a good fitness-for-use of metadata and of schema indicates it is less likely to faulty interpreting data or being uncertain about its interpretation, a poor quality indicates a higher risk.

In addition, we proposed a general framework that can be used by interveners to make appropriate decisions about the risks that may occur during the interoperability process of geospatial datacubes. The framework is based on a hierarchical top-down that corresponds to the proposed categorization of semantic heterogeneity (cube-to-cube, measure-to-measure, dimension-to-dimension, hierarchy-to-hierarchy, and level-to-level). We also demonstrated the usefulness of the proposed approach in environmental applications.

We should remind that the set of the proposed indicators and the proposed framework do not aim being exhaustive or precise but rather at helping the intervener to make appropriate decisions to enhance the interoperability between geospatial datacubes. Such a method is frequently used in several fields like epidemiology and ecology which involve factors that are difficult, even impossible, to evaluate in an exhaustive and precise way.

The example of application showed how the framework as well as the proposed indicators can help interveners to make appropriate decisions about the suspension or the continuation of interoperability process. The framework and the indicators constitute a base for future works dealing with the interoperability between geospatial datacubes, but also with interoperability involving other information systems.

REFERENCES

Agumya, A., & Hunter, G. J. (2002). Responding to the consequences of uncertainty in geographical data. *International Journal of Geographical Information Science*, *16*(5), 405–417. doi:10.1080/13658810210137031

ANSI Z535.6. (2006). *American national standard for product safety signs and labels*.

Bédard, Y. (1986). A Study of data using a communication based conceptual framework of land information systems. *Le Géomètre Canadien*, *40*(4), 449–460.

Bédard, Y., & Larrivée, S. (2008). Spatial database modeling with pictrogrammic languages. In Shekhar, S., & Xiong, H. (Eds.), *Encyclopedia of GIS* (pp. 716–725). New York: Springer Verlag. doi:10.1007/978-0-387-35973-1_812

Bédard, Y., Merrett, T., & Han, J. (2001). Fundamentals of spatial data warehousing for geographic knowledge discovery. In Miller, H. J., & Han, J. (Eds.), *Geographic data mining and knowledge discovery*. doi:10.4324/9780203468029_chapter_3

Bédard, Y., Rivest, S., & Proulx, M. J. (2005). Spatial on-line analytical processing (SOLAP): concepts, architectures and solutions from a geomatics engineering perspective. In Koncillia, W. R. (Ed.), *Data warehouses and OLAP: Concepts, architectures and solutions*.

Bishr, Y. (1998). Overcoming the semantic and other barriers to GIS interoperability. *International Journal of Geographical Information Science*, *12*, 299–314. doi:10.1080/136588198241806

Boin, A. T. (2008). *Exposing Uncertainty: Communicating spatial data quality via the Internet* (p. 197). Unpublished doctoral dissertation, University of Melbourne, Melbourne, Australia.

Brodeur, J. (n.d.). *Interopérabilité des données géospatiales: élaboration du concept de proximité géosémantique* (p. 247). Unpublished doctoral dissertation, Université Laval, Mondreal, Canada.

Cherfi, S. S., Akoka, J., & Comyn-Wattiau, I. (2007). Perceived vs. measured quality of conceptual schemas: an experimental comparison. In *Proceedings of ACM Int.* (Vol. 334, pp. 185-190).

Devillers, R., Bédard, Y., Jeansoulin, R., & Moulin, B. (2007). Towards spatial data quality information analysis tools for experts assessing the fitness for use of spatial data. *International Journal of Geographical Information Science*, *21*(3), 261–282. doi:10.1080/13658810600911879

Few, S. (2006). *Information dashboard design: The effective visual communication of data*. Sebastopol, CA: O'Reilly Media.

Frank, A. U. (2007). Data quality ontology: An ontology for imperfect knowledge. In *Proceedings of COSIT 2007* (LNCS 4736, pp. 406-420).

Genero, M., Manso, E., Visaggio, A., Canfora, G., & Piattini, M. (2007). Building measure-based prediction models for UML class diagram maintainability. *Empirical Software Engineering*, *12*, 517–549. doi:10.1007/s10664-007-9038-4

Kuhn, W. (2005). Geospatial semantics: Why, of what, and how. *Journal on Data Semantics*, *III*, 1–24. doi:10.1007/11496168_1

Levesque, M.-A., Bédard, Y., Gervais, M., & Devillers, R. (2007). Towards managing the risks of data misuse for geospatial datacubes de données. In *Proceedings of ISSDQ 2007*.

Mennis, J. L., Peuquet, D. J., & Qian, L. (2000). A conceptual framework for incorporating cognitive principles into geographical database representation. *International Journal of Geographical Information Science*, *14*(6), 501–520. doi:10.1080/136588100415710

Moody, D. L. (2005). Theoretical and practical issues in evaluating the quality of conceptual models: Current state and future directions. *Data & Knowledge Engineering*, *15*(3), 243–276. doi:10.1016/j.datak.2004.12.005

Rafanelli, M. (2003). *Multidimensional databases: Problems and solutions*. Hershey, PA: IGI Global.

Renn, O. (1998). Three decades of risk research: accomplishments and new challenges. *Journal of Risk Research*, *1*(1), 49–71. doi:10.1080/136698798377321

Rosch, E. (1978). Principles of categorization. In Rosch, E., & Lloyd, B. (Eds.), *Cognition and Categorization* (pp. 27–77).

Sboui, T., Bédard, Y., Brodeur, J., & Badard, T. (2007). A conceptual framework to support semantic interoperability of geospatial datacubes. In *Proceedings of ER/2007 SeCoGIS workshop* (LNCS 4802, pp. 378-387). New York: Springer.

Staub, P., Gnagi, H. R., & Morf, A. (2008). Semantic interoperability through the definition of conceptual model transformations. *Transactions in GIS*, *12*(2), 193–207. doi:10.1111/j.1467-9671.2008.01095.x

Yougworth, P. (1995). OLAP spells success for users and developers. *Data Based Advisor*, 38-49.

ENDNOTES

1. Concepts having a similar representation (e.g., synonyms), generalization, or specialization relationship.
2. *Perceptory*'s Web Site: http://sirs.scg.ulaval.ca/perceptory
3. Hurtado et al. (2005) introduced the notion dimension constraint.

This work was previously published in International Journal of Agricultural and Environmental Information Systems, Volume 1, Issue 2, edited by Petraq Papajorgji and François Pinet, pp. 20-41, copyright 2010 by IGI Publishing (an imprint of IGI Global).

Section 2
Ontologies

Chapter 5
Using Ontologies to Relate Resource Management Actions to Environmental Monitoring Data in South East Queensland

Jane Hunter
The University of Queensland, Australia

Abdulmonem Alabri
The University of Queensland, Australia

Peter Becker
The University of Queensland, Australia

Catharine van Ingen
Microsoft Research, USA

Eva Abal
South East Queensland Healthy Waterways Partnership, Australia

ABSTRACT

The Health-e-Waterways Project is a multi-disciplinary collaboration between the University of Queensland, Microsoft Research and the South East Queensland Healthy Waterways Partnership (SEQ-HWP). This project develops the underlying technological framework and set of services to enable streamlined access to the expanding collection of real-time, near-real-time and static datasets related to water resource management in South East Queensland. More specifically, the system enables water resource managers to access the datasets being captured by the various agencies participating in the SEQ HWP Ecosystem Health Monitoring Program (EHMP). It also provides online access to the statistical data processing tools that enable users to analyse the data and generate online ecosystem report cards dynamically via a Web mapping interface. The authors examine the development of ontologies and semantic querying tools to integrate disparate datasets and relate management actions to water quality indicators for specific regions and periods. This semantic data integration approach enables scientists and resource managers to identify which actions are having an impact on which parameters and adapt the management strategies accordingly. This paper provides an overview of the semantic technologies developed to underpin the adaptive management framework that is the central philosophy behind the SEQ HWP.

DOI: 10.4018/978-1-4666-0333-2.ch005

INTRODUCTION

Climate change, urban development and population growth in South East Queensland are applying significant pressure to urban water supplies. Billion dollar investments are being made in new water projects, such as the South East Queensland Water Grid, with the aim to enhance and secure water supply to the region. However the potential benefits of these massive investments in physical infrastructure will not be fully realized, without the investment in corresponding *cyberinfrastructure*. Scientists, urban planners and policy makers require access to integrated water information management systems that enable them to track water movement, consumption and quality across the "whole-of-water-cycle". They require access to high quality, complete datasets and accurate predictive models on which to base water management decisions. Understanding and satisfying the competing needs of water users, requires the integration of data and models that reflect the behaviour of many systems including: climate, rainfall, catchments, reservoirs, rivers, agricultural demand, industrial demand and urban demand. The linking of models requires a detailed understanding of: the questions being asked, the data that is available and the relationships between the datasets and models being employed. In parallel with the demand for more sophisticated querying and decision support interfaces, there is a trend towards wide-spread deployment of sensors capturing real-time information including volume flows, water quality, images and video. These real-time and near-real-time, temporal and visual data streams require more sophisticated, high speed, data processing, indexing and archival services.

This paper describes the Health-e-Waterways Project – a cross-disciplinary collaboration between information scientists and water resource managers and stakeholders that is developing the knowledge management infrastructure for adaptive whole-of-water-cycle management in South East Queensland. It begins by describing the challenges faced by South East Queensland Healthy Waterways Partnership (SEQ HWP) in managing the data associated with their Adaptive Management Strategy and Ecosystem Health Monitoring Program (EHMP). It goes on to describe the ontology-based knowledge management system developed by Health-e-Waterways to enable the integration and analysis of a wide range of data types. More specifically it describes the ontologies that have been developed, the system architecture and the query, reporting and visualization interfaces – that are enabling scientists and policy makers to identify trends in ecosystem health indicators both geographically and temporally and to understand the impact that water resource management actions are having on water quality.

THE SOUTH EAST QUEENSLAND HEALTHY WATERWAYS PARTNERSHIP

South East Queensland Healthy Waterways Partnership (SEQ HWP) was established in 2002 by government, industry and community stakeholders working in close cooperation to develop a whole-of-government, whole-of-community approach to understanding and managing the region's waterways. The key elements of the SEQ HWP strategy include: the implementation of targeted management actions (ranging from upgrades of sewage treatment plants to rehabilitation of riparian vegetation); a multi-disciplinary science, research and monitoring program that underpins the management action program and monitors its effectiveness; and the Healthy Waterways promotional and educational program (Bunn et al., 2007). Moreover the SEQ HWP is underpinned by an "Adaptive Management" approach as illustrated in Figure 1. "This approach recognizes that management intervention can't be postponed until all of the information required to fully understand the situation is available" (Bunn et al., 2007). As a result, monitoring data is continually acquired through a comprehensive Ecosystem Health Monitoring Programme (EHMP) (EHMP, 2010), analysed and assessed to monitor the effectiveness of management actions, that are then adapted as required. The EHMP comprises two sub-programmes – the Estuarine and Marine EHMP and the FreshWater EHMP.

The *Estuarine and Marine EHMP* began in 2000 and includes monthly monitoring of water quality parameters at 260 sites in coastal waterways from Noosa to the New South Wales border. This program monitors a range of physical/chemical and biological indicators including: turbidity, salinity, temperature, pH, dissolved oxygen, secchi depth, total nitrogen, total phosphorus, seagrass depth range and distribution, $\delta\ ^{15}N$ (sewage plume) mapping, Lyngbya and coral. The *Freshwater EHMP* commenced in 2002 and includes twice-yearly sampling at over 120 sites on all of the major streams in the region. This involves the measurement of 5 indicators (physical/chemical, nutrient cycling, ecosystem processes, acquatic macroinvertebrates and fish) that are derived from 16 parameters (e.g., pH, Temperature, conductivity, dissolved oxygen, nitrogen stable isotope $\delta^{15}N$, algal bioassay, algal growth, number of taxa, proportion of alien fish etc) (Smith & Storey, 2001; Smith & Grice, 2005).

The EHMP delivers an objective regional assessment of the ecosystem health for each of South East Queensland's 19 major catchments, 18 river estuaries, and Moreton Bay (Figure 2). One of the major outcomes from the Program is an Annual Report Card which aggregates measured values of water quality to provide a simple grade (A-F) for each of these waterways (Annual Report Cards, 2009). This grade provides an effective way to communicate the relative health of waterways to the local and State Governments and the broader community, and to evaluate the effectiveness of management actions. To date, these annual report cards have taken 6 months of manual effort by a large team of scientists to retrieve, collate and analyze the data and produce graphical reports using Photoshop and Adobe Illustrator. As the size and heterogeneity of the datasets grows and the number of agencies contributing relevant data expands, this process becomes unsustainable.

Hence the aim of the Health-e-Waterways project is to develop underlying data management, integration, reporting and visualization services to streamline the generation of the Annual Report Cards, enable more interactive online access to the data and associated analytical services and in the long term support the *Adaptive Management* approach, shown in Figure 1.

OBJECTIVES OF THE HEALTH-E-WATERWAYS PROJECT

The primary objective of the Health-e-Waterways project is to enable and promote the sharing and collaborative integration and analysis of high quality information concerning water for the partners and stakeholders of SEQ HWP. Figure 3 provides an overview of the components and architecture of the proposed system. A combination of semantic web technologies, scientific data servers, web services, GIS (Geographical Information Systems) visualization interfaces and scientific workflows is enabling the sharing and integration of high quality data and models, through a combined integrated water information management system and Web portal.

The high-level, long term aim of the project is to provide a Web portal to the latest relevant data and models for Queensland scientific, education, government, community and commercial stakeholders with an interest in water research, conservation, and management. A range of security and access-control levels are also under development to enable different stakeholders to access different types and levels of data. For example, the aggregated report card grades are available to the general public but the raw data underpinning these is only accessible by SEQ HWP partners.

Figure 1. The adaptive management framework adopted by Healthy Waterways Partnership

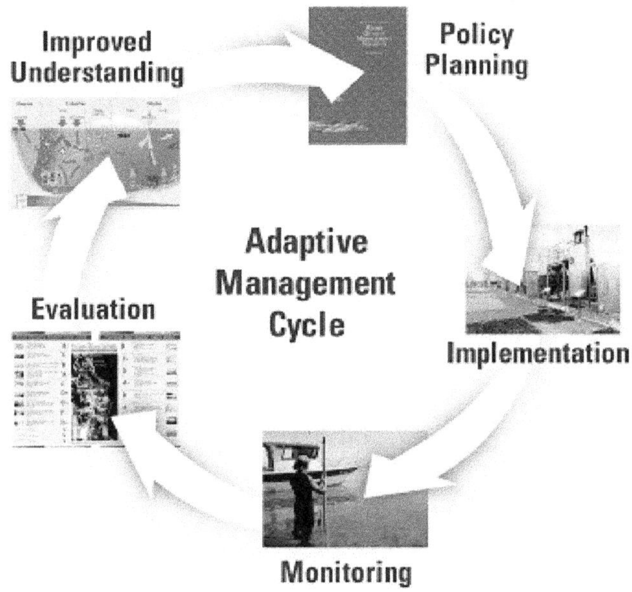

During the first 3 months of the project, it became clear that there were critical components of the SEQ-HWP strategy that required immediate focus by the information scientists involved in the project. These included:

- **Streamlining the generation of the Annual Report Cards:** currently the Report Cards take 6 months to produce and comprise both a 4 page summary Report Card as well as a 150 page Technical Report. These documents contain many graphs and figures that are manually collated and produced using tools such as PhotoShop. Our first objective was to enable the report card grades and associated graphs to be generated dynamically and systematically from the underlying integrated database via a Web interface.
- **Database Registry:** the EHMP databases, Management Action database and other relevant databases are stored on a variety of platforms and servers hosted by different organizations behind their organizational firewalls. Our second objective was to enable these databases to be accessible and discoverable via a single web-based database registry and to be integrated and searchable via a common data model and web server.
- **Semantic Querying, Reporting and Visualization Services:** scientists and resource managers need to be able to easily identify whether specific management actions are having any impact or not. They require decision support tools to help them decide whether and how their management plans should be adapted. Currently this is a complex procedure that depends on manual collation of related data and information. A high priority aim was to provide querying and visualization interfaces that enabled relationships between management actions and corresponding changes in water quality to be identified.

Figure 2. Ecosystem health monitoring program – freshwater and estuarine/marine monitoring sites

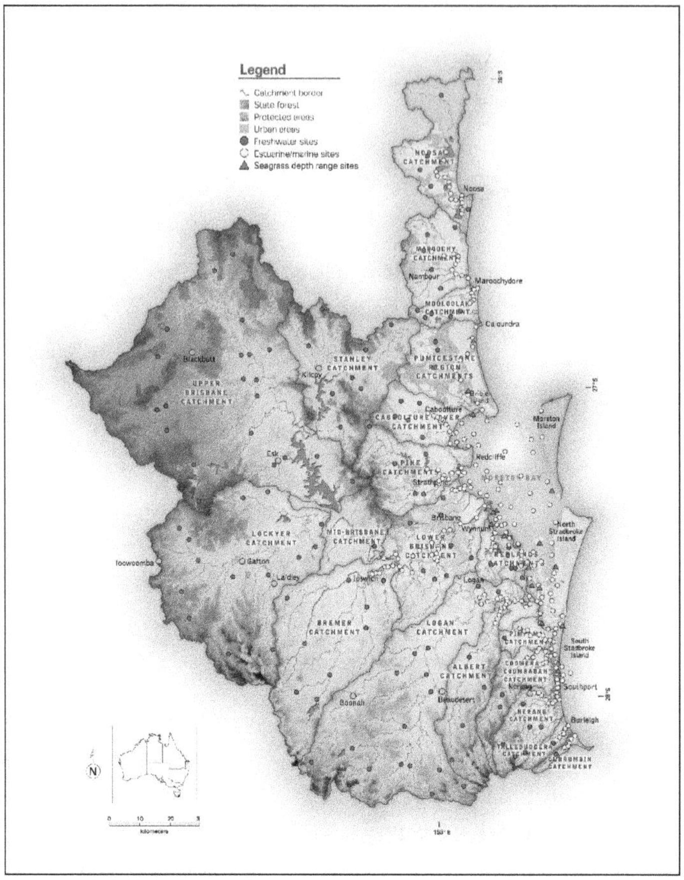

In the remainder of this paper, we describe the methods we have adopted in implementing solutions to these high priority issues and also provide a discussion of the effectiveness of these methods.

RELATED WORK

In developing the Health-e-Waterways system, we first identified a number of existing initiatives and environmental science projects that recognize the critical role of semantic web technologies and Observational Data Models (ODMs) in providing seamless, multi-disciplinary access to observational data. These projects include:

- The Consortium of Universities for the Advancement of Hydrologic Science (CUAHSI) has developed an Observations Data Model (ODM) that is being applied to the storage and retrieval of hydrologic observations data (CUAHSI, 2009) (Horsburgh et al., 2008);
- SPIRE is a research project investigating how semantic web technologies can be used to support science in general and the field of ecoinformatics more specifically (Madin et al., 2007);
- The SWEET (Semantic Web for Earth and Environmental Terminology) has developed several ontologies to provide a com-

Figure 3. Health-e-waterways system - overview and architecture

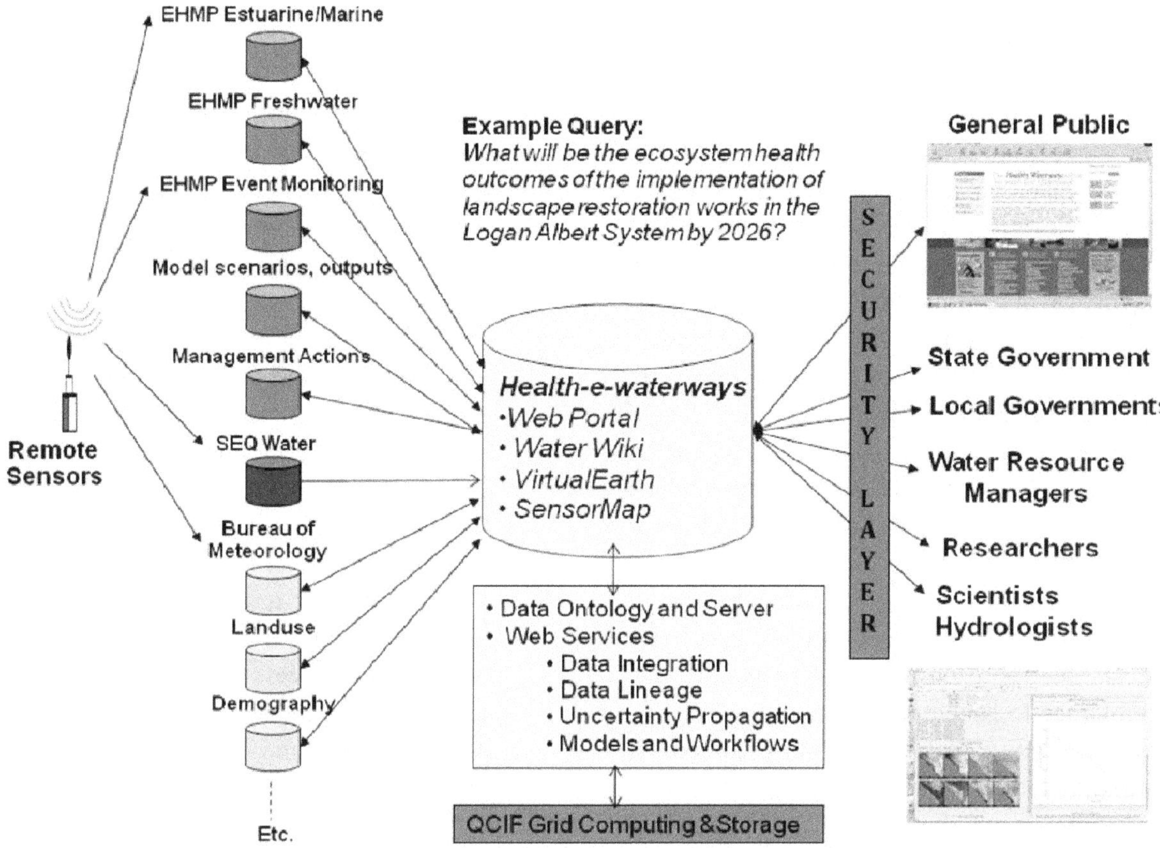

mon semantic framework for various Earth science initiatives (SWEET, 2009);

- The WATERS testbed (Montgomery et al., 2007) is constructing a national-scale network of observational and experimental facilities in the US, to enable systematic water measurements, data storage and curation, modeling and visualization. WATERS is using the CUAHSI ODM as the underlying data model.
- The SNaaS Framework (Sensor Network as a Service) is specifically developing ontologies to describe virtual sensors (Liu et al., 2009).

These previous efforts have primarily focused on ontologies for integrating observational data associated with ecosystems. The EHMP ontology was developed through a 3-pronged approach that involved: a comprehensive analysis of existing database schemas (Hunter & Mauger, 2008); drawing on these existing data models; and inviting input from local domain experts. In addition, we have developed a "Management Action" ontology that models management action data and enables management decisions to be linked to specific targeted ecosystem health indicators. This represents a significant and critical advance on previous efforts.

A number of previous projects and efforts have focused on the application of ontologies, artificial intelligence or machine learning to knowledge management and decision-support in the context of hydrological ecosystems. However the majority of these approaches have focused on computational modeling. For example, Chau (2007) developed an ontology-based system for modeling water flow and quality. This system is based on numerical simulations (not near-real-time observational data) and uses inferencing to select the optimum model and parameters. Other related research efforts (Zhao et al., 2006; Muttil & Chau, 2007; Lee et al., 2002) have primarily focused on statistical and AI methods for identifying the ecologically significant variables associated with environmental models. For example, Lee et al. (2002) applied neural networks to the modeling of coastal algal blooms using long-term water quality data. Although some models have been developed for South East Queensland (Larsen, 2007), there is insufficient long term data available at this stage for scientists and policy makers to be able to reliably depend on the accuracy of these predictive models. Our approach is to apply semantic web technologies to enable scientists to explore and better understand relationships between ecosystem health indicators and management actions in the past – in parallel with the development of more accurate, localized predictive models. The next phase of our work will be to integrate these models to enable predictive queries.

METHODOLOGY

The Health-e-Waterways development strategy comprises seven key components which build on top of a combination of semantic web technologies and web services to access, process and integrate data, maps and models held within both centralized and distributed hydrological databases:

1. Identification and prioritization of the key stakeholder user requirements, queries and datasets;
2. Development of a dataset registry that describes the key datasets, their source, spatio-temporal scope, metadata schema, size, formats etc;
3. Development of the common data models and ontologies to integrate both static and real-time data streams, visual, spatial and temporal data, legacy databases and newly generated datasets;
4. Design and implementation of the semantic interoperability layer on the scientific data server;
5. Development of a Web-based querying, visualization and presentation interface – that integrates GIS technologies with GoogleEarth-type interfaces, ontology-based queries and SensorMaps - to display the latest integrated data sets through a mapping interface;
6. Development of a secure Web Portal and *WaterWiki* that provides different levels of collaborative access to data, models, services, storage and compute power. Standardized authentication and access protocols will enforce controlled access to resources and software;
7. Development of a model registry and scientific workflow tools. This component will enable users to upload and share models as web services through a common model registry, link them using scientific workflows and execute them over grid computing infrastructure.

DATABASE SURVEY AND REGISTRATION

The first phase of the project involved identifying those databases that are most critical and relevant to the water stakeholders of SEQ. In the long term the stakeholders want to be able to integrate the following datasets:

- Ecosystem Health Monitoring Program (EHMP) Database;

- Management Action Database;
- Climate data – rainfall, temperature, humidity from the Bureau of Meteorology;
- Vegetation data – from the Australian Herbarium;
- Biodiversity data – from the Atlas of Living Australia;
- Remote Sensing Satellite Imagery – MODIS (Moderate Resolution Imaging Spectroradiometer) data and images from NASA (National Aeronautics and Space Administration) are providing derived measurements inlcuding surface temperature and soil moisture;
- Atmospheric and Air quality data – available via FluxNet.

In the immediate short term, the users wanted to be able to analyse the data being captured through the Ecosystem Health Monitoring program (EHMP) and relate it to the Management Action Database (MAD). Our first activity was a detailed analysis of these databases and the generation of a report that describes these databases in detail and recommends the best approach for integration. This report recommended the development of a dataset registry (that enables relevant data sets to be easily browsed and discovered) and associated registry services (that enable new datasets to be registered, described, uploaded and published). The report also recommends the development of a common data model/ontology to support the integration of these datasets through a common search interface. The ontologies that we have since developed in response to this recommendation are described in the next section.

THE EHMP ONTOLOGY

The analysis of the existing databases enabled us to develop an Entity-Relationship model that incorporates all of the key top-level classes, properties and relationships. Figure 4 illustrates graphically the model that has been determined by identifying the common top-level entities from the five key databases that we have analysed. We used Protégé to develop the OWL version of the EHMP ontology. We also held numerous meetings with key scientists to interactively explain, refine and validate the ontology.

The advantage of developing a model that is object-oriented and based on a common set of top-level entities is that it is extensible. As new databases become available, it will be relatively easy to incorporate these by refining and extending the existing model. Once the EHMP model was determined and agreed, we represented it as an OWL (Web Ontology Language) ontology. During the development of this common model, we drew on a number of pre-existing water information models of relevance but refined and adopted to accommodate local terms and parameters:

- The CUAHSI Community Observations Data Model (ODM) (Horsburgh, 2008)
- The OpenGIS Observations and Measurements Model (OGC, 2009)
- The SEEK Observation Ontology (OBOE) (SEEK, 2008)
- WRON-RM (Water Resources Observation Network Reference Model) 0.1 (O'Hagan et al., 2007).

THE MANAGEMENT ACTION ONTOLOGY

The South East Queensland Healthy Waterways Strategy for 2007-2012[1] comprises a set of twelve Action Plans:

- Five issue-based Action Plans
  - Point Source Pollution
  - Non-Urban Diffuse Source Pollution
  - Water Sensitive Urban Design
  - Protection and Conservation
  - Coastal Algal Blooms

Figure 4. Graphical representation of the EHMP ontology

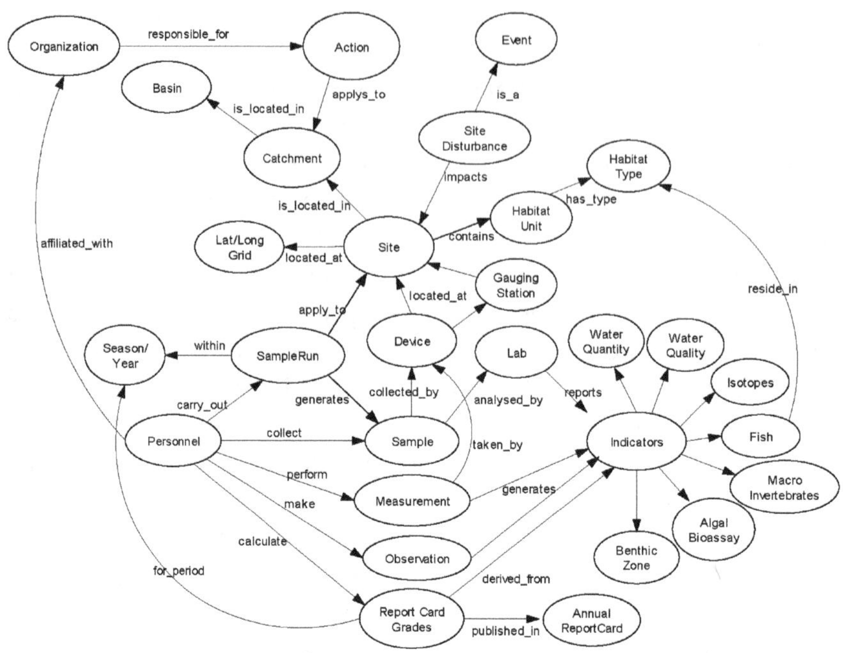

- Three enabling Action Plans
  - Ecosystem Health Monitoring Program
  - Communication, Education and Motivation
  - Management Strategy Evaluation
- Four area-based Action Plans
  - Moreton Bay
  - Northern Catchments
  - Bremer Catchments
  - Logan Albert Catchments

Each Action Plan comprises a number of high level Management Outcomes and sets of Actions that are linked via specific Management Action Targets. Figure 5 illustrates the classes and properties that comprise the MAD ontology.

An illustrative example of an Action Plan (using fictional data) showing its Goal, Target and an associated Action is shown in Table 1.

Over 500 management actions have been stored in an Access database managed by SEQ HWP. One of the key objectives of Health-e-Waterways is to integrate the Management Actions Database (MAD) with the environmental monitoring data in order to identify whether the actions and associated investments are having any impact on the quality of the water – and whether the strategy needs to be adapted or not. At this stage, only the Actions associated with area-based Action Plans are able to be linked to specific regions and indicators. In the next sections, we describe how we integrate these datasets and visualize relationships through a GoogleEarth querying and reporting interface.

SYSTEM IMPLEMENTATION AND ARCHITECTURE

A significant challenge early in the project was the resolution of data access agreements with the custodians of the individual databases. This was a very time-consuming process that required numerous meetings and discussions with stakeholder groups to gain their trust and support. (We are

Figure 5. Graphical representation of the MAD ontology

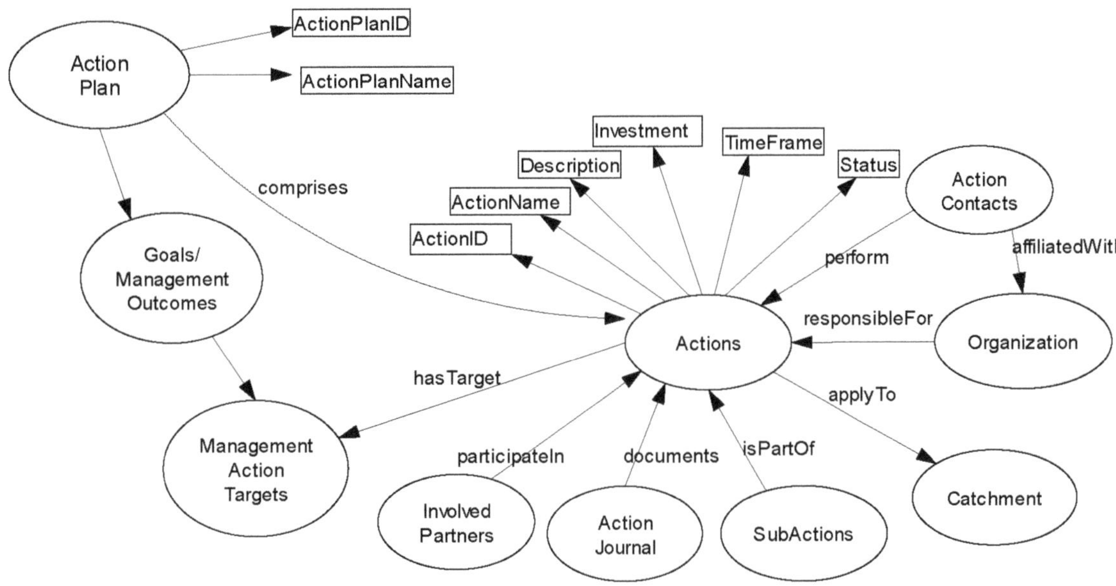

working with the OAKLaw project (Fitzgerald & Long, 2008) to identify licenses that could be applied more generally to enable much easier and streamlined access to the databases via web services interfaces.) Once the data agreements were resolved, we were provided with copies of the datasets for local storage and re-use. We then developed XSLT (XSL Transformations) scripts to automate the mapping of the data from the legacy structure to the common model/ontologies and loaded the mapped data into a common data store. Figure 6 illustrates the ICT components of the system architecture.

Table 1. An example of a management action

Class	Instance
ActionPlan	"Bremer River Catchment Action Plan"
Goal	"Management of Wastewater Treatment Plants"
Action	"Upgrade of Rosewood Wastewater Treatment Plant"
Target	"Total N of 5mg/L and total P of 2mg/L in effluent"
Responsible Organization	"Ipswich City Council"
Involved Partners	"Bremer Catchment Association"
Catchment	"Bremer"
TimeFrame	"2011"
Investment	"$2million"
Status	"Underway"
PercentComplete	"50"
LastUpdate	"20/09/2009"

The Freshwater and Estuarine Marine EHMP data sets, together with the Shape files (outlining the catchments, estuaries and bay zones) and the Management Action Database are stored in a PostgreSQL database. The ontologies described in the previous section are used as the common data model underpinning the database. The Open Source GeoStack (OpenLayers, GeoServer and PostGIS) is used to store, publish and access the data. OpenLayers was chosen because it provides an open source AJAX (asynchronous JavaScript and XML) solution for serving, displaying and layering map data in most browsers, using Open Geospatial Consortium (OGC) OGC Web Mapping Services (WMS). It enables fast development and easy customization and extension of the basic software services.

Figure 6. Architectural components of the Health-e-waterways system

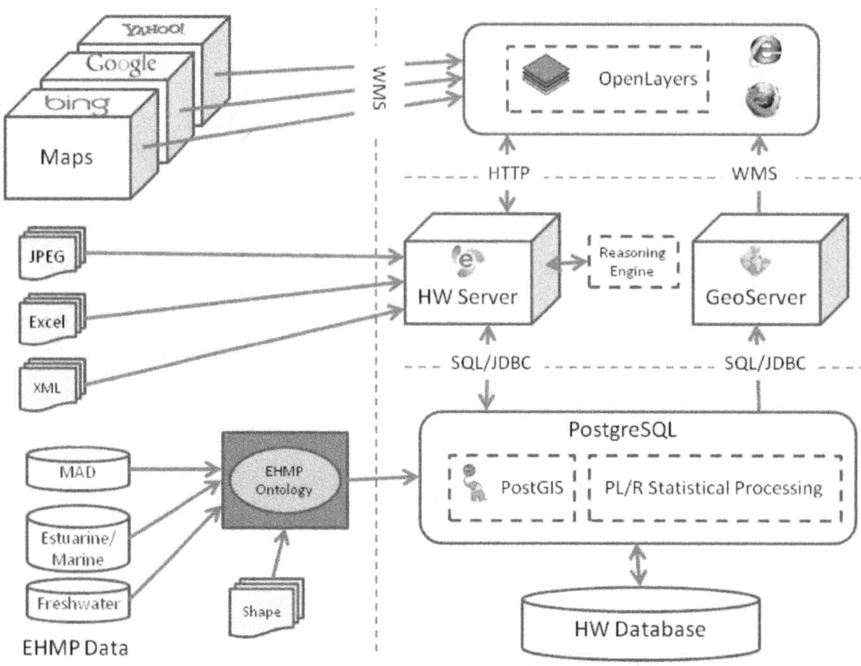

DYNAMIC GENERATION OF ONLINE ECOSYSTEM REPORT CARDS

Initial stakeholder meetings identified the immediate need for a more efficient approach to the production of the annual EHMP Report Cards, published by SEQ HWP (EHMP Report Cards, 2009). Currently hard copies of the EHMP Report Cards (that report the ecosystem heath grades for catchments in SEQ) are printed and published annually. The process takes about 5 months with the most time-consuming step being the manual generation of the graphs and Ecosystem Health (EcoH) plots using Photoshop Illustrator. Politicians and planners from 20 different agencies (including the Queensland Environmental Protection Agency), 4 universities, 18 local councils and CSIRO (Commonwealth Scientific and Industrial Research Organisation), use the report cards to make decisions with respect to land use, water quality, allocations and investments in water recycling plants etc. The report cards provide a standardized method for understanding and comparing the health of catchments between regions and over time. To date, these report cards are largely produced manually, via the following process (as documented in Smith & Grice, 2005):

1. Values for each of the 5 indicators (physical/chemical, nutrients, ecosystem processes, aquatic macroinvertebrates and fish) are calculated from the 16 indices recorded for each site and for two seasons (spring and autumn);
2. Index values are compared against Ecosystem Health Guideline values to derive standardised scores that range from 0 (unhealthy) to 1 (healthy);
3. Standardised scores are averaged across combinations of indices, sites and seasons to provide intelligible summaries for catchments and seasons;

4. Graphical summaries of results are produced to enable the comparison of indices and indicators across reporting areas and between seasons: box and whisker plots, horizontal bar charts, and Ecosystem Health plots (EcoH plots).
5. Summarisations of the EHMP results culminate in a report card grade from A to F.

For the past 6 months, we have been working with the SEQ-HWP staff who are responsible for generating the annual ecosystem report cards. The outcome is a set of software services that enable the report cards to be generated dynamically via R statistical processing tools and a Web-based mapping interface, overlaid on the integrated datasets. Figure 7 shows the results of our collaboration. The interactive map interface enables users to select:

- Spatial regions of interest (e.g., particular catchments or sites) through a mapping interface (GoogleEarth or Microsoft VirtualEarth);
- Seasons and years of interest through a timeline;
- Concepts or indicators of interest through the EHMP Ontology.

A Report Card Grade is generated annually for each catchment, estuary and bay region. Clicking on a grade, displays the corresponding EcoH plots for the current and previous years (dynamically generated from the 5 indicators in the underlying PostGreSQL database) (see Figure 8). Clicking on an EcoH plot, displays the actual raw data (16 indices) used to generate the indicators and plots. The combined use of PostGIS and R statistical processing services enables fast statistical processing and aggregation of datasets and the dynamic generation of graphical mashups.

As users zoom in via the GoogleEarth interface, the locations of individual monitoring sites in each catchment are displayed. Users have the option to view either the EcoH plots or the raw monitoring data/sample data for individual sites (see Figure 9). Metadata and photos for each monitoring site can also be displayed by selecting the *Info* menu item. At this stage, these reporting services are only available to partners of SEQ HWP who can access them from the Health-e-Waterways Web Portal via a secure login id and password. In the next 6 months, we plan to implement more complex multi-level access controls that provide the public with high-level access to report card grades whilst SEQ-HWP partners are able to drill down to view very specific monitoring data for individual sites.

VISUALIZING THE IMPACT OF MANAGEMENT ACTIONS

In the long term, scientists, resource managers and policy makers want to be able to pose predictive queries such as: "If $10million is invested in landscape restoration works in the Bremer catchment today, what will be the ecosystem health outcomes in 2020?" Such predictive queries require the linking of complex models across the length of the waterway (from the upper catchment to middle and lower catchment, the estuary, coast and bay), so users can understand and simulate how changes propagate through the water flow cycle. Although such models are currently under development for SEQ, they are not yet at a stage where they can accurately predict the impact of management actions on ecosystem indicators. More comprehensive and longer-term monitoring data is required for calibration and validation of the models which also require significant time and computational power to execute.

In the meantime however, we are able to quickly and easily display management actions associated with a particular region, alongside the corresponding ecosystem health trends for that region. Figure 10 illustrates such a visualization for the Bremer catchment. The management

Figure 7. GoogleEarth interface to online ecosystem report cards

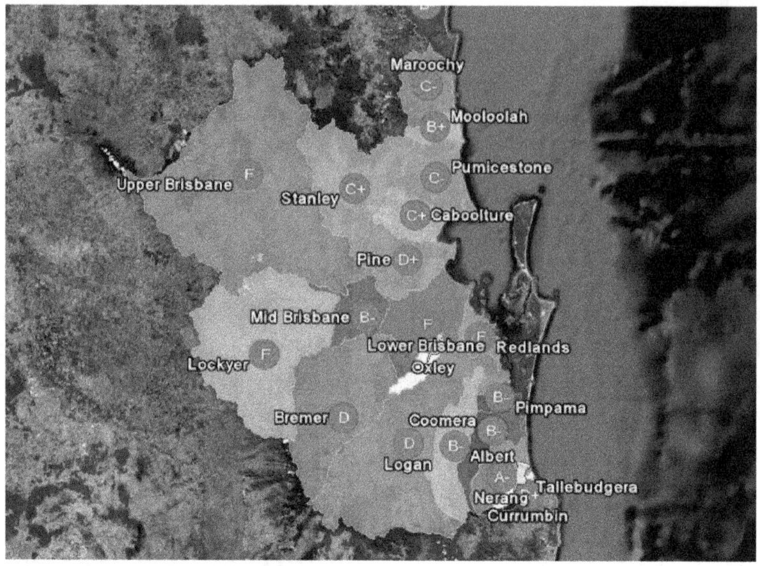

actions are displayed above box plots showing trends for the five freshwater ecosystem health indicators (*physical/chemical*, *nutrients*, *ecosystem processes*, *aquatic macroinvertebrates* and *fish*) over the period from 2004-2007. This display provides sufficient information to answer queries such as "Is the riparian revegetation in the Upper Bremer having any impact on water quality in the Bremer?" Figure 10 shows that in the past 3 years, the grade has improved from F to D due to improvements in the *physical/chemical*, *nutrient cycling* and *fish* indicators - presumably as a result of the riparian revegetation.

Figure 8. EcoH plots for Bremer catchment from 2004-2008

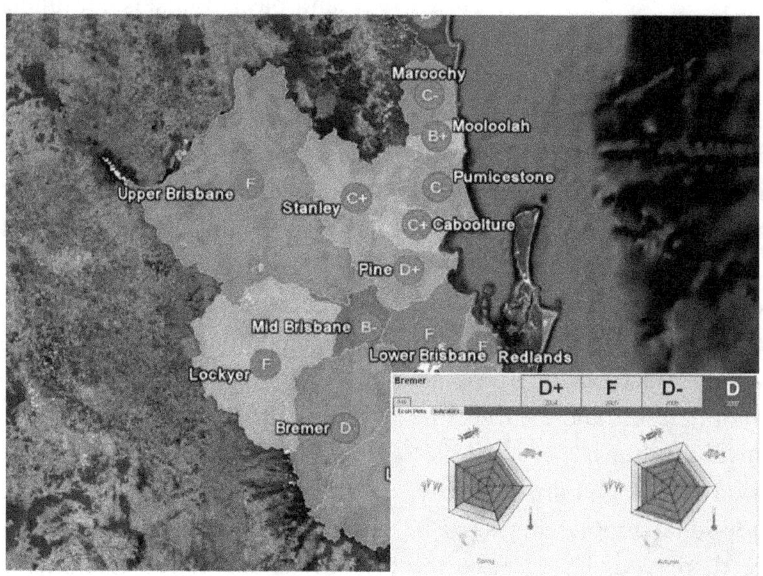

Figure 9. EHMP freshwater monitoring data for an individual site

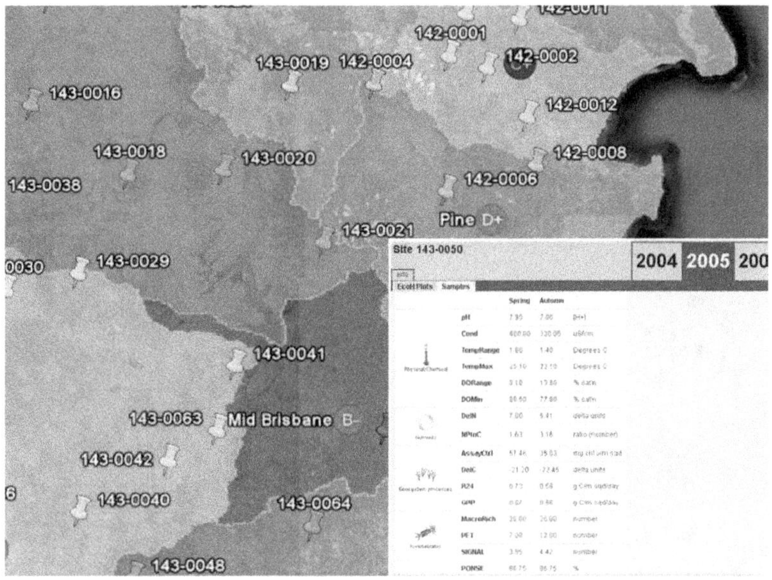

DISCUSSION

The Health-e-Waterways project is still at a relatively early stage (we are currently into the fifth Phase of the seven phases described in the Methodology section), however the outcomes to date include the following components that are either re-usable or can easily be refined, extended and adapted for other regions or similar eco-informatics applications:

- A dataset registry that (that enables relevant environmental and hydrological data sets to be easily, browsed and discovered) and a set of associated registry services (that enable new datasets to be registered, described, uploaded and published);
- A set of OWL ontologies that represent ecosystem health monitoring data and management actions as common data models that are machine-processable and facilitate data integration and reasoning;
- A set of services that automatically map legacy datasets to the OWL data models and store/publish them in the PostGreSQL database;
- A set of optimized reporting services that can dynamically generate a range of graphical plots showing trends in ecosystem health indicators over time and regions (sites, catchments, estuaries and bay zones) at different levels of granularity and resolution;
- Flexible Web Map-based user interfaces (using GoogleEarth, Microsoft Bing Maps or Yahoo Maps) that enable water resource managers to interactively visualize and assess the impact of specific management actions on water quality indicators.

Evaluation trials to date indicate that the technologies and services we have adopted and developed are delivering significant economic, scientific and environmental benefits. We have held a number of stakeholder meetings and demos to acquire feedback from SEQ-HWP users. These sessions have led to numerous refinements of the system functionality and user interface. Rigorous Quality Assurance/Quality Control (QA/QC)

Figure 10. Juxtaposing management actions alongside EHMP trends for the Bremer catchment

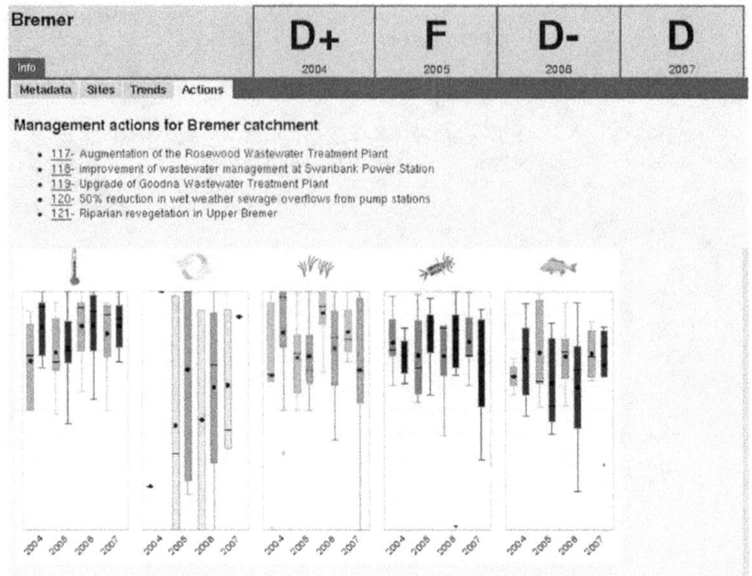

tests have also been carried out. They identified a number of discrepancies between the legacy data that was manually calculated and our automatically computed results. These discrepancies were primarily due to different approaches in handling missing data and have since been resolved.

Feedback from our scientific collaborators within the Environmental Protection Agency (EPA) and Department of Environment and Resource Management (DERM) has been extremely positive. EcoH plots and graphical reports that previously took weeks to generate manually, can now be generated dynamically via the Web interface. Users can immediately access and compare ecosystem health and water quality data over time and across catchments. They can drill down into report card grades to quickly and easily identify which particular parameters are performing badly and the possible causes.

The ability to display Management Actions alongside related Ecosystem Health Indicators for a given catchment, enables policy makers to easily identify the impact of specific actions and investments, and to adapt them accordingly. This capability is critical to support the "Adaptive Management" philosophy that underpins the SEQ HWP approach. However further input from the scientific experts panel is required in order to relate specific management actions and sub-actions to particular water quality parameters at specific sites.

The current system does not yet take advantage of the inferencing capabilities available through technologies such as SWRL. Semantic and spatial inferencing could be usefully applied to automatically infer the anticipated impact of those actions on sites that are downstream e.g., sites that are downstream from a recently upgraded sewage treatment plant should show reductions in $\delta^{15}N$. Ideally the system would intelligently analyse the monitoring data and highlight those management actions that are not having any perceived impact.

The ICT framework that we have developed to date is an ongoing development, but the combined use of OWL ontologies and OpenGeoStack means that the system can easily be extended, refined, customized and adapted – to support the needs of different organizations, different regions, different terminologies or different sets of indicators or indices.

FUTURE WORK AND CONCLUSION

Significant further work and usability studies are required in order to rigorously evaluate the Health-e-Waterways system and refine and enhance its capabilities and functionality. In particular, over the next 12 months we are planning to focus on the following activities:

- Add ontology-based conceptual querying (using the SPARQL query language) to the GoogleEarth interface to enable more interactive and intuitive querying "Which management actions are not meeting their targets in Cabbage Tree Estuary?"
- Add inferencing rules and geo-spatial reasoning (using SWRL) to the system to support richer querying "Which monitoring sites downstream of Rosewood WasteWater Treatment Plant have shown improvements in nutrient cycling in the past year"
- Incorporate new datasets into the system including: near-real-time sensor data, vegetation data, land use data and MODIS (Moderate Resolution Imaging Spectroradiometer) satellite imagery;
- Develop animations that show the changes in water quality parameters over time and waterflow cycles;
- Develop simple services to encourage uploading and quality control of community data – the SPIRE project's Spotter software (SPIRE, 2009) is of particular interest;
- Establish a Water Wiki for exchange of information between the SEQ-HWP partners;
- Incorporate annotation services to enable scientific experts to add comments, explanations and interpretations to the report card results and trend analyses;
- Add support for different levels of access as you zoom in to the data. For example, the specific latitude and longitude of the monitoring sites and the contact details of the individuals collecting the data, should not be displayed to the general public.
- Establish a model registry that enables users to upload and share models as web services, link them using scientific workflows and execute them over grid computing infrastructure.

Numerous state, national and international agencies are advocating the need for standardized frameworks and procedures for environmental accounting. The Health-e-Waterways system provides an ideal model for delivering a standardized approach to the aggregation of ecosystem health monitoring data and the generation of dynamic, interactive reports (that incorporate links back to the raw data sets). The recent development of a "Management Action" ontology, now enables users to link management actions to monitoring data, to identify which actions and investments are having the most impact. The system we have described here will not only save agencies significant time and money in generating environmental accounts, but it can be used to guide and adapt regional, state and national environmental policies and water resource management decisions.

REFERENCES

Abal, E. G., Greenfield, P. F., Bunn, S. E., & Tarte, D. M. (2006, January 16-18). Healthy Waterways: Healthy Catchments –An Integrated Research/ Management Program to Understand and Reduce Impacts of Sediments and Nutrients on Waterways in Queensland, Australia. In *Frontiers of WWW Research and Development – Proceedings of APWeb 2006: 8th Asia-Pacific Web Conference*, Harbin, China (pp. 1126-1135).

Bunn, S., Abal, E., Greenfield, P., & Tarte, D. (2007). Making the connection between healthy waterways and healthy catchments: South east Queensland, Australia. *Water Science and Technology: Water Supply*, *7*(2), 93–100. doi:10.2166/ws.2007.044

Chau, K. W. (2007). An ontology-based knowledge management system for flow and water quality modelling. *Advances in Engineering Software*, *38*(3), 172–181. doi:10.1016/j.advengsoft.2006.07.003

CUAHSI. (2009). *Hydrological Information System (HIS)*. Retrieved March 8, 2010, from http://his.cuahsi.org/

Ecosystem Health Monitoring Program (EHMP). (2009). *Annual Report Cards*. Retrieved March 8, 2010, from http://www.healthywaterways.org/EcosystemHealthMonitoringProgram/ProductsandPublications/AnnualReportCards.aspx

Ecosystem Health Monitoring Program (EHMP). (2010). *Home*. Retrieved March 8, 2010, from http://www.ehmp.org/

Fitzgerald, A., & Long, A. (2008). *OAK Law Project Report - A Review and Analysis of Academic Publishing Agreements and Open Access Policies*. Retrieved March 8, 2010, from http://www.oaklist.qut.edu.au/files/OAK_Law_Review_and_Analysis_v1_20080212.pdf

Horsburgh, J. S., Tarboton, D. G., Maidment, D. R., & Zaslavsky, I. (2008). A Relational Model for Environmental and Water Resources Data. *Water Resources Research*, 44.

Hunter, J., & Mauger, B. (2008). *Health-e-Waterways Database Survey Report*. Retrieved March 8, 2010, from http://www.health-e-waterways.org/reports.php

Larsen, G. (2007). *Modelling hydrodynamic processes within Pumicestone Passage, Northern Moreton Bay, Queensland*. Retrieved March 8, 2010, from http://eprints.qut.edu.au/16634/1/Genevieve_Larsen_Thesis.pdf

Lee, J. H. W., Huang, Y., Dickman, M., & Jayawardena, A. W. (2003). Neural Network modelling of coastal algal blooms. *Ecological Modelling*, *159*(2-3), 179–201. doi:10.1016/S0304-3800(02)00281-8

Liu, Y., Hill, D., Rodrigues, A., Marini, L., Kooper, R., Myers, J., et al. (2009, May 18-22). *A New Framework for On-Demand Virtualization, Repurposing and Fusion of Heterogeneous Sensors*. Paper presented at the Sensor Web Enablement Workshop 2009, 2009 International Symposium on Collaborative Technologies and Systems, Baltimore, MD.

Madin, J., Bowers, S., Schildhauer, M., Krivov, S., Pennington, D., & Villa, F. (2007). An Ontology for Describing and Synthesizing Ecological Observation Data. *Ecological Informatics*, *2*(3), 279–296. doi:10.1016/j.ecoinf.2007.05.004

Montgomery, J. L., Harmon, T., Kaiser, W., Sanderson, A., Haas, C. N., & Hooper, R. (2007). The WATERS Network: An integrated environmental observatory network for water research. *Environmental Science & Technology*, *41*, 6642–6647. doi:10.1021/es072618f

Muttil, N., & Chau, K. W. (2006). Neural network and genetic programming for modelling coastal algal blooms. *International Journal of Environment and Pollution*, *28*(3-4), 223–238. doi:10.1504/IJEP.2006.011208

Muttil, N., & Chau, K. W. (2007). Machine learning paradigms for selecting ecologically significant input variables. *Engineering Applications of Artificial Intelligence*, *20*(6), 735–744. doi:10.1016/j.engappai.2006.11.016

O'Hagan, R., Atkinson, R., Cox, S., Fitch, P., Lemon, D., & Walker, G. (2007). *A Reference Model for a Water Resources Observation Network.* Paper presented at MODSIM07.

Open Geospatial Consortium (OGC). (2009). *Observations and Measurements.* Retrieved March 8, 2010, from http://www.opengeospatial.org/standards/om

SEEK. (2006). *SEEK KR Ontologies.* Retrieved March 8, 2010, from http://seek.ecoinformatics.org/Wiki.jsp?page=KROntologies

Smith, M., & Grice, A. (2005). Ecosystem Health Monitoring Program. In Abal, E. G., Bunn, S. E., & Dennison, W. C. (Eds.), *Healthy Waterways, Healthy Catchments: Making the connection in south east Queensland* (pp. 149–182). Brisbane, QLD, Australia: Moreton Bay and Catchments Partnership.

Smith, M., & Storey, A. W. (2001). *Design and Implementation of Baseline Monitoring (DIBM3): Developing an Ecosystem Health Monitoring Program for Rivers and Streams in Southeast Queensland.* Brisbane, QLD, Australia: South East Queensland Regional Water Quality Management Strategy.

South East Queensland Healthy Waterways Partnership (SEQ HWP). (2009). *Healthy Waterways.* Retrieved March 8, 2010, from http://www.healthywaterways.org/

SPIRE. (2009). *Spire Research Group, Spire Ontology Tool (SPOTTER).* Retrieved March 8, 2010, from http://spire.umbc.edu/spotter/

SWEET. (2009). *Semantic Web for Earth and Environmental Terminology.* Retrieved March 8, 2010, from http://sweet.jpl.nasa.gov/

Wu, C. L., & Chau, K. W. (2006). Mathematical model of water quality rehabilitation with rainwater utilization — a case study at Haigang. *International Journal of Environment and Pollution, 28*(3-4), 534–545. doi:10.1504/IJEP.2006.011227

Zhao, M. Y., Cheng, C. T., Chau, K. W., & Li, G. (2006). Multiple criteria data envelopment analysis for full ranking units associated to environment impact assessment. *International Journal of Environment and Pollution, 28*(3-4), 448–464. doi:10.1504/IJEP.2006.011222

ENDNOTE

1 http://www.healthywaterways.org/TheStrategy.aspx

This work was previously published in International Journal of Agricultural and Environmental Information Systems, Volume 2, Issue 1, edited by Petraq Papajorgji and François Pinet, pp. 1-19, copyright 2010 by IGI Publishing (an imprint of IGI Global).

Chapter 6
A Conceptual Model of Grassland-Based Beef Systems

Guillaume Martin
INRA, UMR 1248 AGIR, F-31326 Castanet Tolosan, France

Roger Martin-Clouaire
INRA, UR 875 UBIA, F-31326 Castanet Tolosan, France

Jean-Pierre Rellier
INRA, UR 875 UBIA, F-31326 Castanet Tolosan, France

Michel Duru
INRA, UMR 1248 AGIR, F-31326 Castanet Tolosan, France

ABSTRACT

Fulfilling the production objectives of a grassland-based beef system requires a robust management strategy to secure the best practicable use of forage resources with regard to the cattle demand. To address the challenging issue of designing such strategies, this article describes the application of an ontology of agricultural production systems to the generic conceptual model SEDIVER, which supports the representation and dynamic farm-scale simulation of specific grassland-based beef systems. The most salient and novel aspects of SEDIVER concern the explicit modeling of (a) the diversity in plant, grassland, animal and farmland, and (b) management strategies that deal with the planning and coordination of activities whereby the farmer controls the biophysical processes. By using the SEDIVER conceptual framework, part of the subjective and context-specific knowledge used in farm management can be captured and, in this way, enable scientific investigation of management practices.

INTRODUCTION

In temperate less-favored areas, beef farming involves the management of a wide range of semi-natural grasslands. Such systems are increasingly threatened by rising input/output price ratios and the growing uncertainty surrounding production due, for example, to year-year weather variability. These problems add to the known difficulties in managing grassland-based systems. In such systems, herbage production is very heterogeneous and variable in time and space (Parsons, 1988).

DOI: 10.4018/978-1-4666-0333-2.ch006

This is partly because of the variation in vegetation types in relation to management intensity and environmental factors, mainly soil conditions and topography, and partly because of weather variability within and between years. The management challenge is thus to make efficient use of grassland production, and to secure the feeding of the herd in accordance with desired, attainable and currently usable herbage production. Of primary importance for farmers is the development of greater flexibility in farm management, enabling them to take advantage of opportunities, reduce vulnerability to adverse events, or cope with their consequences, in order to preserve the sustainability of their production enterprise.

Farmers have long relied on their intuition and on lessons resulting from analyzing other farmers' experiences to make strategic decisions (Jiggins & Röling, 2000). Now, when management processes must change or be adapted, the ability to use experience and history to discern patterns is still helpful, but given the limitations of human intellect and the increase in the pace and scale of change and uncertainty, it can hardly be used to shape robust decisions that perform consistently across a range of possible situations (McCown, 2002). The traditional reductionism of agronomic analysis, which examines a production system by taking it apart and understanding its constituent elements, is also inappropriate (Antle et al., 2001). Indeed the parts interact in complex and non-linear ways in response, in particular, to the manager's actions that are inherently discrete. These interactions are highly significant in the overall functioning and performance of the system. They might give rise to phenomena such as a bottleneck on some resources. Understanding the mechanisms and consequences of these emergent phenomena is of key importance in devising a management strategy that complies with the farmer's objectives and constraints.

The idea that farming systems should take greater consideration of plant, grassland, animal, and farmland diversity, both biological and functional, is generally agreed (e.g. Altieri, 1999; Andrieu, Poix, Josien, & Duru, 2007; White, Barker, & Moore, 2004). Such an approach encourages more flexible and efficient use of natural resources including herbage production. For instance, it enables functional complementarities and synergisms to be promoted between grassland plots that are suitable for different and sometimes multiple uses that depend on context-specific grassland production, and the feeding requirements of different animal categories (e.g. cows vs. heifers) characterized by different and fluctuating animal intake rates (Duru & Hubert, 2003; White et al., 2004). In addition, all four types of diversity constitute a potential source of flexibility that can be used in management choices to cope with uncertainty about uncontrollable factors such as weather. For instance, on a farm scale, farmland and grassland diversity bring organizational flexibility into farm management, i.e. freedom in the implementation and modification of a management strategy, e.g. a switch in the type of grassland use on a field, depending on the actual conditions encountered. On a field scale, plant species diversity makes it possible to take advantage of timing flexibility in grassland management (Martin et al., 2009a), i.e. the extent to which the use of a given grassland may be brought forward or deferred at various times of year. More generally, plant and/or animal diversity enhances operational flexibility, i.e. the farmer's ability to modify the target performance or the state of the plant and/or animal material.

The properties and behavior of agricultural production systems which exhibit such an organized complexity may be studied through modeling and computer simulation. Simulation does not replace intuition or lessons learned from other farmers' experiences but rather supplements it by revealing emergent behavior. Research has produced several simulation-oriented farm models for designing beef systems (e.g. Andrieu et al., 2007; Jouven & Baumont, 2008; Romera, Morris, Hodgson, Stirling, & Woodward, 2004). These models suffer from two main limitations. None of them integrates plant, grassland, animal and farmland diversity, and its consequences on the dynamic heterogeneity of biophysical processes into a single all-embracing model. None considers the flexible and dynamic exploitation of this diversity with sufficient emphasis on the planning and situated, i.e. situation-dependent, coordination over time and space of the farming activities. The contribution of the present work lies in a more detailed and explicit integration of these aspects into a new model. It raises an interdisciplinary challenge of knowledge integration (Bammer, 2005) which concerns among others agronomists, animal scientists, production management scientists and artificial intelligence modelers.

Consistent with this special issue on Intelligent Systems for Engineering Environmental Knowledge and EcoInformatics, this article addresses particularly complex and challenging ecological systems, i.e. agroecosystems in temperate European areas. It presents how an ontology general to the domain of agricultural production systems (Martin-Clouaire & Rellier, 2009) was employed to engineer knowledge particular to grassland-based beef systems. The resulting conceptual model supports the SEDIVER (Simulation-based Experimentation on livestock systems with plant, grassland, animal and farmland DIVERsity) simulation project that aims to design flexible grassland-based beef systems (including radically different ones) and management strategies capable of coping with a wide range of system states and conditions through increased consideration for plant, grassland, animal and farmland diversity. A conceptual model is a non-software description of a computer simulation model (Robinson, 2004). Conceptual modeling is about moving from the recognition of a problem situation, through model requirements to the determination of what is going to be modeled and how. This process makes explicit the link between science and design (Nassauer & Opdam, 2008) thereby constituting the common ground for the interdisciplinary approach. First, it involves identifying relevant concepts and scientific and expert knowledge and determining the appropriate level of detail of the model through the entities, behavior and interactions to be included in it. Then it requires finding areas of overlap to articulate this knowledge in a coherent systemic representation. Finally, it implies clarifying assumptions and simplifications that have to be formulated given the project rationale. In summary, it consists of making choices regarding complexity, uncertainty and imperfection (Bammer, 2005) of the model. The design of the conceptual model is very important because it affects all aspects of a simulation study, in particular the data requirements, the speed with which the model can be developed, the speed of experimentation and the confidence that can be placed in the model results. The approach to beef farming taken in this research is introduced. We then outline the conceptual model of the biophysical and the decision systems and discuss the key modeling choices. We also briefly present how to move from such a conceptual model to simulation-based experimentation.

BEEF FARMING AS THE EXPLICIT MANAGEMENT OF BIOPHYSICAL ENTITIES AND PROCESSES

A Systemic Dynamic Approach

This research relies heavily on a systemic dynamic view of beef farming. The production system is a structured organization to transform system resources, i.e. land, animals, equipment, labor and know-how, into animal products, i.e. meat and milk. It responds to internal purposeful drivers, mainly the farmer's interventions and external factors such as weather conditions. The production system is complex and can be broken down into interacting or interdependent subsystems made up of interacting or interdependent entities, in particular the fields, the forage stocks and the animals, which form a complex whole.

Controlling these interactions by appropriate and timely interventions to achieve his goals despite environmental and biological variation is the main concern of the farmer. Production management is thus considered as a flexible dynamic process rather than solely a function. Indeed, in making decisions, the farmer is able to take a variety of directions which define different system configurations and dynamic resource allocation, given the potentialities and limitations of his farm.

As has long been done in industrial production management (Schneeweis, 1995), production process design involves planning, coordinating and controlling biophysical and decision-making processes at the production system, i.e. farm, level, whilst most existing investigation approaches tend to address one at the expense of the other. To be relevant to real world decision-making and day-to-day management, one must therefore capture and analyze both the sequence and timing of decisions and feasible actions, and their consequences on the state of the system's entities and their relationships. Emphasizing a farm level approach might reveal the flexibility available to a farmer faced with environmental and biological variation, and might even uncover emergent phenomena or system properties. This is especially needed, given that most experimental field research assumes idealized conditions. The data obtained are informative about the potential production of biological entities, e.g. grasslands, but are of limited practical interest due to the management importance of other factors such as, in particular, weather variability or resource restriction aspects.

The Biophysical System as a Set of Managed Entities and Interacting Processes

The biophysical system is composed of a well-structured and ever-changing set of interacting entities. The changes result from biophysical processes and actions taken by the farmer. On the production year scale, the achievement of production objectives calls for skillful management decisions about types of actions, their timing, frequency and extent as a means of influencing the biophysical processes and the interactions between them. As an example, herbage growth on a plot might be affected by the displacement of animals on this plot for grazing to an extent determined by the number of animals, their intake rate and the duration of their stay. For each type of entity, within- and between-entity biological or functional diversities can be distinguished. For instance, grassland diversity exists within a plot through a range of plant species, and between plots through different grassland communities. These within- and between-entity diversities lead to a dynamic heterogeneity of biophysical processes within a farm, in particular for herbage dynamics.

Semi-natural grasslands are complex agrosystems composed of a wide variety of plant species. Their composition depends mainly on management intensity, soil moisture, and nutrient availability (Grime, 1973; Lavorel & Garnier, 2002). Grassland plant species and consequently grasslands exhibit a variety of growth and senescence patterns (Duru et al., 2009), determined by local (i.e. plot-scale) weather conditions, mineral and water nutrition and timing, frequency and extent of herbage use. These patterns have strong practical consequences on the kinds of versatile and profitable herbage use achievable. The trade-off between growth and senescence, which depends on phenological stages and leaf life span (Duru et al., 2009), has major consequences on herbage digestibility (Duru, Cruz, & Theau, 2008) and the related nutritive value, i.e. fill value, intake rate, and energy content, all closely correlated with digestibility (INRA, 2007). In practice, grassland use by grazing or mowing can hardly be synchronized on all plots at a time to exploit the optimum trade-off, but there is a range around this optimum trade-off that does not result in great losses of one property at the expense of another (Ansquer, Duru, Theau, & Cruz, 2008). At the grassland plot scale, there is a time window for using herbage whose duration depends on herbage availability and nutritive value and herd feeding objectives. Inappropriate grazing such as late use can lead to immediate undesirable effects such as low herbage nutritive value, and indirect delayed effects on availability and nutritive value of herbage for subsequent growth cycles. Inappropriate grazing can even compromise the persistence of the grassland over the long term by affecting the dynamic equilibrium between plant species.

The herd is a set of animals characterized by a category (e.g. heifer or cow), a physiological status (e.g. gestating or non-gestating cows), a productive function (e.g. fattened heifers for slaughter or replacement heifers) possibly with a production target and a breed. Feeding requirements and intake and performance patterns vary according to such characteristics. Intake and induced animal performance are regulated by the farmer's decisions and actions. For example he might occasionally opt to underfeed in a situation of acute shortage of feeding resources. The animal component of the production system can further be structured into herd batches that group together animals that are managed in the same way. The determination of the size and composition of the herd batches is a key management decision.

The spring herbage production peak constitutes about half of the yearly production. However, cattle feeding requirements do not follow this trend and display a more regular pattern over the year. Thus, despite the wide literature on the topic, we believe that in grassland-based beef systems, stocking rate is less of a key management issue than the planning and situated coordination over time and space of the encounters between cattle and standing or conserved (e.g. hay) herbage, and the associated distribution of grassland use for grazing or mowing consistent with production potentialities and the likely growth pattern of grasslands, the allocation of animals to appropriate fields according to their feeding requirements, and the frequency and intensity of defoliation likely to result in favorable conditions for the rest of the year. As a consequence, the dynamic heterogeneity of biophysical processes has strong practical implications in the planning and coordination of activities by the farmer. It generates constraints and opportunities for herbage use and animal feeding, which have strong consequences on system performance.

The Manager as an Explicit System that Produces Decisions and Implements Actions

The dynamic system view of the beef system includes the farm manager. Indeed, the farmer who controls the biophysical processes should not be considered as standing apart from the production system but rather as being a main subsystem. As a subsystem he produces decisions and interacts with the biophysical system through control and data collection interventions.

The manager has a management strategy that drives the in-situation management decisions that are both plan-based (anticipatory) and reactive. A strategy reflects the farmer's personal practices, which can be seen in his monitoring and observation behavior, in his understanding of the way the biophysical entities should be operated throughout the season, and in his appreciation of what events are important and how they should be reacted to. As farmers have accumulated experience and advice, they have learned to develop their own temporal organization of farming activities consistent with the overall objective and peculiarities of their production systems. In any particular production system the manager must repeatedly:

- Monitor the occurrence of new events and scrutinize salient aspects of the current state of the production system;
- Revise the management strategy in situations recognized beforehand to require such adaptations;

- Determine the sets of activities that are deemed relevant for execution according to the plan included in the strategy.

As a system, the manager has states and internal dynamics that respond to the passing of time and the influence from other parts of the production system as well as from exogenous factors. The decision-making behavior can be scrutinized, in particular with respect to its ability to cope with uncertainty. This concerns in particular the robustness of the strategy, defined as its capacity to satisfy the multidimensional objectives of the farmer across a range of external factors.

Analyzing the sequence of actions performed on the main biophysical entities e.g. grasslands, may reveal failure or potential improvement in the organization of the production activities over time and space. Such analyses constitute the basis of plan design which focuses on the interdependence between the production activities, whose coordination should achieve the desired outcome, and the feasibility periods and contextual conditions enabling their implementation. The timing of any management operation on biophysical entities, mainly grasslands and animals, depends on their current state. In practice, for any operation there is an ideal time window for its execution. As pointed out by Kemp and Michalk (2007) "farmers can manage more successfully over a range than continually chasing optimum or maximum values". Thus, in contrast with most of the literature on the topic, we believe the vagueness of a plan is not a fault, but, on the contrary, provides the flexibility needed to cope with the huge number of actual circumstances unfolding during its situated enactment, especially in agriculture where uncontrollable factors play a key driving role. Despite its flexibility, a plan may encounter situations where the initial intention is beyond its bounds as particular events occur (e.g. drought); the production system may then require some adjustments affecting for instance the configuration of some herds or the feeding policy, or call for changes of part of the initial plan of activities (e.g. switching the use of a field from hay-making to grazing when there is a shortage of grazing). Such adaptation capabilities need to be incorporated in the strategy. The necessity to expand or modify the strategy in this respect can be revealed by evaluating the application of the strategy in individual scenarios.

Towards the Conceptual Model

The conceptual modeling effort aims at structuring and organizing the pieces of knowledge about the system under study in a coherent systemic representation enabling the issue at stake to be tackled. The key decision regards the granularity required in the model to ensure predictive efficiency rather than fundamental mechanistic certainty (Antle et al., 2001). We do not need to model everything, but to include in the model only those events, processes and processed entities essential to efficiently predict the behavior of the studied system at the time and spatial scales of analysis required by the problem. In this case, because our focus is on the role of temporal aspects and timeliness of the decisions, it consists of evaluating at the farm scale the behavior over years of a farming system configuration and the associated management strategy by simulating the operational level. Day-to-day management is the appropriate level to represent realistically a flexible activity plan faced with environmental variability and its consequences on biophysical processes and on the performance of the farm.

In the present project, the design of the conceptual model is greatly supported by the use of the modeling framework DIESE (DIscrete Event Simulation Environment) that itself relies on a generic conceptual model of agricultural production systems (Martin-Clouaire & Rellier, 2009). This modeling framework provides a specific set of constructs that help to represent and articulate the structural, behavioral and dynamic features of the target system. It involves regarding an agricultural production system as an entity situated in and influenced by what is called the external environment. It can be divided into three interactive subsystems: the manager (decision system), the operating system and the biophysical system (Figure 1). The operating system is simply the system that transforms the manager's decisions into actions and executes them using the available resources. Three fundamental concepts are used for the modeling of this dynamic system: entity, process and event. These represent the structural, functional and dynamic aspects of a system respectively (Rellier, 2005). An entity describes a kind of material or abstract item in the area of interest. The state of a system at a given moment in time is the value of the slots of the entities it comprises. A process is a specification of the behavior of a system, i.e. of the entities composing it. A process causes a change in state when a particular event occurs. Thus, events determine the timing of process triggers.

A CONCEPTUAL MODEL OF THE BIOPHYSICAL SYSTEM

Overview

The model of the biophysical system (Figure 2) is composed of four main types of entity: farmland, herd, food storage units and stable. Stable is just considered a physical location which does not include any process. For reasons of clarity, it is not represented in Figure 2 nor mentioned later in the text. The farmland is a set of groups of grassland plots determined according to the proximity between plots. A plot is made up of two interacting components, soil and herbage, the latter being a set of herbage compartments representing groups of plants within the herbage. Food storage units are a set of food storage units in which food stocks coming from harvested herbage such as hay are stored. The herd is a set of animals structured into herd batches and animal groups. Herd batch is a functional managed entity that can contain different types of animals such as replacement heifers and cows constituting animal groups. The herd is fed based on herbage available on the plots or food stocks contained in food stor-

Figure 1. Production system view in the SEDIVER model

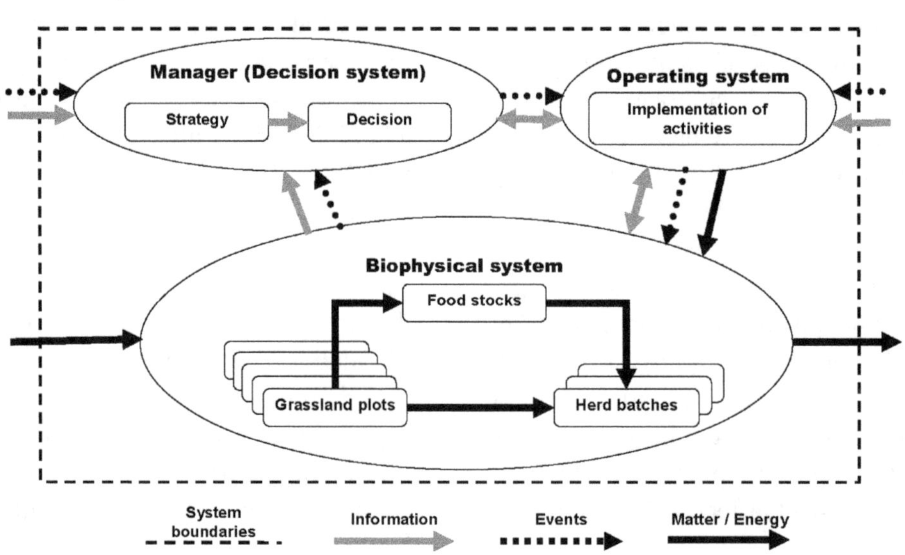

age units. The state of the system entities and the running of the biophysical processes are modified by weather, e.g. temperature and rainfall, and the farmer through his actions. These processes, which determine soil and herbage state i.e. availability and nutritive value, animal feeding and animal state or performance, are modeled at the lowest entity levels, i.e. soil-, herbage- or animal-scale, with a daily time step (Figure 2). Such a level of granularity is required to account for within- and between-entity biological or functional diversity and its consequences on dynamic heterogeneity of biophysical processes such as the variation in time and space of standing herbage. These processes interact: For example, the updating of herbage state on a plot interacts with the animal feeding process particular to each type of animal composing the herd batch grazing on that plot.

Farmland

On the farmland, accounting for groups of grassland plots or land islets is necessary, given that the spatial structure of the farm greatly influences system organization, for example due to the ease of herd batch displacements. The key processes to be modeled on the plot for the purpose of the project are those underlying the dynamics of herbage availability and nutritive value (Figure 2). They depend on so-called defining factors, i.e. weather conditions and plant species characteristics that determine the potential herbage production and nutrient and water availability as limiting factors that determine attainable herbage production (van Ittersum & Rabbinge, 1997). These factors are described here for the case of semi-mountainous grassland-based beef systems such as in Martin et al. (2009a).

Plant species characteristics are defined as a model input including the range of plant species characteristics found in such grasslands. To account for this diversity, the concept of functional diversity defined in plant ecology is quite helpful. In this approach, species are classified into groups, here named herbage compartments, that relate directly to function (primary production) based on shared biological characteristics (plant traits) for plant morphology, physiology and phenology. Functional diversity is based on the identification and measuring of these plant traits in relation to environmental conditions and perturbations (Diaz & Cabido, 2001). A reason for emphasizing a functional ecology approach is that traits are universal whereas species and taxonomic groups are restricted in distribution. Plant traits such as the leaf dry matter content (LDMC) weighted at plant community level are well correlated with agronomic characteristics like herbage digestibility (Al Haj Khaled, Duru, Decruyenaere, Jouany, & Cruz, 2006) and plant phenology (Ansquer et al., 2009), characterized by the beginning of reproductive phase, flowering and leaf life span, all expressed in thermal time units. Plant phenology governs the dynamics of grass growth and nutritive value. LDMC provides a powerful generic (i.e. with a large validity domain) descriptor of a grassland community for characterizing the within- and between-effect of plant functional composition, or relative abundance of each herbage compartment, for the dynamics of biomass availability and nutritive value, the timing of grassland use on a phenological basis, and timing flexibility through time windows for suitable use (Ansquer et al., 2008). Previous studies such as that of Meot, Hubert, and Lasseur (2003) already identified field functions but missed such key vegetation characteristics to qualify what distinguishes fields for dynamically allocating grassland use.

Accessing grassland community characteristics through weighted mean plant traits provides the information to represent variability of herbage state and dynamics between plots (Figure 3), or to characterize the thermal time windows for grassland use. Between-field functional diversity spreads the peaks of herbage dry matter over time. Indeed, differences in phenological stages can reach 400 degree days between two vegetation types (Figure 3, 1st and 2nd columns) which

Figure 2. Main entities and processes in the biophysical system, and interactions between them

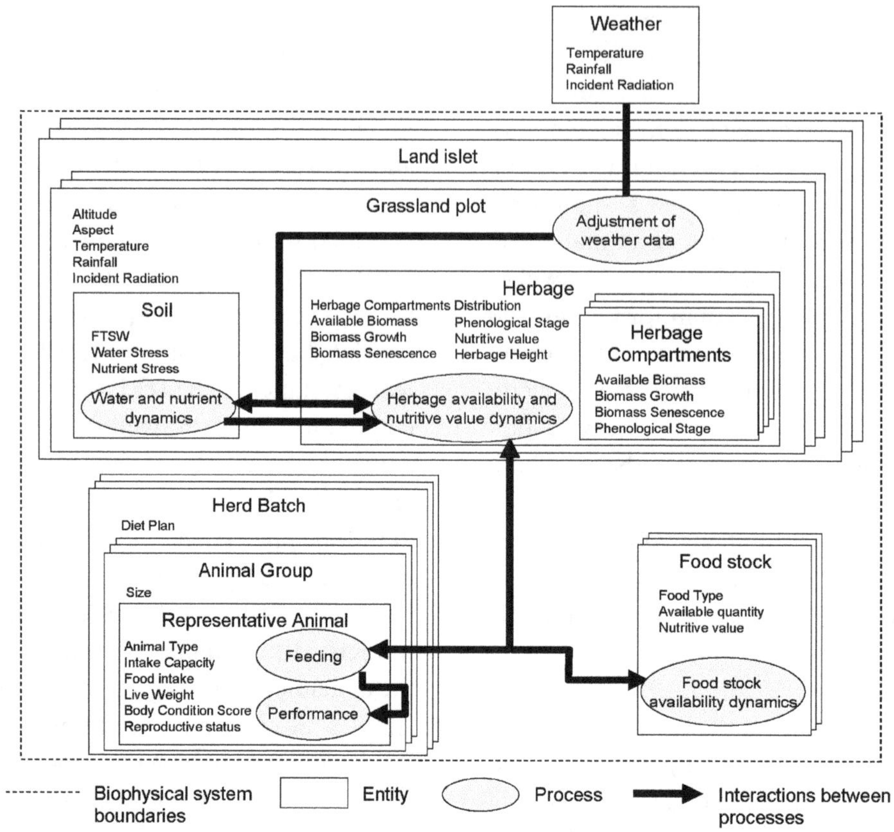

can correspond at a given moment to differences of 50% in above-ground herbage mass (calculation based on Duru et al., 2009) and by 15% in digestibility (calculation based on Ansquer et al., 2008). Within-field functional diversity softens the changes in herbage growth and digestibility decrease rates (Figure 3, 1st and 3rd columns) (Ansquer et al., 2008). Herbage digestibility is needed to properly evaluate herbage nutritive value and animal intake as it determines the fill value of the forage and thus the forage quantity that the animal can eat. This information constitutes the basis for allocating grassland use over time and space.

Weather takes the form of incoming energy and material, i.e. rainfall, radiation and temperature. Given that each plot displays particular topographic peculiarities, upcoming weather data from the reference weather station have to be adjusted in an additional process (Figure 2) according to plot altitude and aspect. These factors can modify temperature by 0.6°C per 100 m of altitude and daily incident radiation by 1% per 100 m of altitude and by -17% when the plot is north-oriented, and by +18% when the plot is south-oriented (Andrieu et al., 2007). Ignoring these corrections to the reference weather data can lead to over- or underestimating daily herbage growth by 50% (Figure 3, 2nd and 4th column; calculation based on Duru et al., 2009). On the other hand we assume that for a given farm, daily rainfall can be regarded as uniform.

Plant water and nutrient stress are driven from the soil component. In the soil, water and nutrient dynamics are governed by inflow from fertilization, excreta or rainfall and outflow - mainly uptake by the plant and drainage. A water stress

Figure 3. Examples of processes occurring at spring (thermal time starts on the 1st of February as recommended by Ansquer et al., 2009) on four grassland communities with different vegetation types (A vs. C vs. A and C mixed; see Ansquer et al., 2008 for further details) or topographic conditions (last column with vegetation type C growing in a north-oriented plot of higher altitude). From the top to the bottom are the dynamics of herbage dry matter accumulation, daily growth rate (calculations based on Duru et al., 2009) and digestibility (calculations based on Duru et al., 2008).

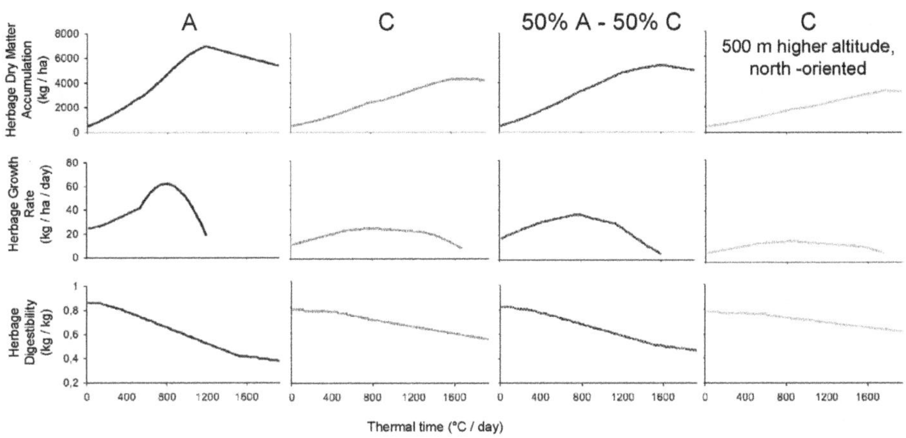

index (Merot, Bergez, Wallach, & Duru, 2008) and a plant nutrient index (Lemaire & Gastal, 1997) representing the extent of water and nutrient limitation experienced by the plant in achieving the potential growth permitted by local weather conditions can be calculated with simple characterization of soil properties such as the fraction of transpirable soil water (FTSW). Based on such indices, 0.2 water and nutrient stresses (both on a scale from 0 to 1) induce 25% and 27% shortfalls in the potential growth (calculations based on Duru et al., 2009) and 5% to the potential digestibility (calculation based on Duru et al., 2008). Available mechanistic models for nutrient dynamics in grasslands have been calibrated for particular areas (e.g. Scholefield, Lockyer, Whitehead, & Tyson, 1991) and would require considerable effort for a recalibration.

The models of Duru et al. (2008, 2009) include all the above-mentioned factors and have been slightly modified, in particular to account for within-plot diversity of plant functional types, or relative abundance of each herbage compartment, and for the modeling of the biomass in the ungrazable strata, e.g. below 5-6 cm for cows. Indeed, indicators used in day-to-day management such as herbage height have to be modeled accurately. Residual centimeters after grazing should not be neglected given the remaining grazing days they could provide (between 100 and 500 kg of dry matter per cm and per hectare according to the type of grassland community) and given that they are essential to ensure the quality of later herbage regrowth (Parsons, 1988). When residual height after grazing is too low, growth is reduced due to a leaf area index that is too low for capturing most of the incident radiation, but beyond this threshold, reducing the intensity of use, either by lengthening the interval between defoliation or grazing, or by increasing the residual height after grazing, results in greater losses from senescence and hence reduced stocking density (Duru & Hubert, 2003).

Interactions between the dynamics of herbage availability and its use take into account day-to-day interactions between processes. Dynamics of herbage availability are modeled on a daily time step as available herbage growth models are not accurate enough to consider shorter time scales. The effects of grazing animals on standing herbage are also modeled on a daily time step. We assume that vegetation characteristics of herbage are uniform on the plot scale.

Food Storage Units

Food storage units contain food stocks, i.e. hay, bale-silage, silage and sometimes concentrates. In grassland-based beef systems, food stocks play a key role in the feeding of the herd. According to the area (e.g. semi-mountainous vs. temperate) they fulfill different functions such as winter feeding for several months or a buffer to compensate for feed shortages at grazing due, for instance, to drought. The importance of food stocks is also related to the farmer's desire for forage self-sufficiency. As a consequence, choices related to food stock management greatly influence the organizational and day-to-day management of grazing, given that the timing, frequency and intensity of grazing depend on the concomitant hay and/or silage-making and distribution. To tackle this issue it is necessary to model food stock dynamics (Figure 2) taking into account the storage capacity of the food storage equipments, which can be a limiting factor. Nutritive value of the food stocks also has to be considered (Jouven & Baumont, 2008) given that different types of food stocks are suited to different animal categories. As with grazing, it is possible to match the availability and quality of the food stocks distributed to the feeding requirements of animal categories.

Herd

Accurately evaluating animal intake and performance, i.e. growth, weight variation, milk production, etc. requires accessing the diversity of animal types encountered on the farm, since different animal categories have different intake rates (INRA, 2007). Animal groups composing herd batches bring together animals of similar category, physiological status, productive function or production level target and breed. Intake can be modified by up to 400% between young grazing calves and cows, up to 30% between lactating and non-lactating cows, 9% per point of body condition score, about 30% between replacement heifers and heifers for slaughter and 8-10% between breeds (calculations based on INRA, 2007). Two processes, animal feeding and performance, are then modeled for a representative animal of each group (Figure 2).

A diet plan is assigned to each herd batch, defining the content of the diet over the seasons. At the animal scale, we do not mechanistically model the act of prehension (grazing), bolus mastication during rumination or the rumen motility cycle. We only model the daily intake process. Indeed, since we consider a representative animal for a group, we cannot simulate individuals' feed preferences through animal behavior at grazing and the related spatial variation consisting of patches of ungrazed vegetation and nutrient excretion. The corresponding quantities of material are then deducted from food stocks or standing herbage mass of grassland plots assuming cattle is not selective at grazing. Given the size of the animal group, animal intake at animal group level is evaluated and finally aggregated at herd batch level. In previous models (Andrieu et al., 2007; Jouven & Baumont, 2008), animal group and herd batch were not distinguished, making it impossible to model different animal categories mixed within a single herd batch. It is especially useful to the issue tackled, given that herd batch re-composition can be used as a regulatory process to adjust stocking rate for coping with variability in herbage availability. As varying herd size is another way of adjusting herd intake rate, accounting for population dynamics over years is essential to take into account the flexibility it constitutes, as regulated by reproductive events such as mating and calving date, and marketing aspects such as target state at sale.

Intake rate is dependent on the animal side on the energy invested in milk production, on live weight and body condition score, and on the fill value of the diet which is related to food digestibility (INRA, 2007). The partitioning of energy intake between displacement, maintenance, pregnancy and lactation and its consequences on live weight and body condition score is modeled as in Jouven and Baumont (2008) and Romera et al. (2004). As a consequence, the reproductive status of the animals and its impact on animal physiology are also modeled.

A CONCEPTUAL MODEL OF THE MANAGEMENT SYSTEM

Rationale

The need for process flexibility in the workflow and process technology communities as a critical quality of effective business processes to adapt to changing market demands and other business circumstances has long been recognized (Reijers, 2006). Several authors (e.g. Sawhney, 2006) suggested that manufacturing flexibility should be included as a key dimension of a firm's manufacturing strategy; researchers have posited that firms increase their manufacturing flexibility to allow them to respond to uncertainty in the environment, and that an appropriate match between business strategy and flexibility improves performance. Agricultural production systems are similarly concerned, although uncertainty does not affect production in manufacturing as much as in agriculture, where some key production drivers such as the weather are out of control.

There is an urgent need for more effective whole-farm analyses to address issues of productivity, profitability and sustainability, and the role played by flexibility. A key requirement for such an analytical framework is the capacity to accommodate weather variability and various kinds of uncertainty. Mathematical models of farm management based on static equilibrium conditions fail on several criteria. Dillon (1979) reflected on the scientific discipline of Farm Management research in Australia and concluded that it had lost touch with practical farming because of "logically attractive but largely inapplicable theory". He emphasized the management challenges created by uncertainty and dynamics, and the failure of existing mathematical models of farm management to capture these features adequately. McCown, Brennan, & Parton (2006) attempted to attract wider recognition of Dillon's criticisms of the mathematical model approach embodied in farm management. Following Dillon, they listed a number of reasons why static equilibrium models cannot be applied to the practice of farm management:

- Farm systems are complex and dynamic;
- Farming is conducted under conditions of uncertainty and performance depends very much on how uncertainty is dealt with;
- Individual farms are unique and farmers have different practices and preferences. A wide variety of so-called farming styles (van der Ploeg, 1994) exist.

Simulation-based approaches that explicitly incorporate production management processes seem to provide a more promising framework to take into account these aspects or highlight the issues they raise. The management system model should explicitly represent the decision-making process and the implementation of the technical actions resulting from this process.

As for the biophysical system, the conceptual model of the management system relies on the DIESE framework that includes specific constructs to represent various aspects relevant to the management functions (Martin-Clouaire & Rellier, 2009). Fundamental to our conceptual model is the commitment to understand things from a farmer's point of view. To be effective, management behavior must be specified by using constructs and language that are intelligible and conceptually close to those actually used in an agricultural setting. The typical approach to representing decision-making in simulation models is to express decision behavior through a set of decision rules. This approach is ontologically limited and becomes cumbersome as the number of rules grows beyond a certain limit; the meta-knowledge about the proper use of the rules, for example which should be applied first when several are applicable, is hard to represent and makes the rule-based approach hard to maintain and reuse.

Hence the basic unit of analysis in our approach is work activity, which is a common high level concept in production management. An activity is a purposeful engagement driven by certain needs to achieve a certain purpose. Activities are contextual in the sense that actual circumstance condition their relevance and greatly affect the way the intended objective is achieved. Activities usually involve the use of resources (equipment, labor). Whenever a combination of activities must be undertaken with a view to achieving a pre-conceived result, a plan is needed to express how those composite activities should be coordinated. A work plan is the result of reflection on prior experiences and in anticipation of particular goals and likely occurrences of important events. Because of this, plans are not rigid in the sense of a definite and precise specification of the execution steps. Plans are flexible and adaptable to circumstances. A slightly more formal and encompassing conceptual description is given in the next subsection.

A Conceptualization of Production Management

In its simplest form, an activity, which we will call a primitive activity, denotes something to be done to a particular biophysical object or location, e.g. a herd batch, building, by an executor, e.g. a worker, a robot or a set of these. Besides these three components, a primitive activity is characterized by local opening and closing conditions, defined by time windows and/or predicates (Boolean functions) referring to the biophysical states or indicators. An indicator is a contextual piece of knowledge or information invoked, assembled, or structured to substantiate a decision-making step, e.g. appraisal of remaining forage amount on a field to decide withdrawal of the herd from it. The opening and closing conditions are used to determine at any time which activities are agronomically eligible for execution; they play a key role in defining the timing flexibility.

The "something-to-be-done" component of a primitive activity is an intentional transformation called an operation, e.g. the harvest operation. The step-by-step changes to the biophysical system as the operation is carried out constitute a functional attribute of the operation. These changes take place over a period of time by means of a process that increases the degree of achievement at each step of the operation until it is completed. An operation is said to be instantaneous if its degree of achievement goes from 0 to 1 in a single step, otherwise the operation is durative which implies that its execution might be interrupted. An operation may require resources such as a mower in case of cutting. In addition, the execution of an operation is constrained by feasibility conditions that relate to the biophysical state. Objects on which an operation is carried out can be individual objects, e.g. a field or a set of fields, or objects having numerical descriptors, e.g. an area. Speed is defined as a quantity e.g. number of items or area, which can be processed in a unit of time. The duration of the operation is the ratio of the total quantity to the speed. In order to have the effect realized the operation must satisfy certain enabling conditions that refer to the current state of the biophysical system, e.g. the field to be processed should not be too muddy, muddy being an indicator. The ability to reap the benefits of organizational and timing flexibility depends on execution competence determined by the involved resources (both operation resources and the executor). Careful representation of the resources and their availability might therefore be essential to get a proper understanding of the situation under study.

Activities can be further constrained by using programming constructs enabling specification of temporal ordering, iteration, aggregation and optional execution. To this end, we use a set of non-primitive or aggregated activities having evocative names such as *before*, *iterate*, and *optional*. Others are used to specify choice of one activity among several (*or*), grouping of activities in an unordered collection (*and*) and concurrence of some of them (e.g. *co-start*, *equal*, *include*, *overlap*). Formally a non-primitive activity is a particularized activity. As such it might also be given opening and closing conditions as well as other properties such as a delay between two activities involved in a *before* aggregated activity. In particular, it has a relational property that points to the set of the other activities directly involved in it (or constrained by it). In addition, it is equipped with a set of procedural attributes that convey the semantics of the change in status specific to each non-primitive activity. The opening and closing of a non-primitive activity depend on their own local opening and closing conditions (if any) and on those of the underlying activities. All the activities are connected; the only one that does not have a higher level activity is the plan. In addition to the timing flexibility attached to the opening and closing conditions of its activities a plan is made flexible by the use of composed activities that enable optional execution or choice between candidate activities. Whether an optional activity is executed and which alternative activity is chosen are context-dependent decisions relying on indicators or priorities between activities.

Besides the flexibility of the timing of the execution of activities it may be necessary to adapt the plan when particular circumstances occur. Indeed a nominal plan conveys the rough course of intended steps to go through under normal circumstances. The specification of when and what changes should be made to a nominal plan is called a conditional adjustment. The trigger for a conditional adjustment is either a calendar condition that becomes true when a specific date is reached, or a state-related condition that becomes true when the current circumstances match this condition. The adjustment can be any change to the nominal plan such as the deletion or insertion of activities. It can also affect the resources used in some activities. Actually a conditional adjustment can also specify a change to be made to conditional adjustments themselves. By this means, the management can be reactive, e.g. by modifying the management objective, and thus cope with unexpected (though still feasible) fluctuations of the external environment and various contingencies. For instance, a target beef production level might be reviewed in case of drought, and materialized by voluntary underfeeding of cattle.

Examples of Operations, Primitive and Composite Activities and Conditional Adjustments

The model includes a wide range of activities corresponding either to daily routine work e.g. distributing stored food, moving a grazing herd batch, etc. or to seasonal work, e.g. mating, weaning calves, selling animals, etc. Let us take the example of hay-making on a farm. This is composed of two steps, i.e. two primitive activities, first cutting the herbage of a grassland plot and, once it is dry enough, storing this new-mown hay. For the cutting activity, the object operated by the cut (the something-to-be-done component) is a plot, in particular the component herbage, and the executor is the farmer equipped with his tractor and mower. The speed of the something-to-be-done component is a harvestable area per unit time. Its effect is the creation of a harvested herbage, the initialization of a drying process on this harvested herbage, and the re-initialization of the herbage component of the plot with its descriptors updated (leaf area index, dry matter, growth cycle age, digestibility, etc.). For the storing activity, the object operated by the storage (the something-to-be-done component) is the harvested herbage, and the executor is the farmer equipped with his tractor, round-baler and trailer. The speed of the storage is a storable quantity of hay per unit time. Its effect is the crediting of the amount of hay stored in the barn by the harvested quantity minus some losses to the yield associated with the whole hay-making process. Storing of harvested herbage can occur only once cutting is complete. Thus hay-making is a sequence of two primitive activities which can be written:

hay-making = *before* (cutting:
operation: cut with mover
operated object: plot
performer: farmer
storing:
operation: store with tractor, round-baler and trailer
operated object: harvested herbage created in cut
performer: farmer)

The opening of any hay-making activity, and consequently of the cutting activity, has to occur within a particular time range delimited by an earliest and a latest beginning date. In addition, the opening predicate refers to a minimum harvestable yield and a given phenological stage for the corresponding herbage, i.e. between stem elongation and flowering, to ensure a compromise between harvested quantity and quality. Once the opening predicate of the hay-making activity has been verified the feasibility conditions attached to the cut operation are examined. These feasibility conditions concern the bearing capacity of the grassland plot, sufficient free space in the barn to store additional hay, and a satisfactory expected air saturation deficit and rainfall in the coming days to ensure proper drying conditions in the field. No closing conditions are specified in this case to ensure completion of the hay-making activity.

To summarize the hay-making aggregated activity is represented as follows:

hay-making = *before* (cutting, storing)
earliest beginning date = a date
latest beginning date = a date
opening predicates concerning:

- minimum yield to harvest = an amount expressed in kg / ha
- earliest phenological stage = stem elongation expressed in degree days
- latest phenological stage = flowering expressed in degree days

Farmers seldom make hay on a single field at a time. Typically, they do it on a set of fields that are close together, i.e. that belong to the same islet. This practice may be risky if too many fields are cut and long period of rainy weather occurs. A typical risk-limiting attitude is to make small groups of plots and harvest these groups in sequence. Bad weather during drying then harms only the plots in the last group treated. In the example of practice considered in this paper, hay-making on the plots of a group can only start if the last hay-making activity executed in the previous group is complete. Sometimes farmers may defer groups of activities for several days to make time for daily routine work not done on the busy days of hay-making. The grouping of activities enables management constraints to be attached to this set, such as the delay between the processing of the groups {Field1, Field2, Field3} and {Field4, Field5}. Using an *and* to make the grouping gives flexibility in the order of execution of the concerned activities, using for instance yield-based preferences. The sequence of hay-making on the two groups of fields can then be written:

before (*and* (hay-making Field 1, hay-making Field 2, hay-making Field 3),
and (hay-making Field 4, hay-making Field 5)
in-between delay = a number of days)

Due, for example, to particular weather conditions in a given year, such a plan might be unachievable. Conditional adjustments of the plan are then necessary to recover a consistent management situation. For instance, in a showery weather period, the farmer might decide to reverse the order of the groups of hay-making activities in the sequence (*before*) to take advantage of the lower drying requirements of herbage on fields 4 and 5. Another adjustment could be the changing of the delay between the processing of the two groups. This conditional adjustment would then change the above activity into:

before (*and* (hay-making Field 4, hay-making Field 5),
and (hay-making Field 1, hay-making Field 2, hay-making Field 3)
in-between delay = a number of days)

FROM THE CONCEPTUAL MODEL TO A SPECIFIC FARM MODEL AND ITS USE IN SIMULATION-BASED DESIGN

The implementation of the conceptual model in the DIESE modeling framework (Martin-Clouaire & Rellier, 2009) amounts to developing particularizations of some classes. For example, from the Entity class, particular cases such as "Plot" or "Herd" were created. From the Activity class, e.g. "Distributing stored food" and "Hay-making" were created. This work led to an implemented simulation model general to the domain of grassland-based beef systems. Going from this general model to a model of a particular beef system was achieved by instantiating the particularized classes, e.g. from the "Plot" class, creating "plot 1", and "plot 2"; or from the "Moving Herd Batch" class, creating "moving Herd Batch A to plot 1". Articulation of the instantiated primitive activities using non-primitive ones then led to "manually" constructed executable plans. The implementation of the conceptual model has been reported in related papers (Martin, Martin-Clouaire, Rellier, & Duru, 2009b; Martin et al., 2009c) with two simulation-based investigations of the merits and limitations of different management strategies applied to three grassland-based beef systems.

The openness and flexibility of the formal language used to represent the management strategies are suited to support a trial and error learning process (Figure 4) by rapidly exploring alternative configuration of the production system and management strategies, their performances and trade-offs between performance criteria at almost no cost. Assume a configuration of the system, both hard (physical layout, dimensioning) and soft (organizational infrastructure, decision logic). Flexibility and adaptability are inherent and built in beforehand in the considered management strategy. In beef systems, "manual" reconfiguration might concern the composition and dimensioning of land and animal resources. Other reconfigurable hard components include technology (e.g. hay drying) and resources (labor). The soft components modifiable in a "manual" reconfiguration are mainly the production plans in which activities can be added or removed, the timing flexibility attributes, the event-based adjustments, as well as the indicators used for decision making. The ability to switch appropriately between options is often more important than offering a large number of options, and sheds light on the key role of temporal aspects and timeliness of the decisions.

CONCLUSION

Keating and McCown (2001) suggested that challenges for farming system modelers are "not to build more accurate or more comprehensive models, but to discover new ways of achieving relevance to real world decision making and management practice." This applies in a wider sense to managed ecosystems. The ontology (Martin-Clouaire & Rellier, 2009) and the con-

Figure 4. The iterative loop based on simulation, evaluation and re-configuration

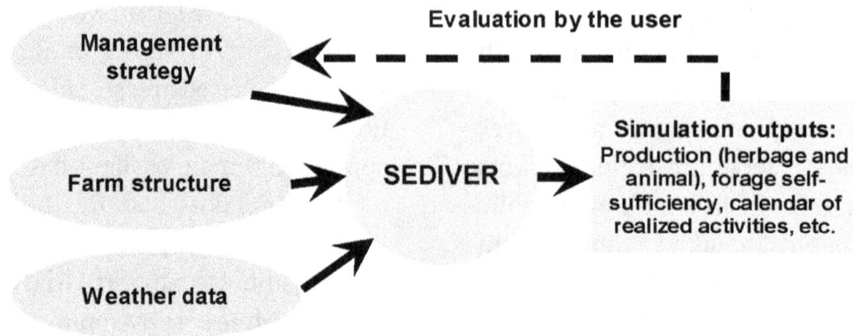

ceptual framework underlying SEDIVER are the result of consistent efforts to improve the representation of management strategies and get closer to the questions raised in practice. Using the integrative conceptual framework that we have described one can develop elaborate simulation models of managed ecosystems. It provides a common structure to help organize and frame monitoring and management activities that can be applied effectively and consistently across any such a system. Running a simulation model under various scenarios of external conditions helps to give a realistic view of the system's behavior and performance, its sensitivity to external factors and the quality of the tested management strategy as regards robustness and flexibility. We can use this approach to give a clearer meaning to the selection and prioritization of management activities by placing the management process in context. The present work and the related applications (Martin et al., 2009b, 2009c) are examples of the merits of ontology-based modeling to engineer knowledge about ecosystems and in fine to address complex management problems.

Alternative modeling approaches such as linear programming models generally assume static equilibrium conditions or rely on farm- or ecosystem-averaged indicators. Thus they ignore the diversity within the system and its consequences on the variability of ecosystem processes in time and space. These approaches can hardly address challenges created by uncertainty and dynamics whereas it is essential that the model be able to reveal how fluctuations might be amplified and how the system may become unstable to large perturbations. Indeed, the complexity of the farmer's management task is not due to the number of components or possible states of the system but rather to the dynamic behavior of the different components which arise from their interactions over time and their dependence on uncontrollable driving factors. The dynamic complexity relates to human difficulty in dealing consistently with feedback effects, and multiple and delayed consequences of interventions. Much of the information about biophysical system functioning and the cognitive process involved in production management resides in the mental models of managers where it remains tacit. By using the conceptual framework set up in the present work, one can expect to capture part of this subjective and context-specific knowledge and, in this way, make it an object of scientific investigation. Improving our ability to make this knowledge explicit and usable for formal modeling and learning can have important effects on both research and practice. Researchers are in a better position to build more complete, accurate and insightful models and practitioners can increase their awareness and mastery of organizational and management issues.

ACKNOWLEDGMENT

This study was partly funded by the French ANR ADD program in the framework of the project TRANS (TRANSformations de l'élevage et dynamiques des espaces, ANR-05-PADD-003) and by the French ANR VMC program in the framework of the project VALIDATE (Vulnerability Assessment of LIvestock and grasslanDs to climAte change and exTreme Events, ANR-07-VULN-011). The authors would like to thank J.P. Theau and O. Therond who, as specialists of beef farming systems, contributed to this research, and the three anonymous reviewers for their relevant comments.

REFERENCES

Al Haj Khaled, R., Duru, M., Decruyenaere, V., Jouany, C., & Cruz, P. (2006). Using leaf traits to rank native grasses according to their nutritive value. *Rangeland Ecology and Management*, *59*, 648–654. doi:10.2111/05-031R2.1

Altieri, M. A. (1999). The ecological role of biodiversity in agroecosystems. *Agriculture Ecosystems & Environment*, *74*, 19–31. doi:10.1016/S0167-8809(99)00028-6

Andrieu, N., Poix, C., Josien, E., & Duru, M. (2007). Simulation of forage management strategies considering farm-level land diversity: Example of dairy farms in the Auvergne. *Computers and Electronics in Agriculture*, *55*, 36–48. doi:10.1016/j.compag.2006.11.004

Ansquer, P., Al Haj Khaled, R., Cruz, P., Theau, J. P., Therond, O., & Duru, M. (2009). Characterizing and predicting plant phenology in species-rich grasslands. *Grass and Forage Science*, *64*, 57–70. doi:10.1111/j.1365-2494.2008.00670.x

Ansquer, P., Duru, M., Theau, J. P., & Cruz, P. (2008). Functional traits as indicators of fodder provision over a short time scale in species-rich grasslands. *Annals of Botany*, *103*, 117–126. doi:10.1093/aob/mcn215

Antle, J. M., Capalbo, S. M., Elliott, E. T., Hunt, H. W., Mooney, S., & Paustian, K. H. (2001). Research needs for understanding and predicting the behavior of managed ecosystems: Lessons from the study of agroecosystems. *Ecosystems (New York, N.Y.)*, *4*, 723–735. doi:10.1007/s10021-001-0041-0

Bammer, G. (2005). Integration and Implementation Sciences: Building a New Specialization. *Ecology and Society*, *10*(2), 6.

Diaz, S., & Cabido, M. (2001). Vive la difference: Plant functional diversity matters to ecosystem processes. *Trends in Ecology & Evolution*, *16*, 646–655. doi:10.1016/S0169-5347(01)02283-2

Dillon, J. L. (1979). An evaluation of the state of affairs in Farm Management. *South African Journal of Agricultural Economics*, *1*, 7–13.

Duru, M., Adam, M., Cruz, P., Martin, G., Ansquer, P., & Ducourtieux, C. (2009). Modelling above-ground herbage mass for a wide range of grassland community types. *Ecological Modelling*, *220*, 209–225. doi:10.1016/j.ecolmodel.2008.09.015

Duru, M., Cruz, P., & Theau, J. P. (2008). Un modèle générique de digestibilité des graminées des prairies semées et permanentes pour raisonner les pratiques agricoles. *Fourrages*, *193*, 79–102.

Duru, M., & Hubert, B. (2003). Management of grazing systems: from decision and biophysical models to principles for action. *Agronomie*, *23*, 689–703. doi:10.1051/agro:2003051

Grime, J. P. (1973). Competition and Diversity in Herbaceous Vegetation. *Nature*, *244*, 310–311. doi:10.1038/244311a0

INRA. (2007). *Alimentation des bovins, ovins et caprins. Besoins des animaux – Valeur des aliments*. Paris: Quae Editions.

Jiggins, J., & Roling, N. (2000). Adaptive management: potential and limitations for ecological governance. *International Journal of Agricultural Resources. Governance and Ecology, 1*, 28–42.

Jouven, M., & Baumont, R. (2008). Simulating grassland utilization in beef suckler systems to investigate the trade-offs between production and floristic diversity. *Agricultural Systems, 96*, 260–272. doi:10.1016/j.agsy.2007.10.001

Keating, B. A., & McCown, R. L. (2001). Advances in farming systems analysis and intervention. *Agricultural Systems, 70*, 555–579. doi:10.1016/S0308-521X(01)00059-2

Kemp, D. R., & Michalk, D. L. (2007). Towards sustainable grassland and livestock management. *The Journal of Agricultural Science, 145*, 543–564. doi:10.1017/S0021859607007253

Lavorel, S., & Garnier, E. (2002). Predicting changes in community composition and ecosystem functioning from plant traits: revisiting the Holy Grail. *Functional Ecology, 16*, 545–556. doi:10.1046/j.1365-2435.2002.00664.x

Lemaire, G., & Gastal, F. (1997). N uptake and distribution in plant canopies. In Lemaire, G. (Ed.), *Diagnosis of the Nitrogen Status in Crops* (pp. 3–44). Berlin: Springer Verlag. doi:10.1007/978-3-642-60684-7_1

Martin, G., Hossard, L., Theau, J. P., Therond, O., Josien, E., & Cruz, P. (2009a). Characterizing potential flexibility in grassland use - An application to the French Aubrac region. *Agronomy for Sustainable Development, 29*, 381–389. doi:10.1051/agro:2008063

Martin, G., Martin-Clouaire, R., Rellier, J. P., & Duru, M. (2009b). A simulation model to design grassland-based livestock systems. *Environmental Modelling & Software*.

Martin, G., Theau, J. P., Therond, O., Carre, J., Cruz, P., & Jouany, C. (2009c). *Bases et premier exemple d'application d'une démarche articulant diagnostic et simulation de systèmes fourragers pour évaluer et améliorer l'efficience d'utilisation de l'herbe*. Fourrages.

Martin-Clouaire, R., & Rellier, J. P. (2009). Modelling and simulating work practices in agriculture. *International Journal of Metadata. Semantics and Ontologies, 4*(1-2), 42–53. doi:10.1504/IJMSO.2009.026253

McCown, R. L. (2002). Changing systems for supporting farmers' decisions: problems, paradigms, and prospects. *Agricultural Systems, 74*, 179–220. doi:10.1016/S0308-521X(02)00026-4

McCown, R. L., Brennan, L. E., & Parton, K. A. (2006). Learning from the historical failure of farm management models to aid management practice. Part 1. The rise and demise of theoretical models of farm economics. *Australian Journal of Agricultural Research, 57*, 143–156. doi:10.1071/AR05051

Meot, A., Hubert, B., & Lasseur, J. (2003). Organisation of the pastoral territory and grazing management: joint modelling of grazing management practices and plant cover dynamics. *Agricultural Systems, 76*, 115–139. doi:10.1016/S0308-521X(02)00105-1

Merot, A., Bergez, J. E., Wallach, D., & Duru, M. (2008). Adaptation of a functional model of grassland to simulate the behaviour of irrigated grasslands under a Mediterranean climate: The Crau case. *European Journal of Agronomy, 29*, 163–174. doi:10.1016/j.eja.2008.05.006

Nassauer, J. I., & Opdam, P. (2008). Design in science: extending the landscape ecology paradigm. *Landscape Ecology*, *23*, 633–644. doi:10.1007/s10980-008-9226-7

Parsons, A. J. (1988). The effect of season and management on the grass growth of grass sward In Jones, M. B., & Lazenby, A. (Eds.), *The grass crop* (pp. 129–178). London: Chapmann and Hall. doi:10.1007/978-94-009-1187-1_4

Reijers, H. A. (2006, June 26-28). Workflow flexibility: The forlorn promise. In *Proceedings of the 15th IEEE International Workshops on Enabling Technologies: Infrastructures for Collaborative Enterprises,* Manchester, UK (pp. 271-272).

Rellier, J. P. (2005). *DIESE: un outil de modélisation et de simulation de systèmes d'intérêt agronomique. Internal report UBIA-INRA, Toulouse-Auzeville.* Retrieved from http://carlit.toulouse.inra.fr/diese/docs/ri_diese.pdf

Robinson, S. (2004). *Simulation: The Practice of Model Development and Use*. Chichester, UK: Wiley.

Romera, A. J., Morris, S. T., Hodgson, J., Stirling, W. D., & Woodward, S. J. R. (2004). A model for simulating rule-based management of cow-calf systems. *Computers and Electronics in Agriculture*, *42*, 67–86. doi:10.1016/S0168-1699(03)00118-2

Sawhney, R. (2006). Interplay between uncertainty and flexibility across the value-chain: towards a transformation model of manufacturing flexibility. *Journal of Operations Management*, *42*, 476–493. doi:10.1016/j.jom.2005.11.008

Schneeweis, C. (1995). Hierarchical structure in organisation: a conceptual framework. *European Journal of Operational Research*, *81*, 4–31. doi:10.1016/0377-2217(95)00058-X

Scholefield, D., Lockyer, D. R., Whitehead, D. C., & Tyson, K. C. (1991). A model to predict transformations and losses of nitrogen in UK pastures grazed by beef cattle. *Plant and Soil*, *132*, 165–177.

van der Ploeg, J. D. (1994). Styles of Farming: an Introductory Note on Concepts and Methodology. In van der Ploeg, J. D., & Long, A. (Eds.), *Born from within. Practice and perspectives of endogenous rural development* (pp. 7–30). Assen, The Netherlands: Van Gorcum.

van Ittersum, M. K., & Rabbinge, R. (1997). Concepts in production ecology for analysis and quantification of agricultural input-output combinations. *Field Crops Research*, *52*, 197–208. doi:10.1016/S0378-4290(97)00037-3

White, T. A., Barker, D. J., & Moore, K. J. (2004). Vegetation diversity, growth, quality and decomposition in managed grasslands. *Agriculture Ecosystems & Environment*, *101*, 73–84. doi:10.1016/S0167-8809(03)00169-5

This work was previously published in International Journal of Agricultural and Environmental Information Systems, Volume 2, Issue 1, edited by Petraq Papajorgji and François Pinet, pp. 20-39, copyright 2010 by IGI Publishing (an imprint of IGI Global).

Chapter 7
An Ontology-Based Framework for Authoring Tools in the Domain of Sustainable Energy Education

Sotirios Karetsos
University of the Aegean, Greece

Dias Haralampopoulos
University of the Aegean, Greece

Konstantinos Kotis
University of the Aegean, Greece

ABSTRACT

This paper presents an ontology-based framework for the production of learning designs, focusing on the domain of sustainable energy education. An ontology of the sustainable energy development domain and educational model designed in compliance with widespread standards constitute the proposed framework to disprove the vagueness of "sustainable development" and enhance reusability and shareability of learning material respectively. The authors envisage this framework both as a means to support the authoring of learning scenarios and as a provisioning of a field for conversation about the appropriate authoring tool in this area. In this sense, an integrating approach for the representation of the learning design domain is kept, focusing on the concept of "mediating artefacts".

INTRODUCTION

The efficiency of the development of learning scenarios highly depends on the ability to reuse existing materials and expertise (Pawlowski & Bick, 2006). Initially the issue of reuse in education was focused in content, in the form of basic independent units (Learning Objects or LOs) (Pasin, 2005), for faster, more economic, and richer coverage of areas of a teaching course, capitalizing experts' work and ability around the world (Mohan & Brooks, 2003).

DOI: 10.4018/978-1-4666-0333-2.ch007

However the simple provision of codified knowledge as data does not safeguard its understanding and using (Pawlowski & Bick, 2006). Therefore there is a tension to move from content-oriented to activity-oriented approach, which promotes a model of the teacher as a designer and is enhanced from the parallel development of tools in the World Wide Web, stimulating the emersion of communities of practice for the exchange of learning scenarios and experiences (Brouns et al., 2005; Pernin & Lejeune, 2006).

As the production and consumption of LOs is distributed, many standards and specifications have been developed for their standardization and generally for data exchange and processes management. A widespread standard for searching, retrieving, evaluating and using LOs, is this of Institute of Electrical and Electronics Engineers Learning Object Metadata (IEEE LOM) (IEEE Learning Technology Standards Committee, 2002), even though, according to Ullrich (2004) it mixes pedagogic and technical information, and does not provide an adequate representation of pedagogical concepts. There are many languages and models for modelling activities in learning processes and is IMS Learning Design (IMS LD) that has been accepted as a specification (IMS Global Learning Consortium, 2003). Even though it provides a general framework for constructing learning designs, its use is hard to non experts. As a consequence, the need for limitation of design options in the sense of provision of specific frames, as patterns, templates, primitives, best practices and/or taxonomies for fast and easy reproduction of learning designs arises (Bailey, Zalfan, Davis, Fill, & Conole, 2006; Griffiths & Blat, 2005; McAndrew & Goodyear, 2007).

In general, to achieve interoperability and shareability, along with being usable, authoring tools must be compliant with specifications, without however being oriented to these but to the pedagogical task, that they have to support (TenCompetence Project, 2006). At any rate:

> *It seems unlikely that one all-encompassing notation system for representing and documenting learning designs will evolve. Instead, a toolkit of a number of representations each used at different times during the educational design process for different purposes seems to be a more plausible option. (Agostinho, 2008, p. 13).*

In the context of learning design, ontologies usage covers a spectrum of roles. Ontologies can support more flexible handling of LOs as to employ just some specific parts of a LO, to reuse the same learning design in different contexts with different LOs, to personalize content of the same LO, and to develop more extensive LO and learning design search and ranking services (Knight, Gašević, & Richards, 2006). According Mohan, Greer, and McCalla (2003) it is necessary to associate LO metadata with domain ontologies since both play an important role in locating relevant LOs for a course.

Based on these concepts the paper introduces an ontology-based framework for the production of Learning Designs (learning scenarios compliant with IMS LD) in the field of sustainable energy education. In the next sections, the rationale, the basic structure and the ontology development of this framework are presented, along with related work, an evaluation experiment, an exemplification of its potential usefulness and some concluding remarks.

MOTIVATION AND RATIONALE

Sustainable energy education is expected to contribute to the solution of a series of challenges – problems related with the dominant developmental models, taking into consideration that the energy production and use should provide "adequate energy services for satisfying basic human needs" while "should not endanger the quality of life of current and future generations and should not exceed the carrying capacity of ecosystems" (Rogner & Popescu, 2000, p. 31).

Regarding the shape of the educational approach that this field should have, the following issues must be taken into consideration:

a. The need for a practical and explicit definition of sustainability in virtue of the vagueness and the polysemous of the term (Robinson, 2004).
b. The necessity for teachers' training, which becomes more urgent in some European countries (e.g. Greece), as relevant reports demonstrate a teacher's lack of basic education on sustainable energy development (Papadimitriou, 2004; Skanavis, Petreniti, & Giannopoulou, 2004; Spiropoulou, Antonakaki, Kontaxaki, & Bouras, 2007).
c. The availability of abundant educational material about sustainable energy education issues, in various languages and forms, requiring however a substantial effort for evaluation, collection, adaptation and distribution.
d. The vast and ever growing literature for learning theories, ideas and approaches with multiple competing claims, which comprises a very difficult and confusing field for educational practitioners, unless they are engaged in full-time research (Dillon & Ahlberg, 2006).
e. The improvement of teaching quality and learning experience, when the teachers adjust and refine materials provided rather than using 100% readymade materials (European Commission Directorate General for Energy & Transport, 2006).

As a result, a reasonable approach to the issue is expected to include the development of an educational authoring tool to support teachers in adaptation of existing educational material and/or in production of new one, according to the needs and the knowledge level of their students.

Accordingly, the issues that should be tackled are reflected in the following research questions:

1. How should the field of sustainable energy development be specified in an explicit way?
2. How the available educational material should be codified, in order to provide information about its content, but also adapted to the local educational environment for it to be reusable?
3. How teachers should be supported to the selection and utilization of the available educational material, its adaptation to the local conditions, in a way that exploits the existing theoretical and empiric pedagogic knowledge and will be, at the same time, attractive?
4. What learning theory or model should be adopted?

THE PROPOSED FRAMEWORK

Based on literature review we introduce a framework that could be used as a base for the development of a prototype authoring tool and for the consequent process of iterated improvements with the participation of practitioners (Figure 1).

The basic framework components are a) the domain model and b) the educational model, which in turn is comprised of a content model and a learning design model. Both models are based on ontologies. According Paquette (2007), a form of domain ontology is needed in order to structure a knowledge domain in e-learning or knowledge management application, since an alternative solution such as a tree organization, although it could reduce significantly the processing, is insufficient to describe the interrelations that tie the concept structures.

In our case the domain model specifies the basic concepts of sustainable energy development and their interrelations. In order to model the domain we rely on the DPSIR indicator-based framework (European Environment Agency [EEA], 2002).

Figure 1. Functions supported by the framework

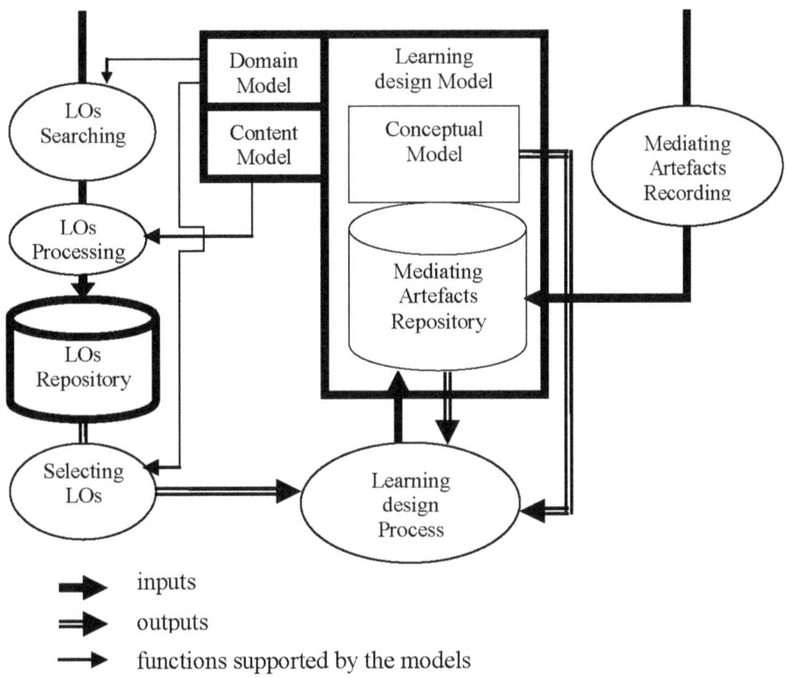

The DPSIR framework depicts the sequence in which the events take place in the energy process, allows identifying the stages where there is a sustainability problem, and connects this problem with a potential response-solution. It can describe the state and the evolution of an energy model from a sustainable development perspective, starting from the demands that the energy sector receives from the society, represented as Driving forces (D). The Pressures (P) are the measures of the effects of energy production and consumption on the environment, the society and the economy, contributing to the change of their conditions (State) and producing an impact on them (Impact). These four steps of analysis of sustainability of the energy sector are followed by the social actions (Responses) that feed back (D), (P), (S), and (I), modify the interrelations of the energy process in order to limit, reduce or mitigate the Impacts (I) and attain a more sustainable energy system (De L. Hierro Ausin & Pérez-Arriaga, 2005).

As it is mentioned in an EEA (1999) document, of particular importance are the links between phases of DPSIR, since they identify dynamic relations between origins and consequences of sustainability problems. These relations show that, in fact, DPSIR does not represent a simple sequential or circular scheme, but a very complex web of interacting factors.

As far as learning design model is concerned we try to keep an integrating approach for the representation of the learning design domain. This is reasonable, as according to Griffiths (2005) most teachers should not be expected to engage with the specification itself but only in the levels of adapting existing units of learning, or mixing and matching parts of them. Also, as mentioned by Conole (2007), practitioners use a range of tools to support and guide their design of learning activities, while according Falconer, Beetham, Oliver, Lockyer, and Littlejohn (2007) multiple representations are necessary as they give the ability to someone to view the designs from different perspectives.

Central point to this approach is the concept of "mediating artefacts" (Conole, 2008), that practitioners use in the authoring process, in order to, according activity theory, bridge the gap between historical state of learning design and their developmental stage.

The learning design model is considered both in terms of supporting practitioners to construct a learning scenario and providing them with ways to represent it. Moreover going up a level the model provides for various alternative solutions for the structure of the learning design tool itself, allowing for the selection of the more appropriate features for a specific context.

In fact, such a model could act as repository of existing artefacts covering different level of abstraction, like pedagogical models, patterns and generic learning designs, case studies, but also learning design processes, learning design conceptual models and representation formats used by other learning design tools.

As far as the content model is concerned, it is based on the IEEE LOM standard.

Mappings between concepts of the above three (domain, content, learning design) models interconnect them. Thus for example the content model is associated with the "Learning content", a subclass of the learning design model while is connected with the domain model via a mapping of LOs with the domain concepts. In addition learning objectives are associated to any domain concept establishing links between learning design model and domain model.

As a conclusion we could note that the specification of causal loops based on the DPSIR framework and related with sustainability problems is very important for the consistency of the approach. Since these loops can act as a backbone of a learning course, they actually connect the domain with the educational model.

ONTOLOGIES DEVELOPMENT

In this section we outline the ontologies' structure and the rationale underlying their development. The tool we use for the ontologies development is the Protégé 3.4 beta.

Afterwards, we are going to use quotation marks when we refer to a *class* of the ontology, and courier new fonts to indicate *slots* (properties of classes).

Learning Design Model

We envisage the learning design ontology both as a means to support the authoring of learning scenarios and in terms of provision of a field for conversation about which should be the appropriate form of an authoring tool in this area. In this sense we attempt to define the basic features that constitute a "Learning design tool". We consider that a tool relies on a "Learning design conceptual model", and potentially can provide a "Learning design process", additional scaffolds in the form of "Guidelines", "FAQs" or "Tips & hints", and other stored artefacts, like "Learning models", or "Learning design solutions", abstract or contextualized. In general, in order to make a structure for storing existent practice in the learning design domain, we rely on the work of Conole (2008), classifying the tools and resources, which are used by practitioners to support the learning design process, as "Mediating artefacts" and the resulting by their aggregation artefacts as "Meta-mediating artefacts" (Figure 2).

The "Learning design conceptual model" specifies the "Learning design elements" and their interrelationships, being also connected with a "Representation format", which defines the way a "Learning design" is represented.

The "Learning design element" is the key concept for the provision of a formal structure for the description or the construction of a learning design.

Figure 2. The classes of the educational model

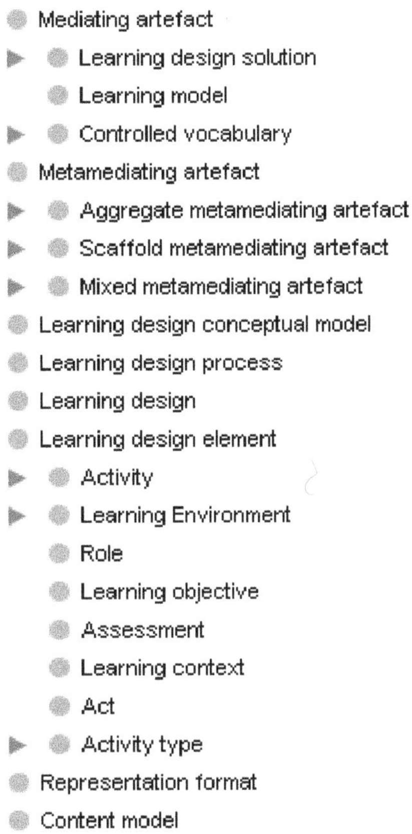

Its subclasses ("Learning objectives", "Learning context", "Activity", "Role", "Environment", "Assessment", "Act") and their interrelationships are following the vocabulary and the basic directions of IMS LD specification. The instantiation of any "Learning design element" is supported by stored "Controlled vocabularies" that are associated with them as potential or default value spaces, providing the ability to choose between a range of solutions.

For example, the revised Bloom's taxonomy (Anderson & Krathwohl, 2001) is defined as default value space for the definition of "Learning objective" regarding the cognitive domain, while a number of similar artefacts constitute the potential value space (Figure 3).

As far as the other subclasses of the "Learning design element" is concerned, we use in most of cases the Dialog Plus taxonomy (DialogPlus Project, 2008) as default value space, in order to facilitate the completion of a learning design, since according to Griffiths and Blat (2005), this taxonomy compresses the definition of the narrative, the representation and coding stages of IMS LD specification. Especially for the "Activity type" we use as alternative solution the Eight Learning Event Model (Leclerc & Poumay, 2005).

We actually use two levels of artefacts' description. The first concerns the artefact itself, whilst the other presents information useful for its localization, selection and use for a specific case.

For example the "Learning design" and the "Learning design solution" refer to the same concept, but the first is defined as instance of the local formalization specified by the "Learning design element" class, while the "Learning design solution" is described by a pattern style form, which is used in order to unify the description of all the range of different approaches are used in the learning design realm (Figure 4). This form includes title, context/audience, granularity, purpose/challenge, solution, example/ generic, focus/approach, creator, references, discipline and related generic solutions, allowing between others the connection among an abstract learning design and its instantiations via the example/generic *slot*.

Learning Content Model

With regard to "Content model" the IEEE LOM standard is followed, so as to take advantage primarily of a complete syntax and semantic created by experts of the Learning Technology and secondarily to ensure the interoperability with similar systems (Capuano, De Santo, Marsella, Molinara, & Salerno, 2000).

The selected subset of the IEEE LOM elements for the description of "Content model" is depicted in Figure 5. Elements' selection was based on their simplicity and their utility for resource localization and sharing.

Modifications that have been made are the addition of the connections between domain

Figure 3. The slot learning objective potential value space

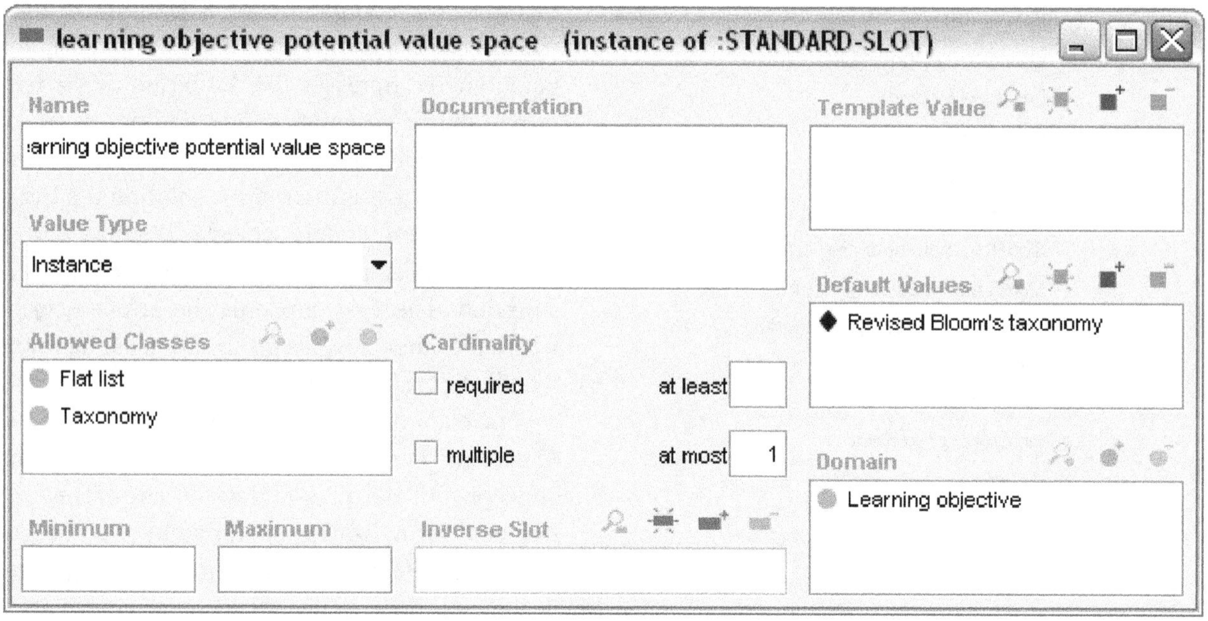

concepts and curriculum disciplines, which are not explicitly mentioned by the standard, as well as discrimination between technical and instructive information in the learning resource type, as suggested by Ullrich (2004). Furthermore we conducted a literature review concerning the learning resource type and we stored relative vocabularies as alternative solutions in order to both interoperability and adaptation to the local conditions are achieved.

Domain Ontology

As indicators can translate sustainability in a practical sense providing a deeper understanding of the main issues and highlighting important relations (International Atomic Energy Agency [IAEA], United Nations Department of Economic and Social Affairs, International Energy Agency [IEA], Eurostat, & EEA, 2005), we actually utilize sets of energy indicators of sustainable development in order to describe the dimensions of the sustainable energy development concept.

We rely on the DPSIR framework, which has been used by the EEA (2002) to connect energy and environment, and extended by De L. Hierro Ausin and Pérez-Arriaga (2005) in order to embody impacts of energy production and consumption not only on the environment but also on the society and the economy. We extend and enrich this last approach with more indicators derived from other sets of energy indicators for sustainability (IAEA et al., 2005; IAEA & IEA, 2001; EEA, 2008), since our aim is not to provide a tool for policy making in a specific context but a broad knowledge base for learning design, including as much as more concepts relative to sustainable energy development.

Except the amount of final energy demand that is the main driver, other indicators can be included, conditioning this amount and its effects on the environment, the society and the economy (Figure 6).

As far as the classes of "Pressure of energy development", "State change of energy development", and "Impact of energy development" are

Figure 4. An instance of the class "Abstract learning design solution"

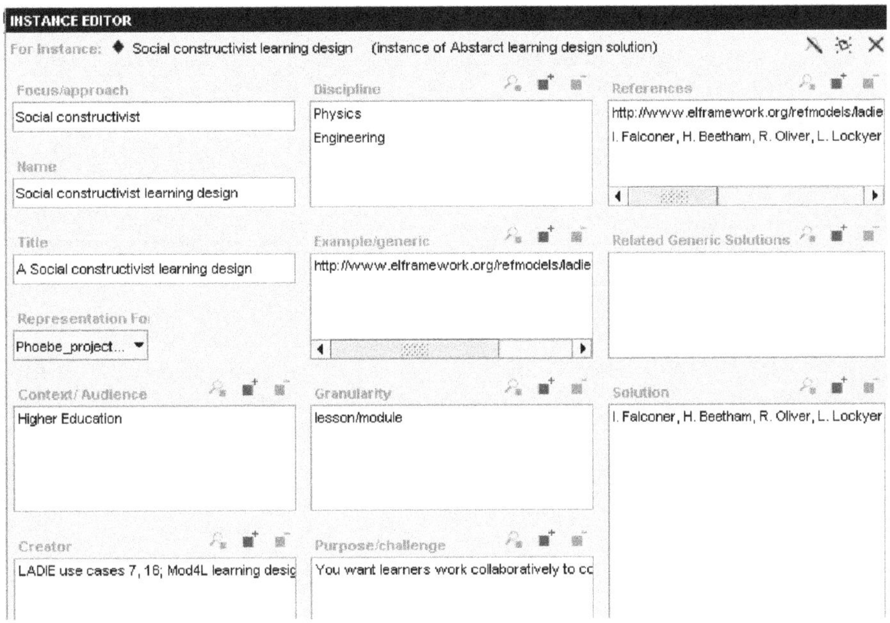

concerned, they are depicted in Figures 7, 8 and 9 respectively.

We should note here that even though we assign any indicator to a concrete phase of DPSIR framework, in fact an indicator can represent more than one DPSIR typology, depending on the context of analysis. A solution could be to define these additional relations via slots in the ontology. For example "Temperature" is assigned as a (D) subclass, since affects the amount of energy demand, but also a slot conditions environment informs that describes the state of environment.

Of particular importance is the class "Response for sustainable energy development" since represents the solution or a mitigation of a sustainability problem. Some responses are considered as negative (D) since other aim to raise the eco-efficiency via the development and penetration of clean technologies (EEA, 1999). We adopt and adapt an approach applied in (IAEA & IEA, 2001) and we define the (R) subclass in terms of the (D), (P), (S), or (I) subclasses that targets and positively affects respectively (Figure 10).

The other slots include: response type, with values based on Environmental Sustainability Index (World Economic Forum, Global Leaders for Tomorrow Environment Task Force, Yale Center for Environmental Law and Policy, Yale University, Center for International Earth Science Information Network, & Columbia University, 2001); response actor, with values based on an approach presented by Zhang and Fujiwara (2007), distinguishing three categories of actors (government, civil society, firm); response level (global, regional, national, city, household, personal); and source.

As far as the other classes of the domain are concerned, a whole view of domain ontology is given (Figure 11).

As a conclusion we should note that the organization of subclasses of the DPSIR types is not exhaustive. More concepts must be embodied, especially in the level of cities and/or households in order to the concept of sustainable energy development is more concrete and familiar to the learners.

Figure 5. An instance of the class "content model"

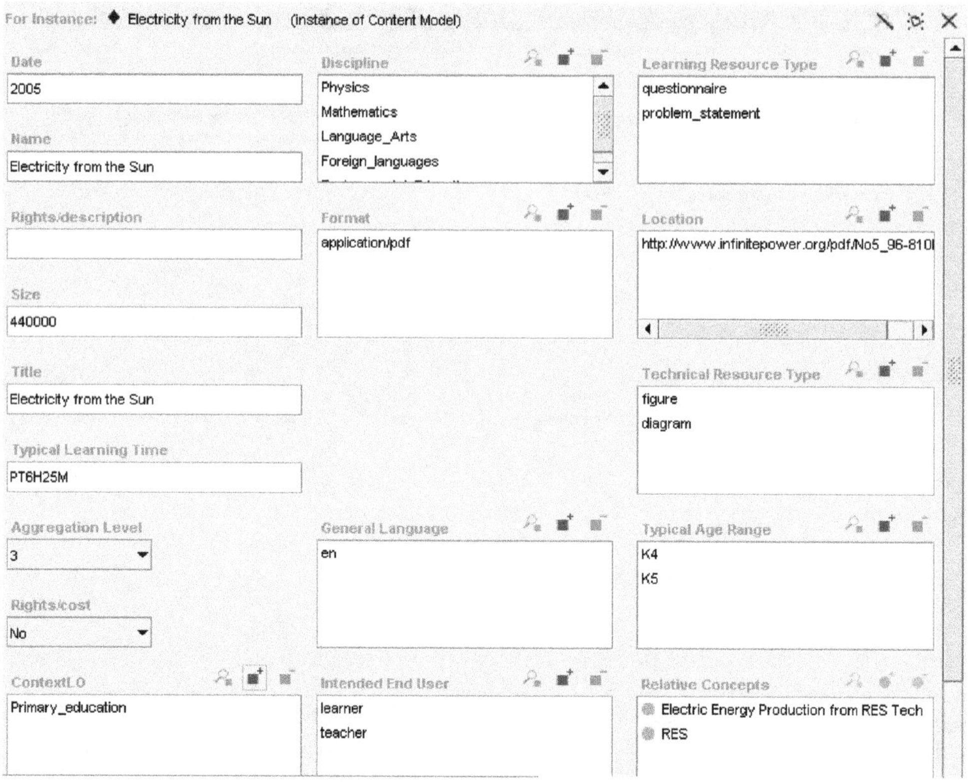

RELATED WORK

Since there are many approaches in the relevant literature that have similarities with one or another part of our approach, it is very difficult to include an analytical state of the art at the present article. Kumazawa, Saito, Kozaki, Matsui, and Mizoguchi (2009) rely on a problem-solving approach in order to model the domain of sustainability and support the extraction of causal conceptual chains, as we also do. However, they conceptualize the domain in a different way and do not focus on the energy concept.

As far as the inclusion of IMS LD specification in our approach is concerned, our aim is not to cover any aspect of the specification as Amorim, Lama, Sánchez, Riera, and Vila (2006) do. Thus we don't define the semantics of the IMS LD concerning its behavioral model since it is not our priority to specify the constraints imposed in a runtime environment during the execution phase, but rather to inspire and facilitate educators in order to design and represent a unit of learning, by defining the conceptual model of IMS LD.

Another related approach is Ontoolcole (Vega-Gorgojo, Bote-Lorenzo, Gómez-Sánchez, Asensio-Pérez, Dimitriadis, & Jorrín-Abellán, 2008), an ontology which describes a tool in terms of a task performed by a specific type of user and the type of artefact that constitutes input and output of the task. Ontoolcole relies to activity theory, as we also do, but it is not orientated towards the representation for storage purposes of potential features of the artefacts, but it is rather orientated towards the provision of the required characteristics for search of Computer Supported Collaborative Learning tools.

Finally, regarding our description layer acting as a manually imposed wrapper to integrate mediating artefact's attached information that

Figure 6. The class "Driving force of energy development"

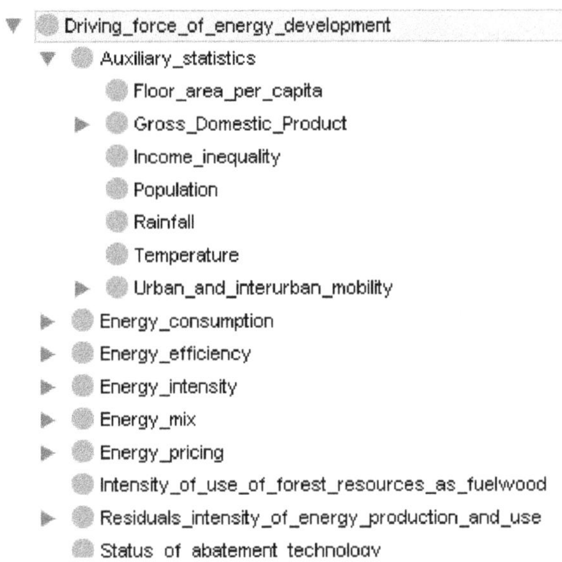

comes from various origins, there may be slight similarities with projects like Edutella (Nejdl et al., 2002), which uses RDF (Resource Description Framework) for annotating materials originated from heterogeneous resource metadata repositories, even though the specific project use these RDF annotations in turn for the purpose of constructing queries for different query languages.

EVALUATION OF THE FRAMEWORK

The technical evaluation of the ontology is carried out in the phases of Conceptualization and Formalization and it comprises of verification, supported via Protégé, and of validation/evaluation, which is carried out with the help of domain experts in a collaborative and argumentative manner (Kotis & Vouros, 2006).

In order to detect if the ontology provides and the adequate expressiveness to practitioners to pose questions and get answers that support them to design learning activities for a concrete learning situation, we conducted an evaluation experiment.

Figure 7. The class "Pressure of energy development"

A group of fourteen (14) teachers of primary and secondary Greek Education took part in this experiment with their valuable comments and contribution. Eight of them were experienced in environmental education while the others were interested to infuse topics of sustainable energy development in their teaching practice.

A Wizard of Oz (Kelley, 1984) technique was applied in order to assess the capabilities of our framework in terms of facilitating teachers to search and select appropriate artefacts such as LOs and abstract representations of learning designs like models, generic learning designs and design patterns, for their educational setting. In this technique a person plays the role of *Wizard* in order to allow users to interact with a system missing some functionality. This way we were able to simulate a search facility supported by the ontology and investigate teachers' expectations of such an approach and their behaviour /reflection in relation to an initial experimental repository of artefacts we developed.

Teachers were asked to formulate one to three questions per category for searching such artefacts in a digital repository, actually containing *instances* of corresponding *classes* of the ontology. These questions were translated from the *Wizard*

Figure 8. The class "State change of energy development"

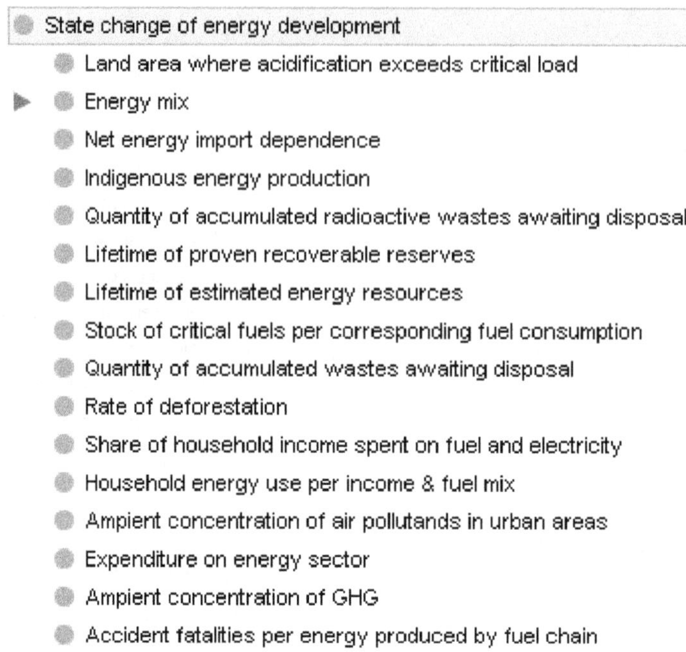

to queries posed to the ontology via the query functionality of Protégé. The translated queries together with the retrieved artefacts were returned to the teachers, which were asked:

a. To evaluate the quality of translation
b. If the retrieved artefacts were consistent to their question and useful for the purpose they needed to.

In order to formulate the questions for searching abstract representations of learning designs teachers had the chance to be supported or not by a list of sentences describing such artefacts in terms of the purpose/challenge they addressed (Table 1), aiming to collect teachers' expectations about representations that are not currently included in the list.

The models that constitute this initial set have been selected to be suitable for the sustainability domain. They are the followings:

a. A way to analyze and teach a sustainability problem suggested by McKeown & Dendinger (McKeown, 2002, p. 19), tied in with the extraction of a causal chain based

Figure 9. The class "Impact of energy development"

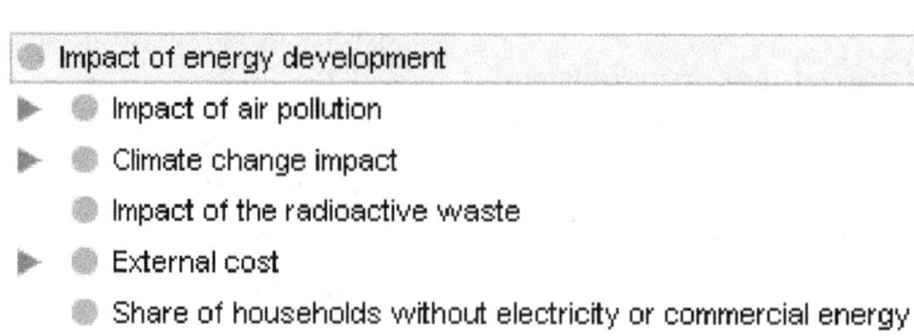

Figure 10. An instance of the class "Responses for sustainable energy development"

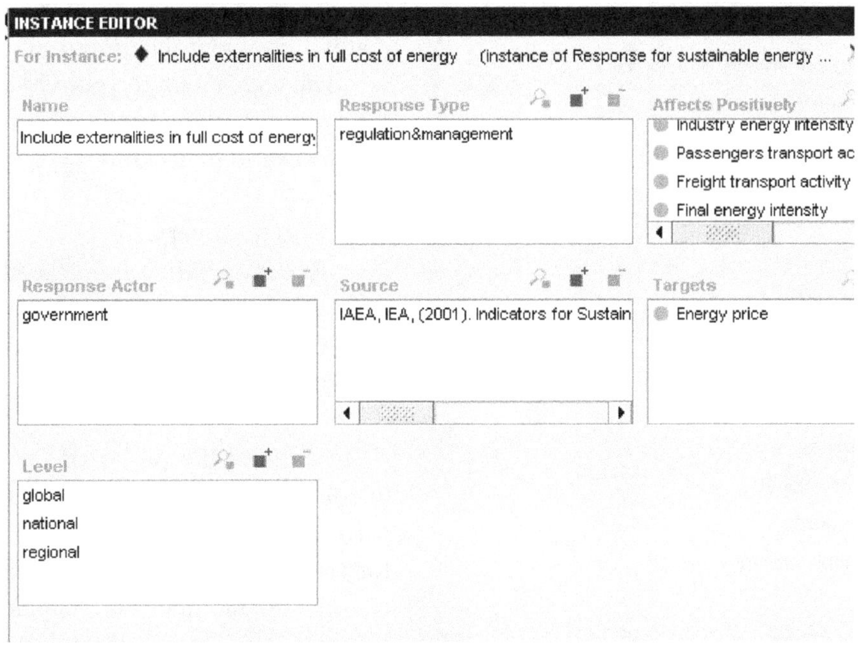

on DPSIR model. Since each one of its steps can accept as input any other learning model, this framework can actually connect the sustainability domain model and the learning design model.

b. Problem based learning.
c. Project based learning.
d. A based on action research model which combines the use of improved Vee Heuristics and conceptual mapping and which has been tested and used in the Education for Sustainable Development (Ahlberg, Aanisma, & Dillon, 2005; Ahlberg & Ahoranta, 2002).

As far as the abstract learning design solutions are concerned, they are mainly derived from Pedagogic Pattern Project (Sharp, Manns, & Eckstein, 2000), Learning Designs Project (2003) and Mod4L and LADIE projects (Falconer et al., 2007).

The results of the study are summarized in the Table 2. They generally indicate a positive response on the part of teachers. Even though the number of the participants is small for safe conclusions, we could remark that there is differentiation in the scoring between LOs and abstract representations of learning designs. This probably reflects the higher level of difficulty that the description of an abstract representation of learning design implies in comparison with the one for a LO-content. Also a score decrease was recorded in assessment from translation quality to consistency and usefulness. This can be partially attributed to the relatively small number of artefacts stored in the experimental digital repository.

Additionally, before the process we described so far, we gave to the teachers, lists of potential criteria for searching and selecting LOs and abstract representations of learning designs. Teachers were asked to select the more appropriate between them, having also the chance to suggest elements out of the lists. Our intention was to identify a LO Metadata profile and a set of description elements for abstract learning designs based on teachers decisions in order to compare them with the corresponding current parts of our

Figure 11. The classes of sustainable energy development ontology

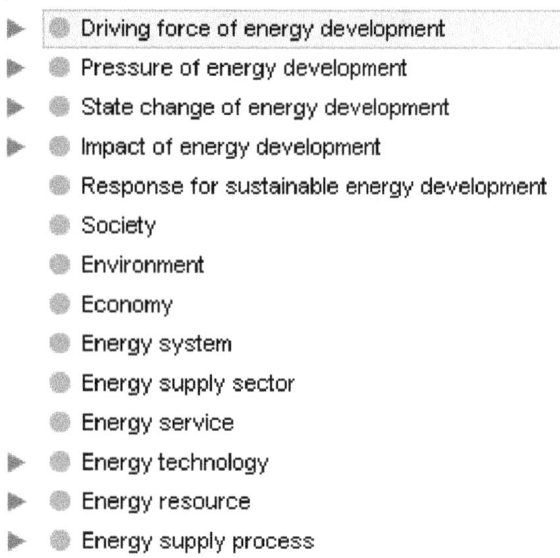

ontology. The teachers were asked both to rank the criteria for searching resources, and to assess criteria for selecting resources as necessary, nice to have or unimportant in order to a search or a results subset of elements be identified respectively. Some quantitative results of this survey are depicted in the Tables 3, 4. In general the results are in accordance with our choices. Some exceptions are concerning mainly the *keyword* criterion regarding the content model, and *subject area (*was scored higher than the contiguous *discipline* we use*), language* and *required technology* regarding the set of descriptors for abstract representations of learning designs.

CONNECTING THE PARTS: HOW THE PROPOSED FRAMEWORK CONTRIBUTES

In this section we connect the fragments of the framework we described so far, in terms of its usefulness and its potential contribution in a learning design case.

The aim of the proposed ontology is to lay the foundations of an energy-environment curriculum that will enable teachers and students to familiarize themselves with the current developments in the area of sustainable development and in this framework to be used for:

1. LOs discovery.
2. Recommending about the selection of learning designs.
3. Construction of learning designs with reuse of LOs.

In parallel the proposed ontology is aiming to constitute an appropriate environment for experimentation, by providing alternative solutions for the structure of a learning design tool itself in order to facilitate its adaptation to different contexts. The *class* "Learning design process" is the key concept for such a function. Any "Learning design process" has step and any "Learning design step", has step number, is related to a "Learning design element" and can be mediated by a "Mediating artefact".

This last *slot* can give quick access to relevant potential solutions for a learning design situation. An example is given in Figure 12, where the step <arrange activities> of an experimental learning design process is highlighted. It is shown that it can be mediated by a number of artefacts (*instances* of "Mediating artefact").

Afterwards we refer some scenarios/ examples for the framework's uses mentioned above.

In the first case a user selects values for one or more *slots* of "Content model" as criteria for searching LOs in a local repository and retrieves a LO as *instance* of "Content model", or starting from one concept browses a graphical representation of the domain's ontology and is guided indirectly to search locally or on the Web for LOs associated with an area of concepts interrelated to the initial one. The value added by our approach to the semantic annotation and searching of a LO is the adoption of object properties as the related

Table 1. A sample of Purposes/Challenges addressed by stored abstract representations of learning designs (models, generic learning designs and patterns)

*You want to engage your students in complex open ended problems
*You want to engage students in action research which includes concept mapping
*You want to teach a sustainability problem
*You want your students to find a solution to a problem which have been set by them
You want to focus on the whole picture
You want students to contribute their fare share on the resolution of a problem
You want to maximize learning by engaging
You want to take different skill levels and interests into account
You want your students to acquire domain-specific or practical skills
You want your students' exposure to the idea that a problem can have many solutions
You want to bridge the gap between the educational and real world
You want your students to develop conceptions
You want to provide to learners a simulated environment for experiential learning
You want your students to think about the past and argue about the future
You want your students to develop tools that they can use later on
Participants are overwhelmed by theory
Participants don't know how to learn outside the official learning environment
Only a few students participate, the rest are quiet
Participants forget the contents
You want learners to organize prior knowledge and internalize new information

* The first four sentences concern learning models

domain concept that facilitate a domain ontology based description of the LO. For example, in the evaluation experiment we conducted, a teacher wishing to search for a LO described it as '*a text, in French language, that informs students 12-14 years old about how can contribute in personal level, to the reduction of air pollution*'. This was translated into a query, by means of the data type properties learning resource type, language, typical age range, assigning them the values <narrative text>, <fr> and <12-14> respectively, and the object property related domain concept, assigning to it as value an *instance* of the *class* "Response action" with its *slot* actor taking the value <citizen>

Table 2. Summarized evaluation results of ontology's search and retrieval capabilities expressed in a five level quantitative scale.

Questions Posed		Translation	Retrieval	
Category	**Number**	**Quality**	**Consistency**	**Usefulness**
Los	35	4.71	4.48	4.33
Abstract LDs	29	4.63	4.17	4.15
Total	54	4.67	4.32	4.24

Note. Values are mean scores on a 5-point scale (1=*insufficient*, 5=*very good*).
[a] As Abstract LDs are defined learning models, learning design patterns and generic learning designs.

Table 3. Quantitative results of a survey for identification of a LOM application profile

Criteria	Search Subset(%)	Results Subset(%)		
Number*** and title	Ranking**	R[a]	N[b]	U[c]
9.1.Related domain concept*	88.1	100.0	0.0	0.0
1.5.Keyword	74.8	100.0	0.0	0.0
5.2.Learning resource type*	73.3	78.6	21.4	0.0
5.7.Typical age range*	59.5	71.4	28.6	0.0
9.1.Discipline *	54.8	78.6	14.3	7.1
1.2.Title*	54.3	71.4	21.4	7.1
1.3.Language*	51.9	64.3	35.7	0.0
9.1.Learning objectives*	50.0	71.4	21.4	7.1
5.5.Intended End User Role*	49.5	64.3	35.7	0.0
5.6.Context*	42.9	57.1	35.7	7.1
5.9.Typical learning time*	30.9	28.6	50.0	21.4
5.8.Difficulty	23.8	28.6	50.0	21.4
4.4.Requirement	21.9	21.4	57.1	21.4
1.6.Coverage	20.9	28.6	28.6	42.9
5.3.Interactivity level*	15.7	14.3	57.1	28.6
5.4.Semantic density	13.8	21.4	42.9	35.7
2.3.1.Creator-Author*	13.3	7.1	71.4	21.4
6.1.Cost*	13.3	28.6	21.4	50.0
1.8.Aggregation level*	8.6	28.6	21.4	50.0
5.1.Interactivity type	8.1	7.1	50.0	42.9
4.1.Format*	7.6	7.1	50.0	42.9
9.1.Related LO	7.6	7.1	42.9	50.0
4.2.Size*	6.2	14.3	28.6	57.1
2.3.3.Date*	5.7	14.3	21.4	64.3
6.2.Rights/restrictions*	3.3	7.1	21.4	71.4

* Included in the Content Model.
** Expressed as percentage of the max potential score. Only first fifteen participant choices (criteria) get points from 15 down to 1.
*** Expresses the classification of IEEE LOM standard
[a]. Required. [b]. Nice to have. [c]. Unimportant.

and the *slots* targets or affects positively taking the value "Air pollutant emission", in order to be posed to the ontology via Protégé.

With respect to the second mentioned case concerning the selection of a learning design, abstract or contextualized, probably the most critical *slot* is the purpose/challenge that this artefact addresses, and implies the fact that a list of its potential values must be given to the user (e.g. Table 1). An example query via Protégé based on one such purpose/ challenge (i.e. '*You want your students to develop conceptions*') is depicted (Figure 13). The query could be more specialized with other *slots* e.g. focus/approach. The output of such a query is an *instance* of corresponding ontology's *class* including a link to this artefact.

The third exemplifying case concerns a generic learning design process which includes the steps depicted in Figure 12.

Table 4. Quantitative results of a survey for identification of a set of elements for the description of abstract representations of learning designs, like learning models, patterns and generic learning designs

Criteria	Search Set (%)	Results Set (%)		
	Ranking**	R[a]	N[b]	U[c]
Purpose/Challenge*	92.9	100.0	0.0	0.0
Context/Audience*	79.5	92.9	7.1	0.0
Focus/Approach *	76.2	85.7	14.3	0.0
Granularity*	66.2	71.4	28.6	0.0
Subject area	65.2	64.3	35.7	0.0
Language	53.3	64.3	35.7	0.0
Discipline*	51.0	35.7	57.1	7.1
Known uses/examples*	46.2	92.9	7.1	0.0
Interaction type	46.2	21.4	71.4	7.1
Group size	42.9	42.9	42.9	14.3
Title*	42.4	42.9	35.7	21.4
Delivery mode	41.4	28.6	57.1	14.3
Required Technology	38.1	50.0	50.0	0.0
Related artefacts*	30.5	35.7	57.1	7.1
Creator/Author*	15.7	0.0	57.1	42.9
Cost	6.2	0.0	57.1	42.9
Date	4.8	7.1	35.7	57.1
Rights / Restrictions	1.4	0.0	42.9	57.1
Solution*		78.6	14.3	7.1
References*		21.4	71.4	7.1
Comment		14.3	42.9	42.9

* Included in the Model. ** Expressed as percentage of the max potential score. Only first fifteen participants' choices (criteria) get points from 15 down to 1.

[a]. Required. [b]. Nice to have. [c]. Unimportant.

Figure 12. An experimental learning design process

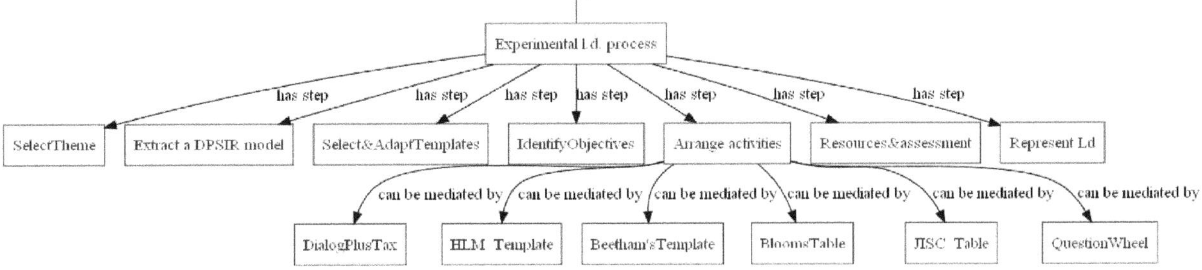

Figure 13. A query via Protégé for the localization of stored "Abstract learning design solutions" which address one purpose/challenge

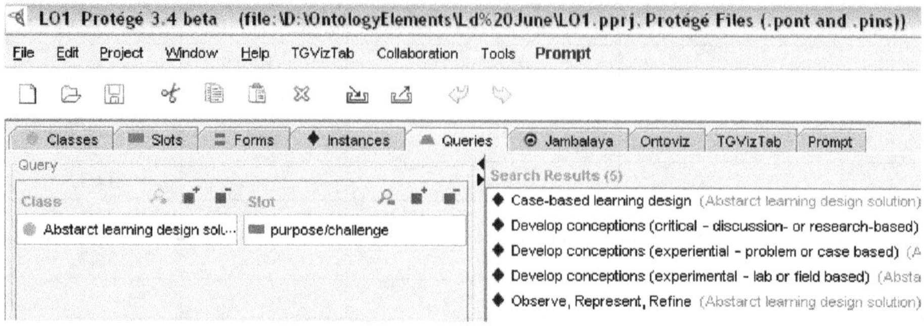

In the first two steps a user selects a sustainability issue/problem, e.g. air pollution, and extracts a causal chain. Thus a number of drivers (e.g. "Urban and interurban mobility", "Energy intensity", "Population", "Pollutants emission intensity", "Abatement technology status" and "Energy pricing"), or potential responses/solutions to the problem (Figure 14) could be identified and support the teacher to assign for example questions/ issues and define tasks for the students. At this point we should note that even though causal chains can be identified via queries based on *slots* like caused by, causes, affects positively or targets, the development of a tool that facilitates the exploration of ontology's visual representation and the extraction of a concept map based on the DPSIR model would be indispensable for the exploitation of the ontology.

Figure 14. A query via Protégé for the identification of "Response actions" that affect positively "Air pollutants emission from energy systems"

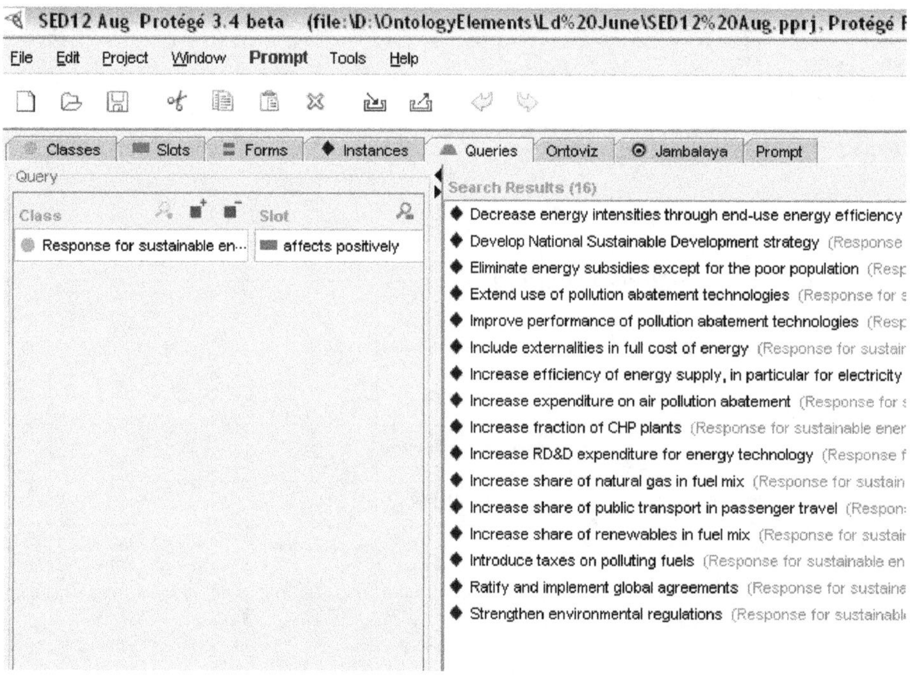

Figure 15. A template of a Vee Heuristic Model

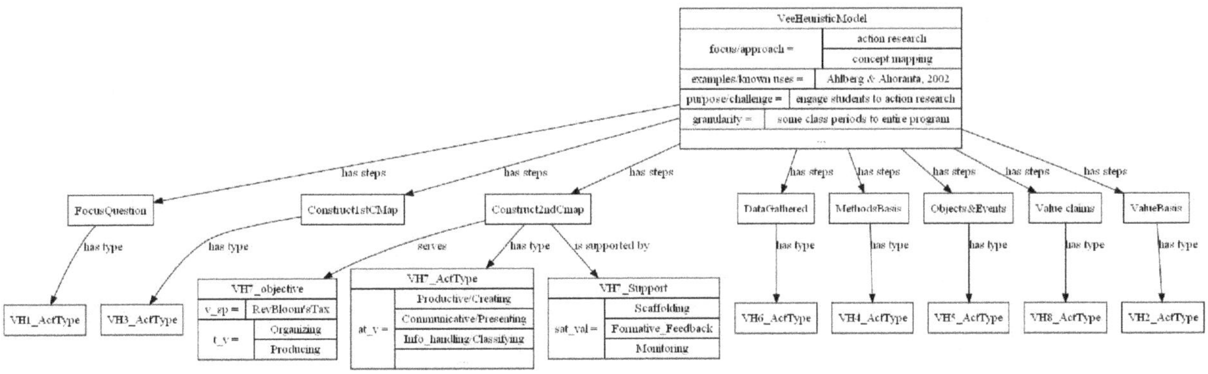

The third step implies the selection and adaptation either of one single template corresponding to a "Mediating artefact" or of a combination of them. Thus for example, in accordance with the previous step the user selects to *teach a sustainability problem* and retrieves via a query based on purpose/challenge the framework suggested by McKeown & Dendinger (McKeown, 2002, p. 19). Then he/she selects to *engage students in action research* and retrieves a version of the improved Vee Heuristic (Ahlberg & Ahoranta, 2002). The first one includes a set of eleven questions, which could be considered as focus questions for the improved Vee Heuristic. Another choice in order to manage the sub-problems identified by this framework, in a way that the students *contribute their fare share on the resolution of a complex problem/task*, could be the retrieving of the Jigsaw collaborative pattern.

Any learning model is expressed via a *slot* has task as a template based on the "Learning design element". This actually means that any task of a learning model is an *instance* of "Act", "Activity structure" or "Learning activity". Learning objectives, learning and support activity types, roles and assessment strategies can be pre-defined for any task of the model. An example concerning the improved Vee Heuristic (Ahlberg & Ahoranta,

Figure 16. Instances of the class "Representation format" that can mediate the completion of the "Learning design step" <Represent Ld>

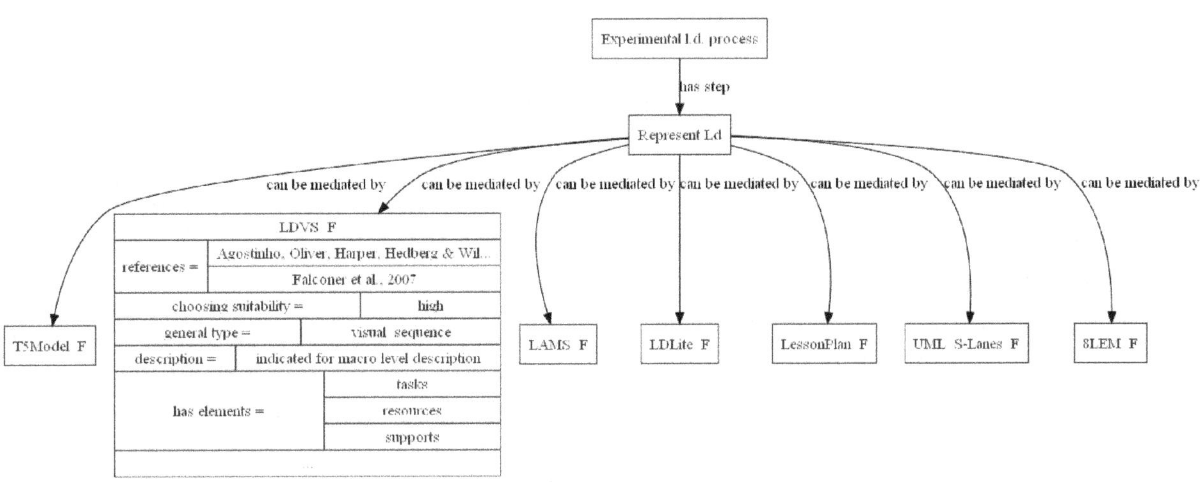

2002) is depicted in the Figure 15, where the task <Construct a second concept map> is highlighted.

This way, the next three steps of the learning design process which are referred to the identification of learning objectives, the arrangement of activities and the selection of assessment strategies, are facilitated. The retrieval and selection of appropriate resources is facilitated via the mapping of domain concepts to LOs and learning objectives we mentioned previously.

Finally a number of artefacts, *instances* of the *class* "Representation format", that could potentially mediate the last step, this of the representation of the learning design, named <represent Ld>, are depicted in the Figure 16. The *slots* that could be used as criteria for such a decision are the general type (text, text/pattern, matrix, visual sequence, xml file), the number and the type of the represented elements, the level of its suitability for different applications of a learning design as are browsing, choosing/evaluating, developing and implementing (Falconer et al., 2007), or the description where hints are provided for the format's indicated use.

CONCLUSION AND FUTURE DEVELOPMENT

In relation with what exists so far in the literature the present work organizes a framework in the field of education for sustainability, focusing to energy-environment and to the solution of related problems. The structure of domain ontology which relies on DPSIR framework (EEA, 1999), provides a base for a systemic approach of sustainability and can contribute to the semantic annotation and retrieval of LOs, acting also as a backbone for the construction of learning courses.

A toolkit approach to the learning design process is followed, integrating actually the relative literature in the structure of the ontology. We conjecture that by providing a variety of supports for the learning design process, ranging from informal solutions, as patterns, up to more constraint structures as the formal representation inspired by IMS LD specification, we allow practitioners to make their own choice of the level of formality that they would like to adopt in their designs.

Also going up a level the framework provides alternative solutions for the structure of the learning design tool itself, supplying the flexibility to design experiments with the participation of practitioners, aiming to the selection of the most appropriate features for a specific context. The provisional users of the presented framework, constituting the participants of such experiments, are teachers of primary and secondary Education in Greece, engaged /experienced in Environmental Education or interested to infuse sustainable energy topics in their teaching practice.

Indispensable for the exploitation of the presented framework is the visual exploration of the domain ontology and the extraction of causal conceptual chains. This actually determines as first priority for the future work the construction of a tool for the visual representation of the ontology in a user-friendly way that overcomes the technical difficulties due to the great number of concepts to be represented. Potential solutions examined are related with techniques based on graphic extension and representation of a defined limited area around a selected concept.

REFERENCES

Agostinho, S. (2008). Learning design representations to document, model, and share teaching practice. In Lockyer, L., Bennet, S., Agostinho, S., & Harper, B. (Eds.), *Handbook of research on learning design and learning objects: Issues, applications and technologies* (pp. 1–19). Hersey, PA: IGI Global.

Ahlberg, M., Aanisma, P., & Dillon, P. (2005). Education for Sustainable Living: Integrating theory, practice, design, and development. *Scandinavian Journal of Educational Research, 49*(2), 167–186. doi:10.1080/00313830500048923

Ahlberg, M., & Ahoranta, V. (2002). Two improved educational theory based tools to monitor and promote quality of geographical education and learning. *International Research in Geographical and Environmental Education, 11*(2), 119–137. doi:10.1080/10382040208667475

Amorim, R. R., Lama, M., Sánchez, E., Riera, A., & Vila, X. A. (2006). A Learning design ontology based on the IMS specification. *Journal of Educational Technology & Society, 9*(1), 38–57. Retrieved from http://www.ifets.info/journals/9_1/5.pdf.

Anderson, L. W., & Krathwohl, D. R. (Eds.). (2001). *A taxonomy for learning, teaching and assessing: A revision of Bloom's Taxonomy of educational outcomes, Complete edition.* New York: Longman.

Bailey, C., Zalfan, M. T., Davis, H. C., Fill, K., & Conole, G. (2006). Panning for gold: Designing pedagogically inspired learning nuggets. *Journal of Educational Technology & Society, 9*(1), 113–122. Retrieved from http://eprints.soton.ac.uk/19642/.

Brouns, F., Koper, R., Manderveld, J., Van Bruggen, J., Sloep, P., & Van Rosmalen, P. (2005). A first exploration of an inductive analysis approach for detecting learning design patterns. *Journal of Interactive Media in Education, 3*, 1–10. Retrieved from http://jime.open.ac.uk/2005/03.

Capuano, N., De Santo, M., Marsella, M., Molinara, M., & Salerno, S. (2000, August). A multi-agent architecture for intelligent tutoring. In *Proceedings of the International Conference on Advances in Infrastructure for Electronic Business, Science, and Education on the Internet,* L'Aquila, Italy. Retrieved from http://www.capuano.biz/Papers/SSGRR_2000_P1.pdf

Conole, G. (2007). Describing learning activities: Tools and resources to guide practice. In Beetham, H., & Sharpe, R. (Eds.), *Rethinking pedagogy for a digital age* (pp. 81–91). Oxford, UK: Routledge Falmer.

Conole, G. (2008). Capturing practice, the role of mediating artefacts in learning design. In Lockyer, L., Bennet, S., Agostinho, S., & Harper, B. (Eds.), *Handbook of research on learning design and learning objects: Issues, applications and technologies* (pp. 187–207). Hershey, PA: IGI Global.

DialogPlus Project. (2008). *Taxonomy of learning activities*. Retrieved from http://www.jisc.ac.uk/uploaded_documents/DialogPlus%20taxonomy.doc

Dillon, P., & Ahlberg, M. (2006). Integrativism as a theoretical and organisational framework for e-learning and practitioner research. *Technology, Pedagogy and Education, 15*(1), 7–30. doi:10.1080/14759390500435747

European Commission Directorate General for Energy & Transport. (2006). *Education on energy: Teaching tomorrow's energy consumers*. Brussels, Belgium: European Communities. Retrieved from http://ec.europa.eu/energy/action_plan_energy_efficiency/doc/education_en.pdf

European Environment Agency (EEA). (1999). *Environmental indicators: Typology and overview* (Tech. Rep. No. 25). Copenhagen, Denmark: Author.

European Environment Agency (EEA). (2002). *Energy and environment in the European Union (Environmental Issue Rep. No. 31)*. Copenhagen, Denmark: Author.

European Environment Agency (EEA). (2008). *Indicators*. Retrieved from http://www.eea.europa.eu/themes/energy/indicators

Falconer, I., Beetham, H., Oliver, R., Lockyer, L., & Littlejohn, A. (2007). *Mod4L final report: Representing learning designs*. Retrieved from http://www.academy.gcal.ac.uk/mod4l/

Griffiths, D. (2005). *From primitives to patterns: a discussion paper, UNFOLD Project*. Retrieved from http://www.unfoldproject.net/ providers_folder/papers/ terms/ primitives-patterns.pdf

Griffiths, D., & Blat, J. (2005). The role of teachers in editing and authoring units of learning using IMS Learning Design. *Advanced Technology for Learning, 2*(4), 1–9.

IMS Global Learning Consortium. (2003). *IMS Learning Design specification ver. 1.0*. Retrieved from http://www.imsglobal.org/learningdesign/index.cfm

Institute of Electrical and Electronics Engineers Learning Technology Standards Committee. (2002). *IEEE standard for Learning Object Metadata*. Retrieved from http://ieeeltsc.org/wg12LOM/lomDescription

International Atomic Energy Agency (IAEA). United Nations Department of Economic and Social Affairs, International Energy Agency (IEA), Eurostat, & European Environment Agency (EEA). (2005). *Energy indicators for sustainable development: Guidelines and methodologies*. Vienna, Austria: IAEA.

International Atomic Energy Agency (IAEA) & International Energy Agency. (IEA). (2001). *Indicators for sustainable energy development*. New York: United Nations.

Kelley, J. F. (1984). An iterative design methodology for user-friendly natural language information applications. *ACM Transactions on Office InformationSystems, 2*(1), 26–41. doi:10.1145/357417.357420

Knight, C., Gašević, D., & Richards, G. (2006). An ontology-based framework for bridging learning design and learning content. *Journal of Educational Technology & Society, 9*(1), 23–37. Retrieved from http://www.ifets.info/journals/9_1/4.pdf.

Kotis, K., & Vouros, G. (2006). Human-centered ontology engineering: The HCOME methodology. *International Journal of Knowledge and Information Systems, 10*(1), 109–131. doi:10.1007/s10115-005-0227-4

Kumazawa, T., Saito, O., Kozaki, K., Matsui, T., & Mizoguchi, R. (2009, March). *Knowledge structuring tool for sustainability science based on ontology engineering.* Paper presented at the International Conference on Complex, Intelligent and Software Intensive Systems, Fukuoka, Japan.

Leclerc, D., & Poumay, M. (2005). *The 8 learning events model*. Liege, Belgium: University of Liege, LabSET. Retrieved from http://www.labset.net/media/prod/8LEM.pdf

McAndrew, P., & Goodyear, P. (2007). Representing practitioner experiences through learning design and patterns. In Beetham, H., & Sharpe, R. (Eds.), *Rethinking pedagogy for a digital age* (pp. 92–102). Oxon, UK: Routledge.

McKeown, R. (2002). *Education for sustainable development toolkit*. Knoxville, TN: University of Tennessee, Energy, Environment and Resources Center.

Mohan, P., & Brooks, C. (2003). Learning objects on the semantic web. In *Proceedings of the 3rd IEEE international conference of Advanced Learning Technologies* (pp. 195-199). Washington, DC: IEEE Computer Society.

Mohan, P., Greer, J., & McCalla, G. (2003). Instructional planning with learning objects. In P. Baumgartner, P.A. Cairns, M. Kohlhase, & E. Melis (Eds.), *Proceedings of the IJCAI 2003 Workshop on Knowledge Representation and Automated Reasoning for E-learning Systems* (pp. 52-58). Koblenz-Landau, Germany: University of Koblenz-Landau.

Nejdl, W., Wolf, B., Qu, C., Decker, S., Sintek, M., Naeve, A., et al. (2002, May). *EDUTELLA: a P2P networking infrastructure based on RDF*. Paper presented at the 11th International Conference on World Wide Web, Honolulu, HI.

Papadimitriou, V. (2004). Prospective primary teachers' understanding of climate change, greenhouse effect, and ozone layer depletion. *Journal of Science Education and Technology*, *13*(2), 299–307. doi:10.1023/B:JOST.0000031268.72848.6d

Paquette, G. (2007). An ontology and a software framework for competency modelling and management. *Educational Technology and Society: Special Issue on Advanced Technologies for Life-long Learning*, *10*(3), 1-21. Retrieved from www.ifets.info/journals/10_3/ets_10_3.pdf

Pasin, M. (2005). *Semantic learning narratives* (Tech. Rep. KMI-05-07). London: The Open University, Knowledge Media Institute.

Pawlowski, J. M., & Bick, M. (2006). Managing & re-using didactical expertise: The didactical object model. *Journal of Educational Technology & Society*, *9*(1), 84–96.

Pernin, J. P., & Lejeune, A. (2006, June). *Models for the re-use of learning scenarios.* Paper presented at the IFIP conference, Ålesund, Norway. Retrieved from http://dspace.learningnetworks.org/bitstream/1820/580/1/Models.pdf

Robinson, J. (2004). Squaring the circle? Some thoughts on the idea of sustainable development. *Ecological Economics*, *48*, 369–384. doi:10.1016/j.ecolecon.2003.10.017

Rogner, H.-H., & Popescu, A. (2000). An introduction to energy. In *World Energy Assessment: Energy and the Challenge of Sustainability* (pp. 31-37). New York: UNDP.

Sharp, H., Manns, M. L., & Eckstein, J. (2000). The Pedagogical Patterns Project. In *Addendum to the Proceedings of the Conference on Object-oriented Programming, Systems, Languages, and Applications* (pp. 139-140).

Skanavis, C., Petreniti, V., & Giannopoulou, K. (2004, August). Educators and environmental education in Greece. In *Proceedings of the International Conference for the Protection and Restoration of the Environment VII*, Mykonos, Greece. Retrieved from http://www.srcosmos.gr/srcosmos/showpub.aspx?aa=4353

Spiropoulou, D., Antonakaki, T., Kontaxaki, S., & Bouras, S. (2007). Primary teachers' literacy and attitudes on education for sustainable development. *Journal of Science Education and Technology*, *16*, 443–450. doi:10.1007/s10956-007-9061-7

TenCompetence Project. (2006). *M6.1- Templates of the pedagogical models to be used in authoring environment: Model that combines classical and new forms of assessment*. Retrieved from http://dspace.learningnetworks.org/bitstream/1820/876/4/M6-1.pdf

Ullrich, C. (2004, November). Description of an instructional ontology and its application in web services for education. In *Proceedings of the Workshop on Applications of Semantic Web Technologies for E-learning, SW-EL'04,* Hiroshima, Japan. Retrieved from http://www.activemath.org/pubs/Ullrich-InstructionalOntology-SWEL-2004.pdf

Vega-Gorgojo, G., Bote-Lorenzo, M. L., Gómez-Sánchez, E., Asensio-Pérez, J. I., Dimitriadis, Y., & Jorrín-Abellán, I. M. (2008). Ontoolcole: Supporting educators in the semantic search of CSCL tools. *Journal of Universal Computer Science, 14*(1), 27–58. Retrieved from http://www.jucs.org/jucs_14_1/ontoolcole_supporting_educators_in/jucs_14_1_0027_0058_vega.pdf.

World Economic Forum. Global Leaders for Tomorrow Environment Task Force, Yale Center for Environmental Law and Policy, Yale University, Center for International Earth Science Information Network, & Columbia University. (2001). *2001 Environmental Sustainability Index*. New Haven, CT: Yale Center.

Zhang, J., & Fujiwara, A. (2007). *Development of the DPSIR+C framework for measuring the social capacity of environmental management*. Retrieved from.http://home.hiroshima-u.ac.jp/~hicec/coe/products/DP2007/DP2007-4.pdf

This work was previously published in International Journal of Agricultural and Environmental Information Systems, Volume 2, Issue 1, edited by Petraq Papajorgji and François Pinet, pp. 40-62, copyright 2010 by IGI Publishing (an imprint of IGI Global).

Section 3
GIS-Based Systems and Spatial Analyses

Chapter 8
Application of Support Vector Machines to Melissopalynological Data for Honey Classification

Giovanna Aronne
University of Naples Federico II, Italy

Veronica De Micco
University of Naples Federico II, Italy

Mario R. Guarracino
Italian National Research Council, Italy

ABSTRACT

In this paper, the authors address the problem of the discrimination of geographical origin and the selection of marker species of honeys using Support Vector Machines and z-scores. The methodology is based on the elaboration of palynological data with statistical learning methodologies. This innovative solution provides a simple yet powerful tool to detect the origin of honey samples. In case of honeys from Sorrento Peninsula, the discrimination from other Italian honeys is obtained with high accuracy.

1. INTRODUCTION

In recent years, an increasing attention has been devoted in Europe to the determination of the botanical and geographical origin of honey. This is partially due to allow European product to compete more effectively with cheap honey appeared on the EU market. To cover the high production costs, honey should be sold at much higher prices than those accepted by international markets. High quality standard, traceability and link to the geographical origin by means of strict regulation, such as Protected Designation of Origin (PDO), can give extra value to European honey. These laws protect a regional food, vegetable or fruit defining specific and objective procedures to determine the quality standard of the product and to verify the link to its geographical origin.

DOI: 10.4018/978-1-4666-0333-2.ch008

At present, the procedure to verify geographical origin of honey is not well established; some attempts have been done to overcome this situation, and a promising field of research relies on the application of statistical methods on data about pollen content of honeys (Aronne, 2010).

Floral honey always contains numerous pollen grains mainly from the plant species foraged by honey bees. Pollen analysis of honey, namely melissopalynology, is of great utility to determine and control both botanical and geographical origin. The determination of geographical origin is based on the entire pollen spectrum being consistent with the flora of a particular region and with any reference spectra or descriptions in the literature (Louveaux, 1978; Herrero, 2001; Persano Oddo, 2004; Persano Oddo, 2007; Aronne, 2010).

Analysis of the geographical origin of honey is based on the assumption that the selection of honey species made by bees is influenced by the peculiarity of local vegetation. Therefore, the palynological component of a honey, if correctly analysed, should provide information on the foraging area. This research topic assumes considerable importance when its aim is to safeguard consumer interests and to protect honest producers of honeys labelled with the indication of geographical origin (Aronne, 2008).

Although some computer-aided methods for the classification of honeys have been developed (Battesti, 1992; Scala, 2004a; Scala, 2004b; Aronne, 2008a; Aronne, 2008b; Aronne, 2008c; Aronne, 2010) at the moment the identification of geographical origin depends mainly on the experience and the knowledge of the palynologists who are asked to compare results from specific samples with a hypothetical pollen spectrum from honey producible in the same geographical area. It is therefore evident that elaboration of melissopalynological data requires precise, sensitive analytical tools which go beyond the subjective evaluation, providing the means to correlate data and information which are otherwise elusive.

Starting from the assumption that palynological data contain valuable information on the vegetation characteristics of the area in which bees have foraged, we used analysis tools and techniques that have been successfully applied to other fields, such as genetics, agriculture, economics and computer science (Guarracino, 2008; Mucherino, 2009). We believe that *data mining* techniques and concepts from *statistical learning theory* could provide the methodology enabling analysis of the pollen content of honeys. This analysis is determined by data models, their analysis, implementation and the use of specific algorithms.

In this paper we report the application of these tools to melissopalynological data; our specific purpose was to test and define a new methodological proposal to trace the geographical origin of honeys. To this extent, we used the results of palynological analysis of chestnut honeys (honey produced from the nectar of *Castanea sativa* Mill.) from the Sorrento Peninsula (Southern Italy) and from other areas in Italy. According to Von der Ohe et al. (2004), a honey can be defined "chestnut honey" if at least 86% of its pollen grains are from *Castanea sativa*. As a consequence, in this kind of honey only the remaining 14% of the pollen component is responsible for differences between samples and it is quite difficult to evaluate the geographical origin of the samples. The aim of our work was to find a mathematical model able to distinguish the chestnut honeys produced in the Sorrento Peninsula from those produced elsewhere. In the following sections, starting from initial statistical evaluation of the data, we report a synthesis of the working phases. In next section, data used for experiments are described. In section 3, data preparation is detailed. In section 4, methods used for classification are given. In sections 5 and 6, results regarding classification and variable selection are discussed. Finally, in section 7, conclusions are drawn.

2. DATA EXPLORATION

Melissopalynological analyses were performed according to (Louveaux, 1979) and Von der Ohe et al. (2004). All samples belonged to the category "chestnut honey". For each sample, every different morphologic pollen type was listed. Consequently, the number and the categories of pollen classes were unequal among the different honeys. However, within every sample the relative quantity of each pollen type was expressed as percentage. As consistent data representation is required for data analysis, data needed to be expressed simply and efficiently. A straightforward intuitive way was to use a table, in which each row corresponds to what is called an *observation* and each column to a *characteristic*. Each observation represents a sample of honey uniquely identified by the place of origin and the year; each column is associated to a pollen type and contains its concentration, relative to each observation, expressed as a percentage.

The data set we analyzed, which we will call the *Chestnut* dataset below, consists of 138 observations (honey samples) with 135 characteristics (pollen types). The set is heterogeneous in that it contains chestnut honey samples from 14 different geographical areas, each identified by a label, a positive number, which denotes its class. Data distribution of points into classes is not homogeneous, as shown in Figure 1. Honey samples from the Sorrento Peninsula were collected from hives thoroughly distributed on the territory to be representative of that geographical area. The other samples came from the main areas of production of chestnut honey in Italy.

3. DATA PREPARATION

Our data were collected from experiments which depend intrinsically on behaviour and environmental phenomena that can be neither controlled nor reproduced. This might have introduced errors and distortions into the data. Some observations could therefore be affected by measurement error during sampling, which might produce observations that are unimportant, or worse, misleading. Such *outliers* have out-of-scale values and do not possess the general behaviour of their dataset. They represent a limit situation, and are useless in statistical terms. To discover such outliers the z-index technique was used.

The search for the outliers can be effected either on a subset of the original dataset, or on the complete dataset. In the former case, a local strategy is used, considering each subset as a distinct set, whereas in the latter case a global strategy is used. Splitting the original dataset into subsets can be useful in cases in which the dataset is heterogeneous, as in the case of *Chestnut*, where the samples can be subdivided with respect to their origin. Hence, in this study, outlier search was conducted with a local strategy, based on geographical areas of production, in order to detect unrepresentative samples for each area. As a result of the analysis, a single outlier was detected in the Sorrento Peninsula dataset, while a single highly suspect observation was spotted in data from Avellino. This means that all samples were statistically valid; we therefore decided to use the whole dataset.

4. CLASSIFICATION

To classify a chestnut honey as produced in the Sorrento Peninsula and not from other geographical areas, general and adaptable criteria are needed. In other words, the diagnostic method must be able to compare any chestnut honey sample with a database of chestnut honeys from the Sorrento Peninsula and to establish whether the characteristics of that sample are consistent with those from the Peninsula database. To maintain the high level of reliability of this analysis it is important to use a mathematical method able to update the dataset at any time. For example, it could include yearly results of the new melissopalynological analysis

Figure 1. Frequency distribution of samples according to provenance

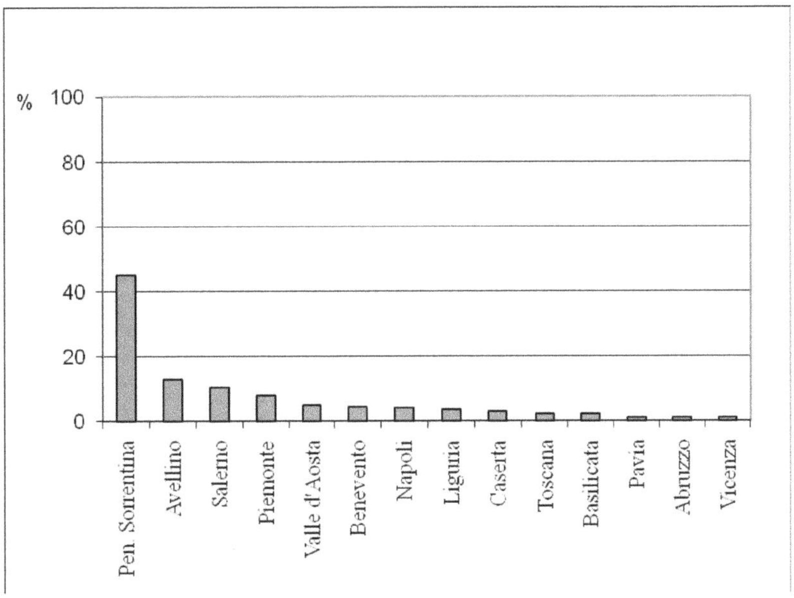

of honeys from that area. By doing so, it would be possible to include in the evaluation the occurrence of quality changes such as those due to the seasonal variation of the flowering period of the honey species or to land use changes. To achieve these aims, various analytical techniques may be used. Each technique has a different discriminatory capacity and range of action.

Below we report a few methods used in this case study. Each was necessary to corroborate the hypothesis to be used for the subsequent steps in the study. Our aim was to find a criterion that could predict the class of each sample, using only the data and the available information from the area of origin. The association of a label to a sample is commonly referred to as *classification*. Automatic classification methods, generally known as *machine learning methods*, are numerous and each has its own peculiarity. We addressed only those based on statistical learning theory because of their strong mathematical background. They belong to the set of *supervised methods*, which refers to the capability of a method to learn from a set of examples. The examples are pairs (*sample, label*) and are used to define a model capable of classifying other samples for which the label is unknown. The classification process consists of two phases: a learning phase and a prediction phase. Among the most widely used classification methods, we recall decision rules and trees, Bayesian methods, neural networks, logistic regression and *support vector machines*.

The classification methods have to numerically detect the discrimination existing in the *Chestnut* dataset between honeys produced in the Sorrento Peninsula and those produced elsewhere.

In the scientific literature, a wide array of algorithmic solutions is available, and for many of them, implementations are freely available. LibSVM (Chang, 2001) is one public domain (GPL license) software library that implements a classification algorithm based on support vector machines (SVMs). SVMs are state-of-the-art classification models based on the determination of a separation plane between two classes of points (Vapnik, 1998). Given two linearly separable sets of points, represented by the rows of two matrices $A \in R^{mxn}$ and $B \in R^{pxn}$ the method finds two parallel hyperplanes $x'\mu - \beta = \pm 1$, where μ is the vector of hyperplane coefficients and $\beta \pm 1$ are the intercepts, such that they leave all points of the two classes from different sides and their distance is maximum. Maximizing the distance is equivalent to solving:

$$\min_{\mu \neq 0} \frac{\|\mu\|^2}{2}$$
$$s.t. \quad A\mu - e\beta \leq 0$$
$$B\mu - e\beta > 0$$

with e unit vector of appropriate dimension. This discriminating plane $x'\mu - \beta = 0$ is defined as the one in the middle of the two planes that maximizes the distance between the two classes. The resulting hyperplane is univocally determined by a subset of points called *support vectors*, as depicted in Figure 2. An unlabelled point belongs to the class that is in the same halfspace with respect to the plane. Data from Chestnut honeys were classified by SVM.

5. RESULTS VALIDATION

The reliability of results obtained with the application of the above method was measured through a validation process. This process uses a criterion to obtain correct and consistent classification results. In *k-fold* cross validation, data are divided into *k* folds of dimension *n/k*. At each step the algorithm is trained on *k-1* folds and the rest is used for the test. Comparison between computed labels and real labels gives the prediction accuracy. The mean accuracy over the *k* steps is therefore the percentage of correctly classified samples over the total number of samples. This method is widely used in assessing the prediction capability of classification algorithms.

In *Chestnut,* data were divided using k=10 and the process was iterated 10 times, each time with a different permutation of the samples, providing 100 train and test sets. For each of the 100 repetitions, train and test sets are disjoint, so that the results are not biased by samples used for both training and testing.

Table 1 shows accuracy results of LibSVM with 10-fold cross validation for the *Chestnut* dataset; in this case, the samples from the Sorrento Peninsula were classified against samples from other areas, taken one at a time. Thus the problem was reduced to a binary classification question with only two classes.

High accuracy values show that honey samples from different production areas may be classified with an error less than, or equal to, 10%. To better assess the prediction ability of the method, *leave one out cross validation* (LOOCV) was also used. In this case, k=1 and the algorithm was trained on all samples except one, which was used for the test. The validation process was repeated for each element in the dataset. This was done to understand how a single sample is described by all the others. The results are shown in Table 2.

The tests were repeated using only two classes, as described above. Our results show that

Figure 2. Discriminating hyperplane determined by SVM

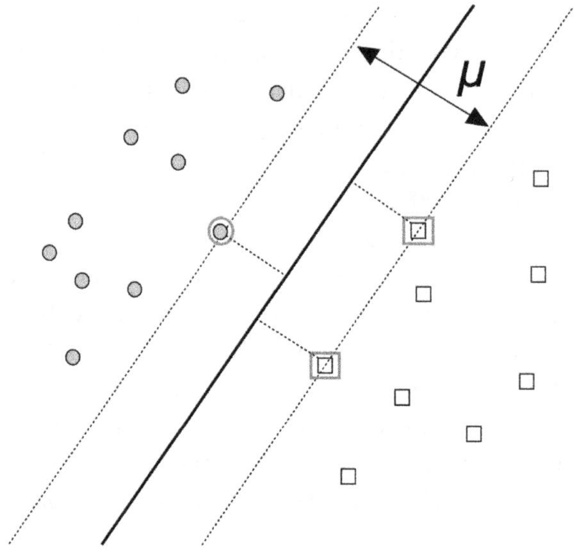

Table 1. Accuracy results of the classification (SVM-k-fold) of chestnut honeys of the Peninsula against samples from other areas

Accuracy			Class for comparison
Min	Max	Mean	
62,50%	100,00%	89,50%	Avellino
37,50%	100,00%	80,50%	Salerno
62,50%	100,00%	92,75%	Piemonte
85,71%	100,00%	98,71%	Valle d'Aosta
57,14%	100,00%	91,14%	Benevento
71,43%	100,00%	89,00%	Napoli
57,14%	100,00%	94,14%	Liguria
71,43%	100,00%	94,14%	Caserta
85,71%	100,00%	97,29%	Basilicata
71,43%	100,00%	95,14%	Toscana
85,71%	100,00%	98,86%	Vicenza
85,71%	100,00%	98,57%	Abruzzo
85,71%	100,00%	97,86%	Pavia

a single sample can be correctly classified with a mean accuracy of over 95%, in most cases.

To the best of our knowledge, this is the first example of the classification methods, based on *support vector machines*, being applied to melissopalynological analysis. Overall evaluation of the results confirms the real possibility of discriminating, from other honeys, those produced in the Sorrento Peninsula with an objective, repeatable method.

6. IDENTIFICATION OF MARKER SPECIES

The last part of the validation was devoted to the identification of species that discriminate among the different areas. This is usually called *feature* or *variable selection*, and many methods are available. Nevertheless, no method is widely accepted as a *de-facto* standard, and therefore we decided to use a well know statistical methodology to devise a variable selection technique. To this extent, we devised a method that we call *z-index Classifier* (ZiC). This index can be interpreted, within a set of samples as a similarity index. Indeed, given a homogeneous set of samples, all belonging to the same area of honey production and hence with the same label, we can compute for each pollen type a z-index (Larsen, 2005). This is defined, for each species as:

$$z_{index} = \frac{x - v}{\sigma}$$

where v is the mean and σ the standard deviation of the selected species.

Taking the maximum value for each type (matrix column), we obtain a set of values, called γ, that can be used to compare different sets.

Taking a set of samples, all with the same label, differing from the previous one, we again compute the z-index using mean and variance from the first set and we compare these z-indexes with those in γ. We can therefore assess to what extent samples belonging to different classes actually

Table 2. Accuracy results of the classification (SVM-LOOCV) of chestnut honeys of the Peninsula against samples from other areas

Class for comparison	Samples correctly classified	Samples not correctly classified	Accuracy
Avellino	72	8	90,00%
Salerno	63	14	81,82%
Piemonte	68	5	93,15%
Valle d'Aosta	68	1	98,55%
Benevento	62	6	91,18%
Napoli	60	7	89,55%
Liguria	63	3	95,45%
Caserta	60	5	92,31%
Basilicata	62	2	96,87%
Toscana	61	3	95,31%
Vicenza	62	1	98,41%
Abruzzo	62	1	98,41%
Pavia	62	1	98,41%

differ. Whenever on one pollen type the value of the second computed z-index is greater than the homologue in γ, the pollen of these samples differs from the other samples of the initial set. Not only does this analysis provide a quantitative criterion capable of discriminating among samples of different classes, but it also gives a qualitative result that can be used to precisely discriminate the pollen differences among honeys from different production areas.

ZiC is applied to samples from only one class. The second set is used for prediction. In this case study, the honeys from the Sorrento Peninsula were used for training, whereas all the others were used for testing. The aim was to find discriminating pollen types between the samples used for training and those used for testing.

The results were obtained considering 10 runs each consisting of 100 permutated subsets, using on each occasion an increasing number of samples for training. It is shown how and which pollen types discriminate a Sorrento Peninsula honey from the others with an accuracy that is over 98%, with a value for an increasing number of samples that is never below 90%. It is also possible to compute the discriminating factor of each of the pollens of the training set. In the *Chestnut* dataset we found a number of pollen types that determine the distinction among honeys. In Table 3 we report the species for which the ratio between the mean number of differences on a single pollen, and the total number of positive occurrences in the test set is 100%. Finally, we observed that some pollen types (*Bryonia, Carex, Cucumis, Potentilla f., Rosmarinus* and *Ziziphus*) can be found only in samples from the Sorrento Peninsula.

The application these methods of *data mining* and concepts from *statistical learning theory* to melissopalynological data from the Sorrento Peninsula showed that it is possible to establish an objective and verifiable link between the honey and its geographical area of origin. To use these methods it is necessary to have an extensive melissopalynological dataset of the different areas.

Table 3. Pollen types useful to discriminate honeys of the Sorrento Peninsula from others of different provenances

Ajuga	Genista	Populus
Albizzia	Geum	Poterium
Alnus	Gleditsia	Pyrus f.
Anemone	Hedera	Ranunculus
Artemisia	Helianthemum	Rhinanthus
Buddleja	Hippocrepis	Rhododendron
Buplerum	Hypericum	Rosa f.
Calluna	Iridaceae	Sambucus nigra
Campanula	Knautia	Saxifraga
Carpinus	Labiatae M	Sedum
Cercis	Lavandula angustifolia	Sempervivum
Cistus	Linaria	Scabiosa f.
Coronilla	Loranthus	Scrophulariaceae
Corylus	Lytrum	Smilax
Cotoneaster	Medicago	Sorbus
Crataegus	Melilotus	Spiraea
Cucurbuta pepo	Mercurialis	Stachis
Cytinus hypocistis	Myrtus	Syderitis
Dorycnium	Morus	Syringa
Euchera	Nigella	Teucrium
Filipendula	Onobrychis	Trixis
Fragaria	Phacelia	Vaccinium
Galega	Pistacia	Verbascum
Geleopsis	Polygonum	Vitis

7. CONCLUSION

Defining the botanical origin of any honey produced and producible in a specific area is undoubtedly a necessary step to improve honey production in that region. However, it is worth defining objective methods to certify not only the botanical but also the geographical origin of the honeys. The identification of geographical origin of honeys is still mainly based on the experience and the knowledge of the palynologists who are asked to compare the pollen spectrum from specific samples with a hypothetical one from honey producible in the same geographical area. Such a comparison is not based on any statistical method; it suffers from the subjectivity of the operator and is not always reliable especially when hundreds of samples have to be considered. Our analysis demonstrated that data mining techniques and concepts from machine learning are appropriate tools to discriminate between honeys collected in the Sorrento Peninsula and those of different origin with an accuracy over 98%. Thus, they represent promising and objectively verifiable methods for geographical origin traceability which surely go beyond the existing lack of other objective tools.

REFERENCES

Aronne, G., & Buonanno, M. (2008). The need to improve bee production in the European Union and Italy. In G. Aronne, M. Buonanno, & V. De Micco (Eds.), *Improving honey production and marketing: new approaches to botanical and geographical characterization* (pp. 9-13). SBR. ISBN 88-901941-3-8

Aronne, G., Buonanno, M., & De Micco, V. (2008). Identification and characterisation of produceable honeys in a geographical area. In G. Aronne, G., Buonanno, M., & De Micco, V. (Eds.). (2008). Improving honey production and marketing: new approaches to botanical and geographical characterization (pp. 32-46). ISBN 88-901941-3-8

Aronne, G., & De Micco, V. (2008). The use of multivariate analysis for honey characterisation. In G. Aronne, M. Buonanno, & V. De Micco (Eds.), *Improving honey production and marketing: new approaches to botanical and geographical characterization* (pp. 24-31). ISBN 88-901941-3-8

Aronne, G., & De Micco, V. (2010). (in press). Traditional melissopalynology integrated by multivariate analysis and sampling methods to improve botanical and geographical characterization of honeys. *Plant Biosystems*.

Aronne, G., De Micco, V., & Buonanno, M. (2008). Analysis of honey variability in different years. In G. Aronne, M. Buonanno, & V. De Micco (Eds.), *Improving honey production and marketing: new approaches to botanical and geographical characterization* (pp. 47-52). ISBN 88-901941-3-8

Battesti, M. J., & Goeury, C. (1992). Efficacité de l'analyse melissopalynologie quantitative pour la certification des origines géographiques et botaniques des miels: le modèle des miels corses. *Review of Palaeobotany and Palynology*, *75*, 77–102. doi:10.1016/0034-6667(92)90151-6

Chang, C. C., & Lin, C. J. (2001). *LIBSVM: a library for support vector machines*. Retrieved November 23, 2009, from http://www.csie.ntu.edu.tw/~cjlin/libsvm

Guarracino, M. R., Cuciniello, S., Feminiano, D., Toraldo, G., & Pardalos, P. M. (2008). Current Classification Algorithms for Biomedical Applications. In *Proceedings of the Centre de Recherches Mathématiques CRM.*

Herrero, B., & Valencia-Barrera, R. M. (2001). Palynological analysis of honeys from Palencia Province (Spain). *Acta Botanica Croatica*, *60*(1), 11–24.

Larsen, R. J., & Marx, M. L. (2005). *An Introduction to Mathematical Statistics and Its Applications*. Upper Saddle River, NJ: Perason.

Louveaux, J., Maurizio, A., & Vorwohl, G. (1978). Methods of Melissopalynology. *Bee World*, *59*, 139–157.

Mucherino, A., Papajorgji, P. J., & Pardalos, P. M. (2009). *Data Mining in Agriculture*. New York: Springer.

Ohe, W. V. D., Oddo, L. P., Piana, M. L., Morlot, M., & Martin, P. (2004). Harmonized methods of melissopalynology . *Apidologie*, *35*, 18–25. doi:10.1051/apido:2004050

Persano Oddo, L., & Bogdanov, S. (2004). Determination of botanical origin: problems and issues. *Apidologie*, *35*(1), S2–S3. doi:10.1051/apido:2004044

Persano-Oddo, L., Piana, M. L., & Ricciardelli D'Albore, G. (2007). *I mieli regionali italiani. Caratterizzazione melissopalinologica* (p. 139). Agnesotti, VT: Ministero delle Politiche Agricole Alimentari e Forestali e CRA.

Scala, M., Aronne, G., & De Micco, V. (2004). Variability in pollen spectra of chestnut honeys from different sited and years. *Polen*, *14*, 403.

Scala, M., De Micco, V., & Aronne, G. (2004). Multivariate analysis of the accompanying pollen for the determination of the geographical origin of honey. *Polen*, *14*, 109.

Vapnik, V. N. (1998). *Statistical learning theory*. New York: John Wiley & Sons.

This work was previously published in International Journal of Agricultural and Environmental Information Systems, Volume 1, Issue 2, edited by Petraq Papajorgji and François Pinet, pp. 85-94, copyright 2010 by IGI Publishing (an imprint of IGI Global).

Chapter 9
Urban Versus Rural:
The Decrease of Agricultural Areas and the Development of Urban Zones Analyzed with Spatial Statistics

Beniamino Murgante
University of Basilicata, Italy

Maria Danese
National Research Council, Italy

ABSTRACT

Until a few decades ago it was very easy to distinguish between city and country: in most cases the edge was defined by defensive barriers. In recent times, the relationships between urban and rural areas completely changed, placing the country in a subordinate position. Consequently, many terms have been coined in order to describe the new phenomena taking place between city and country. The term adopted, "periurban area", despite its large use, does not have a clear and unambiguous definition. Such various approaches are due to the complexity of the phenomenon to be analyzed and to the huge variety of territorial contexts in which it may reveal. The phenomenon is characterized by urban growth with soil consumption generating loss of competitiveness for agricultural activities. This paper defines more precise rules in order to describe the periurban phenomenon, using techniques of spatial statistic and point pattern analysis. This approach has been tested in the case of study of Potenza municipality. Interest in this area comes after the earthquake of 1980, when a large migration of inhabitants began towards the countryside around Potenza.

1. INTRODUCTION

Analyzing the main historical urban functions, Salsano (1998) considered town walls and the market as the first basic elements of the city. First the defensive functions, then the advantages of agglomeration principles and industrial development have led to a long migration process from rural areas to cities. In this period it was very easy to distinguish between city and country.

With the passing of time urban and rural concepts have undergone great changes. Land renting

DOI: 10.4018/978-1-4666-0333-2.ch009

and the expulsion of several typical urban functions outside the city increased pressure on rural areas. The city organized the countryside influencing socio-cultural, economical and functional aspects. Consequently demographic relationships between city and countryside have been changed, generating a reversal trend. New concepts are emerging: urban exodus, new conurbations, "rurbanization", periurban countryside (Charrier, 1994). Therefore, it is not easy to define in a clear and unambiguous way which is the sharp boundary between urban and rural areas (Murgante & Las Casas, 2004; Murgante et al., 2008). Ahlqvist and Ban (2007) developed an ontology considering the degree of urbanization going from rural through exurban, from suburban to urban.

In planning literature, the periurban phenomenon has been defined in different ways. Le Jeannic (1997) describes the population displacement as the need to escape from the dense city in order to have more space and a better environment. Also, growth of periurban belt is due to high costs of flats, the need of individual dwellings and land rents (Guerois & Pumain, 2001). Since 1980s it has been less and less possible to distinguish town from country, denying the concept of two separate entities which was for many years one of the cornerstones of spatial planning (Hidding et al., 2000; Van Den Berg & Wintjes, 2000). Rural areas are more urbanized and an uncontrolled growth of periurban belts has increased the number of inhabitants. At the same time, urban areas have lost resident population gaining population in transit (Alberti et al., 1994), because of the activities concentration in urban areas. All these situations produce a huge commuting phenomenon (Cavailhès et al., 2004). The main feature of this trend is a low density of urbanization which spreads in all directions (Camagni et al., 1998). Growth of these areas is strictly related to urban sprawl, generating negative repercussions on agricultural activities. A great amount of roads have been built to improve dwelling accessibility and car is the only means of transport (Camagni et al., 2002). This is an opposite trend compared to the period after the Second World War, when urban planners used statistical methods to give a dimension of the migratory flows towards towns. This tendency, strictly related to Urban Sprawl, is so complex to analyze, that classical statistics are not enough for a complete understanding of the phenomenon. Settlement location in zones surrounding urban areas takes into account environmental features, accessibility, agricultural losses of productivity. In order to achieve a more complete analysis it is important to analyze each phenomenon according to its spatial location, so that it is possible to consider the concentration of some events in some areas and their possible interactions. Geostatistics can be useful in order to study this problem with an innovative approach compared to the classic socio-economic techniques. This method allows an analysis which may determine the actual trend in one region. This technique has been applied in Potenza Municipality, where a migratory phenomenon began from urban to rural areas after a strong earthquake occurred in 1980. All the informative layers have been combined with a land suitability procedure in order to define a periurban fringe with a certain precision.

2. AN OVERVIEW OF SPATIAL STATISTICS TECHNIQUES

The main aim of spatial analysis is a better understanding of spatial phenomena aggregations and their spatial relationship. Spatial statistical analyses are techniques which use statistical methods in order to determine if data show the same behaviour of the statistical model. Data are treated as random variables. The *events* are spatial occurrences of the considered phenomenon, while *points* are each other arbitrary locations. Each event has a set of attributes describing the nature of the event. *Intensity* and *weight* are the most important attributes; the first is a measure identifying the event strength, the second is defined by the analyst who assigns a parameter in order to define if an event is more or less important according to some criteria. Spatial statistics techniques can be grouped in three main categories: *Point Pattern Analysis*, *Spatially Continuous Data Analysis* and *Area Data Analysis* (Bailey & Gatrell, 1995).

The first group, *Point Pattern Analysis,* considers the distribution of point data in the space. They can follow three different criteria:

- Random distribution: the position of each point is independent of the others points;
- Regular distribution: points have an uniform spatial distribution;
- Clustered distribution: points are concentrated in some building clusters.

The second group, *Spatially Continuous Data Analysis,* takes into account the spatial location and the attributes associated to points, which represent discrete measures of a continuous phenomenon.

The third group, *Area Data Analysis,* analyzes aggregated data which can vary continuously through space and can be represented as point locations.

Spatial analysis aims to identify quantitative and spatial relationships among variables, studying the presence of *spatial autocorrelation.* If some clusters are found in some regions and a positive spatial autocorrelation is verified during the analysis, it can describe an attraction among points. The case of negative spatial autocorrelation happens when deep differences exist in their properties, despite the closeness among events. If it is impossible to define clusters of the same property in some areas, a sort of repulsion occurs. Null autocorrelation arises when no effects are surveyed in locations and properties. Null autocorrelation can be defined as the case in which events have a random distribution over the study area (O'Sullivan & Unwin, 2002). Essentially, the autocorrelation concept is complementary to independence: events of a distribution can be independent if any kind of spatial relationship exists among them.

Spatial distribution can be affected by two factors (Gatrell et al., 1996):

- First order effect, when it depends on the number of events located in one region;
- Second order effect, when it depends on the interaction among events.

If these two definitions seem clearer, it isn't as clear as the recognition of these effects over the space.

2.1. Kernel Density

Kernel density is one of the point pattern analysis techniques, where input data are point themes and outputs are grids. While simple density computes the number of events included in a cell grid considering intensity as an attribute, kernel density takes into account a mobile three-dimensional surface which visits each point. The output grid classifies the Si event according to its distance from the point S and the number of events found inside the three-dimensional surface (Bailey & Gatrell, 1995). The influence function de-fines the influence of a point on its neighbourhood. The sum of the influence functions of each point can be calculated by means of the density function, defined by:

$$\lambda(L) = \sum_{i=1}^{n} \frac{1}{\tau^2} k\left(\frac{L - L_i}{\tau}\right), \quad [1]$$

where:

- λ is the distribution intensity of points, measured in L;
- L_i is the event i;
- K is the kernel function;
- τ is the bandwidth.

L_i, K and τ are the different factors that influence Kernel Density Estimation. L_i is the intensity of point pattern. It could be represented not only by the numerousness of events, but also in terms of strength of a single point.

Another important factor is kernel, characterized by unimodality, regularity, symmetry, finite moments of the first, second, etc. order (Breiman et al., 1977) and it is always not negative. Principal kernel types are uniform kernel, gaussian kernel (Burt & Burber, 1996), quartic kernel (Bailey & Gatrell, 1995), and Epanechnicov kernel (Epanechnikov, 1969). However, according to Silverman (1989) the choice of this parameter is less important than the bandwidth choice.

So the factor which more influences density values is bandwidth: if τ is too big the value of λ is closer to simple density; if τ is too small the surface does not capture the phenomenon (Jones et al., 1996). Approaches used to choose the bandwidth are two: the first one uses one fixed τ for the whole distribution; the second uses instead a variable value. Different methods used in literature are reviewed from Jones et al. (1996).

Finally, like in every grid analysis, a factor influencing density values is cell size. For cell choice, it is known that if the size is less than the bandwidth value divided by a factor between 2 and 5 times the bandwidth, this causes little effects on density estimation (O'Sullivan & Wong, 2007).

2.2. Straight Line Distance

Straight line distance is a function coming from map algebra techniques. It allows to measure the distance between each cell and the nearer source, where the distance calculation is based on the concept of Euclidean Distance (Tomlin, 1990; De Mers, 2002). The source can be in vector or grid format. In the case of grid format some cells will contain information about the source and some others will not, while in the case of a vector theme it will be necessary a previous transformation in grid before determining the distance.

The output of straight line distance is in grid format and the distance is measured between cells barycentre. Also, in this case it is important to estimate some factors such as the maximum distance within which one has to assess measures and sizes of cells.

2.3. Moran Index

Moran index I (Moran, 1948) is a global measure of spatial autocorrelation. This index takes into account the number of events occurring in a certain zone and their intensity. It is a measure of the first order property and can be defined by the following equation:

$$I = \frac{N\sum_i \sum_j w_{ij}(X_i - \bar{X})(X_j - \bar{X})}{(\sum_i \sum_j w_{ij})\sum_i (X_i - \bar{X})^2}, \qquad [2]$$

where:

- N is the number of events;
- X_i and X_j are intensity values in the points i and j (with i≠j), respectively;
- $\bar{X}$ is the average of variables;
- $\sum_i \sum_j w_{ij}(X_i - \bar{X})(X_j - \bar{X})$ is the covariance multiplied by an element of the weight matrix. If Xi and Xj are both upper or lower relative to the mean, this term will be positive, if the two terms are in opposite positions compared to the mean the product will be negative;
- w_{ij} is an element of the weight matrix which depends on the contiguity of events. This matrix is strictly connected to the adjacency matrix.

There are two methods to determine w_{ij},: the "Inverse Distance" and the "Fixed Distance Band". In the first method, weights vary in inverse relation to the distance among events:

$$w_{ij} = d_{ij}^{z},$$

where z is a number smaller then 0.

The second method defines a critical distance beyond which two events will never be adjacent. If the areas to which i and j belong are contiguous, w_{ij} will be equal to 1, otherwise w_{ij} will be equal to 0. *Moran index* I can have values included between -1 and 1.

If the term is high, autocorrelation is positive, otherwise it is negative. Moran index vanishes in very rare cases, but usually the convergence is towards the theoretical mean value E(I), where each value is independent from the others:

$$E(I) = -\frac{1}{N-1}.$$

If I < E(I), the autocorrelation is negative, if I > E(I) the autocorrelation is positive.

The significance of Moran index can be evaluated by means of a standardized variable z (I) defined as:

$$z(I) = -\frac{I - E(I)}{S_{E(I)}},$$

where $S_{E(I)}$ is the standards deviation from the theoretical mean value E(I).

2.4. Local Indicator of Spatial Association and the G Function by Getis and Ord

The Local Indicator of Spatial Association by Anselin (1995) and the G function by Getis and Ord (1992) are indicators at the local scale. Both LISA and G function take into account disaggregated measures of autocorrelation, considering the similitude or the difference of some zones. These indexes measure the number of events with homogenous features included within a distance *d*, located for each distribution event. This distance represents the extension within which clusters are produced for particularly high or low intensity values.

The Local Indicator of Spatial Association (Anselin, 1995) is defined as:

$$I_i = \frac{(X_i - \bar{X})}{S_X^{\,2}} \sum_{j=1}^{N} (w_{ij}(X_j - \bar{X})), \qquad [3]$$

where symbols are the same used in Moran's I, except for $S_X^{\,2}$ which is the variance.

The function by Getis and Ord (1992) is represented by the following equation:

$$G_i(d) = \frac{\sum_{i=1}^{n} w_i(d)\, x_i - \bar{x}_i \sum_{i=1}^{n} w_i(d)}{S(i)\sqrt{\left[(N-1)\sum_{i=1}^{n} w_i(d) - \left(\sum_{i=1}^{n} w_i(d)\right)^2\right] \Big/ N-2}}, \qquad [4]$$

which is very similar to Moran index, except for $w_{ij}(d)$ which, in this case, represents a weight which varies according to distance.

The use of the local indicator is not immediate, because there is the need to make explicit the dataset spatial structure, by defining: *1*) contiguity and distance rules and weight matrix; *2*) homogeneous classes of intensity.

With this aim we left data speak by themselves using a method that integrates the global and the local indicators of spatial association together. More in detail, Moran's I was iteratively used with different values of distance band, with the aim of finding the value maximizing I, that is the degree of cluster level in the distribution. This distance is then used in local indicators.

3. THE CASE STUDY

These techniques have been applied to Potenza municipality. This is located in the southern Apennines area of Italy and it is the chief town of Basilicata, which is a region with a very low residential density.

Its territory, about 18000 ha sized, lays on an area roughly 900 m a.s.l., characterized by a relevant presence of forests, by very steep slopes and by a large number of areas interested by landslides. Although the morphology of this area did not encourage the diffusion of isolated houses, existing houses in the extra urban territory increased by 25% during the 1980-1990 decade. The migration of people from urban to rural areas mainly happened after the earthquake in 1980 (Figure 1), when a huge number of buildings in the urban area were defined uninhabitable and the fear of new seismic events, mixed with the wish of running away from the most damaged places, induced citizens to choose new localizations for building new houses. These reasons coupled with social, economic and cultural aspects generated an inhabitant flow to the countryside around Potenza, so intense that it deeply altered the urban morphology and modified the traditional physical functional relationships between urban and rural areas.

This process caused a decrease of agricultural areas and a development of new settlements with a crown shape all around the town, with no respect for natural morphology and without considering the consequent resources consumption. This kind of uncontrolled expansion, for more than twenty years, generated the following phenomena:

- Diffuse territorial decline, with hydrogeologic phenomena due to anthropic uses (construction of small road infrastructures);
- Abandonment of agricultural soils related to expectations of real estate development, especially in areas close to the town;
- Diffusion of scattered settlements in rural areas, with inadequacy of facilities and infrastructures;
- Pauperization of environmental and landscape valuable components.

On this purpose, it can be useful to apply spatial statistic techniques to define criteria concerning suitability in locating new settlements, considering particular tendencies that some specific areas already manifest.

A first factor considered in this study is density. Periurban area is characterized by a spread of settlements with extensive features, compared to urban area, which has a greater density. Lower density is the first condition distinguishing periurban areas from urban ones.

Rural sites have a strong connection with agricultural activities and the relationship with the urban area is weak. It is also necessary to establish a lower threshold which can distinguish periurban areas from rural ones.

In order to calculate density, all the polygons representing buildings have been converted in points which are the events to take into account in point pattern analysis. The ratio between the number of flats and the number of buildings has been calculated from census data; this value, considered as the intensity of events, has been calculated for each census unit and has been assigned to each building falling inside each unit. Figure 2 compares the density of scattered settlements between 1987 and 2004 and it shows the huge growth of urban sprawl. In the case study, a value of bandwidth of 400 m and a cell size of the grid of 10 m have been used. This bandwidth has been chosen considering a distance between periurban fringe and urban areas that can be travelled on by walking. Consequently the bandwidth becomes representative of accessibility to urban area.

A first rough analysis of periurban fringe takes into account zones with a low density expansion including areas with values of kernel density included between 1 and 18 flats/ha (Figure 3).

Figure 1. Expansion of Potenza Municipality from 1883 to 2004

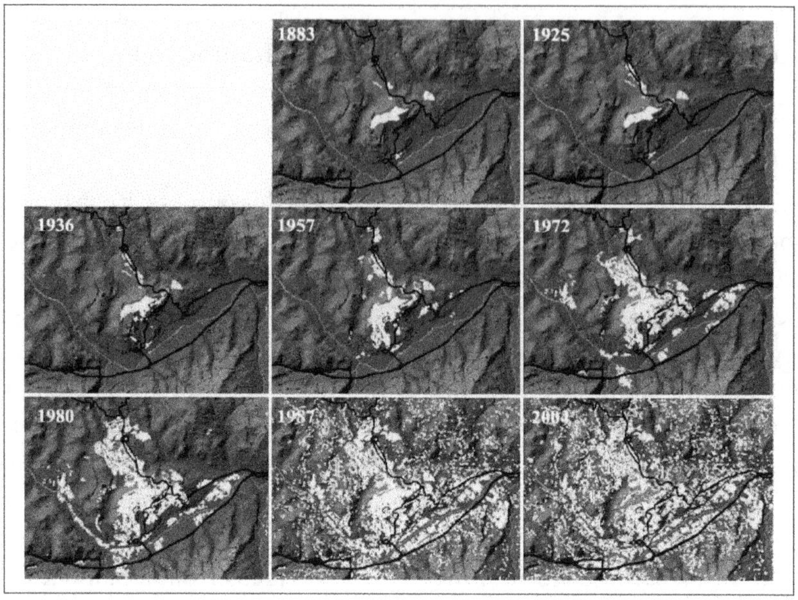

Orography and accessibility define the second factor, which consists of the distance from infrastructures because urban growth is more concentrated along the main line of road network. In order to locate areas with a good accessibility, distance from infrastructures has been defined so that it represents the tendency. Straight Line Distance identified areas with a distance from the main infrastructures within 200 meters. The third factor is the spatial autocorrelation which has been analyzed considering Moran Index, G function as developed by Getis and Ord (1992) and Local Indicator of Spatial Association (LISA).

In this case intensity of events is obtained as the ratio between number of inhabitants and number of buildings in each census zone. Moran index is able to specify if an event is clustered, scattered or with a random distribution. It has

Figure 2. Density of scattered settlements in 1987 and 2004 (flats/hectare)

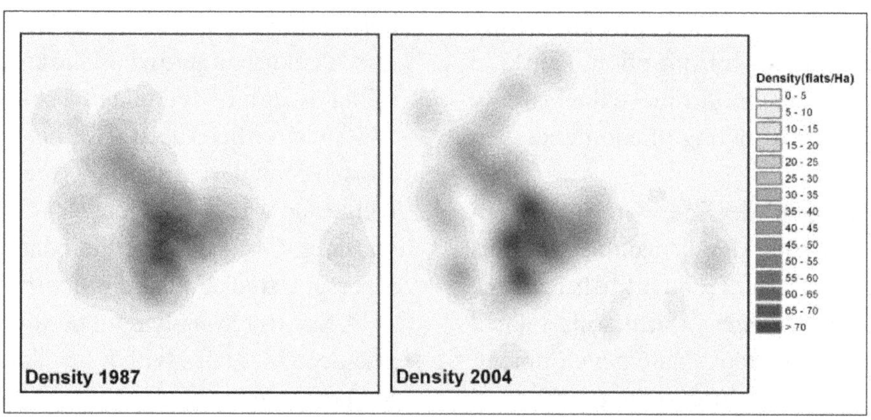

Figure 3. Areas which have Kernel density included between 1 and 18 flats/hectare

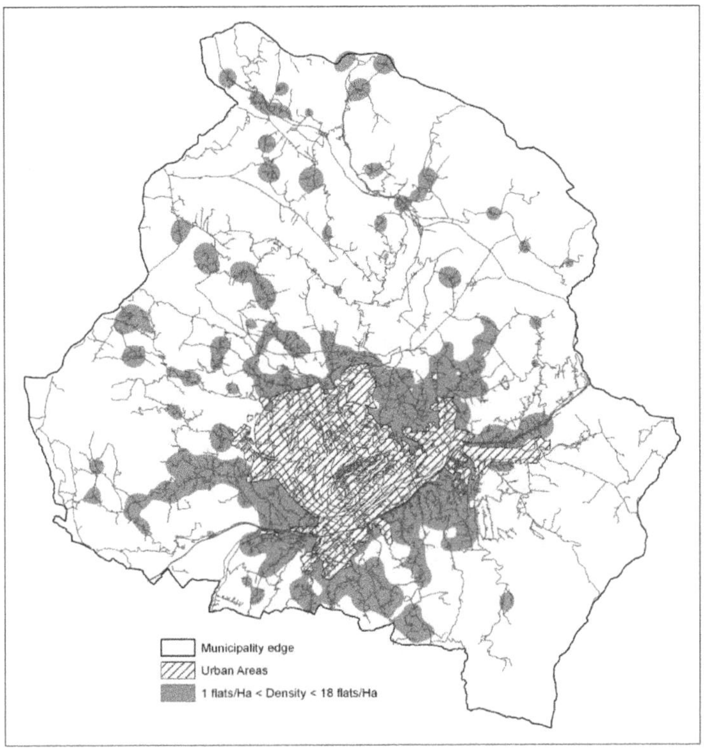

been calculated by means of the inverse distance method considering data in two different periods, 1987 and 2004, to evaluate the variation of scattered rate of settlements. The following values have been achieved:

- Moran Index at 1987: $I_{1987} = 0.0698$;
- Moran Index at 2004: $I_{2004} = 0.0722$.

These two indexes show a low autocorrelation in both cases, and the second one is higher than the first one. These data can be interpreted as growth of settlements concentrated in some particular zones.

The next step of our study was to calculate the contiguity belt considered as the area where the phenomenon grows homogenously and where it will intensify in the future. Moran index depends from the distances among points; it is possible to calculate a distance value which produces an index I with the maximum level of correlation among events, by maximizing the deviation z. This value has been calculated in 1600 m and it has been used as an input parameter in LISA and then in Getis and Ord (1992) function determining zones where events are autocorrelated.

A LISA positive value indicates a positive autocorrelation; obviously a negative autocorrelation corresponds to a negative value.

For the periurban fringe it is important to pay attention to the medium-low level of intensity, so the classes in Table 1 have been considered.

In Getis and Ord (1992) function, highest and lowest values of G mean highest and lowest values of phenomenon intensity.

The classes in Table 2 have been considered.

Figure 4 compares results, showing the similitude of areas with positive autocorrelation achieved with both indicators. In our study case Getis and Ord (1992) function fits the phenom-

Table 1. LISA classes

Class	Autocorrelation	LISA
no correlation	Negative autocorrelation	-106,9 ÷ 0
1 low	Positive autocorrelation among lower bounds	0 ÷ 14
2 low	Positive autocorrelation among low bounds	14 ÷ 28
3 low	Positive autocorrelation among medium-low bounds	28 ÷ 54
4 high	Positive autocorrelation among high bounds	54 ÷ 84.7

Table 2. Getis and Ord (1992) function classes

Class	Autocorrelation	Intensity value X_i (inhabitants/ buildings)	G'
no correlation	Negative autocorrelation	$X_i \leq 18$ $X_i \geq 18$	-1.3 ÷ - 2 -6.3 ÷ 1
1 low	Positive autocorrelation among lower bounds	$X_i \leq 18$	-1.3 ÷ - 2
2 low	Positive autocorrelation among low bounds	$X_i \leq 18$	-2 ÷ - 4
3 low	Positive autocorrelation among medium-low bounds	$X_i \leq 18$	-4 ÷ - 6.3
4 high	Positive autocorrelation among high bounds	$X_i \geq 18$	1 ÷ 11.9

enon better because it is more sensible to intensity changes allowing more accurate classification.

Figure 5 shows in a more detailed map how deep are analyses produced with Getis and Ord (1992) function. Opposite values can occur in contiguous areas. Picture it shows how the highest values of autocorrelation correspond to the highest buildings of the town. Picture b highlights an abrupt transition, in a few metres, from no correlation to high correlation passing from ancient low buildings to high concrete buildings. Picture c shows how very elevated values of autocorrelation correspond to high buildings separate by narrow streets.

Figure 4. Clusters location

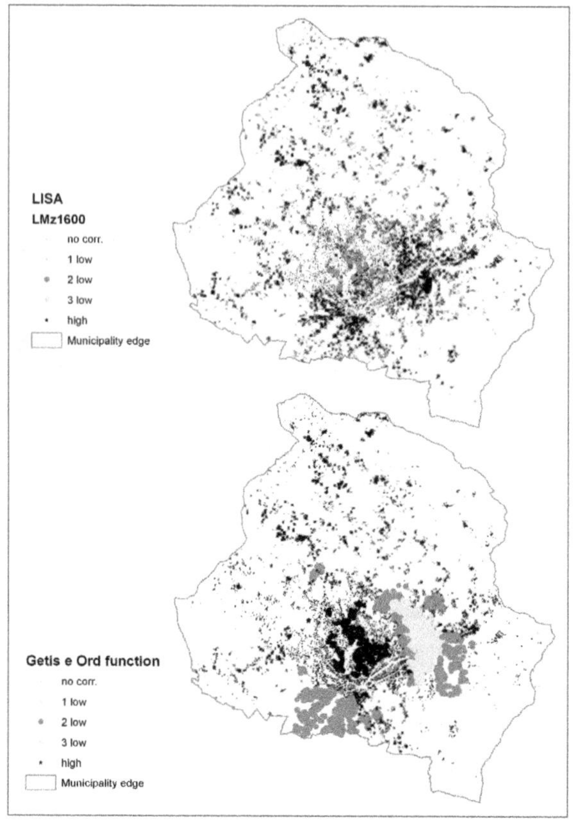

4. RESULTS AND FINAL DISCUSSION

Autocorrelation phenomena included in medium medium-low values have been interpolated thus generating polygons which represent the contiguity belt. These polygons represent the second level of suitability. It is composed by the inclusion rules considered in land suitability procedures reduced considering the global and local measures of autocorrelation. It is obvious that kernel density (Figures 2 and 3) is a rough measure which needs a deeper analysis. Moran index and Getis and Ord (1992) function give a further interpretation of phenomena considering contiguity not in all directions but only in some zones.

The exclusion rules (Figure 6) have been considered in the present study: areas included within a distance of 150 m from rivers, streams

Figure 5. Detailed map with pictures which show autocorrelation difference

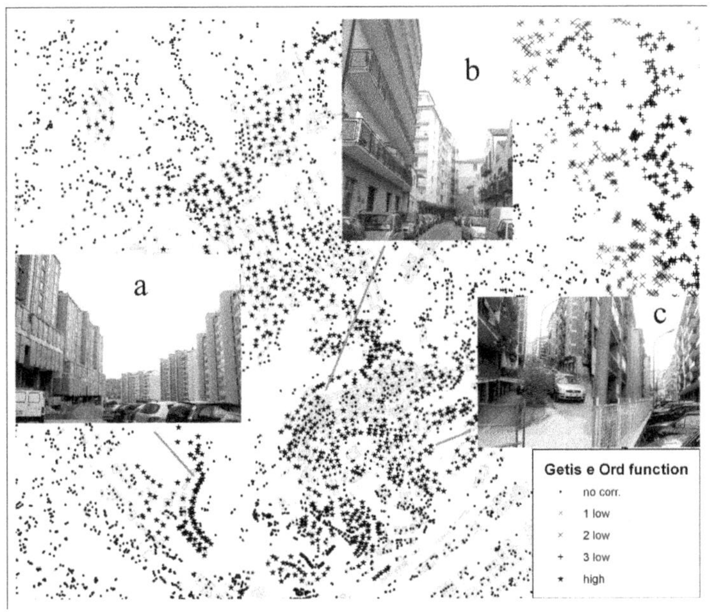

and springs, slopes higher than 35%, Nature 2000 sites, hydro-geological risk zones, areas higher than 1200 m a.s.l., landslides, areas close to railways and road networks.

Figure 6 shows the flow chart of the land suitability procedure for the location of Peri-urban fringe. All these rules have been combined using map algebra techniques. Figure 7 quantifies the reduction of suitable areas achieved after the procedure.

The results are illustrated as geographic components in Figure 8. Location of the contiguity

Figure 6. Scheme of the land suitability procedure for the location of Peri-urban fringe

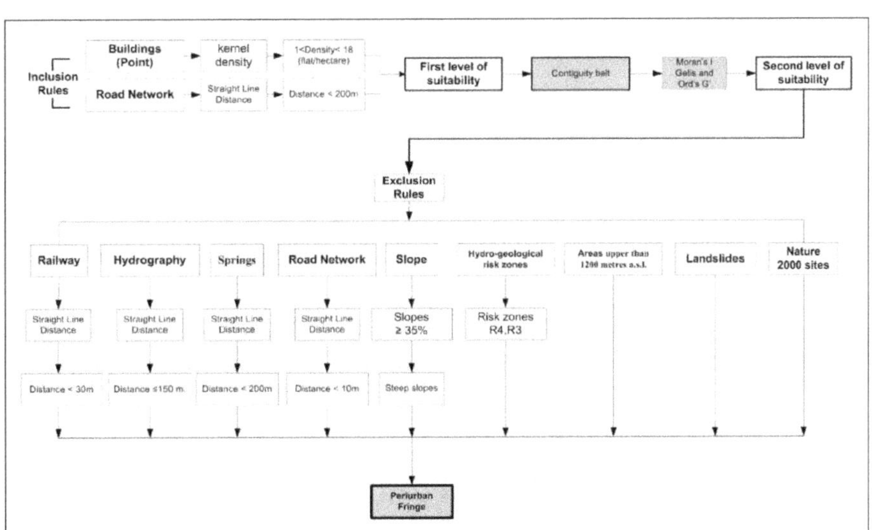

Figure 7. Size of suitable areas

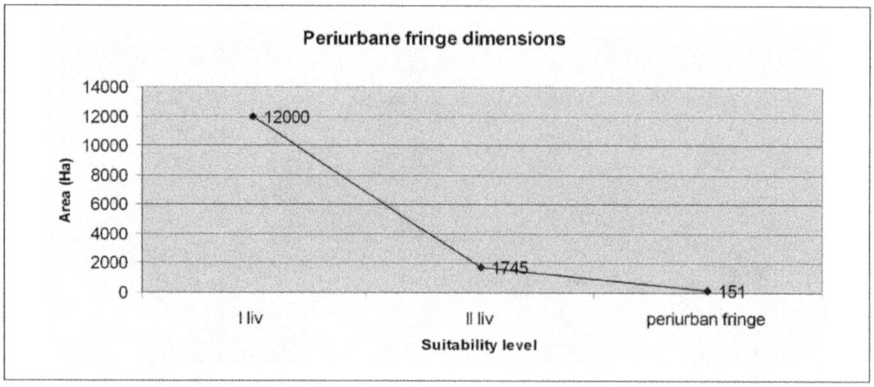

belt is determined by the highway, which determines two gates for the town. In these areas urban sprawl is more intensified, particularly in the eastern part where the other road, which connects the industrial areas of Potenza with FIAT factory, amplifies the phenomenon. Steep slope obstructs urban growth in other zones. Periurban fringe (oblique hatch in Figure 8) considers contiguity belt after the exclusion rules and represents areas suitable for the location of new settlements or for intensifying the existing ones.

After the theorization by Tobler (1970), the first law of geography is reported here: "Everything is related to everything else, but near things are more related than distant things". More experience exists of the use of spatial statistics in geographical analysis; for instance Kernel density has been applied for the location of epidem-

Figure 8. Periurban fringe after the land suitability procedure (oblique hatch)

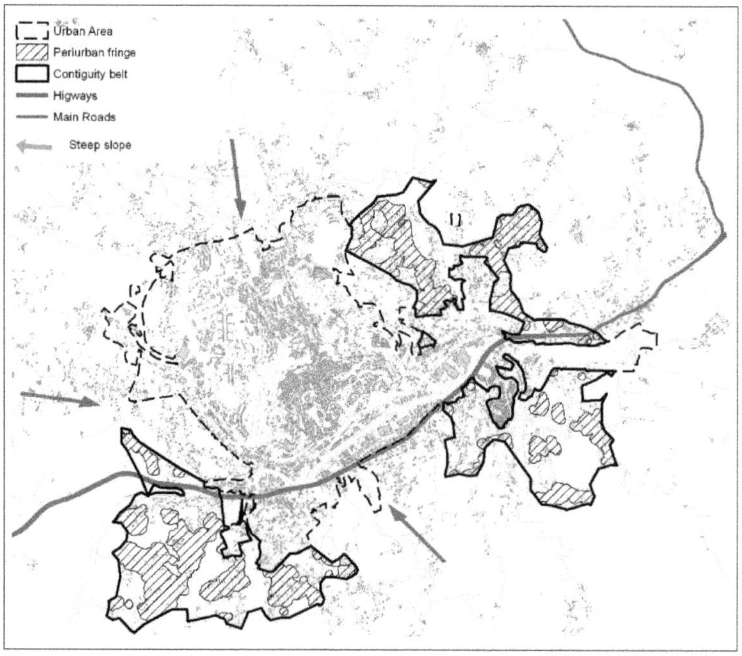

ics (Gatrell et al., 1995), in urban renewal analyses (Murgante et al., 2008), in defining damage scenarios (Danese et al., 2008; Danese et al., 2009) and studies on spreading of city services (Borruso & Schoier, 2004), while these techniques have not been used enough in the field of territorial planning. In this paper several kinds of spatial statistic functions have been applied for a deeper knowledge of territory and to give urban planners a better support for planning choices.

REFERENCES

Ahlqvist, O., & Ban, H. (2007). Categorical measurement semantics: A new second space for geography. *Geography Compass*, *1*(3), 536–555. doi:10.1111/j.1749-8198.2007.00026.x

Alberti, M., Solera, G., & Tsetsi, V. (1994). *La città sostenibile*. Milano, Italy: FrancoAngeli.

Anselin, L. (1995). Local indicators of spatial association – LISA. *Geographical Analysis*, *27*, 93–115. doi:10.1111/j.1538-4632.1995.tb00338.x

Bailey, T. C., & Gatrell, A. C. (1995). *Interactive spatial data analysis*. Harlow, UK: Prentice Hall.

Borruso, G., & Schoier, G. (2004). Density analysis on large geographical databases. Search for an index of centrality of services at urban scale. In A. Laganá, M. L. Gavrilova, V. Kumar, Y. Mun, C. J. K. Tan, & O. Gervasi (Eds.), *Proceedings of the International Conference on Computational Science and its Applications* (LNCS 3044, pp. 1009-1015). doi: 10.1007/978-3-540-24709-8_106

Breiman, L., Meisel, W., & Purcell, E. (1977). Variable kernel estimates of multivariate densities. *Technometrics*, *19*, 135–144. doi:10.2307/1268623

Burt, J. E., & Barber, G. M. (1996). *Elementary statistics for geographers* (2nd ed.). New York, NY: Guilford Press.

Camagni, R., Capello, R., & Nijkamp, P. (1998). Towards sustainable city policy: An economy-environment technology nexus. *Ecological Economics*, *24*(1), 103–118. doi:10.1016/S0921-8009(97)00032-3

Camagni, R., Gibelli, M. C., & Rigamonti, P. (2002). Urban mobility and urban form: The social and environmental costs of different patterns of urban expansion. *Ecological Economics*, *40*(2), 199–216. doi:10.1016/S0921-8009(01)00254-3

Cavailhès, J., Peeters, D., Sekeris, E., & Thisse, J. (2004). The Periurban city: Why to live between the suburbs and the countryside. *Regional Science and Urban Economics*, *34*(6), 681–703. doi:10.1016/j.regsciurbeco.2003.08.003

Charrier, J. B. (1994). *Geografia dei rapporti città-campagna*. Milano, Italy: FrancoAngeli.

Danese, M., Lazzari, M., & Murgante, B. (2008). Kernel density estimation methods for a geostatistical approach in seismic risk analysis: The case study of Potenza Hilltop Town (southern Italy). In O. Gervasi, B. Murgante, A. Laganà, D. Taniar, Y. Mun, & M. L. Gavrilova (Eds.), *Proceedings of the International Conference on Computational Science and its Applications* (LNCS 5072, pp. 415-427). doi: 10.1007/978-3-540-69839-5_31

Danese, M., Lazzari, M., & Murgante, B. (2009). Geostatistics in historical macroseismic data analysis. *Transactions on Computational Sciences*, *6*(5730), 324–341. doi: 10.1007/978-3-642-10649-1_19

Epanechnikov, V. A. (1969). Nonparametric estimation of a multivariate probability density. *Theory of Probability and Its Applications*, *14*, 153–158. doi:10.1137/1114019

Gatrell, A. C., Bailey, T. C., Diggle, P. J., & Rowlingson, B. S. (1996). Spatial point pattern analysis and its application in geographical epidemiology. *Transactions of the Institute of British Geographers*, *21*(1), 256–274. doi:10.2307/622936

Getis, A., & Ord, J. K. (1992). The analysis of spatial association by use of distance statistics. *Geographical Analysis*, *24*, 189–206. doi:10.1111/j.1538-4632.1992.tb00261.x

Guerois, M., & Pumain, D. (2001). *Urban sprawl in France (1950-2000)*. Milano, Italy: FrancoAngeli.

Hidding, M., Needham, B., & Wisserhof, J. (2000). Discourses of town and country. *Landscape and Urban Planning*, *48*(3-4), 121–130. doi:10.1016/S0169-2046(00)00036-0

Jones, M. C., Marron, J. S., & Sheather, S. J. (1996). A brief survey of band-width selection for density estimation. *Journal of the American Statistical Association*, *91*, 401–407. doi:10.2307/2291420

Moran, P. (1948). The interpretation of statistical maps. *Journal of the Royal Statistical Society. Series A (General)*, *10*, 243–251.

Murgante, B., & Las Casas, G. (2004). G.I.S. and fuzzy sets for the land suitability analysis. In A. Laganá, M. L. Gavrilova, V. Kumar, Y. Mun, C. J. K. Tan, & O. Gervasi (Eds.), *Proceedings of the International Conference on Computational Science and its Applications* (LNCS 3044, pp. 1036-1045). doi: 10.1007/978-3-540-24709-8_109

Murgante, B., Las Casas, G., & Danese, M. (2008). Where are the slums? New approaches to urban regeneration. In Liu, H., Salerno, J., & Young, M. (Eds.), *Social computing, behavioral modeling and prediction* (pp. 176–186). Berlin, Germany: Springer-Verlag. doi:10.1007/978-0-387-77672-9_20

Murgante, B., Las Casas, G., & Sansone, A. (2008). A spatial rough set for extracting periurban fringe. *Revue des Nouvelles Technologies de l'Information*, *857*, 101–125.

O'Sullivan, D., & Unwin, D. (2002). *Geographic information analysis*. New York, NY: John Wiley & Sons.

O'Sullivan, D., & Wong, D. W. S. (2007). A surface-based approach to measuring spatial segregation. *Geographical Analysis*, *39*, 147–168. doi:10.1111/j.1538-4632.2007.00699.x

Salsano, E. (1998). *Fondamenti di urbanistica*. Bari, Italy: Laterza.

Sillverman, B. W., & Jones, M. C. (1989). E. Fix and J. L. Hodges (1951): An important contribution to nonparametric discriminant analysis and density estimation. *International Statistical Review*, *57*(3), 233–247. doi:10.2307/1403796

Van Den Berg, L., & Wintjes, A. (2000). New rural lifestyle estates in The Netherlands. *Landscape and Urban Planning*, *48*(3-4), 169–176. doi:10.1016/S0169-2046(00)00040-2

This work was previously published in International Journal of Agricultural and Environmental Information Systems, Volume 2, Issue 2, edited by Petraq Papajorgji and François Pinet, pp. 16-28, copyright 2010 by IGI Publishing (an imprint of IGI Global).

Chapter 10
Coupling Geographic Information System (GIS) and Multi-Criteria Analysis (MCA) for Modelling the Ecological Continuum in Participative Territorial Planning

Mireille Batton-Hubert
Ecole Nationale Supérieure des Mines de Saint-Etienne, France

Marie Bonnevialle
IPAMAC, France

Pierre-Olivier Mazagol
Université Jean Monnet, France

Frédéric Paran
Ecole Nationale Supérieure des Mines de Saint-Etienne, France

Thierry Joliveau
Université Jean Monnet, France

ABSTRACT

This paper presents a research experiment that started in 2007 on the French "blue and green infrastructures." At the prompting of a French local administration, this experiment takes into account ecological stakes and involves actors in territorial planning at a metropolitan scale. Founded on previous works, the goal of the approach builds a participative workshop based on GIS and MCA coupling in order to think collectively and better understand the initial foundation of the method, the problems and proposed solutions, and the theoretical and the practical prerequisites. The final goal suggests methodological solutions to facilitate its generalization and its diffusion. The objectives are: (1) to analyse as objectively as possible the initial approach, methodological and technical choices, constraints, limits, and potential improvements; (2) to take into account questioning about the participative workshop conception and functioning; (3) to test new methodological proposals.

DOI: 10.4018/978-1-4666-0333-2.ch010

INTRODUCTION

Taking into account the natural environment and landscape in territorial planning is often done formally in compiling preparatory documents concerning the territory in question. Then, these documents are discussed and negotiated with multiple committees for development activities regarding environmental and landscape viewing objectives. The approach of BGI (Blue and Green Infrastructures, IVB in French) is interesting for its willingness to rely on the scientific basis of landscape ecology and to use data and tools for digital geodata cartographic representation of the criteria to be taken into account in the decision-making process. This approach is also thought to be participative. It combines government services, public bodies, and local actors in a steering committee. Even if the doctrine of the State on biodiversity, landscape, and amenity for local planning is to be applied, this State doctrine seeks to elaborate a collaborative fashion, involving several areas of expertise. The BGI approach seems a good example for developing a meaningful discussion on the issue of environmental planning of the territory in a collaborative or participatory sense. This mapping process, based on numerical processing, provides "a consistent picture of the territory" which aims to provide easy viewing of the stakes involved and to challenge all the actors. This approach should be used before any development projects as it provides awareness of the ecological and landscape stakes of a territory. Finally, the availability of tools is designed to facilitate the ownership of the method by other research teams to allow its implementation in other areas. The BGI approach should be considered as a consistent spatial information method conducted collaboratively to formalize territorial stakes related to natural environments, landscapes, and natural modes of movement of animal and human communities.

The objectives of this paper are the following: (1) to analyse as objectively as possible the initial approach along with the associated methodological and technical choices, constraints, limits, and potential improvements; (2) to take into account questions related to the participative workshop conception and functioning; (3) to test new methodological proposals.

This approach involves three restraints: (1) *scientific constraints*, because the method needs to develop its prescriptive approach, based on principles of scientific explanation; (2) *organizational constraints*, because the application is done with the data, tools, and skills available in the State services, complemented by those of research consultancy; (3) finally, *socio-political constraints*, because the results obtained by the method must be followed by an approach to study and confirm the results and to define the strategy of the State according to documents, planning, and projects.

Our approach offers an original device at the interface of research (to produce new knowledge) and action (to develop methods and tools used in situations). Faced with a request from a State service, the local offices of the Ministry of Territorial Planning of the Loire department, to improve a method integrating an ecological dimension in a spatial planning process, we chose to respond not by expertise but by building a workshop implementation of the method, tested in the form of role playing in training contexts.

We made the assumption that this approach offered several advantages. First, it allowed the research team to quickly apprehend the method and also the constraints of its application in a simplified case, inspired nevertheless by the real situation. Furthermore, inviting users of the method to participate in the workshop facilitated the implementation of the new approach proposed. Finally, the comments of workshop participants allowed potential problems that the method could cause to be revealed.

This work led to the establishment of a training workshop/research implementing the method as a role playing game. It has been tested with researchers and practitioners in the SAGEO workshops(Batton-Hubert, Bonnevialle, Joliveau, & Paran, 2007a), whose results has been presented at the OPDE conference (Batton-Hubert, Bonnevialle, Joliveau, & Paran, 2008). The method is actually used in a Research Master's course (February 2008, 2009 and 2010). The approach has also been valued during a local communication with the local offices of the Ministry of Territorial Planning (Batton-Hubert, Bonnevialle, Joliveau, & Paran, 2007b).

Our presentation will be set out as follows. First, we discuss the context of the BGI approach in a territorial planning at metropolitan scale in France (The south of the Loire department). Then, we recall the methodological approach of BGI modelling as it was developed by the local offices of the Ministry of the Environment of the Rhône-Alpes region and the local offices of the Ministry of Territorial Planning of the Loire department with the help of the consultancy firm Asconit. Moreover, we perform a critical analysis of this method from the points of view of the findings of both its users and ourselves as outside researchers. We then present the methodological approach, in terms of how the workshop was designed and achieved, by describing how the workshops were held. Finally, we present the elements of the enrichment of the method, and the main questions and perspectives.

BACKGROUND: DEFINITION, HISTORY, INTEREST, DESCRIPTION, AND ANALYSIS OF THE INITIAL METHOD

This section defines the concept of BGIs and the advantage of linking them in terms of local and national stakes. After a brief history of the initial method, it will be quickly described methodologically. Finally the results are presented for the territory that concerns us: the South Loire.

Definition: The BGI Concept in Territorial Planning at a Metropolitan Scale

Blue and Green Infrastructures (BGIs), "green" for the natural land areas and "blue" for the natural aquatic environments, are defined as a set of spaces which are connected and organized into a hierarchy integrating major axes of movement of animals (ecological continuums) and "soft movements" of human being such as walking, cycling, or skiing.

The challenge is to integrate such networks in a local planning whose aim is to coordinate in space and time the planning by various public policies across municipalities. This is a project area that tries to determine the overall balance between urban spaces (to be urbanized) and natural, agricultural, and forestry areas from a perspective of sustainable territorial development.

From National Stakes to Local Stakes

Several activities can be cited as precursors to the BGI approach. Examples include works in Switzerland (Berthoud, Lebeau, & Righetti, 2004) and in France such as the ecological network of the Isère department (Berthoud, 2001) and the biological continuums of the agglomeration of Belfort-Montbeliard (Asconit Consultants, 2003).

The concept of BGI was proposed at the regional level by the local offices of the Ministry of the Environment of the Rhône-Alpes region (Asconit Consultants, 2003) to integrate the objectives of preserving biodiversity, landscape quality, and amenity in approaches to planning. Nationally, the French state has raised the issue in terms of "green and blue frames" envisaged as an instrument of decentralized sustainable development, consultation, and integrated planning. Locally, for example, an experiment was conducted on the South Loire (Asconit Consultants, 2005a, 2005b) at the request of local offices of the Ministry of Territorial Planning of the Loire department and the local offices of the Ministry of the Environment of the Rhône-Alpes region.

The application presented in this paper is a later experiment conducted at the request of the local offices of the Ministry of the Territorial Planning of the Loire department. It is dedicated in part to the South Loire local plan and in another part to the preparation of the North Loire local plan (Figure 1). This application is structured by four major issues: (1) incorporation of an environmental dimension; (2) establishment of a procedure for territorial planning; (3) provision of a digital geodata treatment of the data and the multi-criteria combination; (4) incorporation of a participatory dimension.

The main objective is to improve the methodology of construction of BGI in planning exercises. Indeed, the methods to objectify the BGI are far from being fully stabilized. We will aim to ensure the passage of a decentralized and technocratic experiment to a participatory bottom-up testing based on maximizing the potential of local innovation and research.

The BGI Concept Modelling: The Initial Method

The initial method is based on the work of Berthoud and Righetti (2004) and Berthoud (2001) adapted by the Asconit consultancy firm (Asconit Consultants, 2005a, 2005b). This method is divided into three levels:

- A representation of the stakes of nature by modelling ecological continuums by GIS vector calculation and by classification into three hierarchical levels of remarkable natural environments;
- A representation of the stakes of the human being by classification into three hierarchical levels of recreational areas, "soft movements", and remarkable landscapes;
- A representation of the constraints on previous stakes in four hierarchical levels of barriers to movement.

The Ecological Continuums Modelling

According to the concepts of landscape ecology (Forman & Godron, 1986), a landscape can be described by using three types of elements: (1) non-linear elements: the patches; (2) linear elements, the corridors that connect patches through the matrix; and (3) a dominant and encompassing element: the matrix. Within this matrix, the habitat patches are linked by corridors that contribute highly to the flows within the landscape (Burel & Baudry, 1999). Besides their role in connection, corridors are also important as filters and barriers. Most often, the corridors are studied as "drivers" influencing the dynamics of local populations and consequently of metapopulations (Hanski & Gilpin, 1997) by reducing landscape fragmentation and patches isolation. Within the matrix, these corridors are organized in networks whose preservation is important and which must be taken into account by planners. Many studies have focused on methodologies for identifying these ecological networks (Jongman & Pungetti, 2004). They clearly evidence the diversity of the approaches, the variety of data to integrate, the complexity of the stakes to consider, and generally the difficulty in delineating these networks as a function of the ecological scales and the data available.

The conservation of biodiversity requires the protection of community reservoirs of biodiversity and connections between these environments. This is to protect an ecological network in which animals move. Such a network consists of different elements (Berthoud, 2001):

- Nodal areas or nuclei: natural environments of good quality and sources of biodiversity
- Zones of expansion: lower quality than the corresponding nodal areas but the same types of environments

Figure 1. Location of the Loire department in France: The South Loire (Saint-Etienne, Montbrison) and the North Loire area (Roanne, Feurs) [Data sources: BD Carthage, ArcGIS]

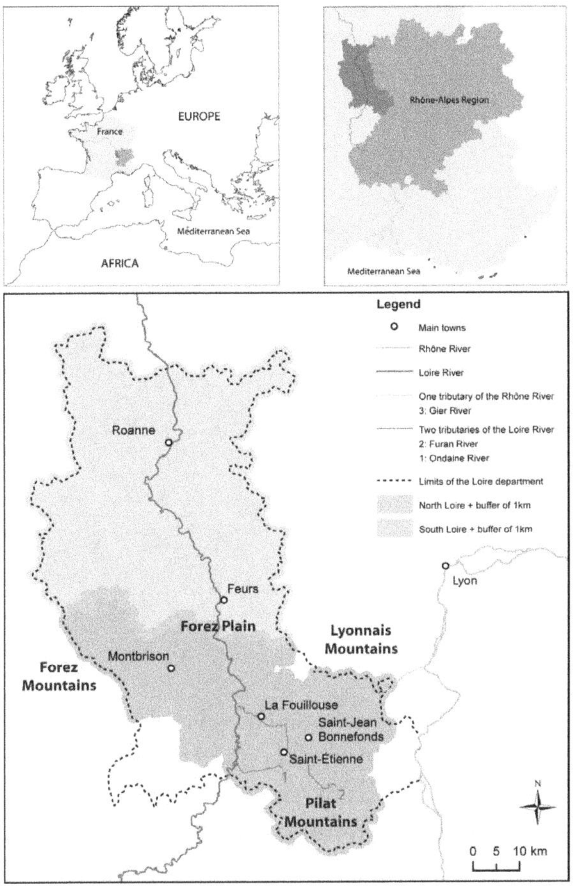

- Continuum: a set of environments favourable to a group of animals, made of continuous elements with nodal areas, areas of expansion, and possibly additional margins
- Corridors: open space providing opportunities for exchange between the nodal areas and development areas
- Development areas: all processed or degraded environments which are potentially favourable

The initial method offers a simplification of the concept of ecological network to build what we call for convenience the *potential of ecological continuums of the territory*. Thus, this method simply distinguishes five types of continuums, which are associated with emblematic species: (1) Woody of Lower Altitude (WLA: deer, wild boar); (2) Aquatic Environment and Wetland (AEW: amphibians, dragonflies); (3) Thermophiles Dry Areas (TDA: Orthoptera, Lepidoptera); (4) Extensive Agriculture and Edges (EAE: partridges, hedgehogs); (5) Woody of Medium Altitude (WMA: roe deer).

Each continuum is split, to simplify, into four groups of environments in which a coefficient of resistance (R) in animal movement is assigned: (1) *Structural environments* (reservoir areas, development of flagship species, R = 0); (2) *Attractive environments* (related and complementary to the structural environment, where animal travel is easy, R = 5); (3) *Little traffic environments* (related to structural and attractive environments, where movement is more rare and penetration harder for animals, R = 30); and (4) *Repellent environments* (that are a priori not frequented by the species and represent an obstacle to the movement, R = 100).

The construction of each of the continuums is performed by spatial analysis in GIS vector mode by aggregating data for Corine Land Cover (2000). The spatialization of a continuum is obtained using Equation [1].

$C = R \times D$ *(1)*

R: coefficient of resistance

D: distance in metres

C: travel cost in points

Knowing the resistance (R) of an environment and the potential initial movement of animals fixed at 3000 points (C), it is possible to calculate the distance (D) that a species can theoretically travel over a territory and to project this distance onto a map to obtain the potential ecological continuum. According to Equation [1], for example, animals cannot travel more than 30 m in an environment characterized as repellent (Figure 2).

Figure 2. Calculating the distance (D) travelled by an animal as a function of the coefficient of resistance (R) of the different environments and the limit movement points (C) set to 3000

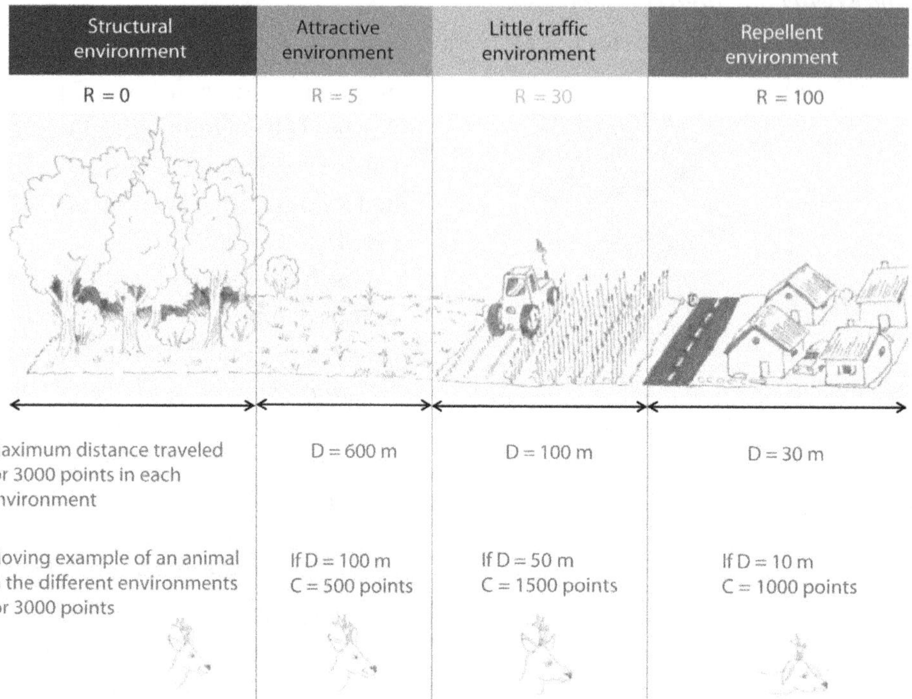

Representation of Other Information Layers

In the original method, the *Remarkable environments* aspect is approached from the perspective of environments enjoying a protected status, graded according to their ecological value (classification based on expert knowledge), and often depending on the severity of regulations. These are the sites of proven rich heritage (fauna, flora, habitats). They are reservoirs or cores of biodiversity and reflect the knowledge and goals of land conservation (e.g. nature reserve, Regional Nature Park). Depending on their ecological value and sustainability, the importance of these environments for the continuums is variable: no importance, significant importance (SI), high importance (HI), or major importance (MI). With this information Asconit generated maps where these three classes can be visually identified.

The territory is crossed by several routes of soft types (soft movements) as small roads (paths) for hiking trips, walking, riding or biking, with the trails marked as more or less known and used. They are used for simple Sunday walks or the discovery of strong points of tourist interest in cultural or environmental heritages. Thus, the initial method assumes that soft movement of humans in a mesh network of leisure is a continuum of amenity that can be mapped. Three categories of data are taken into account: the tourist and heritage sites (e.g. classified sites, monuments, parks, and gardens), the paths, and the remarkable landscapes. The various components of these categories are classified according to their degree of protection. Asconit obtained a map of *Leisure, paths, and landscapes* where we can visually distinguish these classes: no importance, significant importance (SI), high importance (HI), or major importance (MI).

The theme of *Barriers* corresponds to the analysis of landscape fragmentation by mapping natural and artificial barriers of territory. The barriers are all elements of the territory that may cause a rupture in the natural environment and especially in ecological continuums: roads, urban areas, dams, and so on. They are ranked by the degree of discomfort that they create: impermeable barrier, essential disturbance, important disturbance, and less disturbance). This gives an obstacle map where we can visually distinguish these different classes.

Aggregation and Overlay of the Different Maps

In the original method only the final map of ecological continuums comes from a combination of criteria. Other maps (*Barriers*; *Remarkable environments*; *Leisure, paths, and landscapes*) are superimposed. The synthetic map of ecological continuums was obtained on the basis of fixed weights based on expert knowledge: 100 for WLA and AEW and 10 for EAE.

Presentation of the Study Area and of the Main Results of the Initial Method

The study area (Figure 1) is located in France in the Rhône-Alpes region, in the department of the Loire (42). The prefecture of the department is Saint-Etienne, a town about 70 km south-west of Lyon and 500 km south of Paris. The area concerned occupies almost all the southern half of the department, an area of 179 110 ha with 510 000 inhabitants and 117 municipalities. It has the distinction of being in the watershed of both the Rhone River and the Loire River. Four geographic entities largely influenced by human activities structure this territory:

- Three forest areas at the edges: the Pilat Mountains to the south, the Forez Mountains to the west, and the Lyonnais Mountains to the east,
- A great area of agricultural plain in the centre: the Forez Plain crossed from south to north by the Loire River,
- Three highly urbanized valleys converging at Saint-Etienne: the valley of the Furan River from the Pilat Mountains, the valley of the Ondaine River from the Haute-Loire department (43), and the valley of the Gier River from the Pilat Mountains towards Lyon.

The study area is characterized by a rich and fragile natural heritage mixing of sites with high ecological values and ordinary environments at the gate of urbanization. This wealth is now threatened by urbanization, leading to loss of ecological continuums, and by the lack of development sites generating use conflicts.

On four maps of the potential ecological continuums (WLA, AEW, EAE, TDA) of the territory of the South Loire, the woody medium altitude environments (WMA) are absent, and have been generated by Asconit with the vector method. A global map of potential ecological continuums was obtained by weighted aggregation of only three continuums (WLA, AEW, and EAE), because the TDA is spatially extremely limited. The final map is obtained by superposition of the global map of potential ecological continuums and other criteria: *Remarkable environments*; *Leisure, paths, and landscapes*; and *Barriers*.

THE NEW METHOD

Figures 3 and 4 show the key elements of the adaptation of the original method (Asconit consultants, 2005a,, 2005b). On these two figures, the initial method is described on the left and the new method on the right. The main changes are the following:

- Some criteria are modified and renamed;
- The calculation is made in raster mode and not in vector mode;
- The calculation is automated by the dynamic models used with ArcGIS to generate maps;
- The final BGI map is obtained not by a simple overlay mapping but by the use of a combination model associating GIS and MCA. For this, we create a tool capable of evaluating preferences and calculating weights, founded on the Saaty (1982) HAP method. These weights are the input data used in the weighted sum calculated with the ArcGIS software.

We worked from data supplied by the local offices of the Ministry of Territorial Planning of the Loire department and the original files from the application of the Asconit method to the South Loire. A model is defined here as any formalism or algorithm used to construct an indicator representing space or several criteria.

The new method is presented in several steps. First, we adapted the initial modelling methodology of the ecological continuums and redefined the criteria involved. Secondly, we propose an effective method of combination of criteria. In addition, the workshop used is part of the method and led to its development.

Adaptation of the Ecological Continuums Modelling Methodology and Redefinition of the Criteria

Ecological Continuums Modelling

The adaptation of the initial method is based on a reflection on the tools used, the data entry, and the calculation (Figures 3 and 4). The new calculation of the ecological continuums uses a raster mode function, based on a spread/dissemination, named "cost-distance" (ArcGIS 9.2). The construction of the ecological continuums in vector mode was abandoned in favour of the raster mode, which, besides giving a time saving, is more suited to the representation of continuous phenomena, such as spatial ecological continuums (movement of animals). It is possible to adjust the number of points (C in Equation [1]) awarded to the animal's travel and to consider the whole range in the map representation (Figure 9).

The theme *Barriers* remains initially unchanged and superimposed on previous maps. But after the workshop session, barriers were included in the calculation of the ecological continuums.

The Other Information Layers

INSEE (French National Institute of Statistic and Economical Studies) data of human population density and accommodation capacity (number of second homes, campsites, and hotel rooms) per town for the theme *Landscapes heritage* and also the data from the local offices of the Ministry of the environment of Auvergne region for the theme *Protected nature and zoning regulation* have been used.

The theme *Remarkable environments* (Figures 3 and 4) has been modified with additional data (new criteria) and renamed in *Protected nature and zoning regulations*. However, some data have been updated and those concerning Natura 2000 added. The initial classification into three hierarchical levels has been preserved. Figure 5 shows the map of the new criteria aggregated according to the initial classification in the South Loire without weighting.

It is apparent that the theme *Leisure, paths, and landscapes* should be split into two separate models (Figures 3 and 4). The first, the *Landscape heritage* model, is founded on the idea of attractiveness and value associated with a site, for

Figure 3. Major changes from the initial method to the new method: overlay mapping to combination model

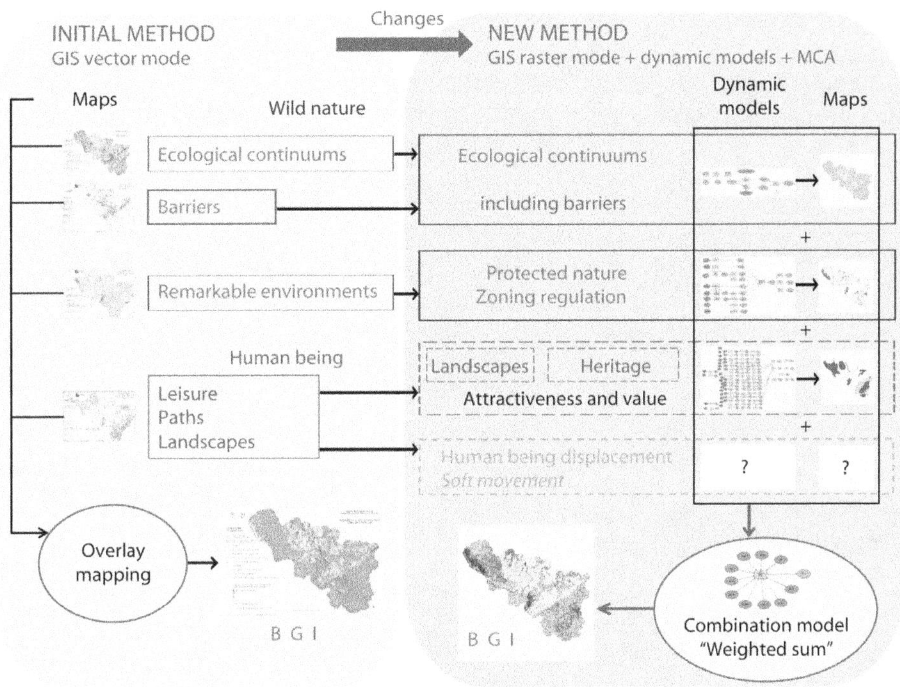

example, area of heritage or landscape interest, density of housing, rates of attendance, and so on. In the present experiment, the test is based on the zoning of remarkable landscapes used in the original method, and on additional data (new criteria) such as rates of touristic attendance and value of touristic sites. Figure 6 shows the map of the new criteria aggregated according to the initial classification in the South Loire without weighting. The second model, *Soft movement* (Figures 3 and 4), could be constructed from a reflection on the concept of network and human displacement (e.g. travel mode, purpose). This model has unfortunately not been implemented during the study period and was therefore omitted. In the future, we have to build this model and to find good criteria. A map of these criteria does not actually exist.

From Split Criteria Analysis to a Multi-Criteria Analysis

In the initial method, the weighted sum allows only a continuum map showing the potential displacements of flagship species to be obtained. Three questions can be considered to improve the previous method:

- How can the ecological continuums and the other criteria like *Protected nature*, *Landscape heritage, Soft movement*, and *Barriers* be aggregated?
- How can the degrees of actor preference that will be considered in the weight choice be taken into account?
- How can the CPU time taken to compute the solution be decreased and use of the method be made easier for the decision makers?

Figure 4. Major changes from the initial method to the new method: technical and criteria changes

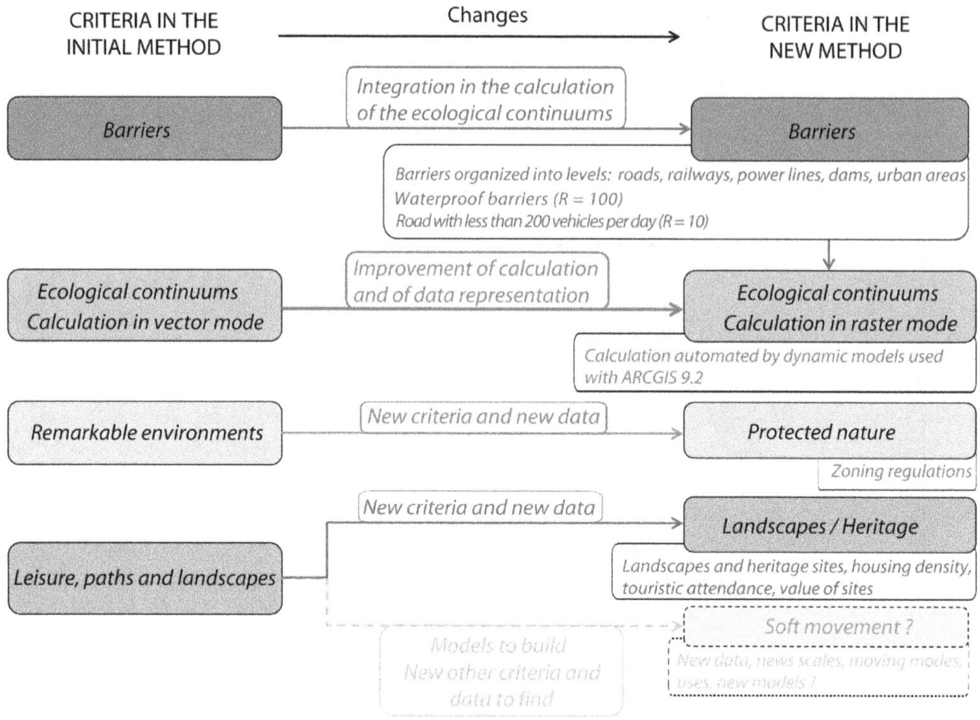

The GIS facilities allow some difficulties like the CPU time and the building of the different data layers to be solved. In fact, the raster data form, with the matrix structure, allows the continuous phenomena (like a gradient of animal displacement) to be solved and the computation time to be decreased. The GIS interface and the tools of ArcGIS (with the dynamic toolboxes) automatically generate the output covers. A spatial algebra, associated with the raster data, allows the weighted sums to be integrated and used. But the last problem concerns the weight attribution.

MCA-GIS methods linking GIS databases and multi-criteria analyses for decision-making are usually used in different applications in landscape management in particular in order to find the shape of a linear highway network (Molines, 2006). This method linking MCA and GIS is also applied, for example, to a public transportation network problem (Tavares-Pereira, Figueira, Mousseau, & Roy, 2007) or corridor identification problem (Chakhar & Mousseau, 2008).

The Analytic Hierarchy Process (AHP) introduced by Saaty (1982) allows us: (1) to describe the current decision object, which is the ecological potential of each raster mesh; (2) to define the different sets and elements of complex system (Saaty & Shih, 2009); and (3) to propose a hierarchical structure containing each element according to the attribute functional properties. The preferences and values defined (Keeney & Raiffa, 1976) are extended and applied by Saaty (1982). This method based on the two-two comparison between alternatives or criteria allows us to define the preferences of each one to another one according to the decision object. The three steps are the following:

- The first step allows us to build a hierarchical structure, a tree (Figure 7), split into elements (criteria) and sub-elements (sub-criteria). The last level is composed of leaves (alternatives). In this tree, each stake defines a criterion, and each criterion has a weight which is computed with the preferences matrixes.

- The second step is the comparison between each element having the same from-node in a level of the tree, in order to obtain the importance of each one in the decision object. For each preference matrix (instances of the importance of sub-criteria are given by the decision players) a coherent weight is computed for each criterion and each alternative. An example of obtained weight is given in Figure 8.
- The last step consists in obtaining the aggregated value from each partial contribution of different levels of the tree with a weighted sum.

This method introduces several advantages. It allows us to resolve complex problems to take into account the interrelationship of the elements of a system and the relating preferences of the factors of a system. Finally, the solution is not the optimum solution but it works out a representative compromise of various opinions. The method allows us to assess the logical coherence of the preferences, refining the definition of a problem by repetition and establishing preferences. It allows us to obtain the total valuation of the desirable character applied to each solution and can be easily interfaced with a GIS. Outranking methods, a family of multi-criteria methods, may be useful in spatial decision problems, especially when ordinal evaluation criteria are implied (Chakhar, 2006).

Nevertheless, the implementation of this approach is difficult in a multi-actors context or when the number of criteria is large. In this last case the risk of obtaining an incoherence ratio associated with preference matrixes can be greater than 10%.

Furthermore, this method asks the decision-maker to consider all the possible comparisons, implicitly making the assumption that all the considered criteria are perfectly comparable. The workshop established an interesting tool to

Figure 5. Protected nature and zoning regulations criteria

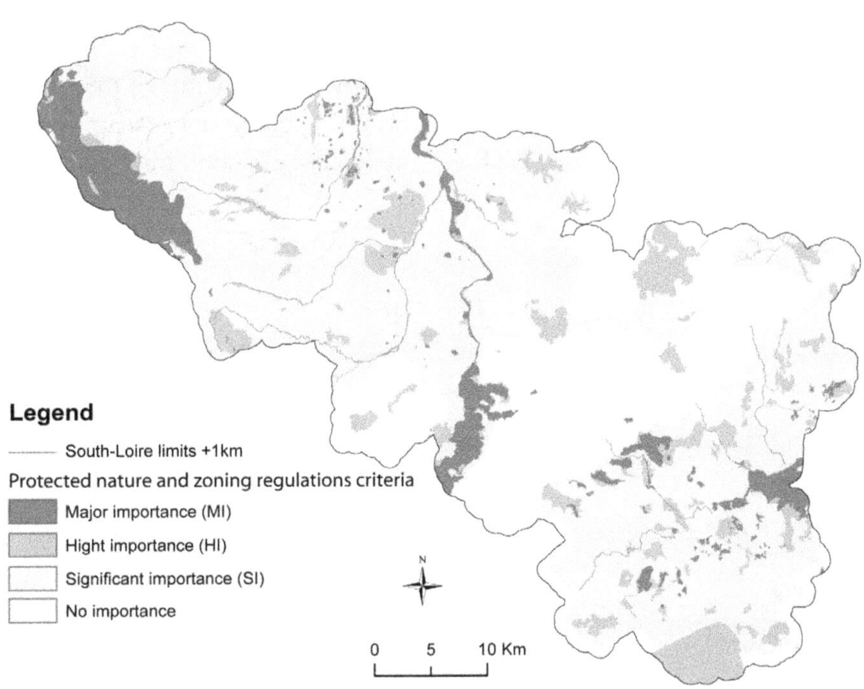

Figure 6. Landscapes heritage criteria

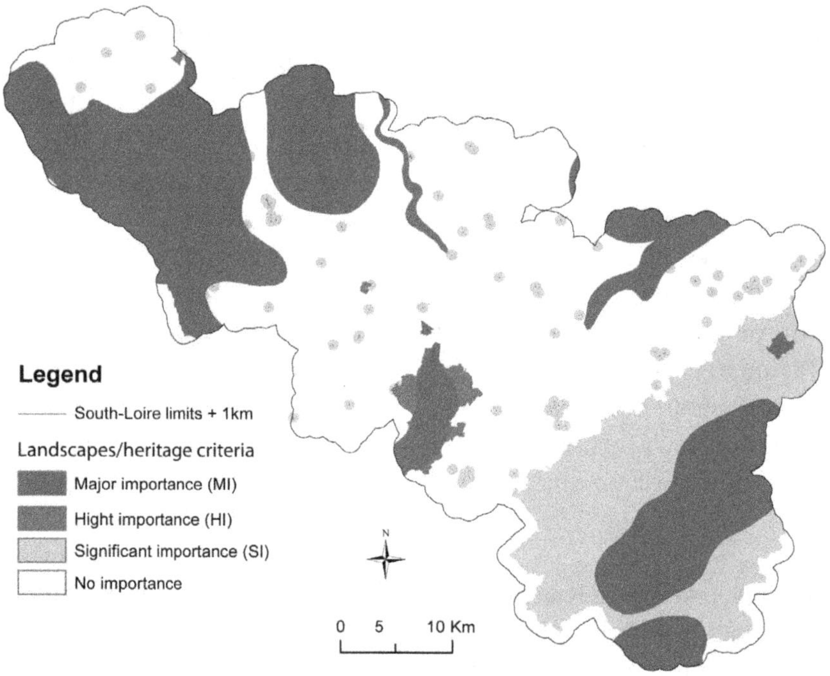

validate the hypothesis that the AHP method was adapted well to the BGI initiative. A state of art of this method can be found in Figueira et al. (2005).

Construction of the Participatory Workshop Combining GIS and MCA

The new methodological approach of BGI construction has been tested at the 2007 SAGEO International Symposium (Batton-Hubert et al., 2007a). These workshops were organized to share experiences about the use of participatory tools for territorial decision-making and to test these tools with a group of researchers and actors (or practitioners) in order to disseminate them more widely. The principle of these workshops was to bring together young researchers and stakeholders interested in action training in the use of spatial

Figure 7. BGI stake

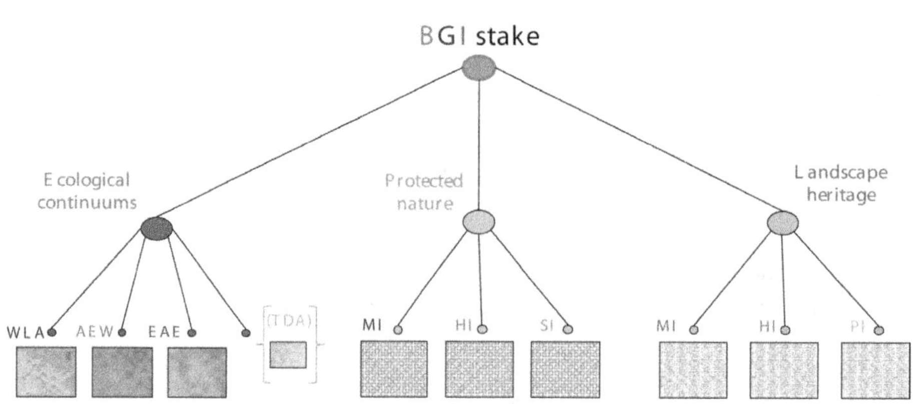

Figure 8. Two-two comparison to obtain preference matrix and to obtain the coherent weights (Saaty, 1982): Interfacing between GIS/MCA weighted sum

The left row (goal) is to compared with the given right column

	Preferences / Continuums			
	WLA	AEW	DTA	EAE
WLA	1	5	4	4
AEW	0,2	1	3	2
TDA	0,25	0,33	1	2
EAE	0,25	0,5	0,5	1

Here, the WLA criteria is to be preferred 5 times over the AEW criteria

Here, the WLA criteria is to be preferred 4 times over the DTA criteria

Here, the WLA criteria is to be preferred 4 times over the EAE criteria

Here, the DTA criteria is to be preferred 0,25 times over the WLA criteria

Criteria	Weight
WLA	0,56
AEW	0,21
TDA	0,13
EAE	0,01

modelling tools that can help actors in their territorial planning decisions or development. Our workshop was repeated in a Master's course integrating students of Saint-Etienne University and of the Ecole des Mines de Saint-Etienne (School of Mines of Saint-Etienne).

Organization and Conduct of the Workshop

The purpose of the workshop is to generate, in a participatory mode, a synthetic map of the potential BGI in the South Loire territory. For this, each participant has an educational device: activity organizer (host), ArcGIS and Excel software, digital and paper documentation, and so on. The three-hour session is divided into several stages:

- Introduction to the different concepts involved (BGI, biodiversity, and ecological network) and to the GIS methods of BGI construction in vector and raster modes, and learning by using an application to build an ecological continuum. As a result, participants are testing an approach using real analysis and spatial reasoning, which goes beyond visual reasoning.
- Presentation of the participatory stakes followed by an introduction to the fundamentals of the MCA (Saaty method).
- An application in which each participant assumes the role of an actor of the territory based on an identification sheet describing this actor. He then introduces the actor preferences in the tool and weights the continuums. Finally, he applies these weights to the ArcGIS tool "Weighted sum" to aggregate the different criteria and obtain a map of the BGI.
- Simulation of a negotiation between participants to choose collective preferences and get a synthesis map of BGI.
- Interpretation and comparison of maps obtained by different actors in negotiation and collective review of the workshop (discussion of the method, selection of criteria, weights, available data, etc.).

The Role Playing Game

To better simulate the negotiation, each of the actors is described in an identification sheet. These actor sheets were based on those used in earlier work on simulations of a negotiation (Paran, 2005).

Figure 9. WLA ecological continuum: Raster versus vector representation

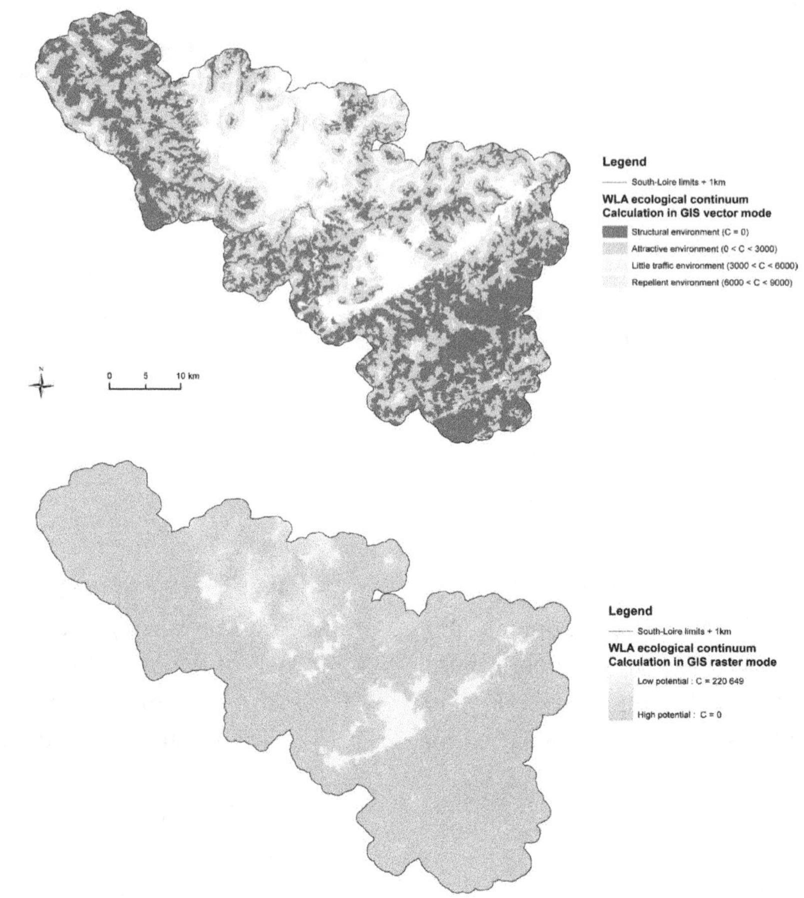

They are a starting point for the interpretation of roles by participants. These actor sheets are deliberately caricatured and simplistic to facilitate the rapid apprehension of the role and expression of preferences and clear positions. They are still operative to reconstruct a coherent spatial context. These sheets describe seven actors involved in the negotiation: a nature protector, a hunter, a fisherman, a farmer, an employee of the State services (local offices of the Ministry of Territorial Planning of the Loire department), an engineer of a Regional Nature Park (the Pilat Mountains), and a tourism promoter. Each sheet describes the actor succinctly according to four specific topics: name, scale of activity, main activities, and specific objectives. The category *main activities* lists the main stakes, objectives, and actions of the actor, while the category *specific objectives* identifies the specific stakes, objectives, and actions dedicated to the negotiation. The *scale of activity* allows us to spatialize the stakes and goals of the actors in the territory of the South Loire.

Improvements of the Modelling Approach and Combination of Criteria

Following the tests we conducted during the workshops with researchers and technicians and during the courses with pupils and students, and on the basis of their comments, refinements of the method have been proposed, and some have been tested or implemented: (1) Sensitivity of the method to initial data used to build the ecological continuums; (2) Interest of an field approach to refine the ecological continuums modelling; (3) Integration of barriers in the calculation of the ecological continuums.

These three tests for the refinements of the method have the same objective: trying to improve the building of the ecological continuums by the addition of different or new data. With the first test we want to know what changes there will be if we use new land cover initial data at a different scale: Spot Thema (ST, scale 1/25 000ᵉ) instead of Corine Land Cover (CLC, scale 1/100 000ᵉ). The second test, founded on a field approach, should allow us to gain a better idea of what is not written on the land cover maps.

With the third test we try to include barriers in the calculation for a more rigorous and elegant method. For this we award a coefficient of resistance to barriers. For roads, it depends on traffic density and width. A highway, for example, is considered as an impermeable barrier (R = 100), whereas a road with less than 2 000 vehicles a day is more permeable for animal movement (R = 10). In addition, coefficients of resistance were attributed for railways, power lines, and dams as a function of their importance, and for the urban areas according to their continuous or discontinuous nature. At this stage it is also possible to integrate the area of the highway projects A47 and A89 of the department. In the calculations each barrier is considered as a buffer zone whose size is proportional to its importance. For a highway, for example, the buffer allocated is 100 m. For a small road, it is 20 m.

RESULTS

Descriptions of each of the maps obtained for each criterion with the two methods would be long, tedious, and altogether unhelpful. Indeed, our results reflect those obtained by Asconit. The difference is mainly methodological. Our method based on raster technology and MCA allows:

- A real-time use of the tool because it is much faster;
- A best representation of the criteria and a better fit of the parameters;
- An objectification of weighted criteria.

These three points are essential for the proper conduct of the workshop. This workshop is the heart of our work and our main results. Figure 9 shows an example of these results. We can see the WLA ecological continuum calculated in GIS vector mode (top) in which the representation in different categories is set by the calculation. We can see also the WLA ecological continuum calculated in GIS raster mode (bottom) in which we can choose a different category after calculation. In that case it is represented in shades of green.

Criteria Modelling and Participatory Workshop

The maps obtained by each of the actors are very different. For example, the map of the farmer overweighs the continuums EAE and TDA, the map of the tourism promoter highlights the *Protected nature* and *Landscape heritage*, while the map of the fisherman emphasizes the continuum AEW. The map generated after consultation (Figure 10) shows important BGI stakes in the mountains, gorges, valleys, and the plain crossed by the River Loire. Following the workshop and discussions with participants, trainees, and students, the following points emerged.

Lessons Learned from the Experiment

Regarding the BGI method, it appears that:

- The concept of BGI is not easy to explain and the construction method is complicated,

Figure 10. Potential BGI in the South Loire SCOT after weighting (Workshop 2008)

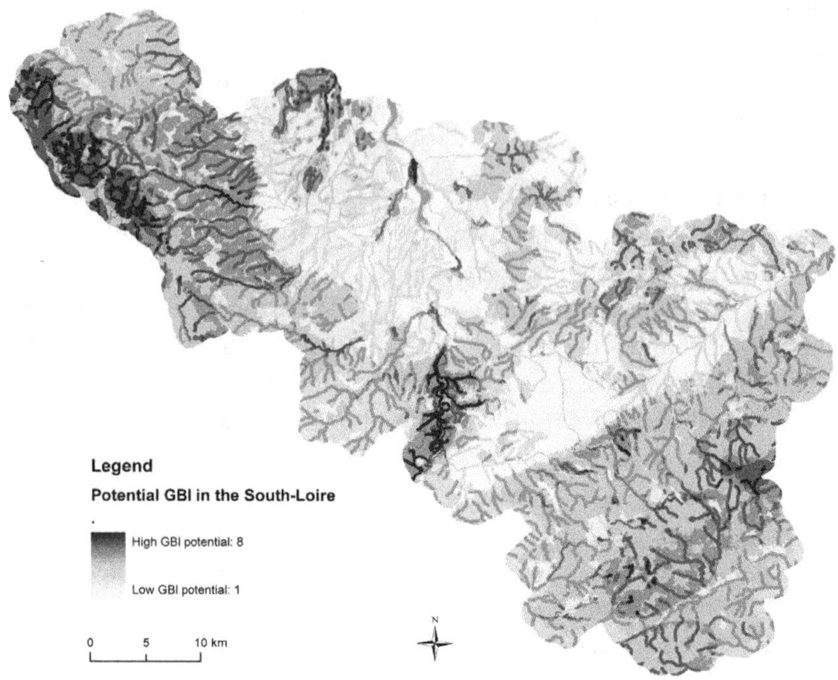

- The ecological validity of the model is questionable (choice of flagship species and of the 3000 points for animal movement),
- The CLC data are not precise enough and an ecological land cover map with barriers, from a field approach, would be necessary.

Regarding the MCA method proposed:

- The Saaty method requires technical support for participants,
- The binary comparison criteria simplify decision making, but do not retain contradictory preferences. Furthermore, the negotiation is disconnected from the spatially defined territorial zones, and does not anticipate the conflicts that occur when decisions directly affect a specific area. One solution would be to weigh spaces (areas) directly rather than specific topics. The difficulty of reaching consensus in a real situation with real actors and finding collective preferences has been addressed. But even if the dialogue fails to produce a consensus, the proposed approach allows everyone to express themselves, to better understand the interests of the others, and to realize the consequences of choices.

Regarding the workshop, we think it is very dense because participants have to assimilate a lot of information and to master various tools quickly. It is necessary to increase its duration.

Outcome Report of the Workshop Approach

From the point of view of the local offices of the Ministry of Territorial Planning of the Loire department which ordered this work:

- Our approach is a concrete improvement of the Asconit method with a direct applicability.
- Many questions about issues, principles, implementation, and presuppositions of the method need to be better explained.
- From the academic point of view:
- The ownership of the method and of the application constraints is rapid.
- The construction of the teaching case is very close to a real case, and it would be necessary to simplify it.
- The tool offers the ability to test the methodological assumptions on participation, the use of the MCA, and the spatial modelling.
- The multi-criteria logic seems to be obvious to the students, and perhaps too obvious. Indeed, they tend to forget, behind the simplicity of the computer interface, the conceptual basis of the MCA method, its complexity, and its limits.
- The host of the workshop has considerable power in the way he tries to create a consensus.
- From the point of view of the other participants:
- Our approach is mobilizing but remains abstract.
- Our new approach facilitates implementation in other places, and allows its potential problems to be shown at full scale.
- Students quickly spotted the arbitrariness that exists in the construction of the continuums, in particular the fact that the structural environment does not allow animals to restore displacement points. The criticism of the ecological validity of this application did not occur in the first workshop bringing together researchers and practitioners.
- The proposed multi-criteria approach seems very abstract since it leads to combining continued criteria throughout the territory and indicating a regionalization retrospectively. Actually, it is possible to overweight only criteria, not areas. This is something that appeared questionable to some participants at the workshop, who would have preferred to work according to zoning determined by experts.

None of the participants doubted the interpretation of the CLC data used both in the original method and in our method. However criticisms were made by the actors themselves when the method was used by the local offices of the Ministry of Territorial Planning of the Loire department for the North Loire area. The same CLC type named "grazing areas" corresponds, in fact, for the actors, to two types of agricultural practices. The idea of asking the actors to reinterpret the CLC data did not appear in the workshop, probably because there is no real knowledge of the field among participants.

Improvement of Our Method

The results derived from experiments in the South Loire validated the GIS raster approach and led to working on smaller areas like La Fouillouse, Montbrison, and Saint-Jean-Bonnefonds (Figure 1) to test the improvement. These improvements were used for a study of the North Loire (Figure 1) area not described in this paper. The adaptation of the method is based on a reflection on the input data, the method of calculation (parameters), and the expert field.

The first improvement consists in observing the sensitivity of the method for initial data (Figure 9). Observation shows that the continuums generated are not identical with data from Corine Land Cover (CLC 1/100 000[e]) and Spot Thema (ST, 1/25 000[e]). The two maps displayed in Figure 11 show a zoom of the study area around La Fouillouse (South Loire): on the left we can see the map generated with CLC data, and on the right the map obtained with ST data. The difference is easily visible. The WLA ecological continuum obtained with Spot Thema data is denser and takes up more space. The structural areas (C = 0, in dark green) are smaller but the rest of the continuum (0 < C < 3000, in light green) is larger. If the difference appears clearly, we need, in the future, to analyse this comparison in more detail. Actually, it appears that a discussion of the CLC and the ST classifications is necessary. Moreover, the absence of an environmentally oriented land mapping throughout the territory is a problem as far as this work is concerned.

The second improvement allows the *Barriers* criteria to be included in the calculation of ecological continuums by awarding a coefficient of resistance. Figure 12 displays two maps which are zooms of the study area: on the left the WLA continuum generated including barriers in the calculation, and on the right without including barriers. At first sight these two maps seem to be similar, but they are different if we look at the details. When the calculation of the continuum includes barriers, the structural areas (C = 0, in dark green) are cut by the lines of little roads. Moreover the WLA continuum seems to be a little less dense. As previously, we need, in the future, to analyse this comparison in more detail.

The third improvement shows that a field approach to a small area was very enriching. It reveals information unavailable in the Corine Land Cover or Spot Thema coverage. In reality, we can see that barriers that seemed at first impervious may actually have secured passage for animals (ducts and tunnels for wildlife under the highway) or could be very dangerous (no wire fence, collision). On rivers also, many small obstacles are not mapped (thresholds, walled banks, pollution, etc.). In the future, we need to think how to include these new data in the calculation of the continuum and how to sample these data in the field.

Figure 11. Sensitivity of the method to initial data for the calculation of the WLA continuum: Corine Land Cover versus Spot Thema

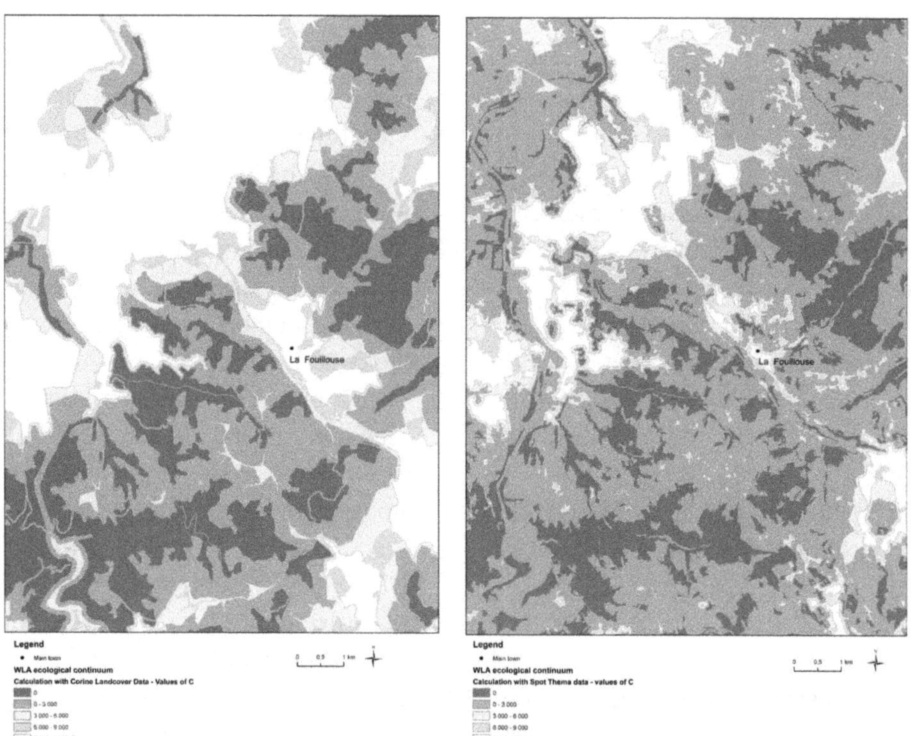

Figure 12. Calculation of the WLA ecological continuum: Including barriers versus without barriers

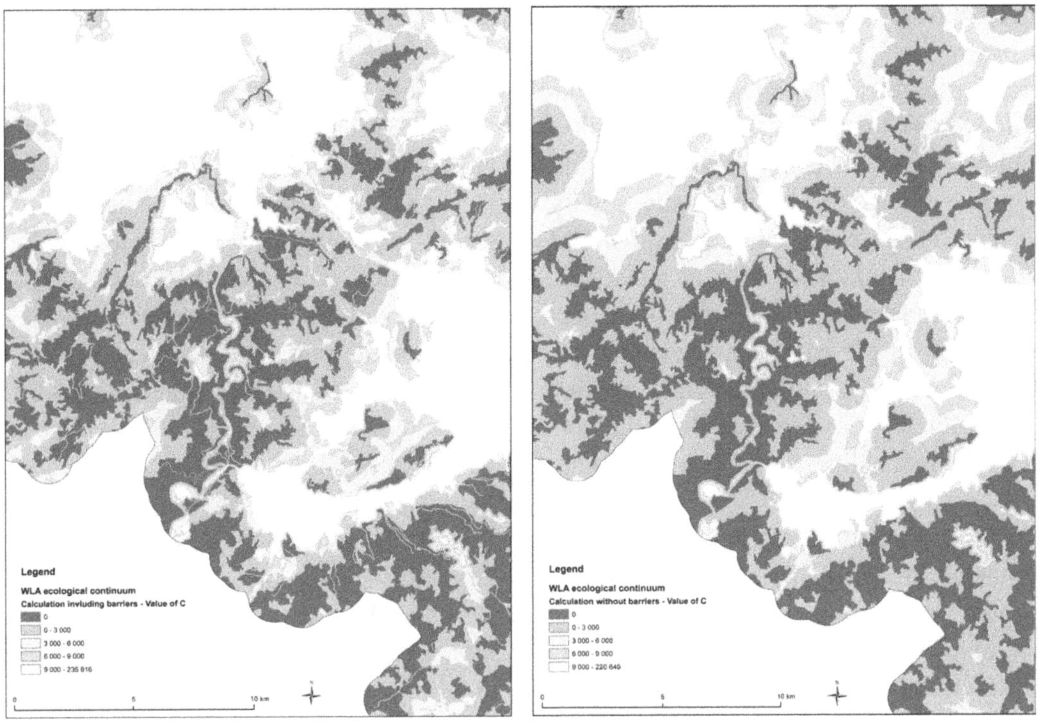

DISCUSSION AND PERSPECTIVES

Actually, the work undertaken raises two main difficulties. Moreover, advantages and drawbacks of the new method are discussed, and perspectives for development in both the research issue and the operational aspects are given.

About the Main Difficulties of BGI Modelling

The first main difficulty is linked with the initial land data and the concept of ecological modelling. On the one side, the absence of an environmentally oriented land mapping is a real gap. On the other side, the allocation of 3000 movement points for all species is questionable. One should therefore turn to more basic research in ecology to improve the modelling of ecological networks (e.g. by distinguishing species, selecting other methods of construction of continuums, and improving knowledge of animal populations). These difficulties concern the initial method and also our method. Actually, the new method can't solve these problems.

The second main difficulty is the integration of barriers in the ecological continuums modelling. Indeed, such a concept is relative to the species considered, which involves problems of change of scale and data availability. For example, if the walls are insurmountable obstacles for some amphibians, this is not true for deer. In this case, it appears that a mapping of these types of barriers (walls) does not exist, and requires a heavy inventory of land. Moreover, their integration at the metropolitan scale remains problematic because of their small size. The integration of barriers requires a more detailed conceptual and methodological study of the problem.

Advantages and Drawbacks of the New Method

The workshop sessions are a good way to test our methodology in a situation that is close to reality. We can check the ownership of the method, its clarity, in spite of the information density, and its working. We can also check its weak points. The issues are only potential stakes which are sometimes dissociated from the reality of the territory. However, we think the workshops sessions are good simulations of the reality. But they are not reality. Actually, the real-life test of our method depends on the real willingness of the politicians and of the administrations.

Actually, our method is able to build the ecological continuums effectively. The calculation is now more elegant because the barriers are included. However, many questions have not been solved. The validity of the model, the consistency of the flagship species, and the initial data sensibility need to be better considered. Furthermore, data are missing because of the difficulty of including them in the calculation. These data are usually non-homogeneous in the territory, such as data on animals run over by cars.

The criterion *Nature protected* seems to fit the problematic because it reflects institutional stakes. In contrast, the models involving data on human beings need to be improved, rethought, or created. The criterion *Landscape heritage* is actually not satisfactory because it takes into account only a small part of the questioning because of a lack of available data. The criterion *Soft movement* does not actually exist. This model is actually in an embryonic state.

About Perspectives

On one hand, it is possible to list the research perspectives:

- More detailed study and development of criteria for a global combination model.
- Enrichment or transform the ecological continuum model with concepts from basic ecology to improve the modelling of ecological network (other methods of calculation, distinction of species, more realistic integration of barriers, etc.).
- Improvement of the knowledge of animal populations (inventories, observations, population genetics, etc.).
- Proposal of a relevant model for the criteria *Landscape and heritage* and *Soft movement*.
- Development of a strategy for participatory enrichment data.
- Consideration of alternative MCA approaches.
- On the other hand, there are operational perspectives:
- Integration of spatial development scenarios: urbanization, territorial planning, and prioritization of actions.
- Change of work scale: from the metropolitan scale to the town scale.
- Mapping of areas of potential conflicts and not only criteria.

Development of both research and operational aspects would allow a diagnosis to be made of functional ecological continuums to identify priority perimeters and prioritize actions and proposed developments. There may be restoration of streams, creation of checkpoints, securing of hazardous areas, improved agricultural practices, and development of improved permeability of the urban areas, or communication actions, animation, or awareness. This diagnosis already seems to be possible at the scale of the town in a local urbanization plan. Moreover, the BGI approach can also be a complementary method to identify geo-historical landscape evolution of a territory.

CONCLUSION

The lessons learned from this experiment are the following. The BGI method is a complex approach which is difficult to communicate. Its ecological validity is questionable because initial data are not well adapted and not validated by expert naturalists in the field. The proposed method requires multi-criteria technical support, which tends to make the negotiation aspatial. The workshop is very attractive but also dense, with a lot of information to absorb.

From the point of view of the local offices of the Ministry of Territorial Planning of the Loire department, our approach is a concrete improvement of the initial method with a test of direct applicability. From the academic perspective the approach and its limitations seem to be easily appropriable. It offers the opportunity to test the methodological assumptions on participation with the use of MCA and GIS spatial modelling. From the perspective of participants, the approach is motivating, but abstract. It facilitates implementation in other areas and shows the potential problems of the method at full scale.

The perspectives of this approach are numerous in terms of both operational and research aspects. In science, it is necessary to deepen or develop the criteria for the construction of a combined global model, to propose new models, to review the calculation of the ecological continuum, to better integrate the barriers, to develop a participatory strategy for enhancement of data, and to consider other multi-criteria approaches. Operationally, it should incorporate scenarios of spatial development, the concept of scale change, and mapping areas of potential conflict through weighting by areas.

The BGI approach is based on the concept of ecological continuum, which is a simplification of the ecological network of scientists. This simplification is appropriate as it provides a potential movement of animal communities in an area. This potential takes into account spatial barriers, land use, and nature protection. To go further, this potential must be confirmed by field work to get a better knowledge of animal communities and ecological land occupation. As we have seen, the data available in government services strongly constrain the accuracy and the type of analysis as well as the nature of the results. In addition, the continuum model could be expanded or replaced in the light of advances in basic research on ecological networks.

Geomatic tools are essential in the BGI process. Those available at the local offices of the Ministry of Territorial Planning of the Loire department, as in many land management agencies, are only capable of processing data in vector format. This complicates and dodges the issue of the calculations related to the simulation of continuous phenomena such as displacement. In the workshop, the treatments are done in raster mode, and participants can test an approach based not only on visual reasoning but also on analysis and spatial reasoning. This kind of analysis is not intuitive or easy because it leads to a relative loss of territorial reference.

One of the main contributions of our work is to consider the BGI approach as based on the combination, ranking, and weighting of different criteria. However, the initial method for expressing and comparing the preferences of various experts of the Steering Committee makes no reference to multi-criteria analysis methods which have been widely tested nevertheless. We propose in the workshop an introduction to the basics of multi-criteria analysis (Saaty approach) that allows participants to test the interest and discuss the boundaries of a formalized approach.

ACKNOWLEDGMENT

This work would not have succeeded without the financial support and the data of the local offices of the Ministry of Territorial Planning of the Loire department, and especially without the help of

Gérard Bol, Christophe Bétin, Nathalie Fournel, and Christine Fournel. Moreover, we are indebted to Bernard Dupuis (University of Saint-Etienne, CRENAM) and Roland Déchomets (Ecole Nationale Supérieure des Mines de Saint-Etienne, SITE) for their valuable help with GIS and their support during the experiment. Special thanks are owed to Georgios Karagiannis for his proofreading work and for his constructive comments and suggestions which helped to improve this paper.

REFERENCES

Asconit Consultants. (2003). *État initial du site et évaluation environnementale du SCoT du pays de Montbéliard*. Paris, France: Direction Générale de l'Environnement de Franche-Comté.

Asconit Consultants. (2005a). *IVB, Utilisation d'un SIG pour l'expression des enjeux de l'Etat dans le cadre d'un SCoT, application au territoire du SCoT Sud-Loire – guide méthodologique*. Paris, France: DIREN Rhône-Alpes.

Asconit Consultants. (2005b). *IVB, Utilisation d'un système d'information géographique pour l'expression des enjeux de l'Etat, application au territoire du SCoT Sud Loire – guide technique*. Paris, France: DIREN Rhône-Alpes.

Batton-Hubert, M., Bonnevialle, M., Joliveau, T., & Paran, F. (2007a). *IVB dans un Scot: Ateliers*. Paper presented at the Rencontres internationales Géomatique et Territoire (Paper presented at the Sageo, Saint-Etienne, France).

Batton-Hubert, M., Bonnevialle, M., Joliveau, T., & Paran, F. (2007b). *Les méthodes d'identification des Infrastructures Vertes et Bleues (IVB) au service des projets d'aménagement et de développement durable des SCoT.* Paper presented at the Semaine régionale du développement durable, Saint-Etienne, France.

Batton-Hubert, M., Bonnevialle, M., Joliveau, T., & Paran, F. (2008). *Intégrer une dimension écologique et paysagère dans la planification territoriale – Méthode et questionnement à propos de la démarche Infrastructures Vertes et Bleues (IVB) dans les SCoT de la Loire (France).* Paper presented at the Les outils pour décider ensemble, Université de Laval, Québec, Canada.

Berthoud, G. (2001). *Les Corridors biologiques en Isère: Projet de Réseau écologique Départemental de l'Isère (REDI)*. Paris, France: Conseil Général de l'Isère.

Berthoud, G., Lebeau, R., & Righetti, A. (2004). *Le Réseau Ecologique National (REN) Suisse Cahiers de l'environnement, Office Fédéral de l'Environnement des Forêts et du Paysage* (p. 373). Berne, Suisse: OFEFP.

Burel, F., & Baudry, J. (1999). *Écologie du paysage: Concepts, méthodes et applications*. Paris, France: Tech. & Doc.

Chakhar, S. (2006). *Cartographie décisionnelle multicritère: formalisation et implémentation informatique.* Unpublished doctoral dissertation, Université Paris Dauphine, Paris, France.

Chakhar, S., & Mousseau, V. (2008). GIS-based multicriteria spatial modeling generic framework. *International Journal of Geographical Information Science*, *22*(11), 1159–1196. doi:10.1080/13658810801949827

Figueira, J., Greco, S., & Ehrgott, M. (2005). *Multiple criteria decision analysis: State of the art surveys* (*Vol. 57*). London, UK: Springer.

Forman, R. T. T., & Godron, M. (1986). *Landscape ecology*. New York, NY: John Wiley & Sons.

Hanski, I., & Gilpin, M. E. (1997). *Metapopulation biology: Ecology, genetics, and evolution.* New York, NY: Academic Press.

Jongman, R. H., & Pungetti, G. (2004). *Ecological networks and greenways: Concept, design, implementation*. Cambridge, UK: Cambridge University Press. doi:10.1017/CBO9780511606762

Keeney, R. L., & Raiffa, H. (1976). *Decisions with multiple objectives: Preferences and values trade-offs*. New York, NY: John Wiley & Sons.

Molines, N. (2006). *Méthode et outils pour la planification des grandes infrastructures linéaires et leur évaluation environnementale.* Unpublished doctoral dissertation, Université Jean Monnet, Saint-Etienne, Université de Laval, Québec, Canada.

Paran, F. (2005). *Représentations territoriales pour la gestion équilibrée d'un patrimoine écologique et anthropique dans le domaine de l'eau.* Unpublished doctoral dissertation, Ecole Nationale Supérieure des Mines, Saint-Etienne, France.

Saaty, T. L. (1982). *Decision making for leaders: The analytical hierarchy process for decisions in a complex world.* Belmont, CA: Life-Time Learning Publications.

Saaty, T. L., & Shih, H.-S. (2009). Structures in decision making: On the subjective geometry of hierarchies and networks. *European Journal of Operational Research, 199*(3), 867–872. doi:10.1016/j.ejor.2009.01.064

Tavares-Pereira, F., Figueira, J., Mousseau, V., & Roy, B. (2007). Multiple criteria districting problems. *Annals of Operations Research, 154*(1), 69–92. doi:10.1007/s10479-007-0181-5

This work was previously published in International Journal of Agricultural and Environmental Information Systems, Volume 2, Issue 2, edited by Petraq Papajorgji and François Pinet, pp. 29-51, copyright 2010 by IGI Publishing (an imprint of IGI Global).

Chapter 11
Spatial Pattern Mining for Soil Erosion Characterization

Nazha Selmaoui-Folcher
University of New Caledonia, New Caledonia

Frédéric Flouvat
University of New Caledonia, New Caledonia

Dominique Gay
Orange Labs, France

Isabelle Rouet
University of New Caledonia, New Caledonia

ABSTRACT

The protection and the maintenance of the exceptional environment of New Caledonia are major goals for this territory. Among environmental problems, erosion has a strong impact on terrestrial and coastal ecosystems. However, due to the volume of data and its complexity, assessment of hazard at a regional scale is time-consuming, costly and rarely updated. Therefore, understanding and predicting environmental phenomenons need advanced techniques of analysis and modelization. In order to improve the understanding of the erosion phenomenon, this paper proposes a spatial approach based on co-location mining and GIS. Considering a set of Boolean spatial features, the goal of co-location mining is to find subsets of features often located together. This system provides useful and interpretable knowledge based on a new interestingness measure for co-locations and a new visualization of the discovered knowledge. The interestingness measure better reflects the importance of a co-location for the experts, and is completely integrated in the mining process. The visualization approach is a simple, concise and intuitive representation of the co-locations that takes into consideration the spatial nature of the underlying objects and the experts practice.

DOI: 10.4018/978-1-4666-0333-2.ch011

INTRODUCTION

New Caledonia is an archipelago located in an exceptional environment. For example, a big part of the Caledonian barrier reef is referenced in the world heritage list by the UNESCO. The protection and the maintenance of this environment are a major goal in New Caledonia. Because of important mining projects on this territory, a global approach for environmental monitoring is necessary. Several national and local scientific plans finance different works about the climate change and the human effects on the environment to save.

In New Caledonia, erosion has a strong impact on terrestrial and coastal ecosystems (mountains, alluvial plains, mangroves, coral reefs). Bush fire, deforestation and/or human activities accelerate erosion in mountainous areas especially on lateritic soils. Frequent and intense precipitation events (tropical depressions, cyclones) regularly strip particles from these sensitive soils. Any eroded sediment is quickly transported to coastal plains and to the sea along the main drainage lines, with immediate and recurrent impact on human activities such as open-cast mining, farming and fishing. There is therefore a need to identify key components of erosion processes for sustainable environmental management and planning. Particularly, hazard assessment is required to plan new developments with limited erosion impacts. However, both mapping of active erosion and assessment of hazard at a regional scale is time-consuming, costly and rarely updated. The volume of data can be considerable and they are generally heterogeneous (images, physical measures, qualitatives observations). Understanding and predicting environmental phenomenons require advanced techniques of analysis and modelization.

The interest of data mining methods for the soil erosion problem has been highlighted in (Gay, Rouet, Mangeas, Dumas, & Selmaoui, 2007; Rouet et al., 2009). GIS raster data was used to describe and predict erosion hazard at regional scale. Nevertheless, the spatial relationship between GIS vector layers was not considered while it could more accurately help to describe areas with erosion phenomena.

We propose in this paper a spatial approach based on spatial pattern mining and GIS to improve the knowledge of the spoiling level of a geographical area. One of the classical task in spatial pattern mining is the extraction of interesting co-locations in geo-referenced data (Koperski & Han, 1995; Shekhar & Huang, 2001; Appice, Ceci, Lanza, Lisi, & Malerba, 2003; Huang, Shekhar, & Xiong, 2004; Lisi & Malerba, 2004; Bogorny et al., 2006; Yoo & Shekhar, 2006; Ceci, Appice, & Malerba, 2007; Celik, Kang, & Shekhar, 2007; Malerba, 2008).

Considering a set of Boolean spatial features, the goal is to find subsets of features often located together. In other words, we aim to characterize relations between interesting features. However, the interpretation of the extracted patterns by domain experts is difficult. One of the major challenges in spatial data mining is to discover and deliver actionable knowledge to domain experts (Cao, 2008). Although many algorithms and techniques for spatial data mining have been proposed, it still remains an open problem to successfully provide useful and interpretable knowledge to experts. Indeed, existing interestingness measures for co-locations can lead to interpretation problems, and co-locations are presented in a textual form.

Our system provides useful and interpretable knowledge to domain experts based on a new interestingness measure for co-locations and a new visualization of the discovered knowledge. Our interestingness measure better reflects the importance of a co-location for the experts, and is totally integrated in the mining process. Note that interestingness measures can be used to quantify some specific relations and to develop environmental indicators. Our visualization approach is a simple, concise and intuitive representation of the co-locations that takes into consideration the spatial nature of the underlying objects and the experts practice. These propositions have been integrated in a prototype with a GIS, experimented on a real geological dataset, and validated by a domain expert.

SPATIAL PATTERN MINING

The aim of spatial data mining is to analyze large geographical databases and to extract implicit informations from spatial data. In other words, spatial data mining can be viewed as the search for interesting, useful and unexpected, but implicit spatial patterns. This is why, classical data mining methods are not sufficient, spatial relationships must be considered. These spatial relationships are often implicitly defined and represent the neighborhood influence. Currently, many works have been realized in the field of spatial data mining (Miller & Han, 2009). There are two types of spatial data mining methods: descriptive and predictive methods. Predictive methods, such as decision trees, can be used to forecast explicit values, based on patterns determined from known results. Descriptive models, such as clustering or pattern mining, describe patterns in existing data, and are generally used to extract implicit relations in the data (such as clusters or association rules).

In this paper, we focus our work on spatial pattern mining techniques and their utilization to describe the erosion phenomenon.

Related Works

Spatial pattern mining finds non explicit spatial relationships in "geo-referenced" data. Several works have studied spatial pattern mining to discover patterns that describe strongly associated features in the data. The discovered patterns are typically represented in the form of implication rules or features subsets (Tan, Steinbach, & Kumar, 2005). To deal with this problem, two families of spatial pattern mining approaches may be identified: multi-relational approaches and co-location-based approaches.

When spatial data is made of various tables describing objects and spatial relationships between objects, multi-relational data mining techniques may be applied to extract spatial interesting patterns such as association rules (Lisi & Malerba, 2004) or emerging patterns (Ceci et al., 2007; Appice et al., 2003; Malerba, 2008).

On the other hand, two types of co-location-based approaches are identified (Huang et al., 2004): transaction-based approaches and event-based approaches.

The first one focus on transforming spatial data into transactional data where classical itemset mining algorithms could be used (Koperski & Han, 1995; Bogorny et al., 2006). In Koperski and Han (1995), authors presented a method for mining association rules in geographic information databases. This method enumerates neighborhoods to "materialize" a set of transactions around instances of the reference spatial feature. The goal is to find co-location of relevant features to the reference feature. Bogorny et al. (2006) extends this work by introducing knowledge constraints in a preprocessing step. These constraints are association previously known as non-interesting for experts.

The second approach focus on the event and their neighborhood relationships (Shekhar & Huang, 2001; Huang et al., 2004; Yoo & Shekhar, 2006). Shekhar and al. have defined the co-location concept based on Koperski's work. The goal is to find all subsets of spatial features likely to occur together. Unlike the first approach, this one considers all features together and original data are not transformed. A prevalence measure has been introduced to filter the most important co-locations. Thanks to the monotonic property of this predicate, a classical levelwise algorithm has been used to extract the solutions. Notice also the recent work (Celik et al., 2007) which extends the notion of co-location to zonal co-location pattern (i.e., intuitively a local zone-scale co-location pattern).

In our application, we are interested in finding subsets of spatial features likely to occur often together. Since we have several types of features to analyze, we decided to base our experience on this last approach. In the following subsection, we will present the co-location concept.

Co-Location Concept

Co-location patterns are subsets of Boolean features associated with spatial objects. In Shekhar and Huang (2001) and Huang et al. (2004) co-locations represent features often located close together.

Definition: *Let* $F=\{f_1, f_2, \ldots f_k\}$ *a set of spatial features,* $O=\{o_1, o_2, \ldots o_n\}$ *a set of spatial objects.* $\Theta: O \rightarrow F$ *is the function that associates an object with a feature.*

This function Θ formally defines the association between objects and features. To simplify notations, the relation between an object $o_i \in O$ and a feature $f \in F$, i.e. the pair $(o_{i,f})$, is noted f_i and represent an object, Figure 1 represents an example where $O=\{o_1, o_2, \ldots o_{12}\}$ and $F=\{A,B,C,D,E\}$. A_9 corresponds to the relation $\Theta(o_9)=A$.

Definition: *A co-location C is a subset of features of F such that there exists a set of neighbor objects having these features.*

For example, on Figure 1, $\{A,B,D\}$ is a co-location w.r.t. a fixed Euclidean distance threshold (represented by circles on Figure 1). However, not every co-location is interesting. There is only one set of three neighbor objects having the features *A*, *B* and *D*. Consequently, we need other concepts to determine if a co-location is interesting or not.

A row instance of a co-location *C*, w.r.t. a neighborhood relation *R*, is a set of objects *O*, such as it has the features of *C* and satisfies the spatial relationship. Note that the neighborhood relationship *R* is reflexive and symmetric. To simplify notations, we write "instance" instead of "row instance" in the rest of the paper. Figure 1 shows that the set of objects $\{A_1, B_8, C_7\}$ is an instance of the co-location $\{A,B,C\}$ $\{A_1, B_8$ and C_7 are neighbor according to *R*). An instance of a co-location *C* satisfies the following property:

Definition: *A row instance* $I \subseteq O$ *of a co-location* $C \subseteq F$ *according to a neighborhood relationship R is a set of neighbors objects associated to each of the features of C, i.e.:*

- $|\{f \in C \mid o \in I \text{ and } \Theta(o) = f\}| = |C|$
- $|I| = |C|$
- $\forall o,p \in I, R(o,p) = true$

A set of objects *satisfies* a co-location *C* w.r.t. the neighborhood relation *R*, if it is an instance of *C*. For example, the set of objects $\{A_1, B_8, C_7\}$ satisfies the co-location $\{A,B,C\}$ (w.r.t. the fixed Euclidean distance threshold) while $\{A_1, B_4, C_7\}$, $\{A_1, B_4, D_{10}\}$ or $\{A_1, B_4\}$ do not. Identically, an object $o \in O$ *participates* in co-location *C* if it belongs to an instance of *C*. For example, objects A_1, B_8 and C_7 participate in the co-location $\{A,B,C\}$ while objects B_3 or E_{12} do not.

A table instance of a co-location *C*, denoted $TI_R(O,C)$ is the set of all its row instances. The table instance of $\{A,B,C\}$ is $TI_R(O, \{A,B,C\}) = \{ \{A_1,B_8,C_7\}, \{A_5,B_6,C_2\} \}$ and the table instance of $\{B,D\}$ is $TI_R(O, \{B,D\}) = \{ \{B_4,D_{10}\} \}$ (Figure 1).

Figure 1. Spatial objects and their features

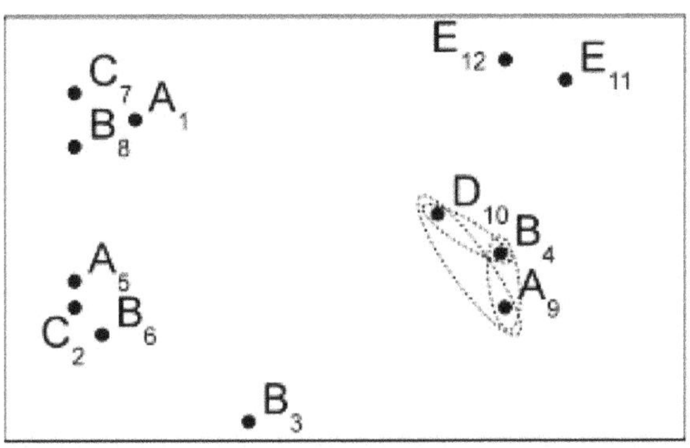

Definition: *Given $C \subseteq F$ a co-location, O the set of spatial objects and R a neighborhood relationship. The table instance of C is defined as:*

$TI_R(O,C) = \{ I \subseteq O \mid I$ is an instance of C *w.r.t.* $R \}$

The participation ratio $pr_R(O,C,f)$ for a feature f of a co-location C, is the fraction of objects from O having the feature f and participating in C, to the total number of objects having the feature f. In Figure 2, $pr_R(O,\{A,B,C\},A)=2/3$, $pr_R(O,\{A,B,C\},B)=1/2$ and $pr_R(O,\{A,B,C\},C)=1$.

Definition: *Given $C \subseteq F$ a co-location, O the set of spatial objects, R a neighborhood relationship and $f \in C$ a feature. The participation ratio of f in the co-location C is equal to:*

$$pr_R(O,C,f) = \frac{|\{o \in I \mid I \in TI_R(O,C) \ and \ \Theta(o) = f\}|}{|TI_R(O,\{f\})|}.$$

Based on above definitions, Shekhar and Huang, (2001) and Huang et al. (2004) have proposed the concept of participation index. This measure is used to estimate the co-location frequency. Two definitions have been proposed:

The first one (Shekhar & Huang, 2001), denoted $pi1_R$ (Figure 2), represents the probability to have the co-location C among all objects that participate in C (supposing that features are independently distributed).

$$pi1_R(O,C) = \prod_{\forall f \in C} pr_R(O,C,f).$$

The second one (Huang et al., 2004), denoted $pi2_R$ (Figure 3), represents the minimal probability to have an object that participates in the co-location C w.r.t. all objects having this feature.

$$pi2_R(O,C) = \min_{\forall f \in C}(\ pr_R(O,C,f)\ ).$$

Based on these definitions, the problem to solve is:

Co-location mining problem: *Given F a set of features, O a set of spatial objects, R a neighborhood relationship and $\alpha \in [0,1]$ a threshold. The problem is to find the set of co-locations $\{C \subseteq F \mid pi_R(O,C) \geq \alpha\}$ where pi_R the function used to compute the participation index.*

These two measures are anti-monotonic (Shekhar & Huang, 2001; Huang et al., 2004). Thanks to this property, these measures can be used as a predicate to prune the search space. Thus, the authors used in their work a levelwise pattern mining algorithm (Mannila & Toivonen, 1997) to extract interesting co-locations.

Figure 2. Example of participation index with pi1

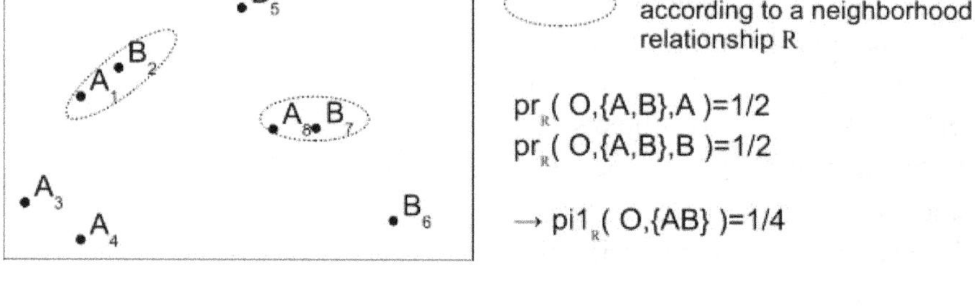

TOWARDS A BETTER INTERPRETATION OF CO-LOCATIONS FOR EXPERTS

We propose in this section a new interestingness measure and a new visualization approach such that the discovered spatial patterns can be easily understood and directly usable by experts.

A New Interestingness Measure for Co-Locations

As we have seen in the previous section, two measures have been defined in Shekhar and Huang (2001) and Huang et al. (2004) to filter interesting co-locations. Nevertheless these two measures can lead to interpretation problems for the experts.

The first measure $pi1_R(O,C) = \prod_{\forall f \in C} pr_R(O,C,f)$ represents the probability to have the co-location C among all objects that participates in C assuming that features are independent. This hypothesis is not usually true in our context since the objective is to describe (and/or find hidden) relation between spatial features. Consequently, this measure is not adapted to represent the importance of a co-location. Figure 2 shows an example where the participation index of *{A, B}* is equal to 1/4 while half of objects having the feature *A* or *B* participate in this co-location.

The second measure

$$pi2_R(O,C) = \min_{\forall f \in C}(\, pr_R(O,C,f)\,)$$

is more relevant than the first one, but its interpretation is difficult for experts. Actually, the importance of a co-location C depends only on the feature from C which has the lowest probability of participating in C. As a consequence, this measure may not represent the global importance of a co-location. The co-location *{A,B}* has the same participation index in two different cases, although this co-location is more important in Figure 3 on the right side.

To deal with these problems, we define a new measure called the **co-location ratio** denoted $cr_R(O,C)$. The aim of the co-location ratio is to have a better interpretation of results by experts. This measure is the fraction of instances participating in the studied co-location, to the minimal

Figure 3. Example of particpation index pi2

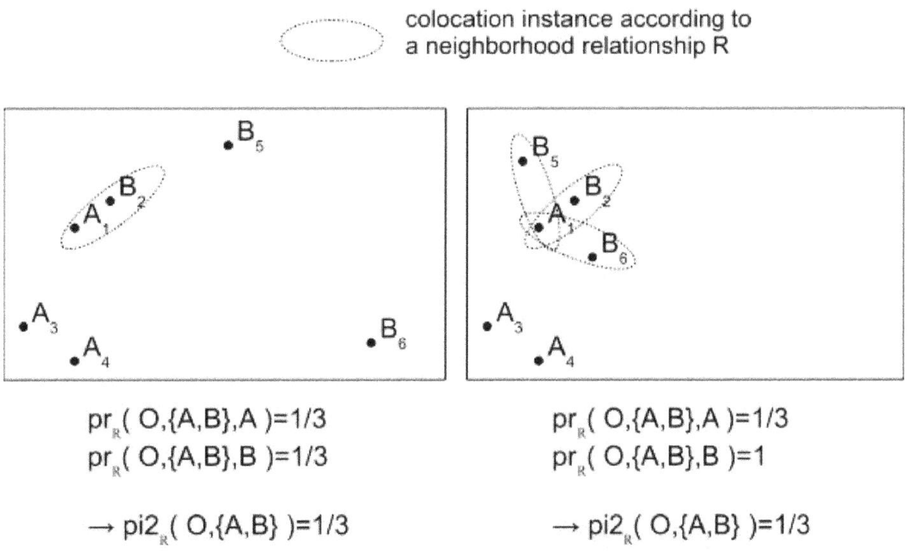

number of instances that we would have if the co-location were always true.

Definition: *Given $C \subseteq F$* a co-location, O the set of spatial objects and R a spatial boolean relationship. The co-location ratio of C is equal to:

$$cr_R(O,C) = \frac{|TI_R(O,C)|}{|TI_R(O,C)| + M},$$

where $M = \max_{\forall f \in C} \left|\{o \in TI_R(O,f) \mid o \notin I, \forall I \in TI_R(O,C)\}\right|.$

Figure 4 shows the value of the co-location ratio for the co-location *{A,B,C}* on the example of Figure 1.

On this example, the co-location ratio is *1/2* since the number of *{A,B,C}* instances is 2 and the number of instances if this co-location were always satisfied would be 4. Indeed, there is two instances satisfying this co-location, and two objects with feature *B* and one with feature *A* that do not.

If the co-location were always satisfied, these objects will be in an instance of *{A,B,C}*. Actually, they would be at least two new instances with B_3, B_4 and A_9, such as $\{A_9,B_4,C_7\}$ and $\{A_5,B_3,C_2\}$, or $\{A_9,B_3,C_2\}$ and $\{A_1,B_4,C_7\}$, depending on the location of these objects.

Note that these objects can generate more instances if they were located elsewhere (or with another neighborhood relationship), e.g. $\{A_9,B_8,C_7\}$, $\{A_9,B_3,C_7\}$, $\{A_1,B_3,C_7\}$ and $\{A_5,B_3,C_2\}$.

Based on the same examples, Figure 5 shows that co-location ratio gives a better interpretation

Figure 4. Example with the co-location ratio

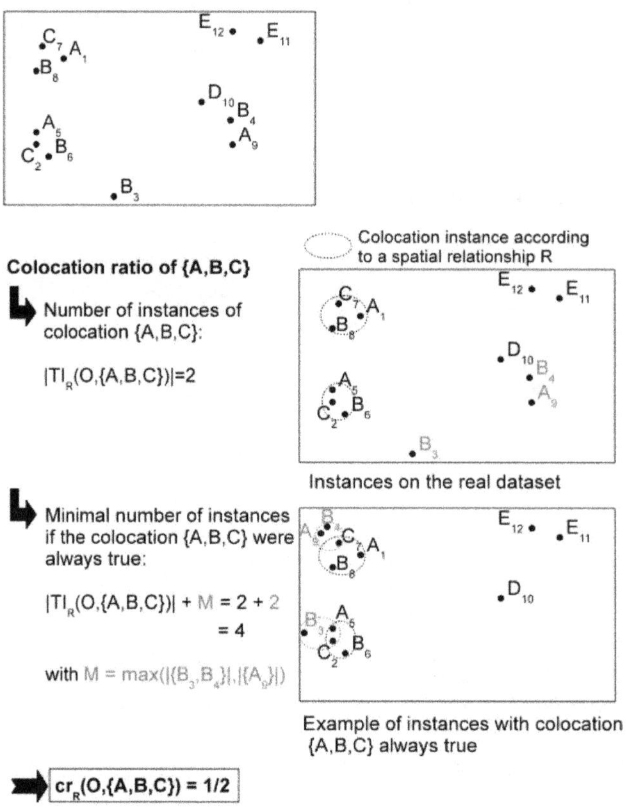

than the two participation index measures above. Our measure provides a more accurate information about the co-location representativeness. Moreover, its definition is more intuitive which leads to an easier interpretation for experts.

Another advantage of our measure is that it can be processed during the mining algorithm without any overhead. Actually, informations used to process the co-location ratio are the same than the ones used to process the participation index and filter interesting co-locations.

A Spatial Visualization of Co-Locations Integrated in a GIS

One of the major issues in data mining is the representation of the discovered knowledge such that it can be easily understood and directly usable by experts (Han & Kamber, 2006). Nevertheless, most of the data mining methods return the results in a textual form. Finding a simple, concise and intuitive representation of co-locations is not that simple.

Several visualization systems have been developed to represent raw data or mining results (Brunk, Kelly, & Kohavi, 1997; Keim, 2002). For example, MineSet (Brunk, Kelly, & Kohavi, 1997) is an interactive system for data mining integrating data visualization. Different kinds of visualizer (statistics, scatter, map, tree) are available according to the type of result to visualize. In Andrienko and Andrienko (1999) authors were interested in representing spatial data and providing a visualization of data mining results. In particular, their system makes a dynamic link between the map and data mining reports. For example, when a cursor is positioned on a tree node or rule in a report, the corresponding objects are highlighted in the map (and vice versa). Recently, Leung, Irani, and Carmichael (2008)

Figure 5. Co-location ratio on datasets leading to interpretation problems with the participation index

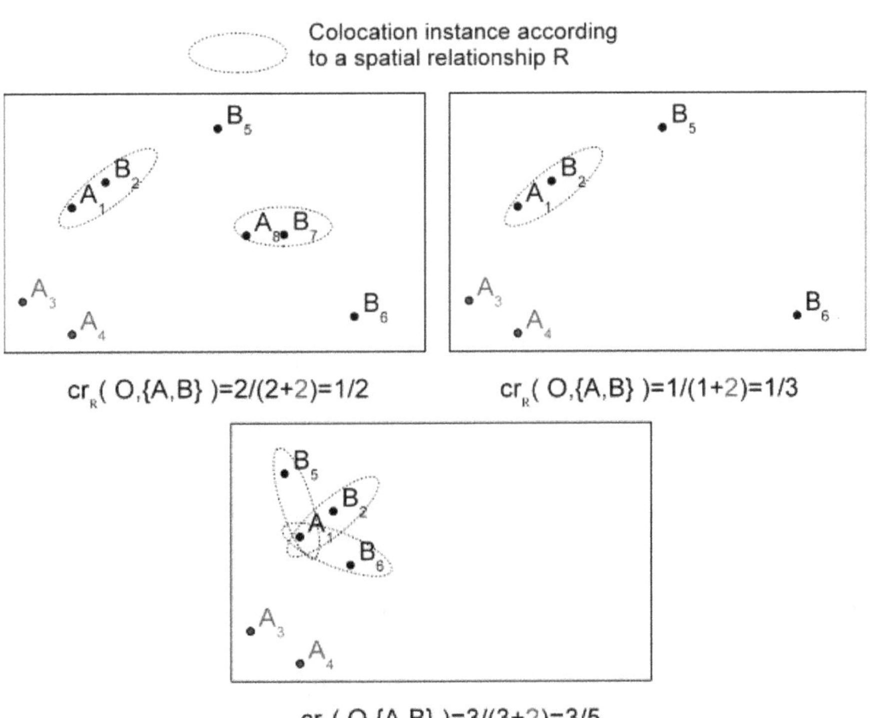

deal with the visualization of frequent itemsets. The authors developed a system, WiFIsViz, for visualizing frequent itemsets based on orthogonal graphs (wiring-type diagrams).

However, to our knowledge, none of the solutions proposed in the literature were designed to display spatial patterns in a simple, concise and intuitive way for experts.

In domains manipulating geographical data, Geographic Information Systems (GIS) are classical tools used by experts for storing and visualizing the data. A main characteristic of GIS is the cartographic visualization of the information in thematic layers. In this context, our objective is to find a spatial visualization of the co-location mining results integrated in the GIS. However, showing on a map all the instances participating in a co-location is unreadable and listing the co-locations in a textual form leads to the loss of the spatial information of objects. To deal with these problems, we propose a new cartographic visualization of co-locations in a GIS. The spatial information of *the instances participating in the co-locations* is summarized and these results are integrated in a "co-location layer" of the GIS (Figure 6).

In our visualization system, a co-location is represented by a set of new spatial objects (generated and stored in a specific layer of the GIS), linked together by lines and positioned on the map. More precisely, each feature f of a co-location C is represented by the centroid of objects participating in C and having the feature f (see an example Figure 7). In other words, if we note by $\{o'_1, o'_2, ..., o'_k\}$ the set of spatial centroids representing the objects of the co-location $C = \{f_1, f_2, ..., f_k\}$, we have:

$$o'_i = (x_i, y_i)\text{, such that } x'_i = \frac{\sum_{\forall o=(x,y) \in \Omega_{f_i,C}} x}{|\Omega_{f_i,C}|},$$

$$y'_i = \frac{\sum_{\forall o=(x,y) \in \Omega_{f_i,C}} y}{|\Omega_{f_i,C}|},$$

and

$$\Theta(o'_i) = f_i,$$

where

$$\Omega_{f_i,C} = \{o \in I \mid I \in TI_R(O,C) \textit{ and } \Theta(o) = f_i\}.$$

Our visualization system also integrates thematic aspects by painting each feature in the color of the corresponding theme. In the same way, the intensity of the links color is proportional to the value of the co-location ratio associated to the co-location.

Figure 8 illustrates the visualization of co-locations *{A,B,C}* and *{A,B,D}* (Figure 8 top) based on their instances (Figure 8 bottom). Each feature color corresponds to a theme. The color of *{A,B,C}* links is darker than the color of the *{A,B,D}*, since the co-location ratio of *{A,B,D}* is higher than the co-location ratio of *{A,B,C}*.

Advantages of Our Proposition

This visualization approach has the following advantages:

- It is totally integrated in the GIS,
- It provides spatial information about the objects behind the co-locations,
- It is integrated in the mining algorithm (not a post-processing step).

Firstly, we get a spatial visualization of the co-location totally integrated in the GIS, and thus adapted to experts' needs and practices. The original data is not affected by our approach, only an additional layer is added. Moreover, it can take advantage of the GIS functionalities. For example,

Figure 6. Principle of our approach

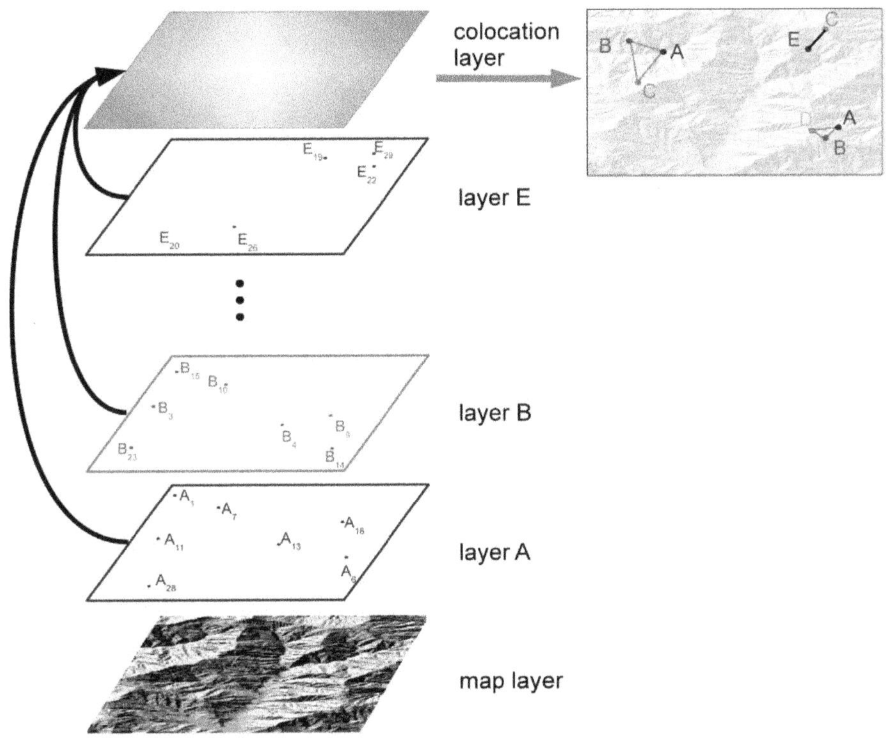

the user can zoom on the map in order to have either a general view of all co-locations (fFgure 15, in the middle), or a detailed view of one or several co-locations (Figure 15 on the right).

Secondly, this representation gives additional informations on co-locations. Actually, it allows visualizing the global location of instances participating in a co-location, i.e. to know in which part of the studied area these instances are generally located. For example, in Figure 9 (top), most of instances participating in the co-location *{A,B,C}* are located in the north west of the map. Thus, the representation of this co-location in the co-location layer (Figure 9 bottom) is also located in the north west.

Our approach also allows to visualize the average distance between objects instantiating the co-location. Figure 9 shows for example that

Figure 7. Example of construction of a co-location representation

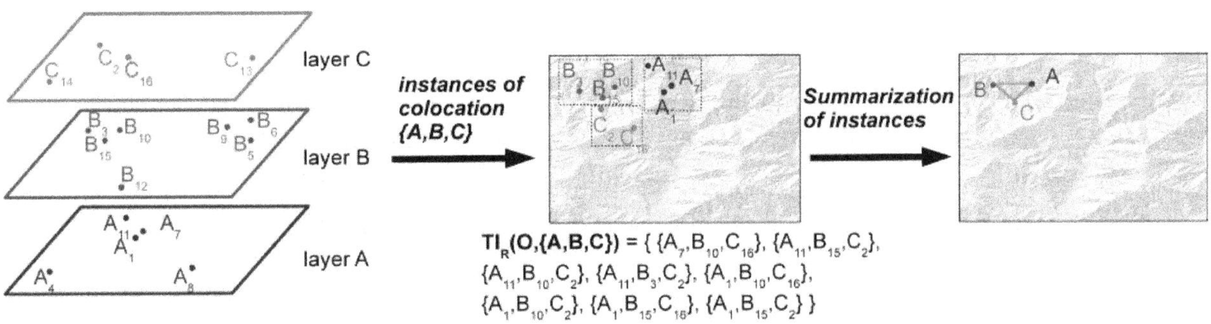

Figure 8. Visualization of two co-locations of size 3

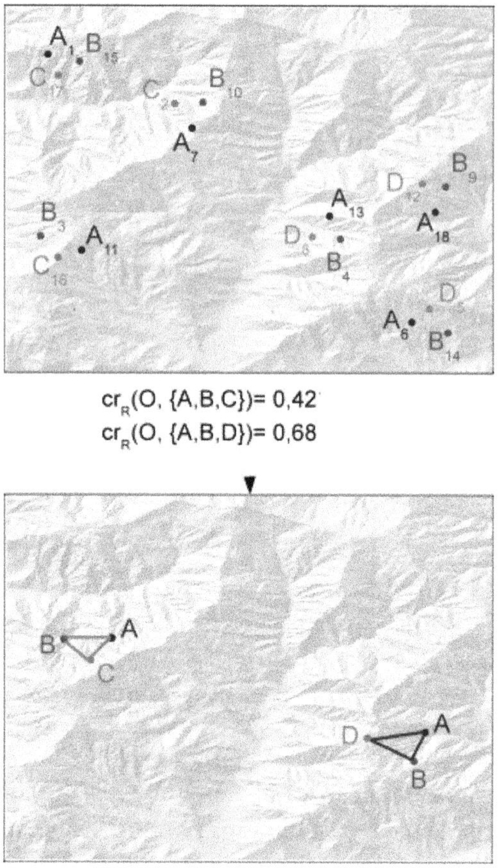

Figure 9. Visualizing spatial informations on co-locations: location, distance and orientation

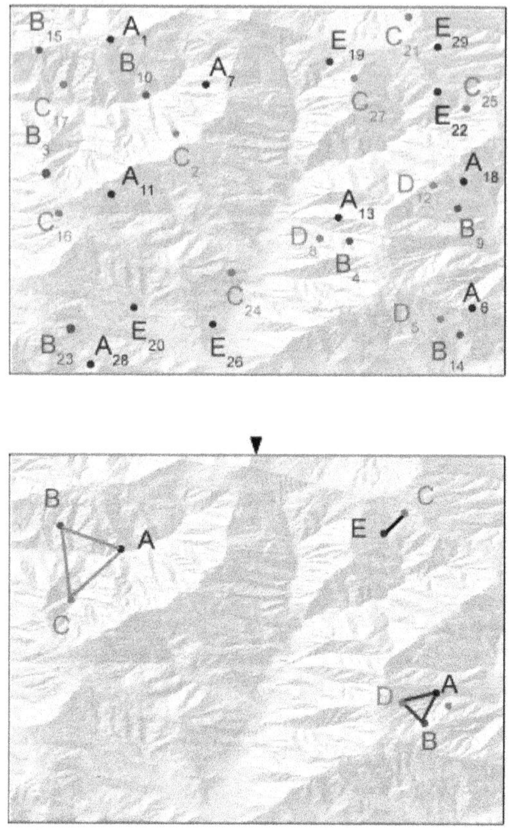

instances of co-location *{A,B,D}* are generally closer than the ones of the co-location *{A,B,C}*.

In the same way, the global position between objects instantiating the co-location can be visualize by our solution. For example, in Figure 9, instances of the co-location *{A,B,D}* have the following configuration: objects having feature *B* are below the ones having features *A* and *D*, and objects having feature *D* are generally located on the left of the ones having feature *A*. Furthermore, note that experts can easily visualize the importance of a co-location and its themes thanks to the color system.

Finally, this visualization approach do not require additional processing, since it is done during the mining algorithm using informations processed for co-location mining.

Main Limit of Our Approach and Solution

However, this visualization approach can lead to an interpretation problem appears when the representation of the co-location is located in the middle of the studied zone. Indeed, instances of such co-locations can be located either in the middle of the area or uniformly distributed all over the area. A solution to deal with this problem is to use clustering in order to have several centroids, and not only one, for each feature of a co-location. Thus, each co-location will have several representations w.r.t. locations of its objects.

Integration of Our Propositions in the Mining Algorithm

Algorithm 1 gives the mining steps to extract co-locations. This algorithm is an adaptation of the levelwise algorithm proposed in Shekhar and Huang (2001), which is based on the *Apriori* algorithm of Agrawal and Srikant (1994). It has been modified to process the co-location ratio (in addition to the participation index) and to construct the visual representation of each pattern.

In line 3, it generates the candidate co-locations level by level (i.e. w.r.t. their size) by using a generalized version of the $apriori_{gen}$ method (Agrawal & Srikant, 1994). Note that this generalized algorithm is also described in Mannila and Toivonen (1997). For each candidate co-location (line 4-13), the algorithm generates its table instance using a SQL query on the database (line 5). The SQL query used in this step is the same than the one described in Shekhar and Huang (2001). Then, the algorithm processes the participation index (line 6). Note that the co-location ratio is processed at the same time since it uses the same informations without requiring additional processing. Then, if the co-location is interesting (i.e. its participation index is higher than the participation index threshold α), we construct its visual representation (line 8). The visual representation with its co-location ratio are stored using a SQL query in the co-location table of the database (line 9). Thus, the participation index is used to prune the search space and the co-location ratio is used for the interpretation by domain experts.

The complexity of this algorithm is in the same order of magnitude than the algorithm of Shekhar and Huang (2001). Our algorithm only processes an additional measure based on the table instance (which is already generated for the participation index), and calculates a sum of coordinates for each interesting co-location. In the worst case, the time complexity of pattern mining algorithms is an exponential function of the number of input features. Indeed, in the worst case, there is 2^n interesting patterns to be generated and tested, where n is the number of features. Most of pattern mining approaches, such as levelwise (the one used in this paper) or depth first strategies, have this complexity. To our knowledge, only the LCM algorithm (Uno, Asai, Uchida, & Arimura, 2003), a frequent itemset mining algorithm, has a linear complexity depending on the number of frequent closed itemsets. However, the notion of closure doesn't exist for co-locations. Moreover, when the number of closed patterns is equal to the number of frequent patterns and all the patterns are frequent, this algorithm also explores 2^n patterns. However, in practice, these pattern mining algorithms still efficient since they take advantage of theoretical property, such as the monotonicity, to prune the search space, and use optimized data structures, such as prefix-trees (Bodon, 2004) or FP-trees (Han, Pei, Yin, & Mao, 2004), to efficiently store and generate the patterns.

Note that we used a prefix-tree (or trie) data structure to store and generate the co-locations. For example, Figure 10 represents the prefix-tree data structure associated to the set $\{\{A,C\},\{A,D,F\},\{A,D,G\},\{A,E\},\{D,E\},\{E,F,G\},\{E,G\}\}$. They have not only a power of compression by factorizing common prefix in a set collection, but are also very efficient for candidate generation. Moreover, prefix-trees are well adapted for inclusion and intersection tests, which are basic operations when considering sets.

APPLICATION TO SOIL EROSION CHARACTERIZATION

Architecture of the Prototype

The proposals discussed in this paper have been integrated in a prototype coupled with a GIS.

This prototype is based on a data mining tool called *iZi* (Flouvat, De Marchi, & Petit, 2009). This tool is used to solve interesting pattern mining problems as defined in the formal framework

Algorithm 1. Co-location mining algorithm

```
Require:
  geographical database D,
  set of features F ,
  boolean spatial relationship R (reflexive and symmetric),
  user defined participation index threshold α
Ensure: spatial representations of co-locations with participation index ≥ α and their co-location ratio
1: Initialize size 1 candidate with F and construct their table instance
2: while the set of candidates is not empty do
3:        Generate candidate co-locations using the generalized apriori gen algorithm
4:        for each candidate co-location generated do
5:                Generate the table instance of the candidate co-location using an SQL query on D
6:                Process the participation index and the co-location ratio using the table instance
7:                if the participation index ≥ α then
8:                        Construction of the visual representation of the co-location based on the table
                          instance (i.e. for each feature of the co-location, process the centroid of the objects
                          stored in the table instance)
9:                        Store the visual representation and the co-location ratio in a table of the
                          geographical database
10:               else
11:                       Prune the co-location of the candidate co-locations
12:               end if
13:       end for
14: end while
```

of (Mannila & Toivonen, 1997), by providing a set of generic algorithm implementations e.g. *Apriori* (Agrawal & Srikant, 1994) and *ABS* (De Marchi, Flouvat, & Petit, 2005). This tool has been extended by two modules:

- The first one allows mining interesting co-locations. It takes as input parameters the spatial relationship studied and the threshold used to select interesting co-locations.
- The second one allows the access and storage of the co-locations in a *PostGis* (http://PostGis.refractions.net) geographical database. *uDig* (http://udig.refractions.net) (a free desktop application framework) is used as an interface to visualize data and co-locations stored in the GIS.

The data are stored in several tables of PostGis. Each table corresponds to a layer of the GIS, i.e. a theme. For example, in our experimentations,

Figure 10. Example of trie data structure

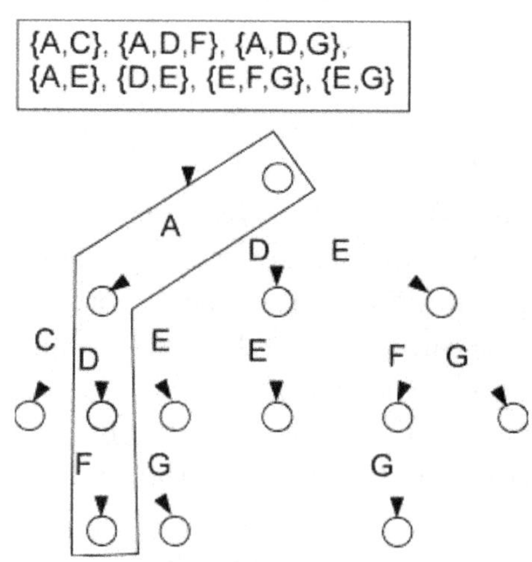

the database is composed of one table for the vegetation, one table for the nature of the ground and one table to store informations about erosion. Each tuple represents a spatial objet. Tables share common attributes such as the id of the objects, its feature (one feature for one object, but several features for one theme), and its "geometry" (its position and shape). Spatial indexes have been defined on the geometry attributes to improve the performances of the spatial join queries.

For each generated co-location (line 4 of algorithm 1), the predicate component of the prototype constructs its table instance using a spatial query on the corresponding tables. The following query represents the computation of the table instance of the colocation $C=\{f_1, f_2, ..., f_k\}$.

SELECT $theme_1$ .id, $theme_2$ .id, ..., $theme_k$ .id

FROM $theme_1$, $theme_2$, ..., $theme_k$

WHERE $theme_1$ .feature = f_1 AND $theme_2$ .feature = f_2

AND ... $theme_k$ .feature = f_k

AND R($theme_1$ .geometry, $theme_2$ .geometry) AND

R($theme_1$ .geometry, $theme_3$ .geometry) AND

... R($theme_k$ .geometry, $theme_{k-1}$ .geometry)

where $f_i \in$ schema($theme_i$)

Note that $theme_i$ can be equal to $theme_j$, $\forall i,j \in [1,k]$, if the co-location is composed of features of the same theme.

The table instance generated is stored in main memory. If the co-location is interesting, the prototype uses the table instance to process the clique representing the co-location. The co-location and its coordinates are stored with the co-location ratio and the participation index, in a specific table of the GIS. The schema of this table is *schema(Colocation)={colocation, feature, theme, coordinate, edge}*. Each tuple of this table store the name of a co-location, one of its feature, its corresponding coordinates, and an edge to draw the clique. For example, the tuples *<{A,B,C}; A; Vegetation; (1000,500); (1000,500,400,500)>*, *<{A,B,C}; B; Erosion; (400,500); (400,500,800,700)* > and *<{A,B,C}; C; Ground; (800,700); (800,700,1000,500)>* represents the informations about the co-location *{A,B,C}* in Figure 11.

Experimentations

Experimental Protocol

We applied our prototype on a mountainous watershed of about 9km^2 in New Caledonia. In this area, natural erosion takes place as well as erosion related to mining activities. When studying soil erosion, three important thematic layers are considered:

- The *erosion* layer (Rouet et al., 2009) represents information about the erosion process (1:50,000 scale). It has been constructed based on a map of bare grounds (6 types) using remote sensing. The state of the soil can be *not bare ground, trail bare ground, active sedimentary zone, river erosion, hillslope erosion* or *mine erosion zone*.
- The *nature of the ground* layer contains information about lithology. It is extracted from a geologic GIS (1:50,000 scale) (DIMENC/SGNC & BRGM, 2005). It is a vectorial data, in polygon type for which 13 different kinds of lithology were detected.
- The *vegetation* layer described the spatial repartition of the 13 different types of vegetation system in the studied area. The layer was generated using remote sensing (DTSI/SGT, 2008).

The studied area is composed of more than 9000 geographical objects. These three layers were analyzed in different ways:

Figure 11. Architecture of the prototype

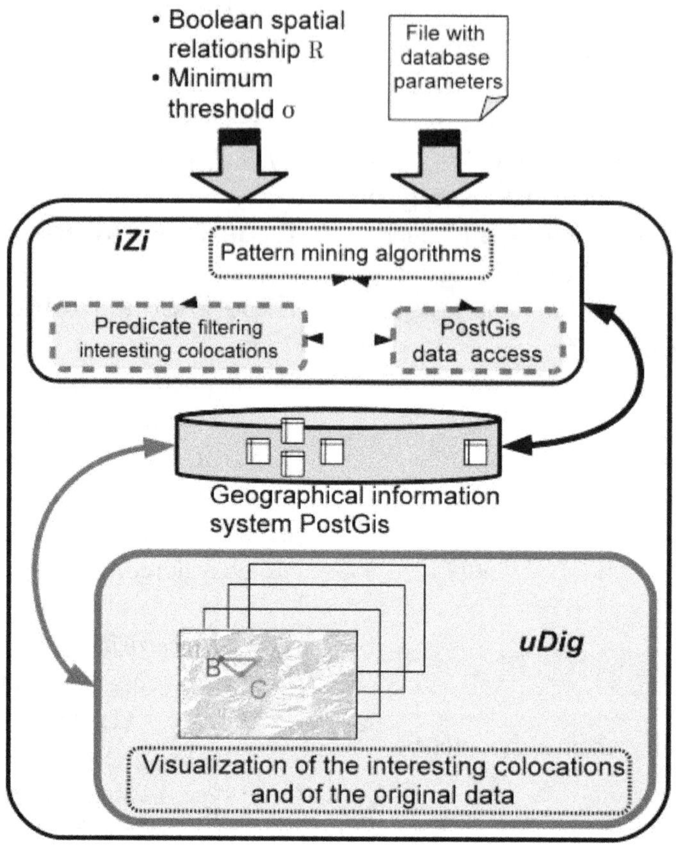

- Process A: because of the large number of polygon available for the studied area, the computation time was very long. This is why, at the beginning, we use the centroid of the polygon to represent a geographic area. Several distances were tested (50 m, 100 m, 200 m) with different co-location ratio thresholds.
- Process B: intersections between polygons of three themes are tested. The three kinds of intersection (point, line, polygon) are also tested.

Experimental Evaluation

Figure 12 shows the performances (execution time) of the original co-location mining (*V0*) w.r.t. our approach with the centroid visualization (*Centroids*). The distance studied is an euclidian distance threshold w.r.t. the centroids of the geographical areas (process A). As shown by this figure, the performances of our approach are of the same order of magnitude than the original algorithm of Shekhar and Huang (2001). Nevertheless, thanks to our new measure and our visualization approach, our solution provides to experts value-added spatial informations about interesting co-locations.

To complete these results, Tables 1 and 2 show the number of co-locations extracted and their length for each relationship and several participation index thresholds. For example, for a distance threshold of 200m and a participation index threshold of 0.3, the algorithm extracts 53 co-locations of length 2, 79 co-locations of length 3,

Table 1. Number of co-locations extracted for the studied area (process A)

	distance 50		distance 100			distance 200				
Threshold	nb coloc size 2	nb coloc size 3	nb coloc size 2	nb coloc size 3	nb coloc size 4	nb coloc size 2	nb coloc size 3	nb coloc size 4	nb coloc size 5	nb coloc size 6
0.3	6	1	19	10	2	53	79	68	30	5
0.5	2	0	10	9	2	32	53	64	30	5
0.7	0	0	4	2	2	17	30	44	28	5
0.9	0	0	1	0	0	4	8	13	13	5

Figure 12. Performances of our approach with the 9000 objects and the 32 features of the studied area (distance based neighborhood relationship, i.e. process A)

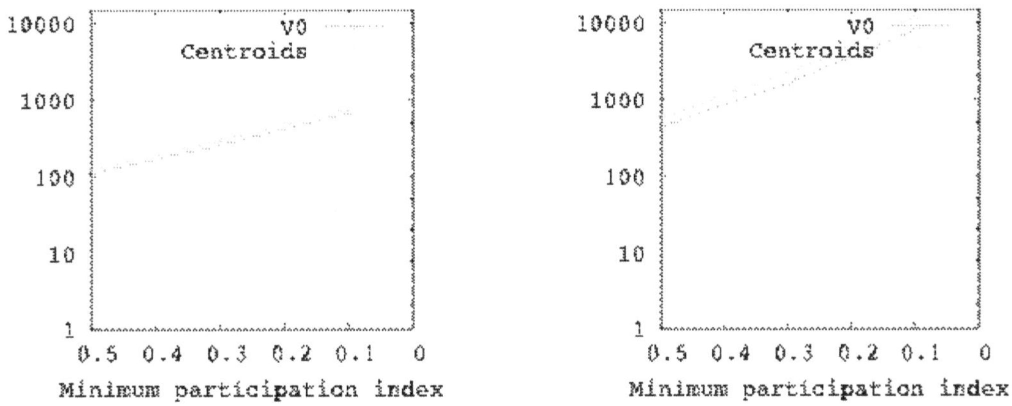

68 co-locations of length 4, 30 co-locations of length 5, and 5 co-locations of length 6.

We also study the performances of our approach on two other datasets:

- The first one has 5802 objets and 19 features
- The second one has 1840 objects and 13 features

Figures 13 and 14 show the execution time of our prototype on these datasets for two different distance thresholds (200m and 300m). These experiments show the efficiency of our approach with different number of features and objects.

Table 2. Number of co-location extracted with intersection (process B)

Threshold	nb coloc size 2	nb coloc size 3
0.3	53	18
0.5	23	2
0.7	7	0
0.9	0	0

Expert Analysis

Figure 15 illustrates the visualization of co-locations on the studied dataset. We can see the spatial objects (left screenshot), the corresponding co-locations (screenshot in the center), and a zoom on a specific area (right screenshot). These

Figure 13. Performances of our approach with 5802 objets and 19 features (distance based neighborhood Relationship

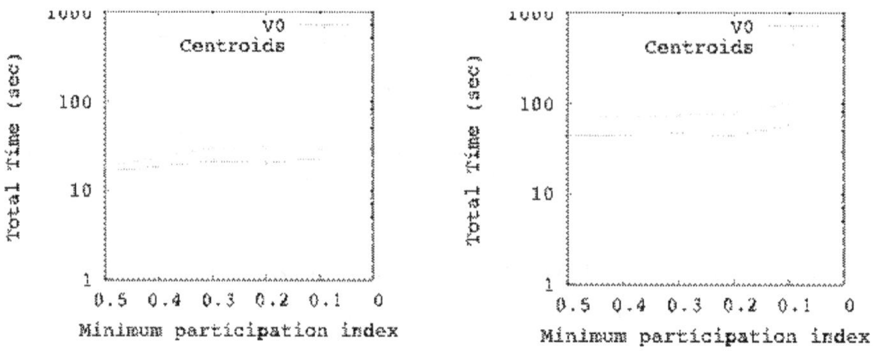

Figure 14. Performances of our approach with 1840 objects and 13 features (distance based neighborhood relationship

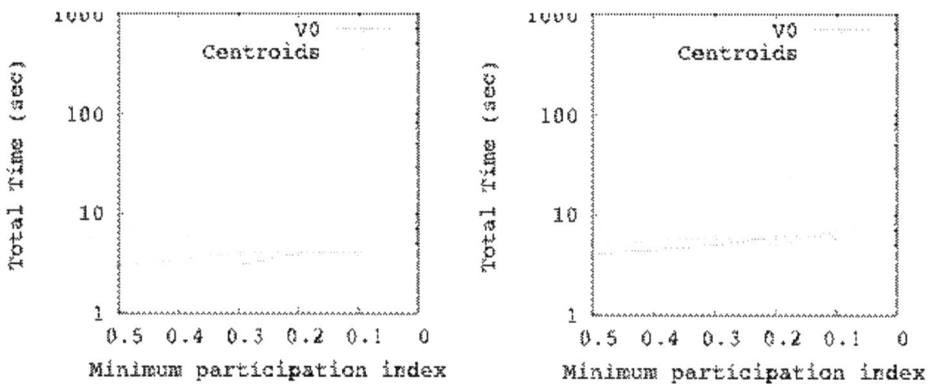

Figure 15. Visualization of co-locations on soil erosion data (threshold: 0.3, distance: 200m)

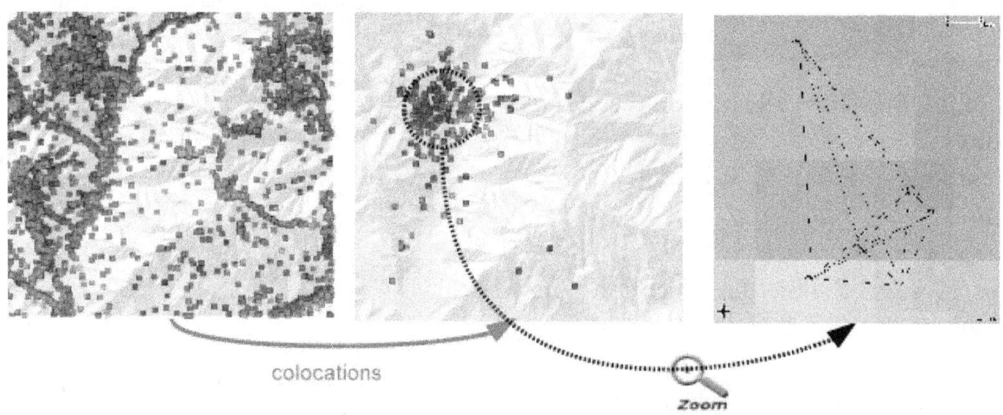

results were analyzed and validated by a geologist, specialist of soil erosion in New Caledonia. They are discussed in the rest of this section.

Results of process A (Table 1) show the co-location at 50 m, 100 m and 200 m. The intra-theme co-locations are mainly related to the theme "state of the soil", with a co-location ratio higher than 0.4. The more significant co-locations point out the fact that the mine erosion zones are close to hillslope erosion zones (0.5 at 50 m, 0.7 at 100 m, 0.9 at 200 m) and also the association between sensitive trails, mines, hillslope erosion and river erosion (0.57 at 200 m). Many co-locations in the layer vegetation appear from the distance of 100m between the centroids. These co-locations have the size 2 or 3. They reveal associations within the vegetation layer (shrubby vegetation and forest on volcano-sedimentary substrate with 0.5 at 100 m, 0.6 at 200 m for example). No intra-theme co-location has been identified about the "nature of the ground". This variation between the results comes from the content of the theme. The "state of the soil" and the "vegetation" result from analysis of the same satellite image being more accurate than the "nature of the ground". In process A, inter-themes co-locations are highlight only from a distance of 200m. The more significant co-locations are associations between sensitive trails, mining zones, river erosion and sparse vegetation (0.37 at 200 m) and between mines, hillslope erosion, woody-herbaceous scrub and sensitive trails (0.32 at 200 m) or river erosion (0.35 at 200 m). They highlight the environmental damage near the areas where humans have used the soils.

Results of process B (Table 2) show another kind of information. Co-locations are computed with the notion of intersection and can have a size of 2 or 3. In this case, intra-theme co-locations are highlighted with the three themes, 7 with the "nature of the ground", 9 with the "state of the soil" and 10 with the vegetation. For example, results show that mining zones and hillslope erosion are often close to each other (0.48, theme "state of the soil"). This co-location already highlighted in process A allows us to say that most of these areas are connected to each other. The mining dumps and debris flows and the mines are closely related (0.83, theme "nature of the ground"). The woody-herbaceous scrub vegetation is contiguous to either sparse vegetation (0.49) or a woodland dense scrub (0.86). This co-location shows that the plant systems can also be related to the environment degradation. The inter themes co-location show also that the sensitive trails are really linked to the lateriting nature of the soil (0.52), to hillslope erosion (0.56) and to the mines (0.52). The savannah and the vegetation on volcano-sedimentary substrate seem to be close to alluvia (respectively 0.64 and 0.5 ; 0.47 with the size 3 co-location) although they are not made up of typical substrate. The woody-herbaceous scrub or the woodland dense scrub and the forest on ultramafic substrate are closely related to harzburgites (respectively 0.61 and 0.48), this can be explained by the high rate of endemism of these species.

These preliminary domain results suggest that our approach could be very relevant for quick and pertinent analysis of environmental GIS data. Common GIS analysis can't give equivalent results in a so short time and the full integration of data mining results as layer in a GIS guarantees a best appropriation of the tools by domain experts. Note that interestingness measures could be used to quantify some specific relations and to develop environmental indicators. For example in our experimentations, intra-theme and inter-theme relations between erosion areas and kind of vegetation all around are warning light environmental condition and these can be used to create a new environmental degradation indicator.

CONCLUSION

This paper shows the added value of co-location mining for soil erosion characterization. This approach provides to experts a formal and intuitive tool which automates the analysis and quantifies the importance of correlations. For example, this kind of approach significantly improves the knowledge of the spoiling level (or conservation level) in the studied area. In a biodiversity hot spot, this issue is especially crucial and experts would have better informations to describe the environmental conditions of an area.

In this context, our contribution is twofold. First, we define a new interestingness measure, the co-location ratio that better reflects the importance of co-locations. The definition of this measure is more intuitive for experts, and improves their interpretation. Its processing is done during the execution of the mining algorithm without any overhead. Second, we propose a new intuitive visualization approach based on a cartographic representation of co-locations. Our system is totally integrated in a GIS and can benefit from its functionalities. Moreover, it provides spatial informations about objects participating in co-locations. We implement these propositions in a prototype coupled with *PostGis* and *uDig*. Experimentations have been done on a mountainous area of 9 km^2 described by three thematic layers (erosion, lithology and vegetation) and 32 features. The co-location mining results were analyzed by a specialist of soil erosion in New Caledonia. They highlight well known relations about soil erosion in this area, such as environmental damage near the areas with human activities or the link between plant systems and environment degradation.

This work has several perspectives. The first perspective is to study more features and themes (e.g. meteorology), and/or other areas in New Caledonia. We plan to develop new environmental indicators, based on spatial data mining, for decision makers and for sustainable environmental management. It would also be interesting to apply and adapt this approach to other environmental phenomena's (e.g. climate change, el niño phenomenon). From a theoretical point of view, a challenging issue is to improve the visualization system to provide more meaningful and condensed informations to experts. Finally, we also plan to study domain driven spatio-temporal correlations in soil erosion data. Most of existing works on spatio-temporal pattern mining focus on mining trajectories. Trajectory mining considers data involving objects moving over time. As an example, trajectory mining studies the movement of cars in a town (Cao, Mamoulis, & Cheung, 2005), or the behavior of flocking sheep for the seasonal migration (Gudmundsson, Laube, & Wolle, 2008). However, in our application, trajectories are not clearly identified. We only have pictures of a studied area over time, and no informations about the "movments" of soil erosion.

ACKNOWLEDGMENT

This work was supported by the CNRT "Nickel and its environment".

REFERENCES

Agrawal, R., & Srikant, R. (1994). Fast algorithms for mining association rules in large databases. In *Proceedings of the 20th International Conference on Very Large Data Bases* (pp. 487-499).

Andrienko, G. L., & Andrienko, N. V. (1999). Knowledge-based visualization to support spatial data mining. In D. J. Hand, J. N. Kok, & M. R. Berthold (Eds.), *Proceedings of the Third International Symposium on Advances in Intelligent Data Analysis* (LNCS 1642, pp. 149-160).

Appice, A., Ceci, M., Lanza, A., Lisi, F. A., & Malerba, D. (2003). Discovery of spatial association rules in geo-referenced census data: A relational mining approach. *Intelligent Data Analalysis, 7*(6).

Bodon, F. (2004). Surprising results of trie-based fim algorithms. In *Proceedings of the Workshop on Frequent Itemset Mining Implementations* (Vol. 126).

Bogorny, V., Valiati, J., Camargo, S., Engel, P., Kuijpers, B., & Alvares, L. O. (2006). Mining maximal generalized frequent geographic patterns with knowledge constraints. In *Proceedings of the IEEE International Conference on Data Mining* (pp. 813-817). Washington, DC: IEEE Computer Society.

Brunk, C., Kelly, J., & Kohavi, R. (1997). An integrated system for data mining. In *Proceedings of Knowledge Discovery in Databases* (pp. 135–138). Mineset.

Cao, H., Mamoulis, N., & Cheung, D. W. (2005). Mining frequent spatio-temporal sequential patterns. In *Proceedings of the 5th International Conference on Data Mining* (pp. 82-89). Washington, DC: IEEE Computer Society.

Cao, L. (2008). Domain driven data mining (d3m). In *Proceedings of the International Conference on Data Mining* (pp. 74-76). Washington, DC: IEEE Computer Society.

Ceci, M., Appice, A., & Malerba, D. (2007). Discovering emerging patterns in spatial databases: A multi-relational approach. In J. N. Kok, J. Koronacki, R. Lopez de Mantaras, S. Matwin, D. Mladenic, & A. Skowron (Eds.), *Proceedings of the 11th European Conference on Principles and Practice of Kowledge Discovery in Databases* (LNCS 4702, pp. 390-397).

Celik, M., Kang, J. M., & Shekhar, S. (2007). Zonal co-location pattern discovery with dynamic parameters. In *Proceedings of the IEEE International Conference on Data Mining* (pp. 433-438). Washington, DC: IEEE Computer Society.

De Marchi, F., Flouvat, F., & Petit, J.-M. (2005). Adaptive strategies for mining the positive border of interesting patterns: Application to inclusion dependencies in databases. In J.-F. Boulicaut, L. De Raedt, & H. Mannila (Eds.), *Proceedings of the Conference* on *Constraint-Based Mining and Inductive Databases* (LNCS 3848, pp. 81-101).

DIMENC/SGNC & BRGM. (2005). *Carte géologique de la Nouvelle-Caldonie au 1/50000.* Retreived from http://dimenc.gouv.nc/portal/page/portal/dimenc/telechargements/tele_geologie/temp_3.pdf

DTSI/SGT. (2008). *Cartographie de l'occupation du sol de la Nouvelle-Caldonie au 1/50000.*

Flouvat, F., De Marchi, F., & Petit, J.-M. (2009). The iZi project: Easy prototyping of interesting pattern mining algorithms. In *Proceedings of the 13th Pacific-Asia International Conference on Knowledge Discovery and Data Mining: New Frontiers in Applied Data Mining* (pp. 1-15).

Gay, D., Rouet, I., Mangeas, M., Dumas, P., & Selmaoui, N. (2007). Assessment of classification methods for soils erosion risks. In *Proceedings of the International Congress on Modelling and Simulation*, Christchurch, New Zealand (pp. 2659-2665).

Gudmundsson, J., Laube, P., & Wolle, T. (2008). Movement patterns in spatio-temporal data. In Shekhar, S., & Xiong, H. (Eds.), *Encyclopedia of GIS* (pp. 726–732). New York, NY: Springer. doi:10.1007/978-0-387-35973-1_823

Han, J., & Kamber, M. (2006). *Data mining, second edition: Concepts and techniques (the Morgan Kauffman series in data management systems).* San Francisco, CA: Morgan Kauffman.

Han, J., Pei, J., Yin, Y., & Mao, R. (2004). Mining frequent patterns without candidate generation: A frequent-pattern tree approach. *Data Mining and Knowledge Discovery*, *8*(1), 53–87. doi:10.1023/B:DAMI.0000005258.31418.83

Huang, Y., Shekhar, S., & Xiong, H. (2004). Discovering colocation patterns from spatial data sets: A general approach. *IEEE Transactions on Knowledge and Data Engineering, 16*(12), 1472–1485. doi:10.1109/TKDE.2004.90

Keim, D. A. (2002). Information visualization and visual data mining. *IEEE Transactions on Visualization and Computer Graphics, 8*(1), 1–8. doi:10.1109/2945.981847

Koperski, K., & Han, J. (1995). Discovery of spatial association rules in geographic information databases. In M. J. Egenhofer & J. R. Herring (Eds.), *Proceedings of the 4th International Symposium on Advances in Spatial Databases* (LNCS 951, pp. 47-66).

Leung, C. K.-S., Irani, P., & Carmichael, C. L. (2008). Wifisviz: Effective visualization of frequent itemsets. In *Proceedings of the International Conference on Data Mining* (pp. 875-880). Washington, DC: IEEE Computer Society.

Lisi, F. A., & Malerba, D. (2004). Inducing multi-level association rules from multiple relations. *Machine Learning, 55*(2), 175–210. doi:10.1023/B:MACH.0000023151.65011.a3

Malerba, D. (2008). A relational perspective on spatial data mining. *International Journal of Data Mining. Modelling and Management, 1*(1), 103–118.

Mannila, H., & Toivonen, H. (1997). Levelwise search and borders of theories in knowledge discovery. *Data Mining and Knowledge Discovery, 1*(3), 241–258. doi:10.1023/A:1009796218281

Miller, H. J., & Han, J. (2009). *Geographic data mining and knowledge discovery*. Boca Raton, FL: Taylor & Francis.

Rouet, I., Gay, D., Allenbach, M., Selmaoui, N., Ausseil, A.-G., Mangeas, M., et al. (2009). Tools for soil erosion mapping and hazard assessment: Application to New Caledonia, SW Pacific. In *Proceedings of the International Congress on Modelling and Simulation*, Cairns, Australia (pp. 1986-1992).

Shekhar, S., & Huang, Y. (2001). Discovering spatial co-location patterns: A summary of results. In C. S. Jensen, M. Schneider, B. Seeger, & V. J. Tsotras (Eds.), *Proceedings of the 7th International Symopisum on Advances in Spatial and Temporal Databases* (LNCS 2121, pp. 236-256).

Tan, P.-N., Steinbach, M., & Kumar, V. (2005). *Introduction to data mining*. Reading, MA: Addison-Wesley.

Uno, T., Asai, T., Uchida, Y., & Arimura, H. (2003). An efficient algorithm for enumerating frequent closed item sets. In *Proceedings of Frequent Itemset Mining Implementations* (*Vol. 90*). LCM.

Yoo, J. S., & Shekhar, S. (2006). A joinless approach for mining spatial colocation patterns. *IEEE Transactions on Knowledge and Data Engineering, 18*(10), 1323–1337. doi:10.1109/TKDE.2006.150

This work was previously published in International Journal of Agricultural and Environmental Information Systems, Volume 2, Issue 2, edited by Petraq Papajorgji and François Pinet, pp. 73-92, copyright 2010 by IGI Publishing (an imprint of IGI Global).

Chapter 12
Describing Spatio-Temporal Phenomena for Environmental System Development:
An Overview of Today's Needs and Solutions

André Miralles
Maison de la Télédétection, France

François Pinet
Cemagref, France

Yvan Bédard
Université Laval, Canada

ABSTRACT

This paper is composed of two parts dealing with the modeling of environmental phenomena. The first part presents the traditional ER and OO formalisms dedicated to geographic information modeling. These languages focus mainly on representing the spatial and temporal properties of this type of information. Many of these languages express these properties visually by using pictograms. After a quick historical presentation of the languages, the authors show the various types of spatiality and temporality usually encountered in these languages. Often qualified as primitive, some of these spatialities and temporalities are simple. Others, which are more complex, result from combinations of simple spatialities and simple temporalities. Still others are used in very specific situations encountered during the development of geographical information systems. These different spatialities and temporalities are presented via examples provided in the field of environmental dynamics.

DOI: 10.4018/978-1-4666-0333-2.ch012

1. INTRODUCTION

Recently, geographic information has become strategic information. In the agricultural and environmental fields, geographic information can help to improve agricultural practices, food traceability or decision-making. Decisions concerning the environment should be made based on reliable data Collecting quality data requires monitoring of agricultural practices and their impact, and, consequently, any information records linked to these practices. Information used for food traceability is very often geo-referenced. It is important to maintain spatiotemporal monitoring within agricultural and environmental practices.

According to two specific studies (Claramunt, Coulondre, & Libourel, 1997; Gayte, Libourel, Cheylan, & Lardon, 1997), geographic information has:

- Thematic properties describing business concepts within the domain of study (for example *land use* or *hydrographic network*),
- Spatial properties depicting the geometrical features of these business concepts,
- Temporal properties characterizing the state of an instance of a concept at a given date or the evolution of that instance.

Spatial and temporal properties concern a large number of objects: static objects (e.g., fields), moving objects (e.g., vehicles) or evolving objects (e.g., evolutions of the boundary of a wetland across seasons).

A large number of studies have been proposed to model and develop *Geographic Information Systems (GIS)* based applications. The nature of the information in this type of applications is complex and modeling GIS-based applications is a major challenge (Bédard, Larrivée, Proulx, & Nadeau, 2004). The first paper to present and demonstrate the advantages of using pictograms to simplify entity-relationship models in GIS was written 21 years ago (Bédard & Paquette, 1989). Later, this formalism was adapted to be used in the object-oriented paradigm (Bédard, 1999b). The work of (Kösters, Pagel, & Six, 1995) also shows that models of GIS-based applications can be simplified by using formalisms dedicated to geomatics applications. Several authors also addressed this issue as presented the *Encyclopedia of GIS* (Shekhar & Xiong, 2008).

After an overview of the different modeling languages designed by researchers or companies (Section 2), in Section 3, this paper shows how geographic information can be modeled using the specific method proposed in (Bédard, 1999b). This method is currently the richer formalism; it allows modeling numerous spatial and temporal concepts. A conclusion finalizes this paper in Section 4.

2. OVERVIEW OF THE LANGUAGES FOR MODELING GEOGRAPHIC INFORMATION SYSTEMS

A large number of methods and formalisms have been developed during the last twenty years to facilitate modeling of GIS-based applications (Bédard & Paquette, 1989). During the past decade, the number of methods and formalisms has exploded, and we are witnessing a new level of maturity with the availability of tools, tool extensions, web sites, publications and the recognition of this expertise in the *Encyclopedia of GIS (Shekhar & Xiong, 2008)*. Table 1 presents the main methods and formalisms dedicated to geographic information modeling. Most of these methods and formalisms are based on the Entity-Relationship (ER) or Object-Oriented (OO) paradigms. Amongst the latter, several formalisms are based on the *Unified Modeling Language* (UML), today's *de facto* standard that has been adopted by the *Object Management Group*, the *Open Geospatial Consortium* (OGC) and the ISO/TC211 Geographic Information/Geomatics Technical Committee.

Table 1. Historical overview of the methods/formalisms used for modeling geographic information

Year	Formalism	Case Tool	Language / Method
1989	Entity-Relationship		(Bédard & Paquette, 1989) proposes the first formalism to model geographic information in databases. Use of pictograms in models to represent the geographic concepts.
1993	Entity-Relationship	Orion	**MODUL-R.** Method based on a MERISE extension (Quang & Chartier-Kastler, 1991). Specific notations to represent spatial information are introduced into conceptual data models. Pictograms represent different types of spatiality and temporality (Caron, Bédard, & Gagnon, 1993). The methodological approach used is based on the MERISE method. It generates code for Intergraph TIGRIS.
1994	Object-Oriented (OOA – Object Oriented Analysis (OOAD, 1997))		**MECOSIG.** This method is an object-oriented method for the design of Geographic Information Systems and includes a formalism called CONGOO (Pantazis, 1994; Pantazis & Donnay, 1996)
1995	Object-Oriented (UML)	Aigle	**OMEGA.** This method integrates an object-oriented formalism based on a former version of UML. This formalism is mainly based on the formalism implemented in MODUL-R (Lbath & Aufaure-Portier, 1995; Lbath & Pinet, 2000; Pinet, 2002).
	Object-Oriented (OOA)		**GeoOOA.** This extension of OOA proposes a type of class diagram for the design of geomatics applications. This method also recommends the use of pictograms in diagrams (Kösters et al., 1995).
1996	Object-Oriented (OOA)		Integration into **GeoOOA** of: - temporal concepts in order to allow modeling the evolution of objects, - and spatial data concerning networks (road networks, electrical networks and hydrographic systems, for instance) (Kösters & Pagel, 1996; Kösters, Pagel, & Six, 1997).
1997	Object-Oriented		**Geo-OMT.** This method extends OMT1 and uses pictograms in OO diagrams (Das Neves, Gordillo, Mostaccio, & Levato, 1997).
	Entity-Relationship extended to Object-Oriented		**MADS.** A method for the object-oriented design of spatio-temporal data. It uses pictograms (Parent, Spaccapietra, & Zimányi, 2008; Parent et al., 1997; Vangenot, 2001).
	Object-Oriented (OOA)	GeoOOA-Tool, REGIS	Development of **GeoOOA-Tool**, a tool based on GeoOOA.
	Object-Oriented (OMT)		**POLLEN.** A method based on **OMT.** It enables the addition of predefined classes to describe spatial and temporal information within models (Gayte et al., 1997).
	Entity-Relationship		**GISER.** Extension of the Entity-Relationship model for applications handling spatial data (Shekhar, Coyle, Goyal, Liu, & Sarkar, 1997).
	Entity-Relationship		**GeoER.** Extension of the Entity-Relationship model for applications handling spatial data (Hadzilacos & Tryfona, 1997).
	Object-Oriented	Uapé	**GMOD.** A type of spatiotemporal model (Lopes De Oliveira, Pires, & Bauzer Medeiros, 1997).
1998	Object-Oriented (UML)		**GeoFrame.** A set of specific pictograms, similar to those implemented in MODUL-R, are proposed for enriched UML diagrams (Filho, 1998; Filho & Iochpe, 1999, 2008; Rocha, Edelweiss, & Iochpe, 2001).
	Entity-Relationship	CASTER + CASTR	**STER.** The components CASTER and CASTR implement STER. These components were designed with the DBMS Oracle (Tryfona & Jensen, 1998, 2000).
1999	PVL (ER, OO and other technologies)	Software font to be embedded within applications.	Pictogrammic language replacing MODUL-R pictograms and designed as a Plug-in for Visual Languages. It was extended for 3D modeling (Bédard, 1999b; Bédard & Larrivée, 2008; Larrivée, Bédard, & Pouliot, 2005, 2006; Proulx, Larrivée, & Bédard, 2002)

continued on following page

Table 1. Continued

Year	Formalism	Case Tool	Language / Method
1999	Object-Oriented (UML)	Perceptory	Proposes an OO version of Orion. It is multilingual and supported by ISO and OGC standards (Bédard, 1999b; Larrivée et al., 2005; Proulx et al., 2002). It generates reports and code for GIS and Universal Servers.
	Object-Oriented (UML)		**STUML.** Extension of UML for spatiotemporal information (Price, Ramamohanarao, & Srinivasan, 1999).
	Object-Oriented	Cobalt	Presentation of an early tool based on the formalism **MADS.**
	Object-Oriented (OMT)		Geo-OMT becomes **OMT-G.** The specification of this method is enriched (Borges, Davis Jr., & Laender, 2001).
2002	Object-Oriented (UML)	T-Aigle	**T-OMEGA.** Extension of OMEGA and AIGLE for modeling spatial information being shared among several computer programs (Pinet & Lbath, 2003).
2006	Object-Oriented (UML)	STGL**2** Profile for Objecteering Case Tools	**STGL.** Extension of UML for the representation of spatial and temporal properties using PVL pictograms (Miralles, 2006; Miralles & Libourel, 2008, 2009).

The use of specific pictograms to describe spatial information models was proposed for the first time by (Bédard & Paquette, 1989). This approach was based on ER formalism (called in the French language "formalisme individuel") in the MERISE method (Tardieu, Rochfeld, & Colletti, 1983) and improved over time to become MODUL-R formalism in 1993. MODUL-R has been implemented in the first spatially-enabled modeling tool, called Orion (Bédard, Pageau, & Caron, 1992), which also included automatic code generation for spatial databases implemented in an object-oriented GIS (Intergraph TIGRIS). Pictograms were used to represent the spatial and temporal properties of geographic information in models; business concepts annotated with the pictogram ⊡ have a punctual geometrical property, concepts annotated with the pictogram "«" have a polygonal property, and so forth. This approach has been tested in several projects (such as the National Topographic Database of Canada, Montmorency Forest, Urban Database and others) (Bédard et al., 2004). Towards the mid-90's, several other formalisms were introduced. Some were based on Extended-ER (e.g., MADS), but most were based on OO formalisms (MECOSIG, OMEGA, GeoOOA, POLLEN). Additionally, some of these formalisms used pictograms similar to those of MODUL-R, but all were based on a unique and proprietary formalism. In order to use these formalisms, one also had to adopt the underlying ER or OO formalism. Since 1997, we have observed an increase of the number of formalisms and of the number of tools implementing these approaches. During this period, the successor of MODUL-R has been developed using a completely new approach: providing a pictogrammic language that is independent from the underlying formalism, allowing a system designer to keep using his favorite formalism while adding the pictogrammic extensions to his modeling tool. This new solution, called PVL (Plug-in for Visual Languages), was implemented in their new modeling tool called Perceptory (Bédard, 1999b; Larrivée et al., 2005; Proulx et al., 2002) on top of an object-oriented formalism (UML). The PVL was also implemented in other modeling tools (e.g., Oracle Designer and Objecteering) and technologies (e.g., JMap Spatial OLAP).

One can also add to the list in Table 1 the standards produced by the *International Standardization Organization* (ISO) and the contributions of the *Open Geospatial Consortium* (OGC). The texts published by these two international organizations result from a consensus between international experts on geographic information. The work of these two organizations is essentially focused on the definition of standards to facilitate the interoperability of geographic information. (Brodeur & Badard, 2008) provide a focus on these standards.

3. TYPE OF SPATIALITIES AND TEMPORALITIES

There are several types of spatial properties, such as *points*, *lines*, *volumes*, *alternate geometry*, *complex geometry*, and so forth. Examples of temporal property types are the *time point* and the *time period*. Spatial and temporal properties can be combined to model advanced concepts such as *moving points* or *deformable polygons*. The study presented in (Miralles, 2006) classifies spatial and temporal properties using the set theory. This approach makes it possible to model the semantics of concepts and the relations between them. In this section we illustrate the modeling of the different types of spatiality and temporality using PVL formalism as it is used in Perceptory.

3.1. Primitive Spatialities and Primitive Temporalities

There are a number of simple concepts used by designers of geographic information systems. In this section, we describe these basic concepts.

All the research and standardization work proposes the primitive spatial concepts of the *point*, *line* and *polygon*, and the primitive temporal concepts of the *time point* and *time period*. Complex spatial and temporal concepts can be produced by combining these five elementary concepts. These five concepts are those most used in the field of geographic information systems. 3D GIS modeling has been addressed only by PVL and Perceptory, which add the ability to differentiate between 2D and 3D universes and include *volume* as a spatial concept.

Figure 1. Parcel concept annotated with a polygonal spatiality

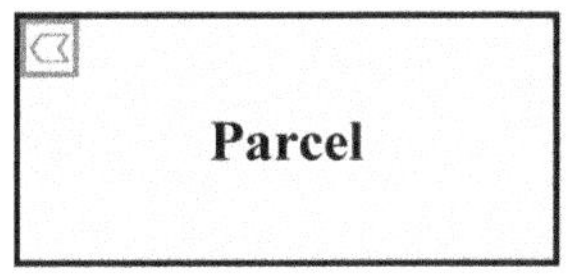

3.1.1. Examples of Primitive Spatiality and Primitive Temporality

A. Example of Primitive Spatiality

All the authors of methods or formalisms for geographic information modeling have adopted the three primitive spatial concepts (Bédard, 1999b; Claramunt et al., 1997; ISO, 2003; OGC, 2005; Pantazis & Donnay, 1996; Parent, Spaccapietra, & Zimányi, 2006; Tryfona & Jensen, 2000) .

For example, the *Parcel* concept is usually related to a polygonal representation of the geometrical property of the concept. In order to introduce this property in ER or OO models, one of the most often used solutions is to annotate the *Parcel* concept with the spatial pictogram "<<" (Figure 1), as proposed by (Bédard & Paquette, 1989).

B. Example of Primitive Temporality

Many authors (Bédard, 1999b; Claramunt et al., 1997; ISO, 2002; Parent et al., 2006; Rocha, 2001; Tryfona & Jensen, 2000) have studied the representation of time in their models. (Bédard, 1999b; Claramunt et al., 1997; ISO, 2002; Parent et al., 2006) use the concepts of the *time point* and the *time period* in their method or formalism. (Rocha, 2001; Tryfona & Jensen, 2000) only use the notion of the *time period*.

The date of an avalanche, for instance, is an important property in the study of natural hazards. In order to represent this property in ER or OO models, one possibility is to add the temporal

Figure 2. Avalanche concept annotated with a time point temporality

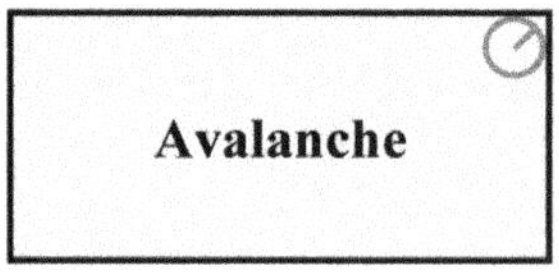

pictogram ⏲. It corresponds to a *time point*. The model in Figure 2 shows the concept *Avalanche*. The distinction between a *time point* and a *time period* depends upon the temporal resolution adopted for the database.

3.1.2. Example of Derived Primitive Spatiality

The notion of derivation is borrowed from UML language (OMG, 2003) and described in the work of (Bédard et al., 2004). They have adapted this notion to the field of geographic information (Proulx et al., 2002). The main goal of the derived concept is to represent the result of a numerical process on spatial and temporal properties. In UML, a derivation is represented by the symbol "/" before a derived attribute or a derived association role. In the pictogrammic language of Perceptory, the derived pictograms are in italics in order to differentiate them from their counterparts.

The model of Figure 3 shows the *Lake* concept. This concept has no spatial property in this diagram. The *Lake* concept could be used in an application which does not need to know the geometric shape of each instance of *Lake*.

We can determine, using simple calculations, the polygon bordering the area covered by the lake because this information exists within the *River Bank* concept. Thus, the pictogram of the derived polygonal spatiality "<<" can be added to the concept of *Lake* (Figure 4).

With this notation, the derived spatial property of *Lake* will be taken into account during the development and the implementation of the information system described by the model. The derived pictogram constitutes exploitable information to remind the designer that a concept must contain an operation, *lineConcatenation()*, allow-

Figure 3. Lake concept without a derived spatiality

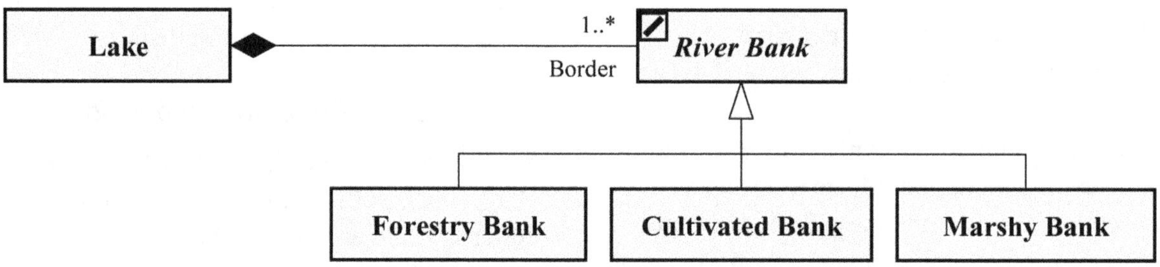

Figure 4. Lake concept annotated with a derived polygonal spatiality

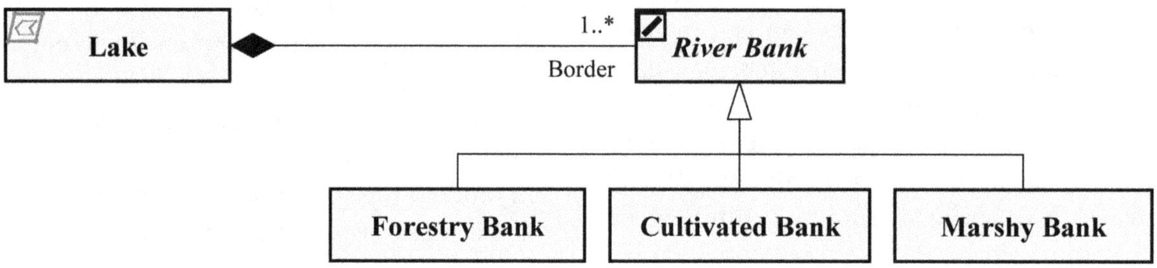

Figure 5. Lake concept with an operation dedicated to the computing of the boundary of lake

ing the program to reconstitute the lake boundary numerically (Figure 5).

3.2. Composite Spatialities and Composite Temporalities

In the previous section, the most often used *primitive spatialities and temporalities* are defined. They are often not sufficient to express all the objects of a GIS. Various research and standardization work has defined *composite spatialities and temporalities*. The latter are always defined as a combination of primitive concepts.

3.2.1. Composite Concepts Involving Only Spatiality or Only Temporality

3.2.1.1. Example of Alternate Spatiality or Alternate Temporality

Alternate spatiality and *alternate temporality* enable us to model concepts in which all instances do not have the same type of spatiality or temporality. The pictogram language proposed in (Bédard, 1998, 1999b) is the only one to implement this mechanism.

A. Alternate Spatiality

The *Farm* concept, annotated with pictogram "□«" (Figure 6), means that instances of this concept can have either a punctual or a polygonal representation. *In this specific example, the choice between these two representations depends on the surface area of the farms*:

- Below a given threshold (e.g., 500 m²), the spatial property of instances of Farm is punctual. It locates the farmhouse with its coordinates.
- Above the threshold, the spatial property of instances of *Farm* is a polygon that represents the shape of the farmhouse.

Figure 6. Farm concept annotated with an alternate spatiality

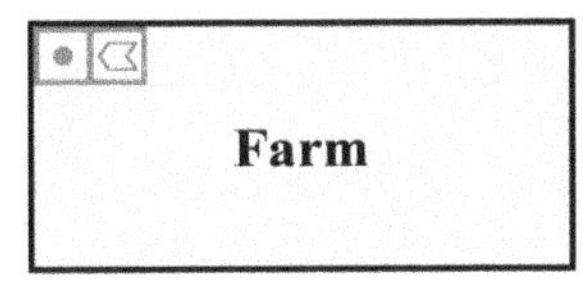

Figure 7. Fire concept annotated with an alternate temporality

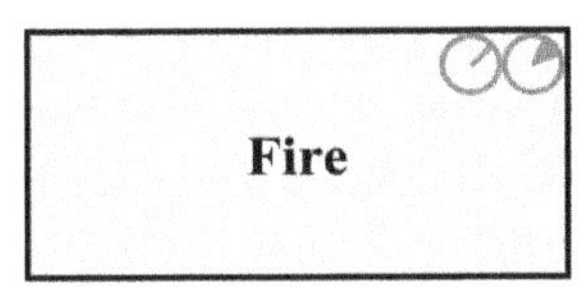

B. Alternate Temporality

In Figure 7, the semantics of the pictogram annotating the *Fire* concept are the following:

- If the duration of the *Fire* is lower than a specific threshold (e.g., 1 hour), only the hour ( ) of the occurrence of the fire is recorded.
- If the duration is longer, the beginning and ending hours ( ) will be recorded.

3.2.1.2. Examples of Multiple Spatiality or Multiple Temporality

Figure 8. Farm concept annotated with a multiple spatiality pictogram

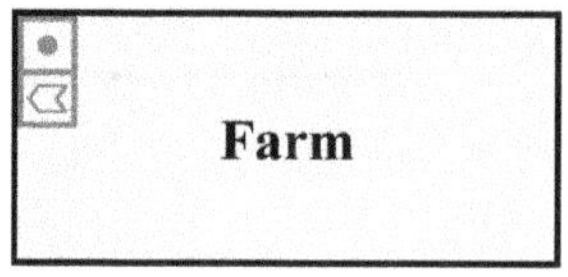

Figure 9. Agricultural Exhibition concept annotated with a multiple temporality pictogram

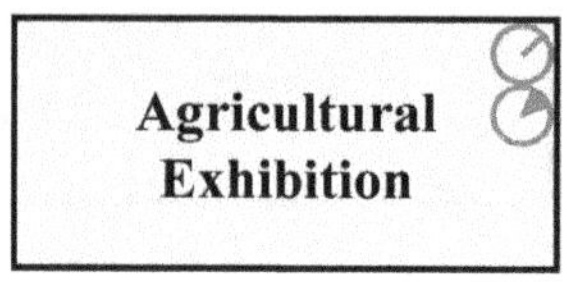

Multiple spatiality and *multiple temporality* allow us to indicate that a business concept has several types of spatialities or temporalities. A number of authors (Bédard, 1999b; Filho & Iochpe, 1999; ISO, 2002, 2003; Pantazis & Donnay, 1996; Parent et al., 2006) have adopted this mechanism in their methods. (Pantazis & Donnay, 1996) uses it only for spatiality.<<

A. Multiple Spatiality

In Figure 8, the *Farm* concept is annotated with the *multiple spatiality* pictogram "⊡<<". It is composed of the primitive pictograms *Point* (⊡) and *Polygon* (<<). This set of pictograms makes it possible to assign two spatial properties to each instance of the *Farm* concept: for example, the location of the headquarters of the farm (punctual representation) and the geographical boundary of the farm (polygonal representation).

B. Multiple Temporality

The multiple temporality pictogram"◷◷<< provides the temporal property of the *Agricultural Exhibition* concept (Figure 9). This property indicates that the organizers of this type of event focus simultaneously on the specific time of the exhibition (◷), for example the award ceremony for innovations in farm machinery and also on the *Agricultural Exhibition* duration (◷). Each instance of the *Agricultural Exhibition* concept will have these two pieces of information.

In some cases, multiple concepts are used to handle the representation of geographic information at different map scales. There is no restriction on the number of spatial or temporal properties that can co-exist in each instance; thus, theoretically the multiplicity can be more than two (although it is very rare in practice).

Figure 10. Punctual Samples concept annotated with multi-point spatiality

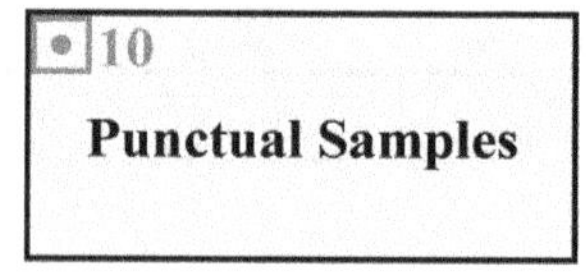

3.2.1.3. Examples of Multi-Spatiality or Multi-Temporality

These concepts are often implemented in the different existing formalisms for GIS (Bédard, 1999a; ISO, 2003; OGC, 2005; Pantazis & Donnay, 1996). The goal of *multi-spatiality* and *multi-temporality* is to express the repetition (*N* times) of the same primitive property.

The concept *Punctual Samples* is annotated with the pictogram ⊡ and the value 10 (Figure 10). This means that each sample is composed of *10 measurements* performed in 10 geographic locations. The punctual coordinates of the different samples must be known.

3.2.2. Composite Concepts Simultaneously Involving Spatiality and Temporality

Composite geographic concepts are the result of a combination of *one (and only one)* spatial primitive and of *one (and only one)* temporal primitive.

Figure 11. Cultivated Parcel concept annotated with spatiality and temporality

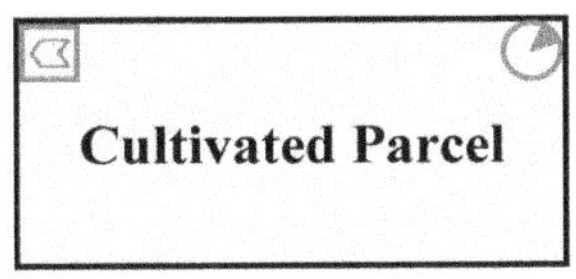

The reasons for this cardinality restriction are explained below:

- The geographic concepts resulting from a combination of spatial primitives and temporal primitives should remain comprehensible.
- It is difficult to explain the semantics of a combination of several spatial and temporal primitives to users who are not familiar with this type of formalism. If such a case happens, Perceptory uses the "!" as a spatial or temporal pictogram. If all possible spatialities or temporalities may happen, Perceptory then uses the wildcard "*" as a pictogram.

3.2.2.1. Example of Simultaneous Spatiality and Temporality

Several modeling languages enable us to represent spatiality and temporality simultaneously (Bédard, 1999b; Filho & Iochpe, 1999; ISO, 2002, 2003; Pantazis & Donnay, 1996; Parent et al., 2006).

In order to illustrate simultaneous spatiality and temporality, Figure 11 provides an example of the *Cultivated Parcel* concept.

The *Cultivated Parcel* concept has a polygonal spatiality ("<<"). In order to optimize their production, farmers sometimes reorganize a contiguous group of parcels. To accomplish this, the duration between two reorganizations must be recorded in order to analyze later the production of the parcels in relation to their spatial distribution. Generally, the duration between two reorganizations is longer than the twice-yearly or the yearly cycle of cereal crops. The pictogram ◷ enables us to record the lifespan of the cultivated parcels, which depends on economic criteria.

3.2.2.2. Examples of Spatio-Temporality

Spatio-temporality combines a primitive spatial property with a primitive temporal property, but, unlike the concepts in the previous section, the temporal property is supported by the spatial property, which annotates a business concept. The annotation is done following a waterfall process. The only methods that take spatio-temporality into account are the pictogrammic language of Perceptory (Bédard, 1999b), POLLEN (Claramunt et al., 1997) and STER (Tryfona & Jensen, 2000).

The *Harvester* and *Epidemic* concepts in Figure 12 are annotated by the spatio-temporality pictograms ⊡◷ and <<◷. The semantics of these pictograms is the following: the spatialities *Point* (⊡) and *Polygon* (<<) are functions of time. The harvesters will have to be equipped with a GPS sensor enabling them to know the successive positions of the machine during work. If this machine is equipped with a sensor continuously weighing the grain, it will be possible to map the variability of the yield of the parcel. This information can be used during the year *N+1* in order to optimize the yield of the parcel.

In the *Epidemic* concept (Figure 12), the polygon representing the extent of the disease can be

Figure 12. Concepts annotated with spatio-temporality

recorded with the date of the measurement. The epidemiologists will have a succession of dated polygons representing the area affected by the disease. It will be possible to evaluate the propagation velocity of the disease.

3.2.3. Macro-Spatiality and Macro-Temporality

Even during the first step of the analysis of a system, the concepts modeled in a diagram can be complex. They can be composed of other concepts exhibiting a smaller granularity (Bédard, 1998, 1999b). For example, a *Rural Bus Network* is a very general concept that aggregates a large number of smaller concepts, *Roads, Bus Stops,* and so forth. The works of (Bédard, 1998) have identified two types of aggregation:

- **Composition,** which consists of aggregating concepts with finer granularity. Consequently, the spatiality and the temporality of the aggregation result from the individual spatialities and temporalities of the finer concepts.
- **Generalization,** which consists of "factorizing" concepts that have common properties.

A macro-concept summarizes a set of concepts and hides the more detailed concepts.

3.2.3.1. Aggregation Macro-Spatiality and Aggregation Macro-Temporality

A. Example of Complex Spatiality

The complex type is not very common in current methods. Only of (Bédard, 1998, 1999b; ISO, 2002, 2003; Pantazis & Donnay, 1996) have integrated the complex type in their respective methods.

The example of the *Rural Bus Network* concept illustrates the mechanism of aggregation. This concept is composed of an aggregation of lines

Figure 13. Rural Bus Network concept annotated with complex spatiality

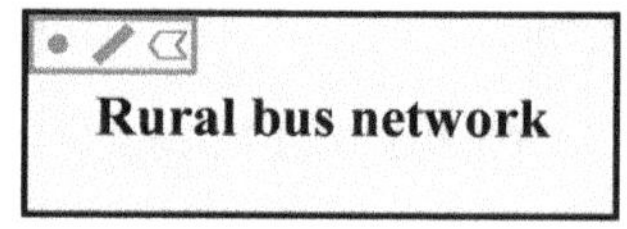

(*Road*), of points (*Bus Stop*) and of polygons (*Terminus*). This description includes two essential pieces of information:

- The first concerns the spatiality of the *Rural Bus Network* concept. Its spatiality includes the linear geometry of the *Roads* (/), the punctual geometry of the *Bus Stops* (⊡) and the polygonal geometry of the *Terminus* (<<).
- The second focuses on the construction of the *Rural Bus Network.* In the description, the use of the word "composed" specifies that this concept is ***an aggregation*** of *Roads, Bus Stops* and *Terminus*. The spatial granularity of these last concepts is lower than that of *Bus network.*

The concept *Bus network* is an aggregation macro-concept. The summarized representation is that of the Figure 13. It is annotated with the pictogram ⊡/<<.

At the beginning of the analysis, the model in Figure 13 may be sufficient to represent the bus network, but at the last steps of the system analysis, the designer would have to refine the *Rural Bus Network* macro-concept. The model in Figure 14 is the result of this refinement.

B. Example of Complicated Spatiality

Only the work of (Bédard, 1998, 1999b) proposes an implementation of *complicated spatiality* and of *complicated temporality* (in the pictogrammic language of Perceptory).

The pictogrammic representation of complex spatiality is not always easy to understand. Thus,

Figure 14. Detailed model of Rural Bus Network concept annotated with complex spatiality

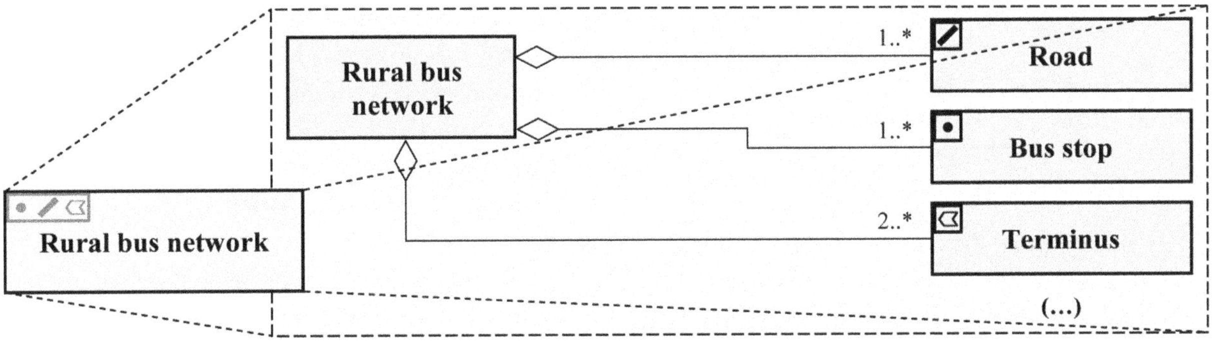

Figure 15. Rural Bus Network concept annotated with complicated spatiality

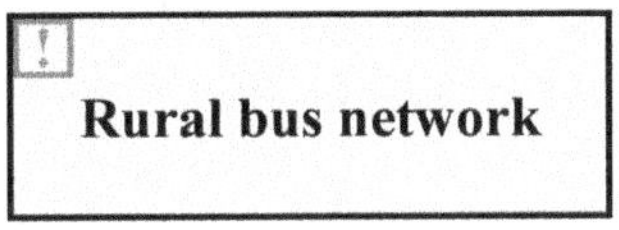

Figure 16. Farm Resources macro-concept without spatiality or temporality properties

(Bédard, 1998) introduced the notion of *complicated spatiality*: e.g., a spatiality not detailed in the diagram. *Complicated spatiality* is represented by the pictogram „!" (Figure 15).

In the first steps of the analysis, the model in Figure 15 may be sufficient. Then, the content of *Rural Bus Network* will be detailed and the complicated spatiality pictogram will be replaced by a complex spatiality pictogram (Figure 13) and then by the model in Figure 14.

3.2.3.2. Generalization Macro-Spatiality and Generalization Macro-Temporality - Example of Multiform Spatiality

Multiform spatiality and *multiform temporality* are only mentioned in the pictogrammic language of Perceptory (Bédard, 1998, 1999b; Bédard & Larrivée, 2008) and in the framework GeoFrame (Filho & Iochpe, 1999). In (Bédard, 1998), they are called "Any possible geometry". This pictogram is only used when there is no limitation on the type of spatialities.

We provide an example constructed around the macro-concept of *Farm Resources* that can be for example *Cereal Silos*, *Road Networks*, *Parcels*, *Irrigation Networks*. The description of *Farm Resources* provides information about:

- Its spatiality; localization (⊡) is an interesting property allowing one to locate the *Cereal Silos*, as is the linear layout (/) for the *Road Network*, the area (<<) for the *Parcels* and complex geometry (/<<) for the *Irrigation Network* (!).
- Its mechanism of construction; the list of the concepts shows that *Farm Resources* is a set of concepts which have no necessary links between them: *Cereal Silo*, *Road Network*, *Parcel* or *Irrigation Network*. The mechanism of generalization is subjacent in this description.

Farm Resources is therefore a macro-concept in which the subjacent structure of the model is based on the mechanism of generalization/specialization. At the beginning of the analysis, the model could be summarized by the *Farm Resources* concept modeled in Figure 16.

Figure 17. Farm Resources macro-concept annotated with multiform spatiality

In this context, the pictogram denoting *multiform spatiality* * can also be used (Figure 17).

A detailed description can be provided by a textual note attached to the *Farm Resources* concept, but it is also possible to refine *Farm Resources* (Figure 18).

The macro-concept of generalized *Farm Resources* is specified by *Cereal Silo*, *Road Network*, *Parcel* and by another macro-concept, called *Irrigation Network*. If necessary, *Irrigation Network* can be also refined. In Figure 18, *Irrigation Network* has a complex spatiality which can be detailed into *Water Supply Line*, *Reservoir* and other possible concepts

The process of refinement shows that the modeling of the macro-concepts *Farm Resources* and *Irrigation Network* is highly conditioned by the points of view of the protagonists of projects.

3.3. Unknown Spatiality or Unknown Temporality

At the beginning of a project, the spatiality or temporality of some concepts may be unknown. The final spatiality or temporality of concepts will be specified later in the project.

Only the pictogrammic language of Perceptory (Bédard, 1998, 1999b) and the framework GeoFrame (Filho & Iochpe, 1999) make it possible for a designer to add *unknown spatiality* "?" and *unknown temporality* "?" to models.

During the general modeling of a hydrological catchment area, it is often impossible to specify the type of spatiality of certain concepts, such as the *River* concept (Figure 19). At a general level

Figure 18. Detailed model of Farm Resources macro-concept annotated with multiform spatiality

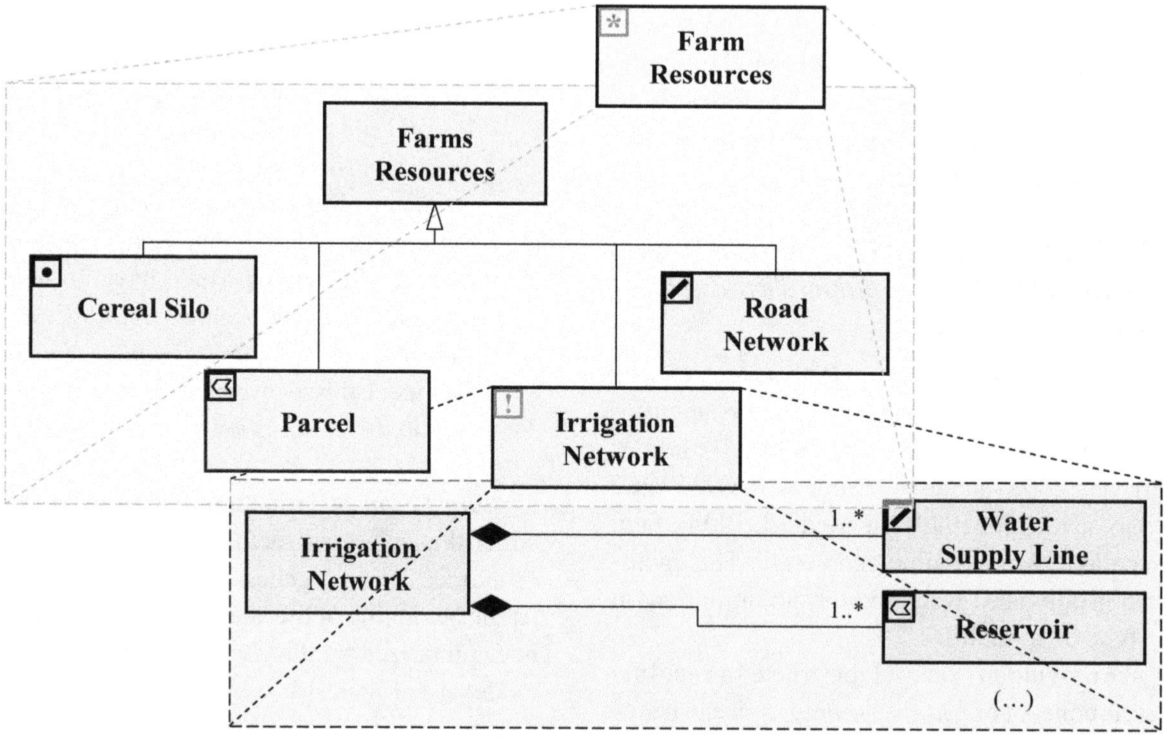

Figure 19. River concept annotated with unknown spatiality

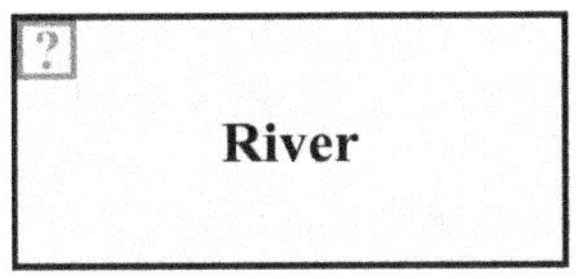

of abstraction, the designer may have no information about the type of final spatiality (linear or polygonal) for this concept. However, it is extremely probable that the type of spatiality will be important later. In this context, the designer can annotate *River* with an *unknown spatiality* "?" while waiting to obtain this information.

If the model of the catchment area is designed to create an information system for management of the river, a linear spatiality could be sufficient. However, a polygonal spatiality will be essential to study the river's behavior during the downstream migration of juvenile salmon (Preusch, 2009). For such studies, biologists often insert tiny radio tags into the belly of the juveniles and follow their movements with a network of antennas (Preusch, 2009).

The *unknown spatiality* pictogram can be used to indicate temporarily a lack of knowledge about the type of spatiality of the *River* concept.

4. CONCLUSION

Currently, there are several formalisms used to represent the spatial and temporal properties of concepts. This paper shows the subtleties proposed by these languages and their power of expression. The visual language of Perceptory has been used here to illustrate different spatio-temporal models.

Recently, most of the languages discussed introduced pictograms to express the spatial and temporal properties of particular concepts. These pictogrammic languages can help the stakeholders of a GIS-based project to understand the models. The languages that allow us to represent the spatial and temporal properties in ER and OO models constitute a solution suitable for the design of GIS. Their use during analytic steps simplifies the resulting models and facilitates the modeling process.

Among the languages cited in our paper, the pictogrammic language of Perceptory currently displays the richest formalism. As presented in this paper, the modeling of numerous types of spatialities and temporalities is made possible with this language. It is the most comprehensive and the most developed from a scientific point of view. Several 3D pictograms have been also introduced in (Bédard & Larrivée, 2008; Larrivée et al., 2005).

An important challenge in the future will be the modeling of spatial and temporal dimensions in multidimensional cubes of data warehouses.

REFERENCES

Bédard, Y. (1998). *Site Perceptory*. Retrieved from http://sirs.scg.ulaval.ca/ perceptory/

Bédard, Y. (1999a, décembre 15). *PERCEPTORY: Nouvel Atelier de Génie Logiciel pour la modélisation des bases de données spatiales et spatio-temporelles.* Paper presented at the Journée scientifique du Cemagref-Engref, Montpellier, France.

Bédard, Y. (1999b). Visual modelling of spatial databases: towards spatial PVL and UML. *Geomatica, 53*(2), 169–186.

Bédard, Y., & Larrivée, S. (2008). Modeling with Pictogrammic Languages. In Shekhar, S., & Xiong, H. (Eds.), *Encyclopedia of Geographical Information Sciences* (pp. 716–725). New York: Springer.

Bédard, Y., Larrivée, S., Proulx, M.-J., & Nadeau, M. (2004, November). *Modeling Geospatial Databases with Plug-ins for Visual Languages: A Pragmatic Approach and the Impacts of 16 Years of Research and Experimentations on Perceptory.* Paper presented at the ER Workshops 2004 CoMoGIS, Shanghai, China.

Bédard, Y., Pageau, J., & Caron, C. (1992, 1-14 August). *Spatial Data Modeling: The Modul-R Formalism and CASE Technology.* Paper presented at the ISPRS Symposium, Washington, DC.

Bédard, Y., & Paquette, F. (1989, 2-7 April). *Extending entity/relationship formalism for spatial information systems, AUTO-CARTO 9.* Paper presented at the Ninth International Symposium on Computer-Assisted Cartography, Baltimore, MD.

Borges, K. A. V., Davis, C. A. Jr, & Laender, A. H. F. (2001). OMT-G: An Object-Oriented Data Model for Geographic Applications. *GeoInformatica, 5*(3), 221–260. doi:10.1023/A:1011482030093

Brodeur, J., & Badard, T. (2008). Modeling with ISO 191xx Standards. In Shekhar, S., & Xiong, H. (Eds.), *Encyclopedia of Geographical Information Sciences* (pp. 705–716). New York: Springer.

Caron, C., Bédard, Y., & Gagnon, P. (1993). MODUL-R: un formalisme individuel adapté pour les SIRS. *Revue de géomatique, 3*(3), 283-306.

Claramunt, C., Coulondre, S., & Libourel, T. (1997). Autour des méthodes orientées objet pour la conception des SIG. *Revue internationale de géomatique, 7*(3-4), 233-257.

Das Neves, F., Gordillo, S., Mostaccio, C., & Levato, A. (1997, April 16-18). *Toward a foundation for Object Oriented GIS Design.* Paper presented at the Joint European Conference on Geographical Information, Vienna, Austria.

Filho, J. L. (1998, November 16-19). *GeoFrame Applying Analysis Patterns in GIS Domain.* Paper presented at the 10th Colloquium of Spatial Information Research Center, Otago, New Zealand.

Filho, J. L., & Iochpe, C. (1999). Specifying analysis patterns for geographic databases on the basis of a conceptual framework. In *Proceedings of the 7th ACM international symposium on Advances in geographic information systems*, Kansas City, MO (pp. 7-13). New York: ACM Press.

Filho, J. L., & Iochpe, C. (2008). Modeling with a UML Profile. In Shekhar, S., & Xiong, H. (Eds.), *Encyclopedia of Geographical Information Sciences* (pp. 691–700). New York: Springer.

Gayte, O., Libourel, T., Cheylan, J.-P., & Lardon, S. (1997). *Conception des Systèmes d'information sur environnement.* Paris: Hermés.

Hadzilacos, T., & Tryfona, N. (1997). An Extended Entity-Relationship Model for Geographic Applications. *SIGMOD Record, 26*(3), 24–29. doi:10.1145/262762.262766

ISO. (2002). *ISO 19108 - Geographic Information - Temporal schema.*

ISO. (2003). *ISO 19107 - Geographic Information - Spatial schema.*

Kösters, G., & Pagel, B.-U. (1996, November 15-16). *The GeoOOA-Tool and Its Interface to Open Software Development Environments for GIS.* Paper presented at the 4th Workshop on Advances in Geographic Information Systems, Rockville, MD.

Kösters, G., Pagel, B.-U., & Six, H.-W. (1995, December 1-2). *Object-Oriented Requirements Engineering for GIS Applications.* Paper presented at the 3rd International ACM Workshop on Advances in Geographic Information Systems, Baltimore, MD.

Kösters, G., Pagel, B.-U., & Six, H.-W. (1997). GIS-application development with GeoOOA. *International journal of GIS, 11*(4), 307-355.

Larrivée, S., Bédard, Y., & Pouliot, J. (2005). *How to Enrich the Semantics of Geospatial Databases by Properly Expressing 3D Objects in a Conceptual Schema.* Paper presented at the OTM Workshops 2005.

Larrivée, S., Bédard, Y., & Pouliot, J. (2006). Fondement de la modélisation conceptuelle des bases de données géospatiales 3D. *Revue internationale de géomatique, 16*(1), 9-27.

Lbath, A., & Aufaure-Portier, M.-A. (1995). AIGLE: Un atelier visuel de génération d'applications géographiques. *Revue Internationale de géomatique, 5*(2), 179-195.

Lbath, A., & Pinet, F. (2000, November 10-11). *The Development and Customization of GIS-based Applications and Web-based GIS Applications with the CASE tool AIGLE.* Paper presented at the 8th ACM Symposium on Advances in Geographic Information Systems, Washington, DC.

Lopes De Oliveira, J., Pires, F., & Bauzer Medeiros, C. (1997). An Environment for Modeling and Design of Geographic Applications. *GeoInformatica, 1*(1), 29–58. doi:10.1023/A:1009704100446

Miralles, A. (2006). *Ingénierie des modèles pour les applications environnementales.* Unpublished doctoral dissertation, Université Montpellier II, Montpellier, France.

Miralles, A., & Libourel, T. (2008). Modeling with Enriched Model Driven Architecture. In Shekhar, S., & Xiong, H. (Eds.), *Encyclopedia of Geographical Information Sciences* (pp. 700–705). New York: Springer.

Miralles, A., & Libourel, T. (2009). Application of a Model Transformation Paradigm in Agriculture: A simple environmental system case study. In *Advances in Modelling Agricultural Systems* (*Vol. 25*, pp. 37–54). New York: Springer. doi:10.1007/978-0-387-75181-8_3

OGC. (2005). Implementation Specification for Geographic information - Simple feature access - Part 1: Common architecture.

OMG. (2003, March). *Unified Modeling Language - Specification - Version 1.5.* Retrieved from http://www.omg.org/cgi-bin/apps/doc?formal/03-03-01.pdf

OOAD. (1997). *Object Oriented Analysis and Design*. Retrieved from http://www.ooad.org/

Pantazis, D. (1994). *Analyse Méthodologique des Phases de Conception et de Développement d'un SIG.* Belgium: Université de Liège.

Pantazis, D., & Donnay, J. P. (1996). *La conception SIG: méthode et formalisme*. Paris: Hermés.

Parent, C., Spaccapietra, S., & Zimányi, E. (2006). *Conceptual Modeling for traditional and Spatio-Temporal Applications: The MADS Approach.* New York: Springer Verlag.

Parent, C., Spaccapietra, S., & Zimányi, E. (2008). Modeling and Multiple Perceptions. In Shekhar, S., & Xiong, H. (Eds.), *Encyclopedia of Geographical Information Sciences* (pp. 682–690). New York: Springer.

Parent, C., Spaccapietra, S., Zimányi, E., Donini, P., Plazanet, C., & Vangenot, C. (1997). MADS, modèle conceptuel spatio-temporel. *Revue Internationale de Géomatique*, 7(3-4), 317–352.

Pinet, F. (2002). *Vers une Méthode de Conception pour la Télégéomatique: Intégration et Validation au sein d'un Atelier de Développement.* Unpublished Thèse de docteur en Informatique, Institut National des Sciences Appliquées de Lyon, Lyon, France.

Pinet, F., & Lbath, A. (2003, April 24-26). *Object-Oriented Modelling of Communicating GIS-Based Applications.* Paper presented at the 6th AGILE European Conference on Geographic Information Science, Lyon, France.

Preusch, M. (2009, July 29). *Acoustic tags help track movement of salmon through dams.* Retrieved February 7, 2010, from http://www.oregonlive.com/environment/index.ssf/2009/07/acoustic_tags_help_track_movem.html

Price, R., Ramamohanarao, K., & Srinivasan, B. (1999). *Spatio-temporal unified modeling language.* Paper presented at the Workshop on Spatio- Models and Languages, IEEE DEXA'99, Florence, Italy.

Proulx, M.-J., Larrivée, S., & Bédard, Y. (2002). Représentation multiple et généralisation avec UML et l'outil Perceptory. In *Généralisation et représentation multiple* (pp. 113–130). Paris: Hermès Sciences Publications.

Quang, P. T., & Chartier-Kastler, C. (1991). *Merise in Practice*. MacMillan Education.

Rocha, L. V. d. (2001). *Framework Conceitual Temporal para Aplicações de Sistemas de Informações Geográficas*. Porto Alegre, Brazil: Universidade Federal do Rio Grande Do Sul.

Rocha, L. V. d., Edelweiss, N., & Iochpe, C. (2001). *GeoFrame-T: A Temporal Conceptual Framework for Data modeling.* Paper presented at the 9th ACM international symposium on Advances in geographic information systems, Atlanta, GA.

Rumbaugh, J., Blaha, M., Lorensen, W., Eddy, F., & Premerlani, W. (1990). *Object-Oriented Modeling and Design*. Upper Saddle River, NJ: Prentice Hall.

Shekhar, S., Coyle, M., Goyal, B., Liu, D.-R., & Sarkar, S. (1997). *Data Models in Geographic Information Systems*.

Shekhar, S., & Xiong, H. (2008). *Encyclopedia of Gis* (1st ed.). New York: Springer.

Tardieu, H., Rochfeld, A., & Colletti, R. (1983). La méthode Merise: *Vol. 1. Principes et outils*. Paris: Éditions d'Organisation.

Tryfona, N., & Jensen, C. S. (1998). *CHOROCHRONOS: A Network for Spatiotemporal Database Systems - Conceptual Data Modeling for Spatio-temporal Applications* (Tech. Rep. No. CH-98-08). European Commission DG XII Science, Research and Development.

Tryfona, N., & Jensen, C. S. (2000, March 19-21). *Using Abstractions for Spatio-Temporal Conceptual Modeling.* Paper presented at the 2000 ACM Symposium on Applied Computing, Villa Olmo, Italy.

Vangenot, C. (2001). *Multi-représentation dans les bases de données géographiques.* Unpublished Thèse de docteur es sciences, Université de Bourgogne, France.

ENDNOTES

1 Object Method Technique (Rumbaugh, Blaha, Lorensen, Eddy, & Premerlani, 1990)

2 SpatioTemporal Geographic Language

This work was previously published in International Journal of Agricultural and Environmental Information Systems, Volume 1, Issue 2, edited by Petraq Papajorgji and François Pinet, pp. 68-84, copyright 2010 by IGI Publishing (an imprint of IGI Global).

Section 4
Decision Support Systems

Chapter 13
Using SocLab for a Rigorous Assessment of the Social Feasibility of Agricultural Policies

Françoise Adreit
Université Toulouse 2 - Le Mirail, France

Pascal Roggero
Université Toulouse 1 - Capitole, France

Christophe Sibertin-Blanc
Université Toulouse 1 - Capitole, France

Claude Vautier
Université Toulouse 1 - Capitole, France

ABSTRACT

This paper presents a theoretical and methodological framework to take into consideration the social dimension in a sustainable development project. To do this, the authors have developed the SocLab software environment, which implements a formalization of a well-established sociological theory, and enables the modeling of social organizations, to analyze their properties and to simulate social actors' behaviors. SocLab was used to assess the social acceptability of new agricultural practices more in line with the preservation of water resources and natural environments, in a well defined context. The paper shows how it was used and presents the main results.

1. INTRODUCTION

The reference model for sustainable development gives an equal importance to the economic, environmental and social dimensions. In practice, little attention is paid to the social dimension: the biophysical aspects and the economic cost of the environmental degradation and its compensation by public policies are considered with much more attention. This situation is due, to a large amount, to the informal and subjective nature of the results provided by sociological studies; they do not

DOI: 10.4018/978-1-4666-0333-2.ch013

feature the properties of quantitative scientific knowledge, and they are difficult to handle and to merge with the results provided by economics and environmental sciences.

As a foreseeable consequence of this lack of attention to the social dimension, many public policies are quite ineffective to change the social practices toward the desirable direction. The social impact of the proposed measures raising some fears, they are not accepted by the actors and thus they are not applied (Vautier et al., 2009).

The *Life* European project *Concert'Eau* was a proof-of-concept project aiming at demonstrating the feasibility to define a set of new agricultural measures proposed and accepted by the stakeholders, following the process shown in Figure 1. In this project, we were in charge, in the step 2, to assess the social acceptability of the measures proposed during the step 1. To provide a quantitative evaluation of the social dimension of each proposed measure, we have used a rigorous theoretical and methodological framework whose generic nature makes it appropriate to model and analyze the social dimension of other cases:

- We base our work on a well-established social theory, the Sociology of the Organized Action (section 2);
- We have formalized this theory as a meta-model for the description of social organizations in a rigorous way (section 3);
- We have implemented a software platform, *SocLab*, which enables to edit the model of an organization, to explore its states and to simulate the behavior of the social actors (section 4).
- Section 5 presents how we have used this framework to estimate the social feasibility of agricultural measures in the Concert'Eau project.

Figure 1. The process of the Concert'Eau project

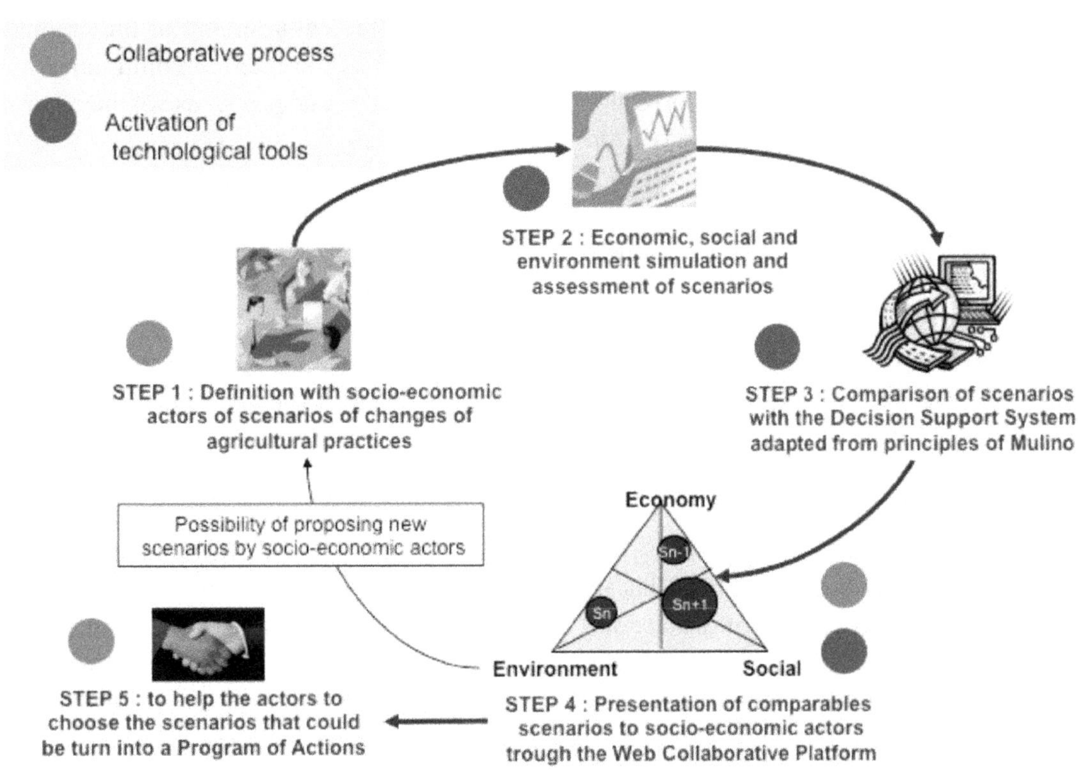

2. THE SOCIOLOGICAL BASE: THE SOCIOLOGY OF THE ORGANIZED ACTION

The Sociology of the Organized Action (SOA) (Crozier, 1964; Crozier & Friedberg, 1977; Friedberg, 1993) defines the organization like "a complex set of intersected and interdependent games through which persons, often provided with very different assets, seek for maximize their gains, respecting the rules of the game which are not written and are imposed by the environment, taking systematically advantage of all their assets and seeking to minimize those of the others [...]. The aim is not to remove power relations – which is an impossible and unproductive task – but on the contrary to recognize them in order to regularize[1] them and to force the partners to directly trade, i.e. to communicate" (Crozier, 1964). These intersected and interdependent games take place in so-called Concrete Systems of Actions (CSA) which can be defined, in a given organizational context, as sets of actors and their alliances, their relations and the regularization of these ones.

If the SOA (formerly called Strategic Analysis) inherits the sociology of the organizations (Clegg et al., 1996), its application covers all kind of "organized" action whatever its level of codification or formalization. A CSA, for instance a firm, an association or a local political system, is composed of "numerous differentiated actors interacting in a non-trivial way" (Crozier & Friedberg, 1977). It is an interaction context which is quite precisely delimited and structures and motivates the cooperation among a set of social actors. These actors are engaged in the long-term achievement of the organization's objectives that may be distinguished from their own ones.

A CSA also includes various resources that are handled by the actors according to rules that comply with the organization's objectives. The rules are admittedly constraining but they always leave some freedom in the way of acting. The behavior of an actor is then neither totally conditioned by the organizational rules that constrain him, nor by purely individual or emotional factors. This behavior is *strategic* and aims at realizing some objectives, would it be deliberated or not. To get the means to achieve his own objectives and those given by the organization's formal rules, each actor seeks, according to his *bounded rationality* capacities (Simon, 1982), to have enough *power* to be able to preserve or increase his autonomy and action capacity within the organization.

This power results from the mastering of one or several "uncertainty zones" (UZ) that enable an actor to behave in a way that is unpredictable by other actors and consequently to set, to some extent, the "exchange rules" in his relations with others. Each UZ involves resources for the action, and thus serves as both a constraint and an opportunity. Each social actor both controls some UZs and depends on some others, so that UZs are the media of the power relationships between the actors. The four main types of UZ that support power relationships within a CSA are based upon: competence or expertise; the control of interactions with the environment of the organization; the control of the internal communication; and the knowledge and proper use of the organization formal rules.

The SOA aims at discovering and explaining the real functioning of an organization beyond its formal rules. It is built upon the idea that organizations are social constructs actualized by and within the relationships among its member actors. It focuses on the regularization phenomenon which ensures at the same time the evolution of such systems and their relative synchronic stability.

The SOA is most often used to study the running of organizations and, more widely, the social processes resulting from the interactions between social entities (public policies, public-private partnerships, governance, and so on). It implies to assume the strategic actor concept, which mainly concerned with the power considered as a mean for the action. Therefore, the SOA is clearly different from the institutionalist theories which focus on the institutional and collective dimension of the organizational phenomena (DiMaggio & Powell, 1991). It also differs from theories which focus on the individuals' action but do not reduce it to its strategic dimension, such as the justification theory (Boltanski & Thévenot, 1991). Therefore, the context of action (more or less institutional, more or less strategic) determines which theory can be applied. In the Concert'Eau project, the design and the evaluation of agricultural measures have been considered as a strategic context of action.

3. THE META-MODEL OF A CONCRETE SYSTEM OF ACTIONS

Purposing to enable the modelling of social organizations, we propose a meta-model as an abstraction that catches the common concepts and properties of social organizations and can be instantiated on specific cases as models of real world social organizations. The model of the structure of an organization is composed of instances of the meta-model's classes which correspond to the constitutive elements of the organization under consideration, and of links between these instances.

The meta-model is graphically represented on Figure 2, as a UML class diagram. It is composed of a class *Actor*[2] and a class *Relation,* linked by the associations *Control* and *Depend*. The *actors* are the active entities who handle the *relations*, by modifying the states of the *relations*. Each *relation* is controlled by a unique *actor*, the only one able to *modify* its state. The *actor* who *controls* a *relation* may also *depend* on this *relation.* Each actor distributes *stakes* on each *relation* he depends on. Thus, he gets some *effect* from this relation, given by the *effect()* function applied to the *state* of the *relation.* Consequently, an actor gets some action *capacity,* as an aggregation of the *stakes* and *effects* he receives from every *relation* he depends on, and some *power,* as an aggregation of the *action capacities* he grants to the *actors* who depend on the *relations* he controls.

The meta-model is described more precisely in the next sections; more information can be found in (Sibertin-Blanc et al., 2006).

3.1. Resources and Relations

The concept of relation is founded on the classical concept of resource. A resource is a concrete or abstract (cognitive) entity whose availability is useful for some actors to achieve actions within the scope of the organization; according to the SOA, this availability is supervised by actors. Each resource is the support of one or several relations that give rise to transactions concerning the resource.

Each *relation* is controlled by an *actor* – one of the supervisors of the *resource* –, whereas some

Figure 2. The meta-model of the structure of Concrete Systems of Action

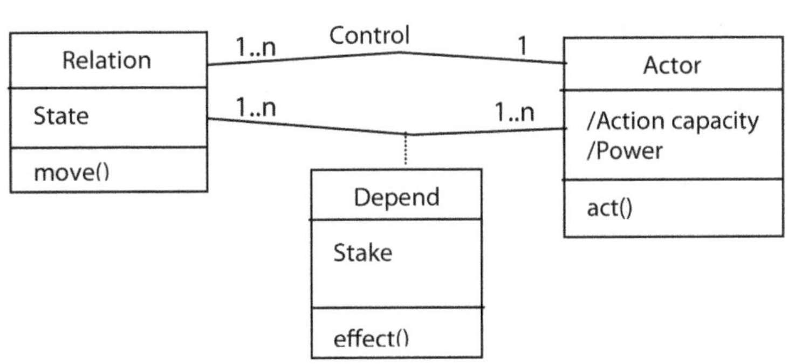

actors who need the resource for achieving their goals are dependent on this *relation*. The *actor* who controls a *relation* is in a position to define how the *resource* is available to the others (in a given transaction concerning the resource), and thereby, he controls to what extent the dependent *actors* are able to reach their goals. Thus, every *relation* is an unbalanced power relation: there is an actor who dominates the others through this *relation* (Friedberg, 1993). But there are many *relations* within an organization and each *actor* controls at least one of them so that, all in all, the actors are mutually dependant on each other. Consequently, every actor is able to negotiate his collaboration.

The *state* attribute of a *relation* represents how it is managed by its controller and stands for the behavior of this *actor* with regard to the supervising of the related resource. Its range of value is a *space of choice*, representing the set of all the behaviors that the actor can adopt to manage the *relation*. Since the SOA does not describe how the actors adjust their behaviors[3], we made the assumption that an *actor* modifies his behavior towards a greater or lower cooperation. In order to keep simple, the space of choice of any *relation* is represented as the interval $SC = [-1;1]$ and this interval is oriented: the lower bound represents the least cooperative behaviors of the controller, the upper bound represents the most cooperative ones, while the zero value stands for a normal, neutral behavior. Therefore, an *actor* who controls a *relation* may modify his behavior by using the function *move()*, that modifies the value of the state of the *relation*, towards less or more cooperation in the management of the *relation*.

The *state* of a *relation* determines the availability of the involved *resource,* i.e., how each actor who depends on this resource is granted to use it according to his goals. So, the *state* of a *relation* produces an *effect* on every *actor* who depends on it. *Effects* measure the capacity to make use of the resource and they take values on a scale from -10 to 10[4]:

the worst access =-10, very bad = -8, …, neutral=0, pretty good=2, …,optimal=10

Effect values are given by *effect()* functions, which are specific to each relation and dependent actor. When the controller of a *relation* r sets the value of its state to s, the *effect* for a dependent *actor* a is given by $effect_r(a,s)$. More formally, the $effect_r()$ function of a relation r is defined as:

$$effect_r : A \times SC_r \rightarrow [-10;10]$$
$$(a, s_r) \mapsto effect_r(a, s_r),$$

where A is the set of actors, SC_r the space of choice of the *relation* r and [-10,10] the range of the capacity to use a resource.

3.2. Actors and Their Stakes

According to the SOA, an *actor* is able to negotiate his collaboration, and this requires the control of at least one *relation*. Therefore an *actor* is defined as someone who *controls* at least one *relation*. *Actors* and *relations* are defined in relation one to another, in a dialogical way: something is a *relation* if and only if some *actors* depend on it, and someone is an *actor* if and only if he controls some *relation*.

In addition, the SOA assumes that the actors are strategic: their behaviors are "motivated by some goals" (Friedberg, 1993, pp. 214-215). To achieve his goals, each actor needs some resources and thus depends on some relations. How much an actor depends on a relation is determined by the necessity of the related resource for the achievement of one of his goals and the relative importance of this goal. This is depicted by the *stake* attribute of the *Depend* association. The more valuable a *relation* is for an actor, the higher his *stake* on this *relation*. *Stakes* are represented by numerical coefficients, on an arbitrary scale from 0 to 10:

null =0, negligible=1,..., significant=5,..., critical=10.

In order to provide every actor with the same degree of investment in the organization, the total amount of *stakes* that each actor can distribute is fixed to 10^5.

3.3. State of an Organization, Action Capacity and Power of the Actors

We define the *state* of an organization as the vector of all the *relation states.* For a given state of the organization and an *actor,* we consider the aggregation of the *effects* of the *relations* the actor depends on, weighted by the *stakes* he puts on the *relations.* Such a quantity depicts the overall ability for an *actor* to gain access to the *resources* he needs to reach his goals, weighted by the relative importance of these resources regarding his goals. It measures, for an *actor,* his *action capacity*, i.e. his ability to reach his goals, by having the means required to actually reach them. More formally[6], the *action capacity* of an actor a, when the organization is in the *state* s, is defined by:

$$action\ capacity(a,s) = \sum_{r\in R} stake(a,r)\times effect_r(a,s_r) = \sum_{r\in R} impact(a,r,s),$$

where stake(a, r) is the stake of a on the relation r and $effect_r(a, s_r)$ is the effect on a of the relation r being in the state s_r. We call *impact(a, r, s)* the product of stake(a, r) by $effect_r$ (a, s).

Since any actor controls some relations, he contributes to the action capacity of the actors who depend on these relations. The global influence of an actor on the action capacity of the others, i.e. how much action capacity he gives by the management of the relations he controls, fits the concept of *power*, a core concept in the SOA. In the same way as the action capacity, the *power* exerted by an *actor* a, when the organization is in the *state* s, is defined by:

$$Power(a,s) = \sum_{r\in R, a\ controls\ r} \sum_{b\in A, b\neq a} impact(b,r,s).$$

3.4. Solidarities between Actors, Satisfaction and Influence

The interactions among actors are not totally framed by operational dependencies or aim dependencies. For example, kinship links or common social conditions or scholarships affect the way actors consider one another and would be hardly represented by *resources* of the organization. Groups or coalitions should also be modeled, as abstract *actors,* who's each component has interests in their satisfaction.

That leads us to introduce the *solidarities* the actors maintain between each other. We represent the solidarities as a function which, for two *actors* a and b, associates the degree of solidarity *solidarity(a,b)* the *actor* a puts on the *actor* b, following the function:

$$solidarity: \ A\times A \to [-1,1]$$
$$(a,b) \mapsto solidarity(a,b),$$

such that ∀a ∈ A, *solidarity (a, a)* = 1, where negative values represent enmities, zero represents disinterest, and positive values real solidarities.

The introduction of *solidarities* leads us to consider two quantities, in addition to the previous *action capacity* and *power,* to depict a social configuration: *Satisfaction* and *Influence*. When an organization is in a *state* s, the *satisfaction* perceived by an *actor* a is the sum of the *action capacities* of all the actors, weighted by his *solidarities* for them:

$$Satisfaction(a, s) = \sum_{b \in A} solidarity(a, b) \times action\ \ capacity(b, s),$$

In the converse way, the influence exerted by an *actor* a on another *actor* b is the sum, over all the *actors*, of the impacts of the *relations* controlled by a, weighted by the solidarities of b:

$$Influence(a, b, s) = \sum_{c \in A} solidarity(b, c) \times \sum_{r \in R; a \text{ controls } r} impact(c, r, s).$$

4. THE *SocLab* ENVIRONMENT

SocLab, for Social Laboratory, is a software environment dedicated to the study of social organizations (Maillard, 2008). As a virtual laboratory, it allows its users to study organizations *in silico* and to test hypotheses on objects for which the experiment *in vivo* is impossible. More precisely, *SocLab* enables its users to build models of social organizations, to make analytical analyses of their properties and to simulate the behavior of their actors. *SocLab* is a Java application, easy to download and to use. It includes several modules. At each step of his work, the user can generate .rtf files that contain all his hypothesizes and comments together with tables, curves or diagrams produced by *SocLab*; thus, the user can keep trace of his work and, e. g., compare different models of the same organization.

4.1. Editing the Model of an Organization

The edition module (Figure 3) enables the user to define the structure of an organization, that is the actors, relations, stakes, effect functions and solidarities, and to annotate each item. This structure is stored in an xml file.

4.2. Static Analysis

The underlying mathematical structure of an organization has enabled us to define many *indicators* that are meaningful to the sociologist. Some of these indicators are structural - as they are not related to a specific state of the organization - and they bear either on a relation or an actor (for instance, the relevance or the strength of a relation, the autonomy or the potential power of an actor) or on the whole organization (for instance the sum of the above indicators). Other indicators are state-dependent and are also related either to a relation or an actor (for instance the capacity or

Figure 3. Editing the stakes of the actors

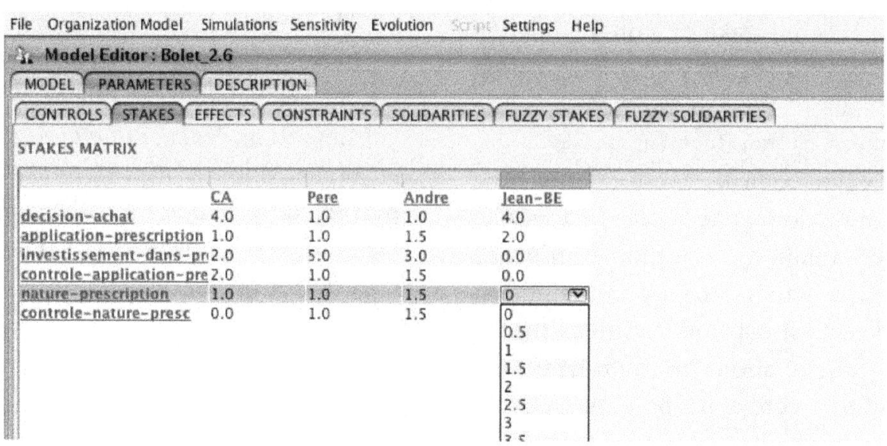

the satisfaction of an actor) or bear on the whole organization.

The actors of an organization are related in many ways, especially by their links of solidarity and of dependency (by means of the relations). For each type of such links, one can give a view of an organization as a social network, where the node actors are connected by these links and, most often, specific indicators may be used to label the nodes or the edges. These networks could be analyzed using the numerous techniques and tools developed for social networks (Jackson, 2008).

The satisfaction function defines, for each state of an organization, a *social configuration* which is the vector of the actors' satisfactions (and the same holds for the capacity of action, the power or the influence). Exploring the possible configurations of an organization is very instructive as it allows to discover all its potential regularizations. Moreover, some of these configurations are particularly interesting because they fulfill some property such as to maximize or minimize the satisfaction of a given actor or the *global satisfaction* (i.e. the sum of all the actors' satisfactions), to be Pareto optima or Nash equilibria and so on. This exploration enables for example to know whether it is difficult for the actors to cooperate, or to detect structural conflicts between two actors when any configuration which provides one of the actors with a high satisfaction also provides the other with a low satisfaction and conversely. *SocLab* computes many such interesting configurations and allows to interactively search all of them (Figure 4); each configuration is detailed by giving the states of the relations and the amount (in value or in proportion) of action capacity and satisfaction given by each actor to others by means of the controlled relations or of the solidarities.

4.3. Simulation of the Actors' Behaviors

From the SOA point of view, an organization determines a *social game* where the actors modify the organization state, by modifying the states of the relations they control, in order to obtain enough action capacity (or to exert some power). Yet, in opposition to classical games in economics, the purpose of social games is not to maximize the action capacity, but to reach a stationary state, a state in which every actor has a satisfactory action capacity, and does not modify the states

Figure 4. State analysis

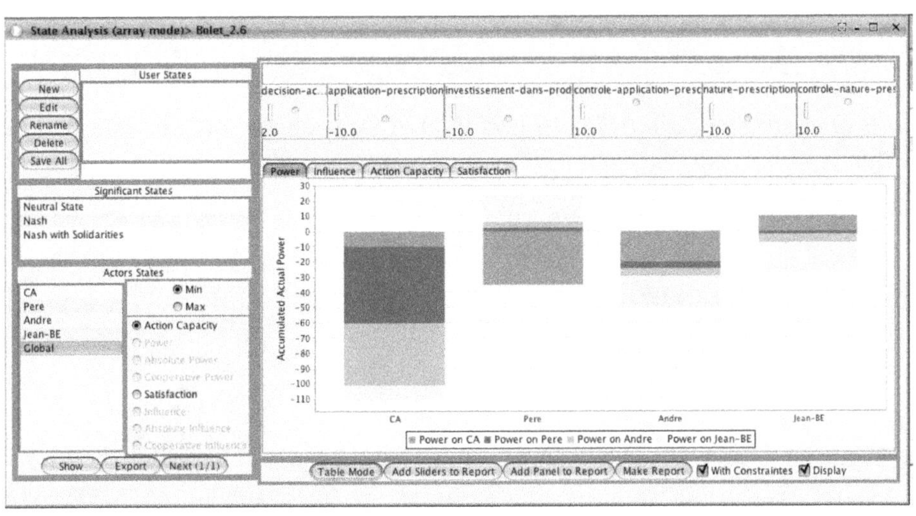

of the relations anymore. Reaching a stationary state means that the actors have adapted their behaviors one to another in a stable equilibrium that corresponds to a feasible regularized functioning of the organization[7]. According to the SOA, the behavior of the actors is strategic: they seek a way to reach their own goals, and therefore, as a meta-goal, to obtain enough action capacity to this end. This strategic behavior is framed by a bounded rationality (Simon, 1981) and is geared toward cooperation.

SocLab implements a model of the actors' rationality as a process of trial and error based on a self-learning rules system (Sutton, 1998). According to this model, the simulator interface enables the user to initialize the simulation parameters (such as the number of runs or the actors' parameters), to play the social game and to visualize the simulation results.

4.4. Sensitivity Analysis

The sensitivity analysis module enables to study the influence of the simulation parameters upon the simulation results (Figure 5). It is possible to make the rationality parameters of the actors vary (their tenacity, discriminability, and so on), as well as their stakes and solidarities (El Gemayel et al., 2009). The development of facilities for the statistical data analysis (PCA, multiple correlations, and so on) of the results of simulations and sensitivity analysis is in progress.

5. USING SOCLAB IN THE CONCERT'EAU PROJECT

In the European Life project Concert'Eau, we have been in charge of the assessment of the social acceptability of new agricultural measures proposed by the stakeholders. We have developed a specific methodology, using Soclab, to produce the demanded results. This section gives a general idea of the methodology and of the interpretation of the results.

5.1. Presentation of the Project

The Concert'Eau project[8] was launched in October 2006 for three years to demonstrate the opportunity to involve agricultural actors of a territory in the design of new practices, more in line with the preservation of water resources and the natural environment, in order to increase the chance of their effective application. The experimentation took place in the upstream part of the Gers river basin inside the Adour-Garonne basin district in the southwest of France, which is classified as a nitrate vulnerable zone for several years. The steps 1 and 4 of the project (Figure 1) gathered about

Figure 5. Results of a sensitivity analysis: the variation of the average satisfaction of the actors depends on their tenacity

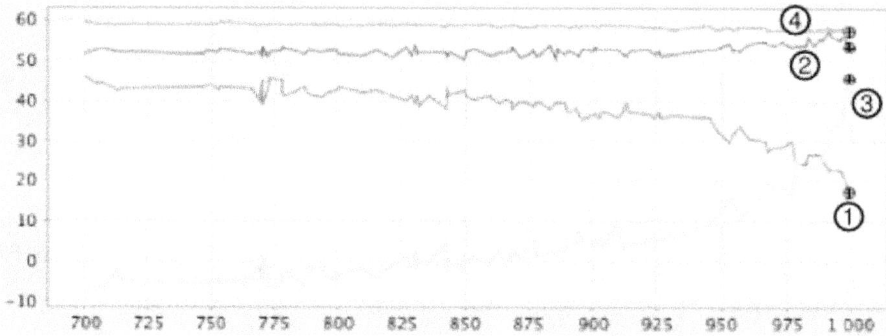

seventy people of the socio-economic system, representing about fifteen organizations or communities, who elaborated about seventy alternative scenarios for new agricultural practices. In step 2, these scenarios have been evaluated by scientists on the environmental, economical and social dimensions. On the environmental dimension, the evaluation consisted in simulating the effects of the proposed measures on the concentration of nitrogen molecules and pesticides in the surface waters of the Gers river and its affluents. The economists had to assess the potential effect of these measures on the working time of the farmers, on their gross farming income and on the direct cost for the society. The sociologists had to estimate the potential acceptability of the proposed measures for the whole social system and just for the farmers (Vautier et al., 2009).

5.2. The SocLab Model of the System

The social system was not a formal organization with well-defined frontiers but a diffuse system of action gathering a great amount of actors, actions, relations and strategies, and having a fuzzy frontier. Taking exhaustively all of them into account would have resulted in a model not simple enough to enable a clear interpretation of the results. However, the limitation of the number of actors who are taken in consideration requires to make choices which oblige to tighten the model, *i.e.* to focus on the essential components. The actors who were considered in the Concert'Eau model are really the ones who, according to experts knowing this context of action and according to the interviews, are fundamental for the regulation of the agricultural use of the water.

5.2.1. Relations and Actors

Our model of the system is composed of eight actors. Each actor is provided with the control of a single relation which combines the resources which are the major means of his influence:

- The *Water Agency: Financing* of the various projects related to agriculture and water.
- The *Department*[9] *Council: Cofinancing* the projects.
- The technical and scientific *institutes:* the *data* about the quantity and quality of water.
- The *Chamber of Agriculture: Consulting* (technical route - the dates of ploughing, of seedling, of fertilization, the doses of products to spread, and so on).
- The so-called *conventional farmers* who are represented by the local units of the FNSEA[10] (or of the CNJA[11]): *Majority lobbying.*
- The *organic farmers* (and the environmental associations): *Anti-establishment lobbying.*
- The *DRAF* and the *DDAF* (Regional and Departmental Directions of the Agriculture and the Forest) and also the ONEMA (National Office of the Water and the Aquatic Environments): *Legal inspection.*
- The *agricultural cooperatives*: *Interface* between farmers and other actors.

5.2.2. Stakes and Solidarities

Table 1 shows the stakes of the actors. The values have been collected with a specific investigation protocol. For each actor of the model, several persons have been interviewed. The values consist in a weighed average of the values given by persons who compose each collective actor.

Similarly we have collected and made the averages of the values given for the solidarities (Table 2).

*Table 1. Stakes (in **bold** for the actor who controls the resource)*[12]

	Water agency	DRAF DDAF	Depart. Council	Chamber of Agri.	Agri. coop.	Tech. Instit.	Conv. farmers	Organic farmers
Financing	**4**	3	2	1	0	1.5	2.5	0.5
Cofinancing	2	1.5	**4**	1	0	1.5	1	1
Data	2	2.5	1.5	1	1	**3**	0	0
Consulting	0	0	0	**2**	2	1	1.5	1
Majority lobb.	0	0	0	2	3	0	**2.5**	1
Anti-establish. lobbying	0	0	0	1	0	0	0.5	**3**
Inspection	2	**3**	2.5	1	1	2	1	2.5
Interface	0	0	0	1	**3**	1	1	1

5.3. Simulation Results about the Current Situation

The simulation module of SocLab computed the levels of satisfaction and power of each actor, when the game is regularized (Table 3).

A value has to be read by comparison with the others on the same row (for example, the satisfaction of the Water Agency is greater than the satisfaction of the organic farmers).

The results generated by SocLab gives information about the level of satisfaction of each actor and, by comparison, the most (and the least) satisfied actors. This information allows the sociologist to give relevant interpretations about the CSA, highlighting some facts that could be remained unseen. In the Concert'Eau context, according to Table 3, the Agricultural Cooperatives, the Chamber of Agriculture and the conventional farmers are the most satisfied actors (71 to 78), the organic farmers are the less ones (38), while the institutional actors, the DRAF-DDAF and the Technical and Scientific Institutes (50 - 52), the Water Agency and the Department Council (61 - 68), stay in-between.

These results suggested the following sociological interpretation: the highest and lowest satisfactions are not surprising in a system which operates on a relative *status quo*: the domination of a productivist conception of agriculture,

Table 2. The solidarities between the actors: The table has to be read by row (for example, the solidarity of the Water Agency for the DRAF is 0.15). The sum of each row equals 1.

	Water agency	DRAF DDAF	Depart Council	Chamber of Agri.	Agri. Coop.	Tech. Instit.	Conv. farmers	Organic farmers
Water agency	0.8	0.15	0.15	- 0.2	- 0.2	0.5	- 0.2	0
DRAF/DDAF	0.2	0.6	0.05	0.05	0	0.1	0	0
Depart. Council	0.2	0.25	0.8	- 0.25	0	0	0	0
Chamber of Agri.	0.1	- 0.1	- 0.1	1	0.2	0.1	0.2	- 0.4
Agri. coop.	0.2	- 0.1	0	0	0.8	0.2	0.2	- 0.3
Techn.instit.	0.1	0	0.1	0.1	0	0.5	0.1	0.1
Conv. farmers	0.1	- 0.1	- 0.1	0.3	0.15	0.2	0.8	- 0.35
Organic farmers	0.15	- 0.15	0.15	0.2	0	0	- 0.35	1

combined with the habit of being subsidized to reach quantitative aims without evolution of the agricultural practices. The satisfaction of the contributors, the Water Agency and the Department Council, is mainly symbolic and concerns their reputation. As for the DRAF-DDAF and the Technical and Scientific Institutes, they are institutional actors who get their legitimacy from outside the social game (i.e. the State) and their action capacity depends only partly on the other actors of the social game.

The simulation results also give information about the level of power of each actor and, by comparison, the most (and the least) powerful actors. According to Table 3, the more powerful actor is the Technical Institutes (apart from the Water Agency which is the main fund-provider with the European Union). According to the sociological interpretation, this is not surprising if we consider that the data concerning the water-agriculture relations are the basis of the public policies, of private actions (agricultural cooperatives, associations) but also of the objections to the current state of affairs; the nature and the quality of these data are at the center of very affirmed struggles between the actors.

The results generated by SocLab give more precise information about the level of cooperation of the actors. To this end, Soclab calculates two indicators: the absolute and the cooperative powers (Table 3). The cooperative one is the sum of the positive contributions of an actor to the satisfactions of others; the absolute one is the absolute sum of all his contributions. The study of the differences between these two powers shows that some actors are very cooperative (they do not prevent others to achieve their goals) while others are not. For example, according to Table 3, the Water Agency, the Department Council and the Technical Institutes are very cooperative (according to their function in the organization or their own interest) while the Chamber of Agriculture is not (according to its lobbying activity). The conventional farmers, like the organic ones, also are not cooperative but for different reasons (the sociological interpretation): the first ones are allied to the cooperatives and the Chamber of Agriculture; they compose an agricultural oligopoly which keeps a leading power (if we add the powers of the three actors) and are more inclined to be self-interested than cooperative. The organic farmers consider that they are isolated and that no other actor can (or wants to) embrace their cause.

Upon the basis of these results and interpretations, we can determine the actors who will have some reason and be able to be opposed to (or to promote) each proposed measure.

5.4. Evaluation of the Social Feasibility of the Proposed Measures

The demand of the Concert'Eau project was to evaluate the social acceptability of the proposed measures for all the social system and for only the farmers. We have revised this acceptability as a feasibility that is the possibility for the system to find a new regularization.

The starting point was the actors' satisfactions at the current situation given by the simulation. We made a second round of investigation to evaluate, for each actor, the acceptability (i.e. the anticipated satisfaction) of each proposed measure in comparison with the current situation.

However, the increase or decrease of satisfaction does not induce the same reaction for all the actors: an actor with a high level of power and a significant increase (resp. decrease) of satisfaction will have the capacity to promote (resp. to prevent) the adoption of the measure, while a low-powered actor will not, or slightly. Therefore, we have merged the acceptability of each actor with his actual power to evaluate how much he could supports each measure. Indeed, an actor undertakes an action when he has a motivation (his satisfaction) and some means (his power).

For each measure, we added up the supports of all the actors to evaluate its feasibility for the whole societal system and the supports of the

Table 3. The satisfactions and powers of the actors (simulation results)

	Water agency	DRAF DDAF	Depart. Council	Chamber of Agri.	Agri coop.	Tech. Instit.	Conv. farmers	Organic farmers
Satisfaction	68.4	50.7	61.4	71.4	78.1	52.0	72.0	38.0
Absolute power	146.6	41.5	74.2	25.0	65.1	141.8	101.7	56.5
Cooperative power	144.2	27.5	74.2	7.9	40.0	141.8	55.4	0.9

Table 4. Feasibility of each measure M_i for the whole societal system and just for the farmers[13]

	current	M1	M2	M3	M4	M5	M6	M7
System	457	468.1	476.2	494	511.4	483	459	501.1
Farmers	90.5	60.3	60.3	90.5	98.2	96.3	88.5	114.6

two kinds of farmers to evaluate its feasibility for the farmers (Table 4) (doing so, we hide the difference between the two kinds of farmers, but this was the demand of the project).

The first column ("current") gives the feasibility at the current situation. Then, we can compare the feasibility of each measure M_i with it. For example, M_1 and M_2 are not supported by the farmers (60.3 is smaller than 90.5) but they are supported by the whole system (468.1 and 476.2 > 457). So, we can expect that the other actors will promote these measures and could make them to be adopted, but in a contestation context from the farmers. We could also have computed the Gini coefficient (which measures the quality of a distribution of values) of the actors' satisfaction for each measure as another parameter of the feasibility; but we did not.

Thus, the analysis indicates the actors who could be opposed to a measure and give an idea of the discussions and the oppositions whether a measure is adopted (even if the measure seems to be globally feasible).

6. CONCLUSION

We have presented a theoretical and methodological framework that allowed us to provide a quantitative assessment of the social acceptability of envisioned agricultural policies more in line with the preservation of water resources and the natural environment.

This kind of assessment is very useful for supporting the decisions of policy makers: whatever the scales of value, quantification allows on the one hand to study the properties of an alternative (by comparing the rates of various indicators applied to this alternative) and, on the other hand, to compare alternatives (by comparing the rate of indicators for these alternatives). These comparisons may be done by considering either the raw values or the relative values, that is the position of a value within its range (value – minimum_value) / (maximum_value – minimum_value), not shown in section 5. In addition, the results provided by this kind of sociological analysis can be aggregated with results provided by other sciences thanks to the use of tools like *Mulino*[14].

We believe that this framework is also relevant for companion modeling processes, where actors have a collective interest in establishing a diagnostic of the running of the CSA (Becu et al., 2003). In this context, this approach allows each actor to express his representation in a formal way which includes no ambiguity and can be discussed by others; the problematic points can be clearly identified and, most often, it is possible to establish a shared representation. Then, SocLab can be used to highlight properties of the current running and to investigate other distributions of the power relationships among the actors.

As the quantitative results are obtained from a model which requires a quantitative analysis of the organization under consideration, we have to wonder about the sociological feasibility and rigor of such analyses. For lack of space, this paper does not describe the inquiry methodology used by the sociologists to obtain necessary data to elaborate the model of the Concert'Eau context. However, once the relevant actors and relations have been identified, the quantitative values of the model (the stakes, the solidarities and the shape of effect functions) have been determined according to the answers given by the ground actors to the sociologists' questions. It appeared that, with an appropriate presentation of the meta-model, the people understand and respond to the concrete questions with firm and quite precise answers that are then easily combined to determine the needed values.

Assuming that the feasibility of a rigorous quantitative assessment is established, we have also to wonder about the validity of the results. First of all, let us recall that a model must not be confused with the modeled object. A model of an object is always a partial view of this object, a simplification intended to improve its understandability. Thus, the proper use of quantitative results always necessitates their interpretation in the field of the social reality. This interpretation requires some competency in the practice of sociology. For that reason, SocLab is a scientific tool that must be used by competent and trusty peoples (as misuses, we can mention the return of results that disturb an organization, or a scientific guarantee given to erroneous assertions). Moreover, considering social affairs, we have to deal with the potentially unpredictable behavior of human beings. Thus, in this domain, forecasts are only plausibilities. All in all, the validity of the results which are provided by our approach is grounded in the validity of the Sociology of the Organized Action to address the considered cases, in the accuracy of our formalization of this theory (Phan et al., 2009) and of its implementation into the SocLab modeling tool, and in the scientific rigor of the methodology followed by the sociologists to analyze and study the organization.

REFERENCES

Becu, N., Bousquet, F., Barreteau, O., Perez, P., & Walker, A. (2003). A methodology for eliciting and modelling stakeholders' representations with agent based modelling. In D. Hales, B. Edmonds, E. Norling, & J. Rouchier (Eds.), *Proceedings of the 4th International Workshop on Multi-Agent Based Simulation III* (LNCS 2927, pp. 131-149).

Boltanski, L., & Thévenot, L. (2006). *On justification: The economies of worth*. Princeton, NJ: Princeton University Press.

Clegg, S. R., Hardy, C., & Nord, W. R. (1996). *Handbook of organization studies*. London, UK: Sage.

Crozier, M. (1964). *The bureaucratic phenomenon*. Chicago, IL: University of Chicago Press.

Crozier, M., & Friedberg, E. (1977). *L'acteur et le système: Les contraintes de l'action collective*. Paris, France: Seuil.

DiMaggio, P., & Powell, W. (1991). *The new institutionalism in organizational analysis*. Chicago, IL: University of Chicago Press.

El-Gemayel, J., Sibertin-Blanc, C., & Chapron, P. (2009). Impact of tenacity upon the behaviors of social actors. In *Proceedings of the 1st International Workshop on Applied agent based simulator engineering for complex system study*, Nagoya, Japan.

Friedberg, E. (1993). *Le pouvoir et la règle: Dynamiques de l'action organisée*. Paris, France: Seuil.

Jackson, M. O. (2008). *Social and economic networks*. Princeton, NJ: Princeton University Press.

Maillard, M. (2008). *MultiAgents formalisation of the sociology of the organized action*. Unpublished doctoral dissertation, University of Toulouse, Toulouse, France.

Mulino. (2004). *MULtisectoral INtegrated and Operational decision support system, or mDSS.* Retrieved from http://www.netsymod.eu/mdss/

Phan, D., & Varenne, F. (2009). Agent-based models and simulations in economics and social sciences: From conceptual exploration to distinct ways of experimenting. *Journal of Artificial Societies and Social Simulation*, 10.

Sibertin-Blanc, C., Amblard, F., & Mailliard, M. (2006). A coordination framework based on the sociology of organized action. In O. Boissier, J. Padget, V. Dignum, & G. Lindemann (Eds.), *Proceedings of the Conference on Coordination, Organizations, Institutions and Norms in Multi-Agent Systems* (LNCS 3913, pp. 3-17).

Simon, H. A. (1982). *Models of bounded rationality: Behavioral economics and business organization* (*Vol. 1-2*). Cambridge, MA: MIT Press.

Sutton, R. S., & Barto, A. G. (1998). *Reinforcement learning: An introduction*. Cambridge, MA: MIT Press.

Vautier, C., Roggero, P., Adreit, F., & Sibertin-Blanc, C. (2009). Evaluation by simulation of the social acceptability of agricultural policies for water quality. In *Proceedings of the International ICSC Symposium on Information Technologies on Environmental Engineering*, Thessaloniki, Greece (pp. 478-490).

ENDNOTES

1 To make regular

2 In this section, every element of the meta-model will be written in *italic* type, in order to distinguish them from the elements of the Concrete System of Action they represent.

3 The SAO addresses only the observed behaviors within an organization and the actors' motivation, not how these behaviors are reached.

4 The scale we chose is obviously arbitrary; yet, since the results of analysis and simulations only make sense in comparison one to another, this does not matter.

5 Anyway, a less invested *actor* could be modeled by adding a constant effect *relation* that he *controls* and whose he is the only one to *depend* on.

6 Under the hypothesis that there are no interferences between resource uses; otherwise, impacts must be aggregated in a specific way.

7 Comparing this state with all the other possible configurations is very informative.

8 It is a european Life Project led by Ecobag-ADERA ("Association pour le Développement de l'enseignement et de la Recherche auprès des universités, des centres de recherche et des entreprises d'Aquitaine") and AEAG (Water Agency of Adour Garonne), assisted by: Cémagref, the Centro de Recursos Ambiantales de Navarra, CNRS, Enfa, Gramip, Infoterra, INPT, INRA, IRH, Silogic and the University of Toulouse 1.

9 an administrative district in France

10 "Fédération Nationale des Syndicats d'Exploitants Agricoles" which is the french national federation of farmers unions

11 "Centre National des Jeunes Agriculteurs": National Center of Young Farmers.

12 The sum of each column is equal to 10 which is the total amount of stakes that an actor distributes.

13 The table is an extract of the 70 proposed measures.

14 Mulino (MULtisectoral INtegrated and Operational) is a multi-criteria decision support system fields for the management of natural resources, used in the Concert'Eau project to integrate data supplied by different (Mulino, 2004).

This work was previously published in International Journal of Agricultural and Environmental Information Systems, Volume 2, Issue 2, edited by Petraq Papajorgji and François Pinet, pp. 1-15, copyright 2010 by IGI Publishing (an imprint of IGI Global).

Chapter 14
Pyroxene:
A Territorial Decision Support System Based on Spatial Simulators Integration for Forest Fire Risk Management

Eric Maillé
Cemagref, Mediterranean Ecoystems and Risks Research Group (EMAX), Territorial Management Department (GT), France

Bernard Espinasse
Aix-Marseilles University, Information Sciences and Systems Laboratory, LSIS UMR CNRS 6168, France

ABSTRACT

Medium term management of territorial risk increase requires decision support tools able to represent and simulate spatial dynamics. For forest fire risk, simulators of spatial dynamics of both fuel zones and vulnerable zones are produced by specialised disciplines, like ecology and geography. In order to integrate simulators of spatial dynamics of different themes at different scale levels, this paper proposes a spatial agent and GIS based software infrastructure called Pyroxene. This infrastructure is designed to implement "models for models integration", specified by domain expert users. To do so, end users have to use a specification framework, specifically designed for spatio-dynamic models integration. The purpose of the implemented platform is to execute and synchronise integrated simulations. To operate the semantic and syntax requirements of such integration, the platform is structured in an HLA-like architecture, and implemented as a multi-agents system for models integration, compliant with the FIPA specification. It is also organised around a GIS. First steps of validation confirm the validity of the system at the functional level and precise the limits of using the approach on the operational level.

1. INTRODUCTION

Nowadays environmental changes, from global climate change up to local land cover change, often lead to natural risks severe increase, like flood risks or forest fire risk. Modelling the interaction between natural systems (ecosystems) and human system dynamics is a key activity to produce decision support tools for environmental risks management. In the case of forest fire, the increase of the risk stems from the interaction between fuel ecosystem dynamic (forest) and

DOI: 10.4018/978-1-4666-0333-2.ch014

vulnerable human system dynamic (discontinuously urbanised zone) occurring on the same territory. Moreover, decision of risk management is a multi-scale approach, and refers to diagnostics and operational actions at different scale levels.

Many models and implemented simulators exist, designed to represent either spatial human systems dynamics (Veldkamp, 1996) or ecosystems dynamics (Coquillard & Hill, 1997). Models can be very heterogeneous, regarding notably their mode of representation of space and time (Ramat, 2004). Moreover, aggregated approaches contrast with individual based approaches (Grimm, 1999), like agents based modelling (Ferber, 1999). These models are usually not specialised in risk dynamic representation, but can be used to represent the interaction between spatial systems sharing the same territory.

Trying to design one unique model of the global « humanised-ecosystem » is highly difficult first because of the complexity of such a global system, and secondly because the required knowledge is distributed in very different disciplines. On the other hand, using existing models, already produced by the different scientific disciplines, requires an integration process.

We propose a conceptual and operational solution for a decision support tool to manage forest fire risk increase due to land cover changes. The tool is based on existing spatial dynamic simulators integration. The integration system is "multi-thematic": it permits to integrate models referring to different ontological fields (Grüber, 1995), like forest ecosystems, urban dynamics and natural risk. It is also a "multi-scale" integration system, as it permits to integrate models describing the same territory at two different scale levels.

In section 2, a general presentation of forest fire risk change modelling based on thematic and scale integration of several disciplinary spatial models is proposed. In section 3, the conceptual framework of the integration system is described. In section 4, the implementation of this framework as an agent-based integration platform called PYROXENE is presented. Finally, in section 5, a description of operational use of the platform for land planning decision support is proposed.

Figure 1. Integration of land cover change models to simulate risk evolutions

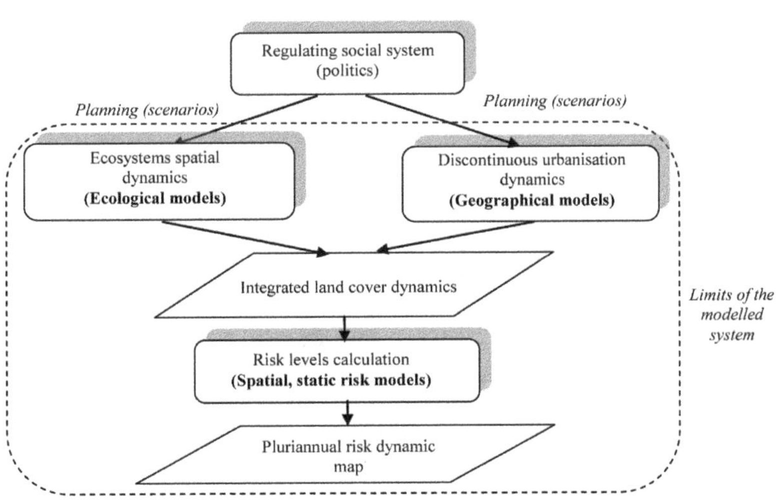

2. MULTI-MODELLING OF SPATIAL DYNAMICS LEADING TO CHANGES IN FOREST FIRE RISK

Forest fire risk increase stems from the competition between two very different spatial dynamics: *fuel spaces* dynamics (forest ecosystems), and *vulnerable* spaces dynamics, in particular discontinuously urbanised territories (Jappiot et al., 2000). Both of them are controlled by the social regulation system, for example through land planning. In Figure 1, we describe the link between land cover dynamics and changes in the map of risk levels, as well as the limits of the integrated system to be modelled.

In order to support spaces management planning decision-making, it is not required to represent the social system of regulation decision making itself. This social system is, in fact, the end user of our specified tool. The end user will elaborate and test different planning scenarios in order to assess their possible effect on the future map of the risk levels during the plan execution period.

Integration of spatio-dynamic models aims to represent the interaction between processes sharing the same spatial resource. We give the example of two land cover change models, i.e. one forest and one urban spatio-dynamic model (Millington, 2007). If both forest sprawl and urban sprawl progress simultaneously on the same territory, interactions between objects of their model will occur randomly all over the territory. The integrated dynamic is represented by the results of these interactions (Figure 2).

2.1. Three Themes

Three fields of thematic knowledge have to be formalised into operating models, and then integrated, in order to represent the risk dynamic: the forest ecosystem thematic (fuel zones dynamic), the discontinuous urbanisation thematic (vulnerable zones dynamic), and the forest fire risk thematic itself.

- Ecology provides models able to represent the vegetal dynamics at different scale levels: Some models like AFFORSIM (Prevosto et al., 2003) or CAPSIS (de Coligny, 2006) aim to represent the "spatial behaviour" of vegetal individuals like trees through individual-based models. They are based on results of functional ecology that

Figure 2. Principle of spatio-dynamic simulations integration

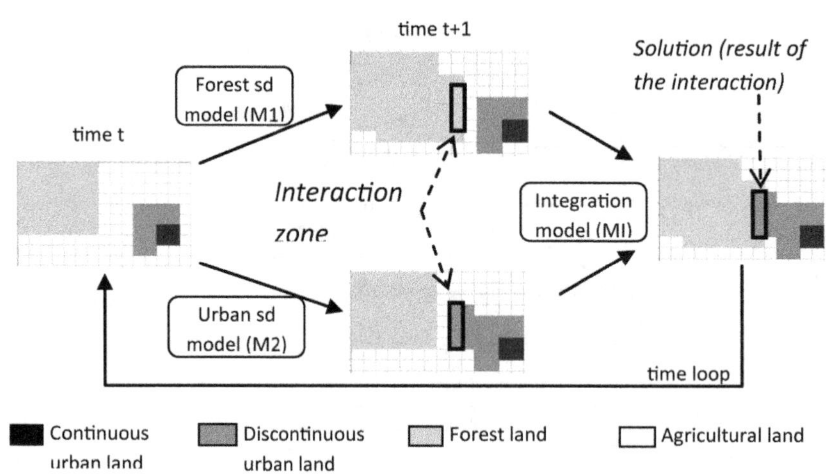

in particular describes the vegetal reproduction processes (seeds widespread, competition, etc.). They are usually applied to very small spatial extensions (forest patches). Other models as SIERRA (Mouillot et al., 2001) or LANDIS II (Scheller et al., 2006) represent the global behaviour of vegetal stands at "landscape" scale. It is based on knowledge provided by landscape ecology, and can be applied to larger spatial extents (whole forest, forest region, etc.).

- Modelling vulnerable zones dynamic and urbanisation process is based on knowledge mainly produced by geographical sciences. Models are often formalised by cellular automata, in order to represent spatial diffusion processes, taking into account spatial attributes of the geographical space on which the diffusion occurs (Batty et al., 2000; Dubos-Paillard et al., 2003; Ellerkamp, 2001).

However, specific works on discontinuous urbanisation process modelling, at local scale, are quite rare. We developed a model called MACROPOLIS. It is a cellular automaton implemented on a raster Geographical Information System (GIS) environment. It is specialised in the representation of dynamic of complex spaces, called *interface spaces* (Lampin et al., 2005) where vegetal "natural" geographical objects (trees) are mixed with human isolated buildings. Transition functions between the different states of the automaton cells, are based on the quantification of the spatial relationship between the different components of the *interfaces spaces*. The spatial relationship is assessed by raster spatial analysis indexes calculation.

Some of the social processes leading to land cover changes can also be represented in the simulation models (Napoléone, 2005), and then formalised using analytical, individual based, or mixed paradigms. We developed a Multi-agents system called MICROPOLIS (Maillé & Espinasse, 2007), in which some social agents representing economical actors of the real world territorial system (owner, buyer, land manager, etc.) interact. These different agents exchange land plots in order to locate new buildings (individual houses). This model aims to simulate the discontinuous urbanisation process of periurban areas.

- Finally, forest fire risk assessment in relation to land cover is based on static spatial risk models, provided by risk sciences. Four main approaches of forest fire risk modelling are considered (Maillé & Espinasse, 2006): (1) *analytic attribute* risk modelling allows calculating the risk level at any geographical point, in relation to its geographical attributes (slope, vegetation etc.) ; (2) *aggregated spatial* risk modelling aims to assess the risk level at any geographical point, in relation to its neighbourhood ; (3) *analytical spatial* risk modelling aims to assess the risk on one given geographical point, in relation to the ignition probabilities and the propagation possibilities in its environment ; finally, statistical risk modelling is based on statistical study of multi-simulation results of forest fires, using specialised forest fire simulators. All these approaches lead to a risk level map related to a given territory. Models are implemented on GIS.

2.2. Two Scale Levels

Both land cover dynamics and forest fire risk mapping can be considered at different scale levels. Two main scale levels are commonly considered for land planning decision support for forest fire management: decision support for *operational* planning requires integration of *small scale* spatial models i.e. based on detailed description, on small surfaces, of elementary geographical entities: plots, individual buildings, vegetal entities (copse, individual tree, etc.). Spatial extension of territory that can be simulated by such models may be comprised between some hectares, up to some square kilometres. Individual-based models are often well adapted to such a scale range. We call this level the "micro-local" scale level.

Strategic decision support uses *larger scale* levels, i.e. able to simulate larger territories, using aggregated modelling approaches. Diffusion models based on cellular automata are common at this scale level. Geographical entities described at these scale levels are for example, an "agricultural zone", a "forest massif", a "town". The geographical extent of the simulation can be comprised between some square kilometres and a whole European region. We call this level, the "macro-local" scale level.

These two scale levels are articulated in two different perspectives: for the final use of the simulation on one hand, and in a purely functioning perspective on the other hand.

- In the end using perspective, the aim is to solve two different groups of problems in a consistent way: problems of the operational planning for the micro-local level, and problems of the strategic planning for the macro-local level. In this perspective, the integration is not functional, but mainly aims to ensure the consistence of the different simulations.
- In the functioning perspective of the integrated simulation, the macro-local simulations define some particular *critical zones* whose dynamic is only known at the micro-local level. These critical zones stem themselves from the interaction between two maco-local simulations. On each macro-local critical zone, one micro-local simulation has to be initiated in order to represent its specific dynamic.

3. THE CONCEPTUAL LEVEL OF THE INTEGRATION

In this section, we describe the general conceptual framework of the integration system as well as a specification framework of some "*integration models*" aiming to make the different models to be integrated *semantically consistent*.

3.1. The General Conceptual Framework

Our purpose is to integrate six simulators, related to three thematics and two scale levels. All are spatial models. Four are dynamic models and two are static.

In order to make these simulators function together, and their *models* "understand" each other, some mediating components, called *models for model integration,* have to be specified (Maillé, 2008). Figure 3 describes the general conceptual framework organising these different components.

Models for models integration aim to make semantically consistent the information provided by the different models to be integrated. Four models for model integration have to be specified:

- Two thematic integration models, i.e. one for each scale level (ITMacro and ITmicro).
- Two scale integration models i.e. one for each thematic (IS1 and IS2).

Models for models integration have to be specified by the expert user, within a specification framework proposed by the integration system. We will describe this specification framework in the next section.

Figure 3. Conceptual framework for model integration

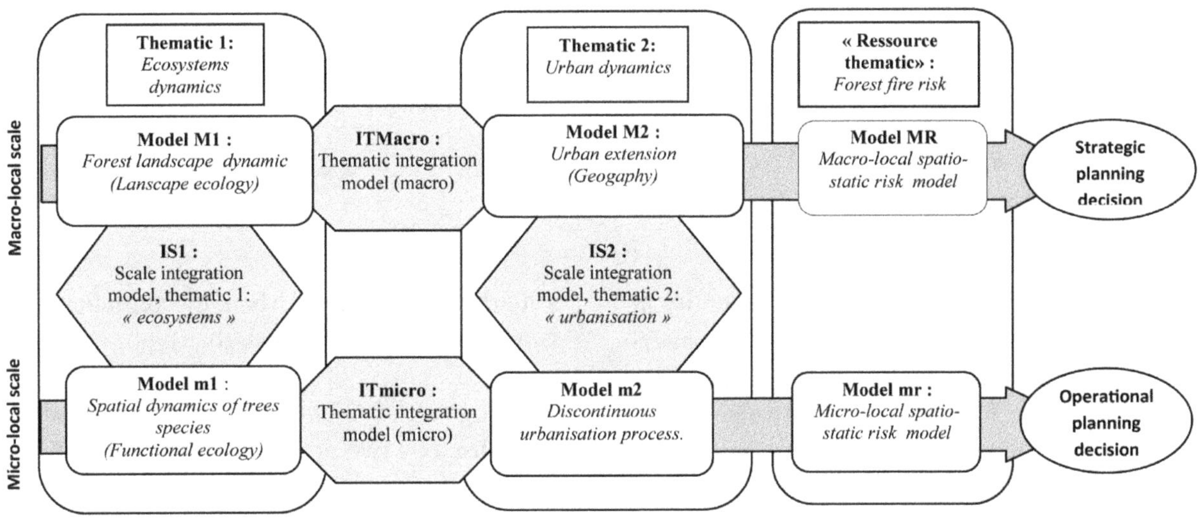

3.2. The Conceptual Specification Framework of Models for Models Integration

A conceptual specification framework of models for models integration is proposed (Maillé, 2008). The framework first distinguishes "thematic" integration specification, that concerns models at the same scale level, and scale integration specification, that concerns models operating at different scale levels. Scale integration specification can be considered as a complex thematic integration, including scale aspects.

For both thematic and scale integration specifications, the framework is structured in relation to the three *views* of the generic conceptual framework for *spatio-temporal* systems representation proposed by D. Peuquet in 1994 (Peuquet, 1994), called the Peuquet triad. The three views of the Peuquet triad are the spatial view, the temporal view and the semantic view.

We first explain principles of thematic integration specification before giving elements on scale integration specification.

3.2.1. Thematic Integration

The purpose of the specification framework is to permit the user to specify the result of spatial interactions between objects produced by spatio-dynamic models (Figure 4).

This scheme is to be interpreted as follows: when a dynamic spatial entity e1 of class C1, produced by the model M1, gets into spatial interaction with an entity e2 of class C2, produced by the model M2, a new entity eI of the class CI is created. Creation of the new entity requires the specification of:

- The *semantic* of the newly created entity, mainly represented by its class, in relation to the semantic of the initial entities. The newly created entity's class belongs to the semantic (the set of classes) of the integrated model (MI). This specification is called the "semantic view" on the integration.
- The *spatial description* of the newly created entity, in relation to spatial description of initial entities. This specification is called the "spatial view" on the integration.

Figure 4. Example of an interaction between spatial objects produced by two models

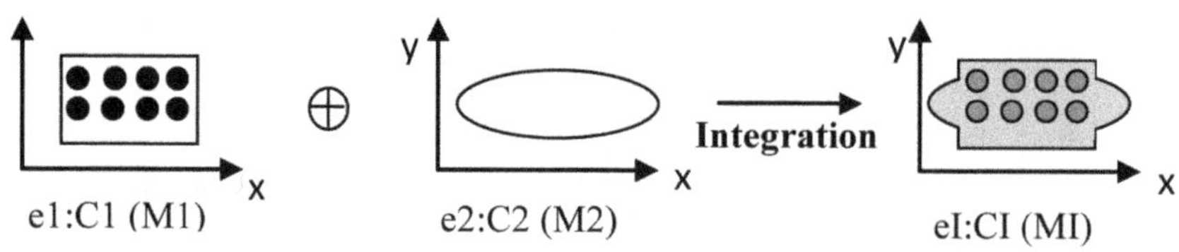

- The *dynamic* of the newly created entity in relation to the dynamic of the initial interacting entities. The dynamic is specified by specifying a relationship between *events* modifying initial entities, and events modifying the newly created entity. Specification of these relationships is called the "temporal view" on the integration.

The "*model for models integration*" is composite, and is based on the specification of these *three views* on the integration, that are components of the model.

Specification of the Semantic View

Specification of the semantic view is illustrated by the following example: it is assumed that a MEco model of wild ecosystems spatial dynamic uses three concepts, plus the « not-active » concept, that represents locations where MEco does not operate (agricultural land, for example). The semantic sMEco of the MEco model is defined as follows [1]:

sMEco = {grassLand, bushyLand, forestLand, notActiveEco} [1]

It is also assumed that the MUrb model of urban sprawl uses three concepts, plus the « not-active » concept. The semantic sMUrb of the MUrb model is defined as follows [2]:

sMUrb = {continuousUrban, discontinuousUrban, wildLand, notActiveUrban} [2]

The "*model for models integration*" can be formalised by two proposition lists defined from the cartesian product sM1 X sM2 on respectively sets sM1 and sM2 [3]:

MIEco: sMEco X sMUrb -> sMEco

MIUrb: sMEco X sMUrb -> sMUrb

So the global model for models integration has two major components:

MI*Eco* is the component of the global "*model for models integration*" that permits to take into account information produced by initial model MUrb in the dynamic represented by MEco.

MIUrb is the component that permits to take into account information produced by initial model MEco in the dynamic represented by MUrb.

The semantic view is then specified using a propositional-like formalism, by defining operators $\wedge_{MIEco}$ and $\wedge_{MIUrb}$, related to respectively models components MIEco and MIUrb, and by explicitly specifying each proposition (Table 1).

The propositional formalism is used to facilitate the future check of consistence.

In the example (Table 1), MUrb is not influenced by MEco (col 2), but it modifies MEco's behaviour (col 1): for example, it forbids the access to any MEco entities to "continuous urban" locations *(1)*.

Table 1. A partial example of the semantic view of thematic integration specification

MIEco	*MIUrb*
(1) grassLand $\wedge_{MIEco}$ continuousUrban -> notActiveEco grassLand $\wedge_{MIEco}$ discontinuousUrban - > grassLand grassLand $\wedge_{MIEco}$ wildLand -> grassland grassLand $\wedge_{MIEco}$ notActiveUrb -> grassLand bushyLand $\wedge_{MIEco}$ continuousUrban -> notActiveEco bushyLand $\wedge_{MIEco}$ discontinuousUrban - > notActiveEco	continuousUrban $\wedge_{MIUrb}$ grassLand -> continuousUrban continuousUrban $\wedge_{MIUrb}$ bushyLand - > continuousUrbancontinuousUrban $\wedge_{MIUrb}$ wildLand -> continuousUrbancontinuousUrban $\wedge_{MIUrb}$ naEco -> continuousUrban discontinuouUrban $\wedge_{MIUrb}$ grassLand -> disconti-Urbandiscontinuous Urban $\wedge_{MIUrb}$bushyLand - > disconti-Urban

Specification of the Spatial View

In addition to this semantic view, the spatial view aims to specify the spatial dimension of the interaction. In particular, the spatial view specifies the spatial class and description of the entity resulting from the interaction, in relation to the spatial classes and descriptions of the initial interacting entities.

This specification is commonly limited to a spatial function that permits to build the spatial entity resulting from the interaction. In the example below, the result of an interaction (here, an intersection) between a polygonal entity e1 of class C1 produced by model M1 and an entity e2 of class C2 produced by model M2 is a polygonal entity eI of class CI, so that the spatial description of eI is the result of the *union* of the spatial descriptions of entities e1 and e2. A simple formal expression of this proposition is provided [4] (Figure 5).

The spatial view on the integration specifies the spatial organisation of the result of the interaction between entities, in relation to their initial spatial description.

Specification of the Temporal View

This last view on the integration models specifies interactions between *changes events* of entities handled by different models. These changes can be class transitions or state transitions. A generic specification of the temporal view is proposed [5]:

$$\text{transition}_{M1}(A, B) \Rightarrow \text{transition}_{M2}(I, J) \qquad [5]$$

An entity e1 of a spatio-dynamic model M1 being in spatial interaction with an entity e2 of a model M2, if a transition from A to B occurs for e1 (if A and B are classes, e1 has to be re-instantiated), then a transition from I to J occurs for e2.

Figure 5. Specification of the spatial view

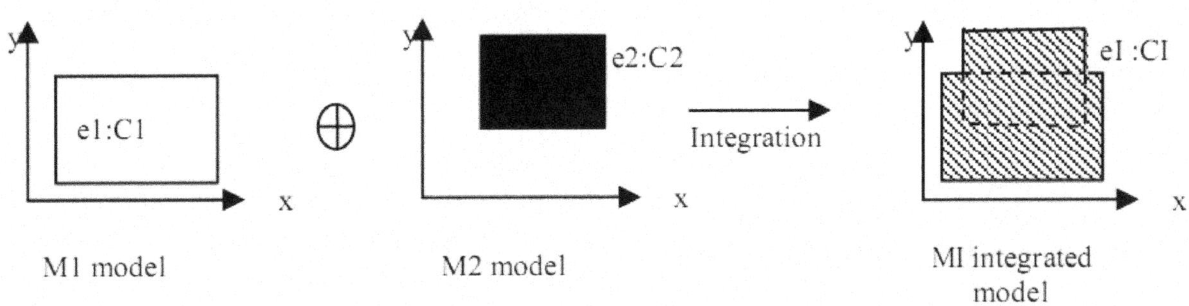

intersects (e1, e2) -> eI , eI = union(e1, e2) [4]

This specification is required in some very particular cases. Although there are no semantic relationships between two sets of concepts of two models, there may be relationships between changes. We give a concrete example, applied to land cover dynamics and forest fire risk management, to illustrate this specification requirement:

Two models of land cover dynamics have to be integrated to properly manage forest fire risk. Model M1 is an ecosystems dynamic model (ecology), model M2 is a stochastic forest fire occurrence model (produced by the *cindynic* discipline).

Semantic of M1: s_{M1} = { NoVegetation, GrassLand, BushLand, ForestLand } [6]

Semantic of M2: s_{M2} = { NeverBurnt, OneTimeBurnt, TwoTimesBurnt, MoreThanTwoTimesBurnt } (7)

Each of the different classes of both models has its own attribute set (vegetal density, burning probability, etc.) and its own atomic spatio-dynamic model representing state changes (attribute value changes).

There are no semantic relationships between classes of both semantics. For example, the semantic view on the integration models specifies:

ForestLand $\wedge_{MI1}$ OneTimeBurnt -> ForestLand [8]

Even if the patch of land burnt one (or more) times, it can be a forest land if these fires occurred a long time ago.

However, the domain expert user may want to specify in the temporal view of the integration model:

transition_{M1}(NeverBurnt, OneTimeBurnt) => transition_{M2}(ForestLand, NoVegetation) [9]

If the patch of land burns, there is no more vegetation.

So the temporal view of the model for models integration specifies the relationships that can exist between transitions related to entities handled by dynamic models to be integrated.

3.2.2. Scale Integration

Scale integration specification is a particular thematic specification taking into account scale constraints. Two scale levels are considered, a macroscopic level and a microscopic level. In order to represent the interaction between macro-level entities and micro-level entities, the first step is to operate a *scale transfer* of entities from microscopic level to the macroscopic level. Scale integration specification is focused on the scale transfer, which is mainly based on aggregation functions.

The Semantic View

The semantic view specifies the *semantic* aggregation functions that relate microscopic entities to macroscopic entities (Figure 6).

The semantic view on scale integration specifies the way to determine the class C1 of the macroscopic spatial entity given the population of microscopic entities it is composed of. These components of the models for models integration may be based on area thresholds, instance average density thresholds, etc. For example, a macroscopic area may be considered as "Forest lands" (macroscopic level) if density of "Trees" overtakes a defined thresholds at microscopic level.

The Spatial View

The spatial view specifies the way to define spatial description of macroscopic entities, given the spatial description of the microscopic entities it is composed of (Figure 7).

Formal models are mainly based on envelope curve determination algorithms.

Figure 6. Graphical illustration of the semantic view of scale integration model

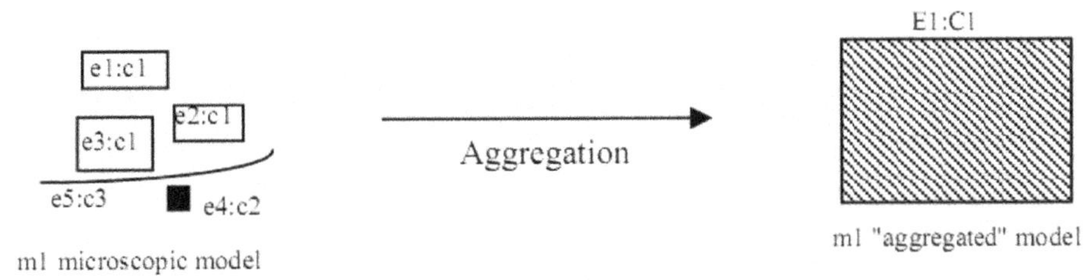

The Temporal View

The temporal view on scale integration model aims to represent the interaction between *processes* running at different organisation levels. It assumes that a change event series at microscopic level can induce a change event at macroscopic level (Figure 8).

The same example as for the temporal view on the thematic integration (§ 3.2.1) can illustrate the temporal view on scale integration. The aim of this concrete example is to integrate a microscopic fire occurrence model and a macroscopic ecosystem dynamic. It is assumed that the user wants to specify that micro-local fire repetition leads to a global (i.e. macroscopic) desertification of the zone. The temporal view component of the model for models integration can be formalised as follows [10]:

$(\Sigma_t^{t+i}((e2{:}FuelLand_{:M2}).numberOfFiresChanged() * (e2{:}FuelLand_{:M2}).getArea()) / i > threshold) \Rightarrow (E1{:}VegetalLand_{:M1}).\ setVegetalDensity(0)$

If the frequency of "microscopic" fire occurrences, weighted by fire areas, during a given (t+i – t) period, exceeds the specified threshold, the entire macroscopic ecosystem is affected and the vegetation disappears.

3.2.3. Synopsis of the Specification Framework

The different components of the specification framework are synthesized in the following six cells table (Table 2), crossing the three views (semantic, spatial and temporal) and two integration types (thematic and scale integration).

For both, thematic and scale integration, the semantic view specifies the interactions between ontological concepts held by classes handled by the different models. The spatial view specifies the spatial interaction between objects. The temporal view specifies the interaction between events.

This framework is simplified, as it is necessary to differentiate "structural" interactions, i.e. interactions between *classes* from "*dynamic*"

Figure 7. Graphical illustration of the spatial view of scale integration model

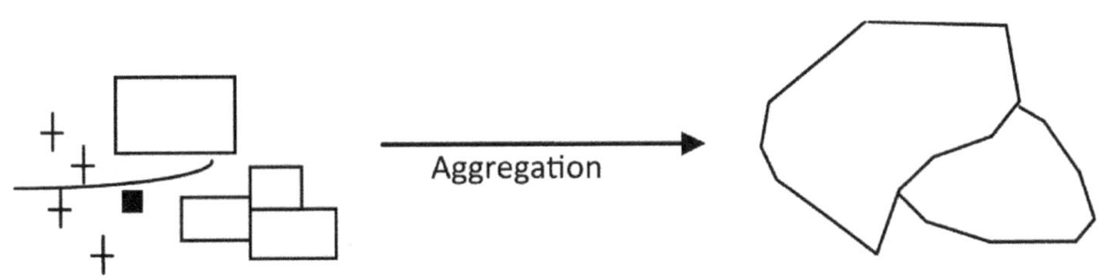

Figure 8. Graphical illustration of the temporal view of scale integration model

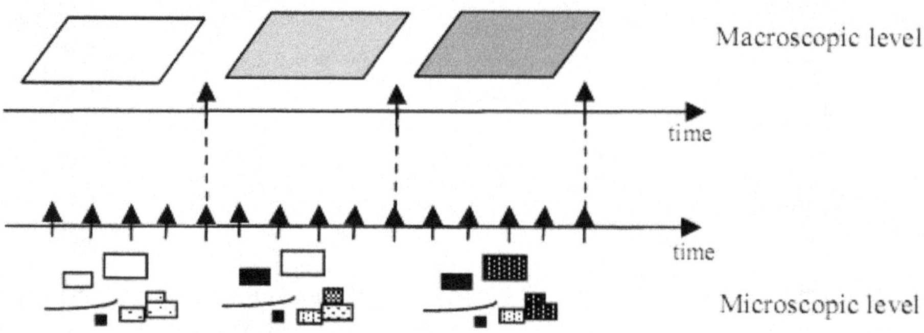

interactions, i.e. relationships between attributes values of interacting objects. The final specification framework has got twelve components.

This conceptual level aims mainly to solve semantic constraints of the integration of spatio-dynamic models. The integration system also has to solve all syntax constraints to the simulators interoperability, in order to make them able to function together synchronously, and produce correct results. We call this part of the integration, which leads to an efficient operational system, "operational integration". The operational integration that allows the interoperability between models is ensured by a specific agent-based architecture, implemented in the Pyroxene platform, described in the following section 4.

Table 2. Simplified synopsis of the specification framework

	Thematic integration	Scale integration
Semantic view	Relationship between *concepts* held by classes	Semantic aggregation of *concepts*
Spatial view	Relationship between "*spatiality*" of objects	Spatial aggregation of geographical objects
Temporal view	Relationships between *events* related to objects	Aggregation of events series

4. ARCHITECTURE AND IMPLEMENTATION

Pyroxene is a prototype of a thematic and scale integration platform of land cover change simulators for the simulation of forest fire risk change. In this section we first precise some specificities of the platform, then we describe its architecture, and finally we illustrate some of its functioning principles through the description of information flow during an integrated simulation.

4.1. Specificities of the Pyroxene System

The main specificities of the Pyroxene system are:

- This is an *agents-based* integration system. It is an integration multi-agents *system* (MAS) developed with the JAVA language, using the cognitive agents development platform Tilab JADE, compliant to the Foundation for Intelligent Physical Agents specifications (FIPA, 2001). The integration is operated by wrapper agents, that make the link between the spatial simulators and the integration MAS. Agents of the MAS are spatial agents (Rodriguez et al., 2002).

- It is a distributed system, partly compliant to the High Level Architecture (HLA) specification (IEEE, 2000). It is composed of a central agent-based mediating infrastructure (RTI-like) and of some peripherical *federated* components, mainly the wrapped simulators and other software supporting the different integration functions (Serment, 2007).
- The integration process is based on the execution of thematic and scale integration *models* on Geographical Information Systems, either raster (GRASS, open source under GNU GPL licence) or vector (ESRI ArcGIS/ArcInfo Workstation).

The implemented Pyroxene tool has got a fixed structure that permits to integrate three themes (ecosystem dynamics, urban dynamics, risk calculation) at two scale levels (micro-local et macro-local) (Figure 9).

Pyroxene can manage the integration of simulators of spatial dynamics that respond to two main constraints:

- They must admit as input and produce as output at least one semantic geo-layer, either raster (classified) or vector.
- They must admit as parameter the number of cycles to be simulated.

4.2. Agent-Based Architecture Functioning

Main capabilities of the *Pyroxene* platform are the selection and the configuration by the user of the different simulators and risk models, the specification by the user of the scale and thematic models for models integration and finally the management of the integrated simulations. Software specialised modules, in their own computer environment, ensure these different functions. Modules are "wrapped" by wrapper-agents, and organised in an agent-based distributed architecture (Figure

Figure 9. PYROXENE GUI and the Tilab JADE SMA module

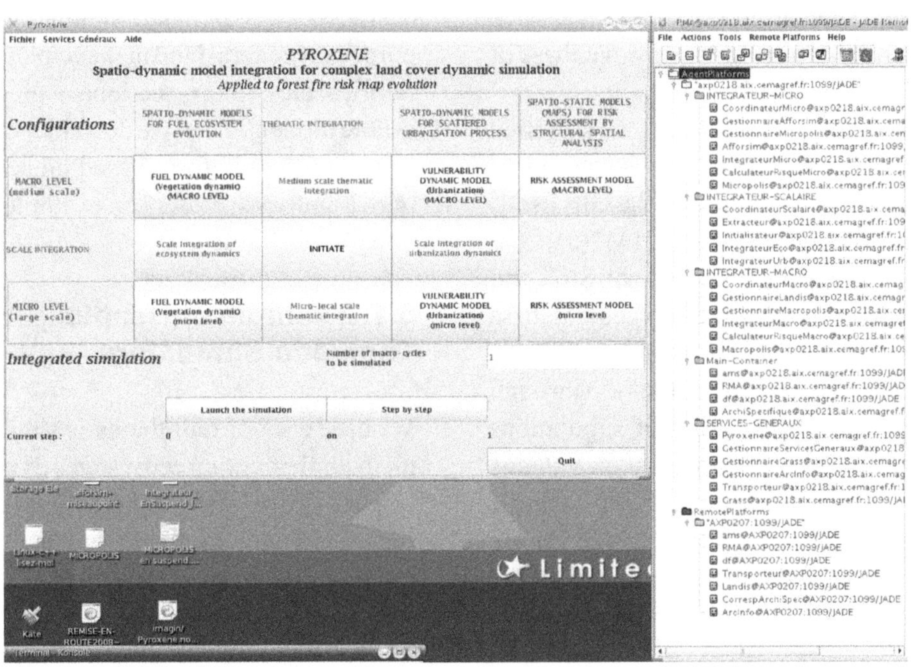

10) where functional modules are coordinated by a central mediation agents based system (the coordinator). This architecture is inspired by the HLA architecture, specialised in simulation integration.

When an integrated simulation is started, the different agents exchange messages in order to synchronise the execution of the different required tasks, by the different software systems: simulations (by the simulators), execution of the integration models (by the GIS), data format conversion and exchanges (transport), by specific software and protocols, etc. (Figure 10).

Models for model integration are executed by the GIS under the control of some integrator agents (one integrator agent for each theme and each scale level).

4.3. Messages and Data Flows

Two main kinds of information are exchanged between the different components of the system: messages are exchanged between agents (wrappers, mediators agents, etc;) and geo-data layers are exchanged between software modules (simulators, GIS, GUI, etc.). All geo-data layers are dated. During an integrated simulation a chronological stack of geo-data layers "grows" in the geo-database of the GIS (Figure 11).

By order of the coordinator agent, the specified model for models integration is executed on its suitable software module (usually the GIS itself) ①. Resulting data layers layerM1M2(t0) and layerM2M1(t0), respectively usable by SM1 and SM2 simulators are stored in the GIS geo-database and then transferred to their respective simulator ②. The coordinator orders to simulators (through their wrappers) a given number of simulations steps (usually one step, if step duration value is the same for both simulators). Results of these simulations (layerM1(t1) and layerM2(t1)) are stored in the GIS ④ and can be integrated by the "integrator" software module. Then the cycle restarts.

Many other not represented general operations, like data transportation, data format conversion, etc. are realised by others software modules, under the control of the mediation agents based system through their wrapper agents.

Although the infra-structure may be interpreted as a kind of Service Oriented Architecture (Nickul, 2005), as it has got SOA's main characteristics (granularity, re-usability, weak coupling, wrapping, distribution, etc.), its implementation remains resolutely object oriented, without being compliant with SOA usual foundations like web-services, XML objects serialisation, etc. On the other hand, the platform can be integrated in a SOA as a particular service provider in an integrated land management multi-risk decision support tool for example.

5. OPERATIONAL USE AND VALIDATION OF THE PYROXENE INTEGRATION PLATFORM

Operational use of the platform for risk management in land planning requires first thematic land cover change simulators. These are usually not designed for risk change simulation but for other purposes (general land management, virtual laboratories, etc). In this section we give examples of usable models. Then we describe some principles of use of the platform based on scenario definition. Finally we present an example of an applied integrated simulation.

5.1. Example of Usable Thematic Simulators

We used a set of four thematic simulators, two of them implementing ecosystems dynamics models respectively at micro-local scale and macro-local scales, and two implementing urbanisation models, at the same two scale levels.

Figure 10. HLA-like architecture of the Pyroxene platform

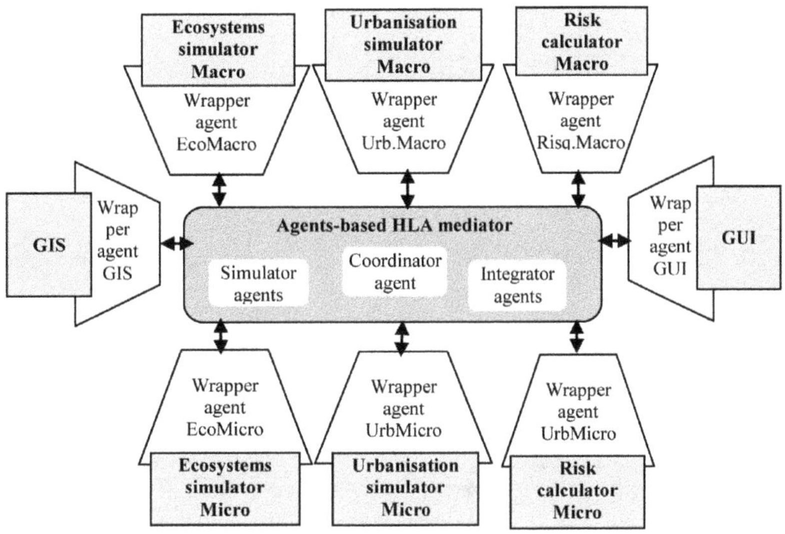

- AFFORSIM (Prevosto et al., 2003) is a functional ecology model designed to simulate trees sprawl on wild open lands at micro-local scale. It is an objects-based model, considered as a virtual laboratory. It is implemented as an autonomous simulator in C++ language. The following figure illustrates an AFFOSIM simulation session (Figure 12).
- LANDIS II (Scheller et al., 2006) is a landscape ecology model designed to simulate forest dynamics at macro-local scale. The simulator is considered as a decision support tool for forest management. It is implemented in Microsoft C# langage.
- MACROPOLIS (Maillé, 2008) is an urbanisation simulator specialised in simulation of interfaces' dynamics at macro-local scale. It is a cellular automaton (Langlois et al., 1997), developed on a raster GIS (GRASS). The following figure illustrates a Macropolis simulation session on GRASS as well as the transition functions definition interface implemented in JAVA language (Figure 13).
- MICROPOLIS (Maillé, 2008) is an agent based model (Wooldridge, 2002) for micro-local urbanisation simulation. It simulates installation of new buildings (houses) on wild land or agricultural lands, using spatial economics models (Napoleone, 2005). Figure 14 illustrates a Micropolis simulation session on a French "commune".

Risk models are static. They are designed to calculate a risk level at any point of a geographical zone, using either vector or raster representation of fuel and vulnerable areas (Jappiot, 2000). Different approaches exist, usually based on spatial analysis indexes calculation. We use two models, TOPOVector, at macro-local scale, and CONTEXTRaster, at micro-local scale, that both assess the spatial relationship between fuel and vulnerable areas through respectively vector and raster spatial analysis indexes (Maillé, 2008). All these risk models are still under construction and validation by researchers (specialised in cindynics) and only prototypes are available.

All these models have their own *semantic*, i.e; their own set of handled object classes. They

Figure 11. Messages and data flows during an integrated simulation

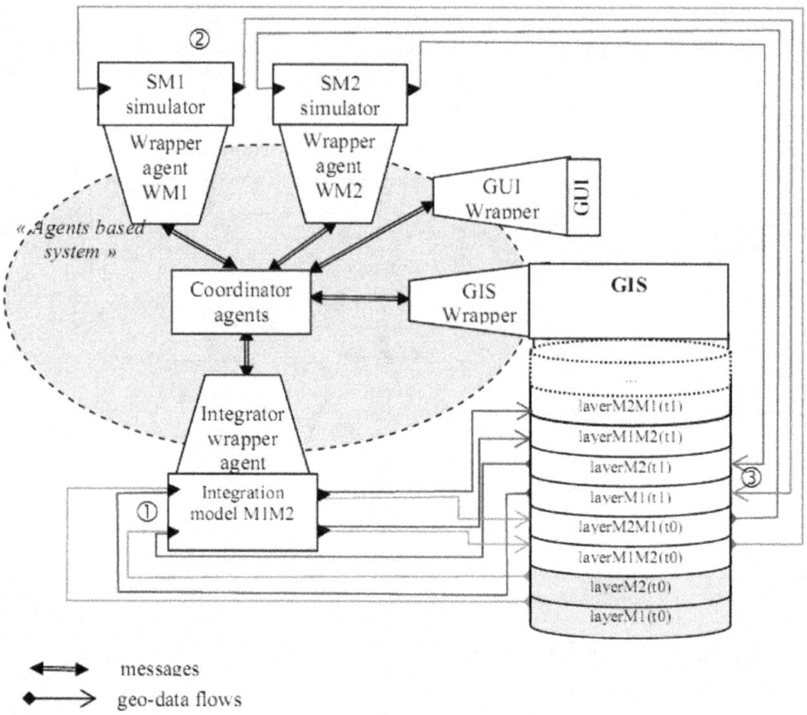

admit as input and produce as output geo-data layers, of different formats. Simulators also admit as parameter the number of cycles to simulate.

5.2. Scenario

Scenarios definition is done by the end user who has to specify two sets of data (Figure 15):

- The land management *plans* are provided to the system as GIS layers in order to *control* the integrated simulation. For each model, these layers specify, for example, excluded zones, where installation of buildings or trees is impossible.

These land management plans elaborated for scenarios definition can aim either to control a given land cover dynamic or to directly reduce the risk level. They periodically change the input values of one (or several) given initial models. Each of their modalities (action) is dated. For example, the plan may specify a spatially described "fuel brake" realization at the given *d* date.

- Parameter values of each model, representing the hypothesis related to the general environment of the simulation (ecological, climatic, economic, etc.), are strictly specific to each simulator.

Each scenario should be simulated several times in order to assess its effects on the risk evolution (scenario assessment). Then it can be modified if results do not correspond to the land manager's objectives.

5.3. Real Terrain Simulations and Validation

Pyroxene is now tested on one experimental zone in the Aix-Marseille conurbation, gathering 43

Figure 12. An Afforsim simulation session (trees appear on the "terrain")

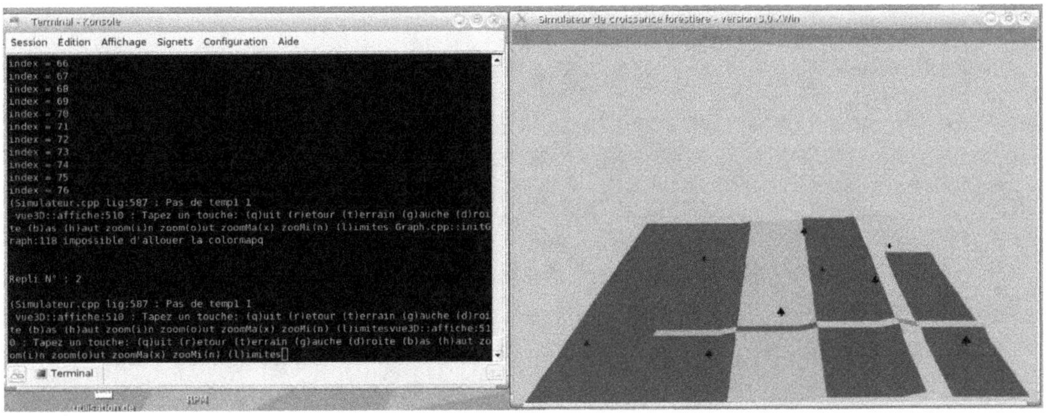

"communes" (elementary administrative entities) of the "*Pays d'Aix Agglomeration Community*".

Validation protocols of the risk level dynamic are still under construction by the risk sciences researchers. The validation is therefore focused on the quality of the integrated land cover dynamic representation. Validation is based on the comparison between the integrated simulations results and past land cover changes measured by diachronic mapping using remote sensing images, satellites or aerial (Maillé & Bouillon, 2005).

Two validations are operated: the initial models validation for each theme and scale level, and then the validation of integrated simulations. Our present objective is mainly to assess the error due to the integration process itself, by subtracting the error due to the initial models from the total error.

Functional validation aims to verify *efficiency, reliability* and *consistence* of the prototype, in order to validate the global conceptual approach as well as the architectural and technical choices. It uses real terrain past data, and compares integrated simulations to diachronic mapping results. Figure 16 presents an example of a serial of dated risk level maps obtained by integrated simulation on the studied area (Maillé & Espinasse, 2009).

The presented simulation example (Figure 16) is limited to scale integration of two urbanisation oriented models of different scales: Macropolis and Micropolis. Both models have different time step values, so the figure represents ten Mac-

Figure 13. A Macropolis simulation session and the transition functions definition interface

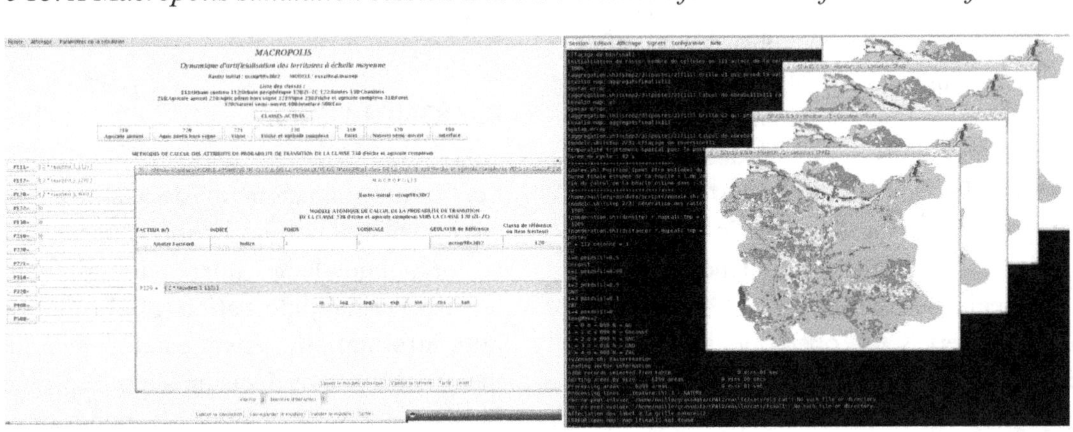

Figure 14. A Micropolis simulation session. New "produced" buildings appear as blue "disks".

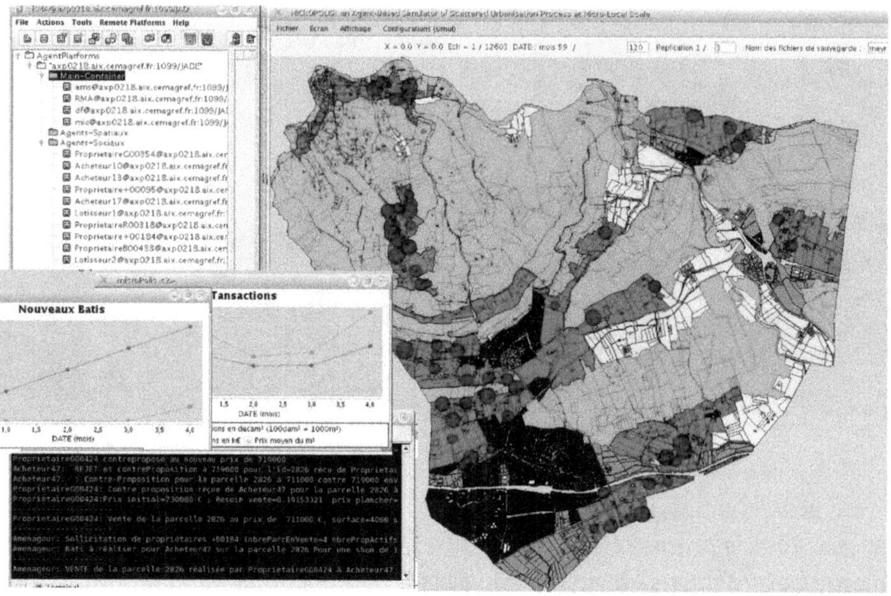

ropolis annual time steps (from 2000 to 2009) and one hundred and twenty Micropolis monthly time steps located on some critical interface zones.

The simulation anticipates a decrease of the *high risk* area (-8%), due to the limitation of pure forest surfaces, and an increase of the *very high risk* area (+19%), due to extension of isolated buildings into the forest (interfaces). The *Low risk area* also decreases (-5%), because initially opened discontinuously urbanised areas (with low risk) get more and more vegetal fuel (trees, shrubs, etc), and so become *high risk* areas. *Medium risk* areas also decrease (-19%): these are often former agricultural areas consumed by both urbanisation phenomenon and forest advance phenomenon after abandonment. Finally, *very low or null risk* areas remain very stable (these are mainly continuous urban areas, neck soils, water surfaces, etc.).

These results validate the *consistence* of the integrated simulator productions in relation to diachronic mapping measures, and their interpretation by specialised experts.

The next step is the operational validation that aims to assess simulations results *accuracy* in representing quantitative real world dynamics, as well as the application domain limits. This is a long term work still in progress, based on many different experimental terrains. First results show important gaps measured in terms of spatial distribution of changes. The integrated simulation does not correctly represent contours of urban or forest extension zones without constraints. However, such limits are mainly due to initial simulators limits, even if they are often increased by the integration process.

At this stage, the tool can be useful for strategic land management planning, but cannot be used as the unique decision support tool for localised operational planning.

6. CONCLUSION

Territories management decision-making for environmental risk limitation is based on multi-sectorial knowledge related to different scale levels. Formal integration of this knowledge into information systems helps guaranteeing the consistence of decision support provided by these systems. In a dynamic perspective, and

Figure 15. Scenarios definition, simulation and evaluation cycles

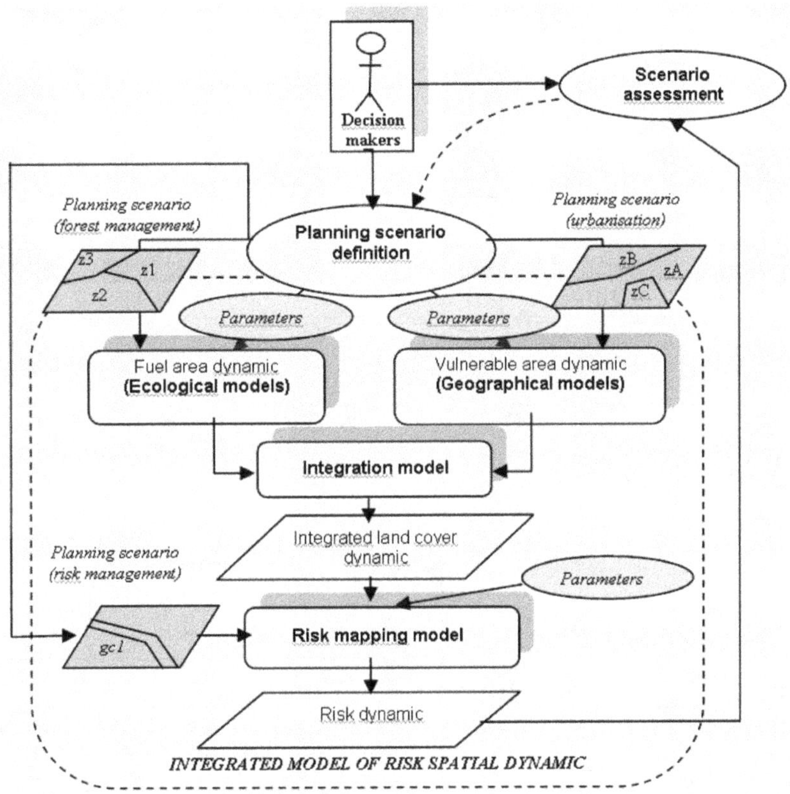

in particular for territorial planning, dynamic models integration permits the representation of the interaction between processes described by different disciplinary fields.

We propose a land cover simulator and risk models integration system, designed to ensure the semantic consistence of models produced by different disciplinary fields, and the syntax interoperability of their software implementation. This integration system has the specificity to be *agents-based.* The main interest of such an agent based mediating technology is that it can solve all integration constraints by using existing software systems (GIS, simulators, etc.). Semantic constraints are managed by using knowledge management and semantic information processing capabilities of cognitive agents (Ferber, 1999). Syntax constraints are managed by agent specialisation and agent-based wrapping: computing environments heterogeneity is over-layered by a unique agent paradigm that ensures all of the required communication between modules. Specific constraints due to spatiality and temporality of territorial systems representation are also managed by spatial agents, closely related to GIS. Temporal issues, in particular synchronisation, are managed through the HLA architecture of the integration multi-agents system, centralised around some coordination agents specialised in that task. Globally, the system enhances agent interactional capabilities that are crucial in the interaction approach.

However the integration is based on the definition of models for model integration. These are

Figure 16. Example of integrated simulation of risk level change

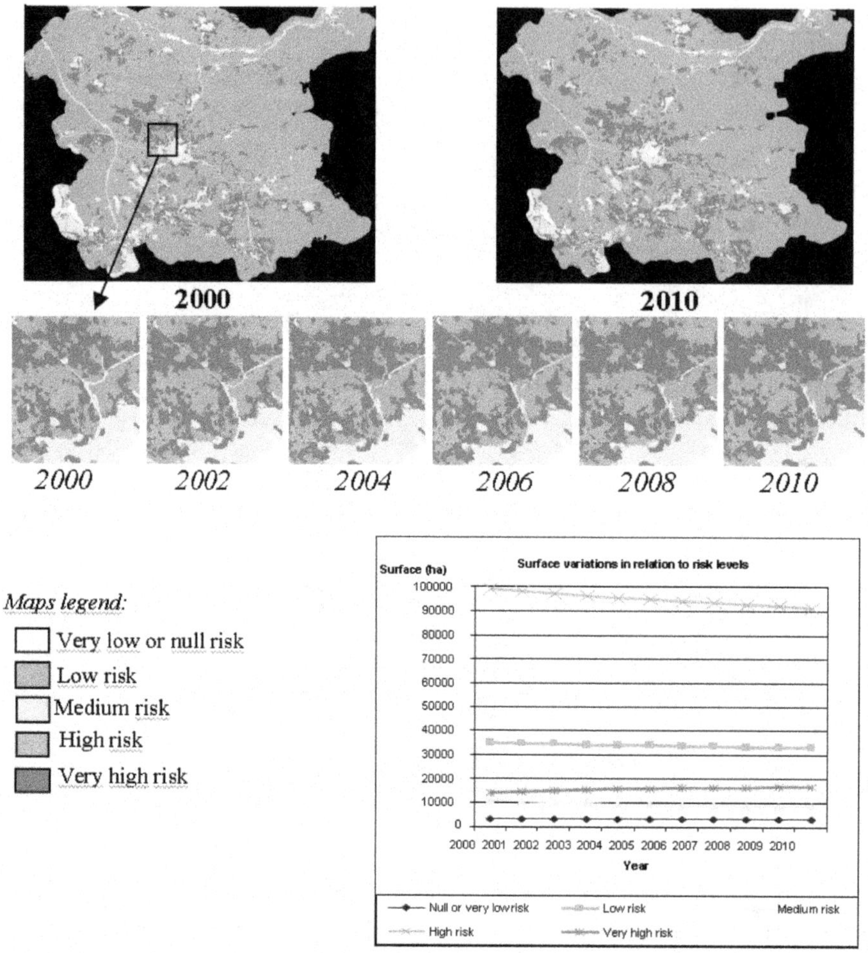

complex models, specified by expert users, and related to several knowledge fields, produced by several scientific disciplines. In order to facilitate such a specification, these knowledge fields could be formalised into domain ontologies that would permit to formally relate them, and easily identify the relationships between their concepts. As the handled information is spatio-temporal, spatial ontologies (Vangenot, 2004) or spatio-temporal ontologies (Spaccapietra, 2004) should be used.

Validation of such a system is a long term work that requires first validation of the initial models, then the validation of the integrated simulations of land cover dynamics, and finally the validation of the integrated dynamic risk model. This work is based on an experimental approach, associating computer scientists together with specialists of the different disciplines that produce the different models.

From an operational point of view, the system has no predictive ambition, but should be considered only as a prospective system. It allows land managers to compare the impact of different land planning scenarios on the future evolution of the risk. It cannot be the unique planning decision support tool used by land managers, but is part of a large range of decision support tools including operational numerical and mapping tools or even textual traditional tools that can be used for a better result interpretation.

REFERENCES

Batty, M., & Jiang, B. (2000). Multi-agent simulation: Computational dynamics within GIS. In Atkinson, P., & Martin, D. (Eds.), *Innovation in GIS VII: GIS and geocomputation* (pp. 55–71). London, UK: Taylor & Francis.

Coquillard, P., & Hill, D. R. C. (1997). *Modélisation et simulation d'écosystèmes: Des modèles déterministes aux simulations à évènements discrets*. Paris, France: Masson.

De Coligny, F. (2006, November 13-17). Efficient building of forestry modelling software with the capsis methodology. In *Proceedings of the Second International Symposium on Plant Growth Modeling, Simulation, Visualization and Applications*, Beijing, China.

Dubos-Paillard, E., Guermond, Y., & Langlois, P. (2003). *Analyse de l'évolution urbaine par automate cellulaire, UMR 6063 IDEES*. Paris, France: Laboratoire MTG Rouen.

Ellerkamp, P. (2001). *Etude de l'extension urbaine autour d'Avignon et de Montpellier de 1936 à 1999 par la simulation de dynamiques spatiales à l'aide de modèles cellulaires d'interaction locale*. Unpublished doctoral dissertation, University of Avignon, Avignon, France.

Ferber, J. (1999). *Multi-agent systems, an introduction to distributed artificial intelligence*. Reading, MA: Addison-Wesley.

Grimm, V. (1999). Ten years of individual-based modelling in ecology: What have we learn and what could we learn in the future? *Ecological Modelling, 115*, 129–148. doi:10.1016/S0304-3800(98)00188-4

Grüber, T. R. (1995). Towards principles for the design of ontologies used for knowledge sharing. *International Journal of Human-Computer Studies, 43*(5-6), 907–928. doi:10.1006/ijhc.1995.1081

Institute of Electrical and Electronics Engineers (IEEE). (2000). Standard for modeling and simulation (MetS), high level architecture (HLA) *(Tech. Rep. No 1516-2000)*. Washington, DC: IEEE Computer Society.

Jappiot, M., Blanchi, R., & Alexandrian, D. (2000, October 25-27). *Cartographie du risque d'incendie de forêt: besoins, méthodes et données, essai de normalisation.* Paper presented at the Journées GIS Incendie de Forêt, Hyères, France.

Lampin, C., Jappiot, M., Long, M., Borgniet, L., & Dumas, E. (2005, June 20-23). *Outils de caractérisation et cartographie des interfaces habitat-forêt*. Paper presented at the International Colloquium of Spatial Analysis and Geomatics, Avignon, France.

Langlois, A., & Phipps, M. (1997). *Automates cellulaires: Application à la simulation urbaine*. Paris, France: Hermes.

Maillé, E. (2008). *Intégration conceptuelle et opérationnelle de modèles spatiaux-dynamique*: *Application à la dynamique du risque d'incendie de forêt.* Unpublished doctoral dissertation, Paul Cezanne University, Marseille, France.

Maillé, E., & Bouillon, C. (2005). *Consommation des espaces agricoles et naturels par le processus d'urbanisation discontinue. Aix-en Provence*. France: Cemagref.

Maillé, E., & Espinasse, B. (2007, June 18-19). *Un cadre de modélisation pour la simulation de dynamiques spatiales complexes.* Paper presented at the International Colloquium of Spatial Analysis and Geomatics, Clermont-Ferrand, France.

Maillé, E., & Espinasse, B. (2009). Spatial agents for scale integration of spatial dynamics models. *European Journal of Geomatics and Spatial Analysis, 19*(4), 523–549.

Millington, J. D. A. (2007). *Modelling land-use/cover change and wildfire regimes in a Mediterranean landscape*. Unpublished doctoral dissertation, London King's College, London, UK.

Mouillot, F., Rambal, S., & Lavorel, S. (2001). A generic process-based simulator for Mediterranean landscapes (SIERRA): Design and validation exercises. *Forest Ecology and Management, 147*, 75–97. doi:10.1016/S0378-1127(00)00432-1

Napoleone, C. (2005). *Prix fonciers et immobiliers, et localisation des ménages au sein d'une agglomération, urbaine*. Unpublished doctoral dissertation, Aix-Marseille University, Marseille, France.

Nickul, D., Reitman, L., Ward, J., & Wilber, J. (2005). *Service oriented architecture (SOA) and specialized messaging patterns.* Retrieved from http://www.adobe.com/ government/pdfs/ SOA-technical-whitepaper.pdf

Peuquet, D. J. (1994). It's about time: A conceptual framework for the representation of temporal dynamics in geographic information system. *Annals of the Association of American Geographers. Association of American Geographers, 84*(3), 441–461. doi:10.1111/j.1467-8306.1994.tb01869.x

Prevosto, B., Hill, D., & Coquillard, P. (2003). Individual-based modelling of pinus sylvestris invasion after grazing abandonment in the French Massif Central. *Plant Ecology, 166*(1), 121–137. doi:10.1023/A:1024404214782

Ramat, E. (2004). *Contribution à la modélisation et à la simulation des systèmes complexes, habilitation à diriger les recherches*. Unpublished doctoral dissertation, Littoral Cote d'Opale University, Brest, France.

Rodriguez, A., & Raper, J. (1999). Defining spatial agents. In Camara, A. S., & Raper, J. (Eds.), *Spatial multimedia and virtual reality research, monograph* (1st ed., pp. 111–129). New York, NY: Taylor & Francis.

Scheller, M. R., & Domingo, J. B. (2006). *LANDIS-II core model description, Landis-II models v 5.1 user guide.* Retrieved from http://www.landis-ii.org/exts/age-only-succession/LANDIS-II%20Age-Only%20Succession%20v2.0%20User%20Guide.pdf

Serment, J. (2007). *Une infrastructure d'intégration à base d'agents logiciels pour l'élaboration de systèmes d'aide à la décision environnementale: Application à la gestion hydraulique camarguaise.* Unpublished doctoral dissertation, Paul Cezanne University, Marseille, France.

Spaccapietra, S., Al-Jadir, L., & Yu, S. (2004, April 4). Somebody, sometime, somewhere, something. In *Proceedings of the International Workshop on Ubiquitous Data*, Tokyo, Japan (pp. 6-13).

Vangenot, C. (2004). Les ontologies spatiales. *European Journal of Geomatics and Spatial Analysis, 14*(2).

Veldkamp, A., & Fresco, L. O. (1996). CLUE: A conceptual model to study the conversion of land use and its effect. *Ecological Modelling, 85*, 253–270. doi:10.1016/0304-3800(94)00151-0

Weinberg, G. M. (1975). *An introduction to general systems thinking.* New York, NY: John Wiley & Sons.

Wooldridge, M. J. (2002). *An introduction to multiagent systems*. Chichester, UK: John Wiley & Sons.

This work was previously published in International Journal of Agricultural and Environmental Information Systems, Volume 2, Issue 2, edited by Petraq Papajorgji and François Pinet, pp. 52-72, copyright 2010 by IGI Publishing (an imprint of IGI Global).

Chapter 15
On the Use of Abduction as an Alternative to Decision Trees in Environmental Decision Support Systems

Franz Wotawa
Graz University of Technology, Austria

ABSTRACT

Although decision trees are frequently used in environmental decision support systems, they have shortcomings. In the case of an available model, decision trees have to be constructed manually from the model. Moreover, not all knowledge is represented in the decision tree. To overcome this issue, the author proposes the use of abductive reasoning directly applied to the available cause-effect model. In particular the abduction problem the author introduces (i.e., the problem of finding a cause for observed effects), shows how this problem can be extended to allow distinguishing between competing explanations, and discusses the integration of testing and repair actions within the framework. The latter is especially important in case of environmental decision support systems.

INTRODUCTION

In the environmental domain knowledge-based systems for simulation, problem solving, and prediction are widely used. This holds especially in the case of environmental decision support systems. It is common to use either rule-based systems or decision trees for representing the knowledge used in such systems. In this paper we argue that neither decision trees nor rule-based systems are very well suited for knowledge representation for decision support systems and in particular diagnostic systems. As an alternative we propose the use of abductive diagnosis, which allows for deriving root causes from effects where the underlying knowledge base represents cause effect relationships directly. Such models are usually available in the natural sciences. In order to show the use of abductive diagnosis in the environmental field, we use a model that was

DOI: 10.4018/978-1-4666-0333-2.ch015

used for diagnosing a waste-water treatment plant (Comas et al., 2003).

We first start with a discussion about the characteristics of decision-tree-based systems. A decision tree comprises knowledge from effects to causes. In order to come up with a single solutions test vertices where the user is asked a question. Depending on the answer a certain branch is taken. Finally, a leaf vertex is reached that corresponds to the cause. Although, this kind of knowledge representation is very effective there are some drawbacks.

Usually available models of the real world describe the cause-effect relationships. Hence, extracting test vertices, which ask the state of effects, is a hard task and has to be done manually. The consequences are: (1) the decision tree is not very flexible. If other tests would be more effective the whole tree has to be generated again. (2) Not all knowledge is used and the focus is more on an effective computation of a result. (3) Model changes usually require lot of changes in the decision tree. (4) Manually extracted decision trees are likely to be error-prone and/or the risk for not accurately representing the model is high.

Although, the resulting decision tree allows for effective computing of a diagnosis, it's use is limited. The coded knowledge can only be used for the purpose someone has in mind when constructing the decision tree. This is might be now direct drawback but consider the case where there are several different tasks to be fulfilled with a model. For each task a new decision tree has to be generated. Hence, decision trees represent knowledge more in a procedural fashion and not in a declarative one.

In order to overcome these drawbacks we suggest the use model-based techniques where a model is directly used for some purpose like diagnosis. In particular we focus on abductive diagnosis. In literature there are many applications of consistency-based diagnosis to the environmental domain reported. See for example the work by Heller and Struss (1996, 1997, 1998), and Struss (1998). However, the use of abductive diagnosis for the same purpose has to our knowledge not been described. Since abductive diagnosis uses the fault behavior directly to extract root causes, it seems to be more appropriated in the domain of environmental decision support systems for plants in cases where the model can be formulated in terms of cause-effect relationships. Other examples of model-based reasoning applied to the environmental domain are articles written by Bredeweg and colleagues. In (Bredeweg et al., 2006; Salles et al., 2003a, 2003b) the authors introduced models for some environmental problems. In contrast to this paper and work by Struss and colleagues the focus was more on qualitative modeling and not on providing a methodology for diagnosis or reasoning in general. However, the used qualitative models also incorporate cause-effect relationships.

This paper is organized as follows. In the next section, we discuss problems regarding decision trees and rule-based systems in detail. Afterwards, we introduce a small running example that is based on a decision support system for a wastewater treatment plant. In the next section we use the running example for introducing the basic definitions of abductive diagnosis. Based on these definitions we discuss the use of abductive diagnosis and probing in the process of extracting appropriate repair actions. In a section on modeling for abductive and model-based reasoning we show how models should be organized in order to support re-use and maintenance activities. Finally, we conclude the paper.

Problems with Decision Trees

The objective behind all different kind of decision support systems is to support a user in drawing conclusions from available knowledge in a particular situation and a given domain. The overall goal is to come up with a decision that is good enough and well elaborated. Another important issue is the explanatory capability of decision support systems. The system should allow for given reason why a particular decision is the best given the available knowledge. Unfortunately in most cases the systems do not allow for explanation beside in a very general form like the one saying that given the factual knowledge and my underlying the model some decision is more likely than another. The reason for this incapability lies in the way decision support systems are constructed. Only in rare cases a model of the system is directly provided. Instead rules or decision trees are used both of them relying on a single on the basic idea of providing single solutions to given problems. In cases of uncertainty probabilities are used to control the overall decision-making process via selecting the most likely path from the initial state to the desired goal state. The reason behind a derived conclusion is hidden behind probabilities, fuzzy values, uncertainties, possibilities or other measures for stating knowledge that is not completely justified.

In this paper we do not want to discuss all the issues related with reasoning methods based on uncertain knowledge and their methodologies. Instead we focus on decision support systems based on either rules or decision trees. In particular, we will raise some problems of these methods and alternatively provide a methodology based on models. We will show that model-based reasoning does not face these problems at the costs of providing models, which turns out not to be that simple. However, we also provide a modeling methodology that can be used in practice bearing great advantages like re-usability and flexibility regarding changes. The decision to ignore uncertainty does not mean that probabilities or other means for representing uncertainty cannot be integrated in the approach we discuss in this paper. De Kleer & Williams (1987) for example introduced probabilities in model-based diagnosis. The idea was to assign a probability value to each diagnosis that can be obtained from known fault probabilities. The same approach can be used in abductive reasoning.

Before talking about model-based reasoning let us discuss the problems arising when using simple rule-based or decision tree based reasoning techniques for decision support systems. Because of the fact that we are interested in environmental decision support systems we start with a simple example from the wastewater treatment-plants domain. For simplicity, we ignore some aspects at the cost of not being a complete representation of such a treatment plant. In particular we use a simplified model of a constructed wetland.

A constructed wetland (see Figure 1) is a very simple wastewater treatment plant where the natural capabilities of plants are used for cleaning water or in this case eliminating or at least reducing substances like phosphor and nitrates from wastewater. Constructed wetlands are used in cases where bigger wastewater treatment plants are economically not reasonable for example in case of small villages. The decision support we have in mind is a diagnosis system, which allows for detecting the root cause of unexpected observations regarding the constructed wetland. Such observations might impact the functionality of the wastewater treatment plant and should be corrected as soon as possible.

In order to come up with such an environmental decision support system for diagnosing constructed wetlands, we first have to formulate the expected behavior of such a system together with possible failures finally leading to undesired effects. In case of constructed wetlands we have several involved components, which have to work together in a correct way. The difference between the inflow and the outflow to the constructed wetland has to ensure the water level being within giving boundaries. The concentration of nitrates and phosphates in the incoming water has to be within a certain boundary. Moreover, the membrane preventing the wastewater to be going into the surface water and thus spoiling it has to have not leaks. Otherwise, again the water level might not be within the right boundary. The last factor is the weather. During the summer time the

Figure 1. A constructed wetland

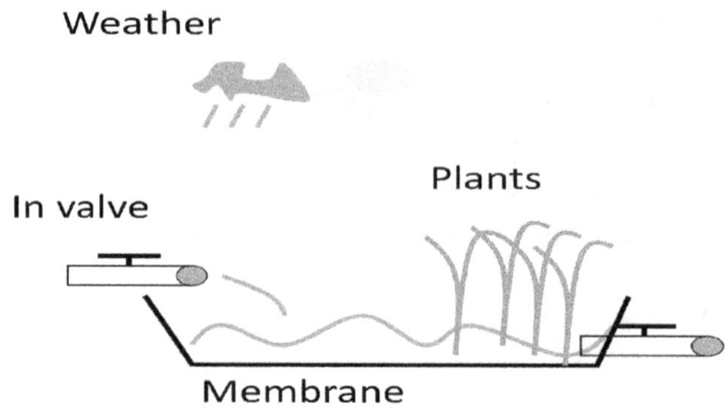

sun might has a negative impact on the water level. Enough water and enough nitrates and phosphates are necessary for plants to grow. As already said this description does not capture all aspects of constructed wetlands. However, it can be used to explain problems with existing approaches.

In order to come up with a diagnosis system for a constructed wetland, we have to discuss what we expect from such a system. In particular diagnosis systems are used to explain a detected misbehavior. The detection of misbehavior or an unexpected situation has to do with observations that are in contradiction with our beliefs of how a system should react in a certain situation. In case of constructed wetlands we might be wondering why the plants are not green any more. This should not happen because it is the manifestation of a state of the system where the desired functionality, i.e., removing undesired substances from wastewater, cannot longer be achieved. In order to bring back the system in a correct state we have to identify the cause of the observation. Note, that even in this simplified example a simple replacement of the plants with new ones might not be enough because the new ones might also be dead after a certain time because the real root cause has not been identified. Hence, the first step of repairing a system has to be the identification of the root cause for a certain undesired observation.

So how can we identify or isolate the real cause of the observation plants to be brown and not green as expected? To answer this question we have to find potential cause effect chains in the model. For example, we know that plants become brown if they have not enough water. We also know that a leak in the membrane might cause the water level to decrease beyond a certain threshold. Hence, the membrane might be the real cause of the observations. Unfortunately, the situation is a little bit more complicated in our example. Not only the state of the membrane, i.e., having a leak or not, has an impact on the water level. The weather or the position of the valves controlling the inflow and outflow has an impact. Moreover, not only the water level is a potential cause of plants dying. Another reason might be the nitrate and phosphate concentration. How to handle all these potential impacts? Another problem is that the used models of the system describe the behavior in terms of an implication from causes to effects, but we know the effect and we are interested in causes. From the point of classical logical reasoning the derivation of causes from effects is not directly possible. We cannot say that the water level is the only reason for plants to be brown because the nitrate and phosphate concentration might also be a reason. Hence, there is no single source of explanation. Even worst it can be the case that different explanations are in contradiction. Using classical logical reasoning does not solve these problems.

In order to solve the questions raised classical AI (for an introduction into AI see Nilson, 1980) came up with rule-based systems and decision trees implementing reasoning from effects to causes. The general idea is simple. Whenever there are two or more possible explanations a test is performed. The output of the test separates possible explanations. For example, if we know only that the plants are brown and not green, a test on the water level separates the nitrate and phosphate concentration cause from the membrane leak cause. Such a test does not allow for separating the membrane leak cause from the other two causes inflow-outflow difference, and the weather. More tests are necessary in order to come up with a single cause if there is any. When following this approach we finally define rules of the form *If test output is x, then either there is a single explanation or we have to apply a new test.*

Such rules can be easily represented as decision trees, where each non-leaf node represents a test. Each directed edge between two nodes represents the output of the corresponding test of the parent node. And each leaf node corresponds to a single explanation. Figure 2 depicts a decision tree for our constructed wetland example. It is worth noting that the shape of the tree depends on the used tests. Hence, there might be smaller or larger decision trees for the same problem. Moreover, the number of necessary tests might vary from tree to tree making one tree more effective in terms of less required tests than others. Hence, the aim is to make a small decision tree that solves the desired problem. Another important issue to mention is that decision trees can be easily represented as set of simple rules. We only have to use the rule used previously to explain decision trees. For our constructed wetland decision tree from Figure 2 we obtain the following rules:

```
start_diagnosis → apply_test(Plant
color green?)
test_result(Plant color green?, YES)
→ output(Correct behavior)
test_result(Plant color green?,NO) →
apply_test(Membrane leaking?)
test_result(Membrane leaking?, YES) →
output(Membrane defect)
test_result(Membrane leaking?,NO) →
apply_test(Not enough water low?)
test_result(Not enough water flow?,
YES) → output(Wrong flow cond.)
test_result(Not enough water
flow?,NO) → apply_test(Hot air tem-
perature?)
test_result(Hot air temperature?,
YES) → output(Bad Weather)
test_result(Hot air temperature?,NO)
→ output(Wrong NP conc.)
```

Rule-based systems (which are also called Production systems or Expert Systems) that are based on forward-chaining theorem proofers directly use such rules to derive results. Such systems have been very effectively used in practice (for an introduction, some applications, and a discussion on Expert Systems see Parsaye & Chignell, 1988; Partridge, 1987; Pham, 1988). One advantage of Production or Expert System is that if a correct and consistent rule base is available, the computation of results is very fast. Hence, decision trees or rule-based systems provide a good basis for commercial products. However, there are some problems that correspond to such systems.

The first problem we discuss is the problem of obtaining a rule-base. Although, the procedure for obtaining rules or decision trees seems to be straightforward it is not. Everything that is known in a domain is formulated either as equation or relational constraint or as cause-effect implication.

Figure 2. A decision tree for diagnosing constructed wetlands

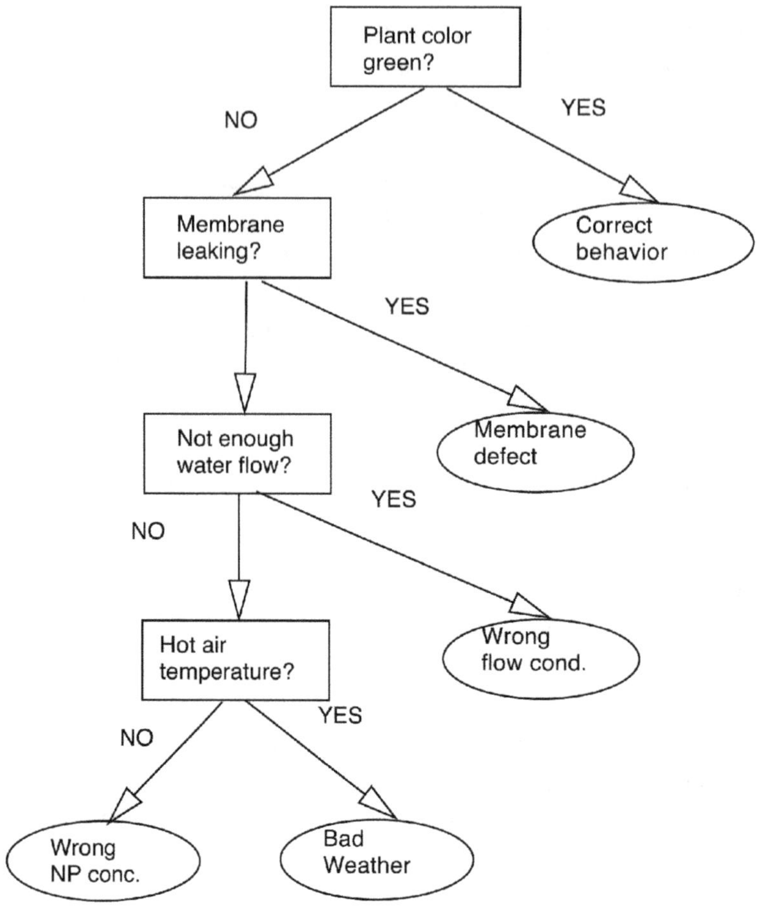

The latter is especially of importance in domains where causes for observed effects are searched for like in diagnosis. What we have is knowledge describing how to derive effects from causes but for diagnosis we need knowledge that enables us to derive causes from given effects. Therefore, the available cause-effect knowledge cannot be used directly for diagnosis if using a rule-based system or a decision tree. As a consequence when using decision trees or rule-based system the knowledge base capturing the effect to causes knowledge, has to be constructed manually using the entire cause-effect knowledge. This procedure is failure prone and any changes in the original model have an impact on the rules. Hence, not using the available knowledge might not be the best idea.

The second problem is due to the fact that the used tests have an impact on the shape and size of decision trees and rule-bases. Moreover, tests also have assigned costs or might not be easy to achieve if even possible. For example, testing whether the membrane has a leak is almost impossible. A visual inspection is very costly and might cause heavy construction work at the constructed wetland going on. Hence, such tests would be performed only if this is really necessary. Other tests, like checking the inflow and the outflow can be easily achieved. Therefore, during construction someone would choose the most cost effective tests. Unfortunately, this cannot always be guaranteed. For example, if a new technology makes one test less expensive, the new one might replace an old test. Hence, the decision tree and the rule-base have to be changed. In the case of the constructed wetland example, Figure 3 shows the new decision tree. In terms of rule we have to remove 2 rules and replace them by 2 new ones. Our example is a small one but approx. 20 percent of all rules has to be changed.

Consider a huge rule-base where the relationships between the different rules are not that obvious. Therefore, maintenance activities for rule-bases and decision trees are a big problem with high associated costs. The maintenance problem occurs not only due to changes in tests but also in cases the underlying knowledge base has to be changed.

The third problem is similar to the second one. If we have a decision tree or rule-base for one system, it is almost impossible to use it or even parts in case of a different systems with structural changes. The same happens in case of merging two systems. The combination of the two corresponding decision trees or rule-bases is not that simple requiring a lot of changes. Hence, decision trees and rule-bases have a severe knowledge re-use problem.

The fourth problem of decision trees and rule-based system is due to the fact that they are not very flexible with respect to the available information. Assume that some observations are available but not the one needed to answer the tests necessary to follow the path from the root node to the leaf node where the solution is stated. In this case both type of systems will fail. It is not possible to replace some information with other. The only way out of this problem would be to construct alternative decision trees or rule-bases capturing different situations at higher costs both in construction and maintenance activities. Hence, we will hardly obtain a decision support system that is flexible and can adapt to different situation when it is based on rules or decision trees.

The fifth problem can be seen as an instance of the fourth one. In case of absence of information a decision tree or rule-based system cannot give you an answer. However, it would be good to get the answer that everything should behave as expected.

In summary, decision support systems based on decision trees or rule-based systems are less flexible on different inputs, are hard to construct and maintain, and hardly handle missing information. Hence, especially in the environmental domain where knowledge changes fast over time and where there are many hidden relationships between different systems and theories, the use of rule-based systems and decision trees seems to be not the right option except for some rare cases.

In order to avoid the mentioned problems of decision trees or rule-based systems we suggest to use abductive reasoning for environmental decision support systems. The idea of abduction is to use logical theories that represent knowledge in the form from causes to effects directly for explaining undesired effects and using observations as soon as they are available. Hence, an abductive reasoning engine provides the user with all possible explanations. Furthermore such a reasoning engine allows for computing the next optimal test. In contrast to decision trees or rule-based systems the tests needs not to be known in advance. Instead the whole currently available knowledge is used. In the next sections we formally introduce abductive reasoning and discuss modeling issues. Afterwards, we present a modeling paradigm that is well adapted for the purpose of model-based reasoning and supports re-use as well as maintenance activities.

An Example

In this section we introduce our running example, which is based on a decision tree for diagnos-

Figure 3. Another decision tree for diagnosing constructed wetlands

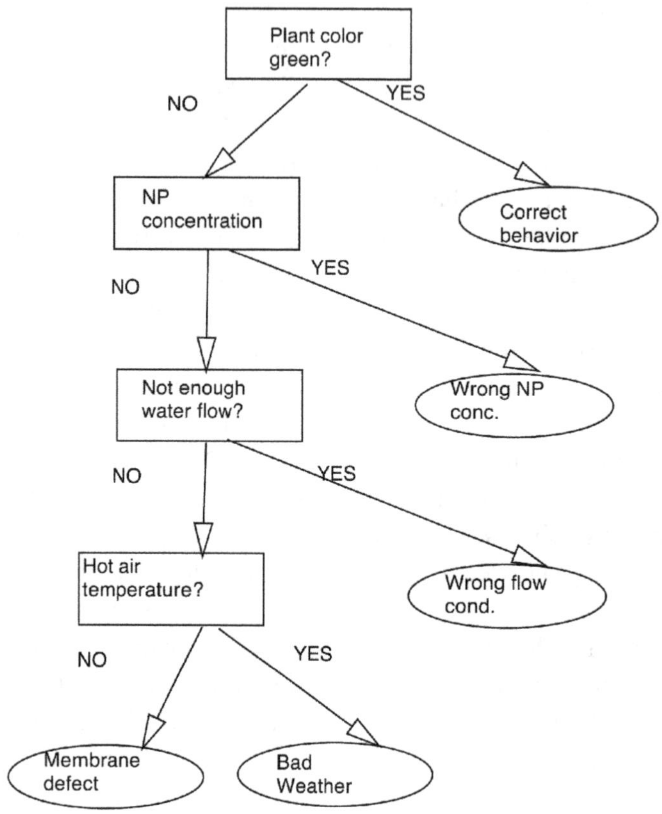

ing a wastewater treatment plant. Comas and colleagues (Comas et al., 2003) introduced the whole knowledge base of the wastewater treatment plant. We only concentrate on a subtree that is depicted in Figure 4 and which is given on the left side of Figure 3 in (Comas et al., 2003). It is worth noting that the given decision tree is not really a tree but a graph because there are vertices having more than one parent. Hence, Comas and colleagues use decision diagrams instead of trees. In the following, we show how to extract an abductive knowledge base from the decision diagram and further in the next section how to use this knowledge base for finding diagnosis and appropriate repair actions.

We can easily extract a rule-based knowledge from decision trees. For this purpose we only need to transform each branch of the tree $t_1, \ldots, t_k, c$ where t_i denote test vertices and c the consequence vertex into a rule. However, a straight forward transformation like $t_1 \wedge .. \wedge t_k \rightarrow c$ would not be sufficient because it would map effects to causes, which would not lead to an abductive knowledge base, i.e., a knowledge base that maps causes to effects. In order to obtain an abductive knowledge base we have to transform each branch into a rule that maps causes to effects, e.g., $c \rightarrow (t_1 \wedge .. \wedge t_k)$. The resulting rules are representing the knowledge in an abductive way but they are not horn clauses. In order to obtain horn clauses the single rule $c \rightarrow (t_1 \wedge \ldots \wedge t_k)$ must be transformed into a set of rules $c \rightarrow t_1, \ldots c \rightarrow t_k$.

Another difficulty arises in cases where we do not have a decision tree but a decision diagram. For decision diagrams the introduced transformation does not work. Consider the decision diagram in Figure 4. There are two paths from the root vertex to the cause *Other*. A direct conversion would lead to contradicting rules, e.g.,

```
Other → free_swimming_cilliates_pre-
dominance = No and
Other → free_swimming_cilliates_pre-
dominance = Yes.
```

The underlying cause of contradiction is that the two paths represent alternatives, which has to be represented by two logical sentences combined with an exclusive or operator. Such a logical construct cannot longer be converted into a representation that comprises only horn clauses. A solution would be to represent the cause *Other* by two different causes *Other*1 and *Other*2.

The transformation, applied to the example decision diagram from Figure 4, leads to the theory Th_0:

```
Th0 = { too_high_fm → no_fswcp, too_
high_fm → fp, sp → no_fswcp, sp →
no_fp, sp → ws, o1 → no_fswcp, o1 →
no_fp, o1 → no_ws, o2 → fswcp, o2 →
nh_fm, ul → fswcp, ul → low_fm }
```

In order to complete the knowledge base we should add completeness rules like saying that the outcome of one testing vertex can only be *Yes* or *No* but never both at the same time. Hence, we add completeness rules to our example theory Th0 and obtain a new extended theory Th_1:

```
Th1 = Th0 ∪  no_fswcp ∧ fswcp → ⊥,
no_fp ∧ fp → ⊥, no_ws ∧ ws → ⊥, nh_
fm ∧ low_fm → ⊥ }
```

In the following section we introduce the basic definitions of abductive diagnosis used to compute explanations for a certain given set of observations.

Abductive Diagnosis

In this section we lay out the basic ideas already described in a formal way. For this purpose, we first define a knowledge base comprising horn clause rules over propositional variables. The restriction of logical formula to be horn is not a real limitation in the context of physical systems since those models usually code the knowledge from causes to their effects. The used definitions are close the ones introduced by Friedrich et al. (1990) and others in the area of abductive diagnosis (Console et al., 1991; Console & Torasso, 1990).

Definition 1 (Knowledge base (KB)): *A knowledge base (KB) is a tuple (A,Hyp, Th) where A denotes the set of propositional variables, $Hyp \subseteq A$ the set of hypothesis, and Th the set of horn clause sentences over A.*

In the definition of KB hypotheses corresponds directly to causes such that for every cause there is a propositional variable that allows to hypothesis about the truth value of the cause. Hence, we use the terms hypotheses and causes in an interchangeable way.

Example: The tuple *({too high fm,sp,o1,o2,ul, no fswcp,fswcp,no fp, fp,no ws,ws,nh fm,*

Figure 4. Partial decision diagram for solving the disperse-growth problem in wastewater treatment plants

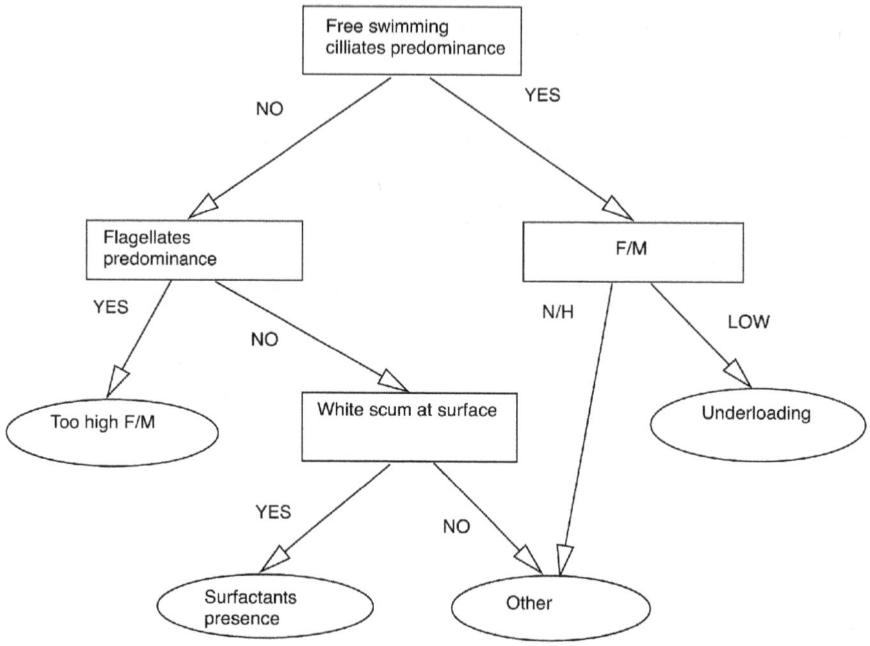

`low fm}, {too high fm,sp,o1,o2,ul},` Th_1`)` `form the knowledge base of the running example.`

Definition 2 (PHCAP): *Given a knowledge base (A,Hyp, Th) and a set of observations Obs ⊆ A then the tuple (A,Hyp, Th,Obs) forms a propositional horn clause abduction problem (PHCAP).*

Example: [cont.] Assume that we observe no flagellates predominance, i.e., *Obs = {no fp}*.

Given a PHCAP we define a solution like in (Friedrich et al., 1990) as follows:

Definition 3 (Diagnosis; Solution of a PHCAP): *Given a PHCAP (A,Hyp, Th,Obs). A set $\Delta \subseteq Hyp$ is a solution if and only if $\Delta \cup Th \models Obs$ and $\Delta \cup Th$ is consistent. A solution Δ is parsimonious or minimal if and only if no set $\Delta' \subset \Delta$ is a solution.*

The definition of abductive diagnosis takes the given theory and the observations we want to explain as input. A diagnosis is a set of hypothesis explaining the observations. In terms of logic we express explanation as the ability to derive the observations logically from the diagnoses and the theory, i.e., $\Delta \cup Th \models Obs$. Moreover, we know that in logic we can derive everything from inconsistent theories. If the diagnosis together with the theory is inconsistent, *Obs* can be derived but such an explanation would not be regarded as the solution we want to compute. Because of this reason, an abductive explanation has to ensure that $\Delta \cup Th$ is consistent.

In the above definition diagnoses need not to be minimal or parsimonious. However, in practice we are almost only interested in minimal diagnoses or minimal explanations for given effects. Therefore, from here on we assume that all diagnoses are minimal diagnoses if not specified explicitly.

Table 1. The propositions used in Th_0 have the following meaning in the decision diagram.

Causes	*too_high_fm*	Too high F/M
	sp	Surfactants presence
	o1	Other 1 (left)
	o2	Other 2 (right)
	ul	Underloading
Effects	*no_fswcp*	Free swimming cilliates predominance = No
	fswcp	Free swimming cilliates predominance = Yes
	no_fp	Flagellates predom. = No
	fp	Flagellates predom. = Yes
	no_ws	White scum at surface = No
	ws	White scum at surface = Yes
	nh_fm	F/M = N/H
	low_fm	F/M = Low

Example: [cont.] For our running example we obtain 2 minimal diagnoses: $\Delta_1 = \{sp\}$ and $\Delta_2 = \{o1\}$ for the observation *no_fp*.

What is worth discussing now is the question whether abductive reasoning avoids the problems mentioned for rule-based systems and decision trees in a previous section of this paper? Consider the first problem that is due to modeling. Usually knowledge about a domain is given in cause-effect form but decision trees and rule-based systems require knowledge in form of effect to cause reasoning. This is definitely not required for abductive reasoning where cause-effect knowledge is used directly. Hence, the first problem is solved when using abduction. The second and the third problem we mentioned is somehow related. Due to the restrictions imposed when constructing a rule-base the structure of the rule-base or the decision tree heavily depends on the intention behind the construction. E.g. which test to be used in order to guide the user through the decision tree or rule-base in order to come up with a solution. In abductive reasoning such considerations are not necessary. The rules should only reflect the given cause-effect knowledge. In case of changes to the knowledge-base abductive reasoning uses the new knowledge and adapts to inconsistencies that may arise. Hence, changes of the system do not impose further problems when using abduction. Therefore, the second and third problem are also at least less demanding in the case abduction.

The fourth problem, i.e., the missing flexibility of the observations used during diagnosis, is also solved now. In abductive reasoning we compute a solution for given observations. Whether we have one or many observations is not a question for the reasoning mechanism behind but influences the computed diagnoses. Hence, a system that is based on abductive reasoning automatically adapt to a given situation. The same holds for the fifth problem, i.e., reasoning in case of missing information. If some information is unknown, it is not used for computing a solution. Missing information never impacts the capability for computing solutions but of course has an impact on the computed solutions, which depend on the available observations.

It is well known that the problem of finding minimal diagnoses for a given PHCAP is NP complete, which means in the case of abductive reasoning that the running time for computing solutions is exponential in the number of hypothesis. See for example Friedrich et al. (1990) for a proof. Moreover, in the same paper the authors describe a general diagnosis and therapy process that make use of a PHCAP directly in order to identify and correct the detected faulty behavior. Unfortunately the introduced algorithm cannot be used in our setting where exactly one diagnosis has to be identified before executing the necessary repair actions. Hence, instead of computing a diagnosis, checking its hypothesis, and if necessary modifying the corresponding PHCAP in order to obtain other diagnoses for further elaboration, in our case we have to compute all possible diagnoses and to reduce this set by adding new observations, which allows for discriminating diagnoses until one diagnosis is left and the corresponding treatment can be applied.

A reason for this change is that in the industrial context checking for correct hypothesis, i.e., mainly via replacing components, might be too expensive. Hence, the identification of one root cause allows for reducing the overall costs. In order to compute discriminating additional observations we adopt the approach by De Kleer and Williams (1987) who introduced an algorithm for measurement selection, which is based on entropy. The original work of De Kleer and Williams is based on consistency-based diagnosis (Reiter, 1987) whereas our version is adapted for the abductive case.

The idea behind measurement selection is do identify an observations that help to restrict the search space as efficiently as possible. This is the case if an observation allows for dividing the set of diagnoses into two disjunctive subsets of almost the same size where one subset predicts the observations and the other does not. In this case search is done in a binary way, which requires only a logarithmic number of steps to find a unique solution.

Definition 4 (Discriminating observation): *Given a PHCAP (A,Hyp, Th,Obs) and two diagnoses Δ_1 and Δ_2. A new observation $o \in A \setminus Obs$ discriminates two diagnoses if and only if Δ_1 is a diagnosis for (A,Hyp, Th,Obs $\cup \{o\}$) but Δ_2 is not.*

Example: [cont.] For our running example knowing the truth state of either *ws* or *no_ws* discriminates the two diagnoses *{sp}* and *{ol}*. All other observations do not allow for distinguishing the competing diagnoses. Note that in practice the knowledge base should not be extracted from a decision tree. Instead a cause-effect model should be obtained from the underlying theory, which is richer in terms of effects to be obtained from a cause or a combination of causes.

The entropy *H* of an event is the product of the event's probability and the logarithm of the event's probability and is a measure of the information gain.

$$H(o) = -p(o) \cdot \log_2 p(o) - (1 - p(o)) \cdot \log_2 (1 - p(o))$$

The $1 - p(o)$ part on the right of the equation is due to the fact that we also have to consider the observation to be not *o*. The probability $p(o)$ can be obtained using the number of diagnoses that predict the observation and the overall number of diagnoses.

$$p(o) = \frac{\left|\{\Delta | \Delta \in \Delta - Set, \Delta \cup Th \models Obs \cup \{o\}\}\right|}{\left|\Delta - Set\right|}.$$

Δ-Set is the set of diagnoses obtained from the original PHCAP, i.e., *Δ-Set* = $\{\Delta | \Delta \in Hyp, \Delta \cup Th \models Obs\}$.

Example: [cont.] For our running example with the *Δ-Set = {sp, o1}* both *H(ws)* and H(no_ws) are 0.5, and the entropies of the other propositions are 0. Hence, only ws or *no_ws* allow for discriminating the obtained minimal diagnoses.

De Kleer and Williams (1987) entropy-based probing approach is similar to ours but there are some differences, which we discuss now. First, in De Kleer and Williams (1987) the used diagnosis technique is consistency-based diagnosis where the correct behavior of a system is modeled. In abductive diagnosis the behavior in case of a fault has to be specified in order to extract the root cause. We refer the reader to (Console et al., 1991) where the relationship between abductive and consistency-based diagnosis is discussed in detail.

Second, in De Kleer and William's approach (De Kleer & Williams, 1987) all diagnoses and not only the minimal ones are used to compute the next best measurement location. In classical consistency-based diagnosis every superset of a diagnosis is itself a diagnosis and therefore minimal diagnoses characterize all possible solutions. This is not the case for abductive diagnosis. For example, consider our wastewater treatment plant model together with the observation *no_fp*. *{sp}* is a minimal diagnosis but *{sp, too_high_fm}* is not because this set is inconsistent with the theory Th_1. Hence, in our case diagnoses do not characterize all possible diagnoses and using them for probing seems to be infeasible. A more detailed discussion about the use of fault models for diagnosis and its consequences can be found in (De Kleer et al., 1992).

Third, the introduced probing is not based on a specific type of model, e.g., a component-connection model. The entropy values for the propositions are directly computed. Since, there are several propositions, which are tightly coupled with a certain test, it makes sense to use such information within the diagnosis and probing cycle. We discuss this issue in the next section.

Integrating Diagnosis, Probing, and Repair

In this section we focus on the integration part of abductive diagnosis, probing, and repair. This part is especially important for environmental decision support systems in order to finally come up with a repair suggestion for a given problem. Given the abductive theory someone is interested in finding a unique root cause for the given observations, which can be eliminated applying one action or a sequence of actions. In the used example from the decision tree is exactly used for the purpose of identifying a root cause, which is directly connected with actuations like "Improve pretreatment". This direct coupling of repair knowledge given a root cause when using decision trees is a consequence of the underlying knowledge representation, which goes from effects via causes to the necessary treatment. Note that this view on repair we also find in Friedrich et al.'s work (Friedrich et al., 1990) where there is a one-to-one correspondence between the abductive explanation and repair.

In this paper we follow this approach and assume the existence of a function that maps root causes or in our case hypotheses *Hyp* to repair knowledge. How this repair knowledge is going to be stored or executed is out of this paper's focus. In particular it is also possible that such a function returns a textual description about how to tackle a problem, which has to be read and executed by the human operator. Formally we introduce such a function ρ as a function mapping *Hyp* on *TEXT* where *TEXT* is the set of all possible strings.

The second issue we discuss before presenting a general solution to the integration problem is the representation of different values associated with variables. For example, in our running example there are two values possible for white scum at surface, i.e., *yes* or *no*, which are represented by the propositions *ws* and *no_ws* respectively. Note that it is also possible that more than two values are possible. Within the theory we are not able to directly say that either *ws* or its negation is true at a time because this would lead the theory to be not horn anymore. What we do instead is to state some rules establishing knowledge about impossibilities. '$no\_ws \wedge ws \rightarrow \bot$' is an example of such a rule.

A consequence of this representation is that at the theory level we are not able to use such information to improve probing. In our running example both propositions *ws* and *no_ws* have the same entropy value. However, in practice every test on white scum on the surface will lead to either *ws* or *no_ws* to be true. In order to represent this knowledge correctly we use the same idea than before with repair knowledge. And indeed both repair actions and probing are actions where someone has to perform a certain task to obtain the intended result. Hence, we introduce a function τ: $A \rightarrow TEXT$, which maps elements from the set of propositional variables to the corresponding test actions. In our example such a mapping may look like: $\tau(ws) = \tau(no\_ws) =$ *'WHITE SCUM TEST'*.

In order to use this knowledge for probing we now do not only compute the information gain for each propositional variable but calculate the expectation value for each test action and choose the test with the highest value. The expectation of a probability variable is given as the sum of products of a value and its probability. In our case the expectation for a test action act is given as follows:

$$E[act] = \sum_{o \in A(act)} H(o) \cdot \frac{1}{|A(act)|},$$

where $A(act) = \{x | \tau(x) = act\}$ is the set of proposition that correspond to the same test action.

One underlying assumption of the computation of E is that the each outcome of the test is equal likely with probability $1 / |A(act)|$.

The functions τ and ρ provide interfaces to a user of the system. The function τ is used to establish a relationship between a test or a question and possible answers to be provided by the user. The function ρ is for returning appropriate actions to the user, which have to be carried out in order to eliminate the detected cause. Hence, a decision support system based on abduction comprises three parts: an abductive theory or model of the underlying system, the function ρ, and the function τ.

Definition 5 (Abductive Decision Support System (ADSS)): *An abductive decision support system is a tuple (KB, τ, ρ) where KB = (A,Hyp, Th) is an abductive knowledge base, τ: A $\rightarrow$ TEXT is a function mapping propositional variables to tests to be answered by a user, and ρ: Hyp $\rightarrow$ TEXT is a function mapping causes to repair actions.*

The following algorithm implements the decision process for ADSS and is based on our previous discussion in the paper.

Algorithm *computeSingleExplanation*

Inputs: An ADSS comprising an abductive KB *(A,Hyp, Th)* and the functions ρ, τ,, and a set of observations $Obs \subseteq A$.

Output: A single abductive explanation for *Obs*.

1. Compute the set of abductive diagnoses *Δ-Set* for the given KB and *Obs*.
2. If there are no diagnoses, then report a conflict and leave the algorithm.
3. If the size of *Δ-Set* is larger than 1, then do the following:
   (a) Compute the expectation value for each test action and select the action *t* with the highest expected entropy.
   (b) Provide the user with the selected action *t* and add the answer *o* to *Obs*. Note that $t = \tau(o)$.
   (c) Go to Step 1.
4. Let *Δ* be the only remaining element of *Δ-Set*.
5. Return the set $\{\rho(h) | h \in \Delta\text{-}Set\}$ as result.

Note that *computeSingleExplanation* always returns only one solution, which might comprise several root causes or hypotheses from *Hyp*, whenever a solution exists. A solution does not exist in cases where either the KB itself is inconsistent or the given observations cannot be explained. The first case should never happen if the KB is design appropriately. However, the latter case can happen in practice if the observations are conflicting. This can be the case because of two reasons. First, the observations are not consistent or there are only explanations explaining the conflicts, which are inconsistent because of the KB. Such conflicts have to be reported and solved by the user.

Another important fact thing about *computeSingleExplanation* and the underlying abductive theory is that there might not only be a single explanation. The set to be returned might comprise several actions coming from several hypotheses necessary to explain Obs. This is different to the usual use of decision trees where the aim is only to provide one explanation with the implicit assumption that a single explanation is good enough for almost all cases. Of course it is also possible to handle the multiple explanation case using decision trees but at a substantial increase of size and complexity. Using ADSS the case of multiple explanations is handled for free and does not need further elaborations.

In order to use the proposed methodology it is necessary to obtain an abductive model. Because of the nature of such a model, i.e., capturing the cause-effect relationship, it should not be difficult to build the model. In our running example we used a model obtained from a decision tree. This model is only for explaining the presented theory. We do not claim that converting decision trees into an abductive model should be done. Because of the fact that decision trees only partially represent the underlying model such a conversion is not appropriate and the loss of knowledge is might be unacceptable. Hence, the recommendation in practice is to start from scratch developing an abductive model for the given problem.

At the end of this section we want to give some more information regarding the computational aspect of abductive diagnosis. It is well known that in principle the problem is NP-complete and therefore has an exponential running time with respect to the number of hypothesis. Algorithms computing abductive diagnoses have a worst case running time exponential in the size of *Hyp*. However, there are algorithms and systems available that lead to a practically acceptable running time. One of such systems is the assumption-based truth maintenance system (ATMS) originally invented by De Kleer (1986a, 1986b). Beside an algorithm (De Kleer, 1988) there are some improvements (e.g., De Kleer & Williams, 1986; Forbus & De Kleer, 1988), available to mention only few of them. An ATMS can be directly used for the purpose of abductive diagnoses because it stores all possible consistent explanations for a given propositional variable. Hence, if we have the set of observations $Obs = \{o_1, \ldots, o_k\}$ we only have to add a rule of the form $o_1 \wedge \ldots \wedge o_k \rightarrow EXPLAIN$ where *EXPLAIN* is a new propositional variable to the ATMS. The corresponding ATMS label of *EXPLAIN* comprises all possible explanations.

Although, the upper bound of the running time required for abductive diagnosis is high, in practice the use of reasoning engines like the ATMS (De Kleer, 1986a) in the most improved version should help to keep the running time for computing a solution in an acceptable range. And indeed, we implemented a first prototype of an abductive reasoning engine based on the ATMS. For theories comprising less than 10 hypotheses and about 5 to 20 rules, the computation of all abductive diagnoses for a given theory and set of observations took less than one second using a today's notebook. Moreover, we tested the system on larger knowledge bases comprising 18 hypothesis and more than 140 rules. In this case the implementation computed all solutions in about 40 seconds. It is worth noting that the implementation is hardly optimized and that the running times are for computing all solutions and not only one. Further improvements can be expected when guiding search using probabilities and restricting search to only a few solutions.

Modeling for Abductive Reasoning

After presenting a solution to the mentioned problems regarding environmental decision support systems based on decision trees or rule-based systems namely abductive reasoning there are still some open questions to discuss. The problems regarding re-use and maintenance have not been completely covered in the previous sections. Although, allow for modeling the cause-effect relationships directly, which definitely supports maintenance activities, the question of re-use have not been tackled so far.

In order to provide a model for abductive reasoning that supports reuse and reduces maintenance costs we follow the component-oriented modeling paradigm. This paradigm has been successfully used in the context of model-based diagnosis (De Kleer & Williams, 1987; Reiter, 1987). Heller and Struss published several papers on the application of model-based reasoning to the domain of decision support systems in ecology. In (Heller & Struss, 1996) the authors present an approach for transforming models into corresponding qualitative models that can be used directly by a reasoning engine. In (Heller & Struss, 2002) the application of model-based reasoning and in particular consistency-based reasoning for environmental decision support has been reported. Both papers are of particular interest even in the case of abductive reasoning because the modeling principles have been given there.

In the component-oriented modeling paradigm every system comprises a set of interconnected components. Each component has an assigned behavior that depends on an internal state. The system can only be observed via the connections between components. Usually, the internal state of a component cannot be directly observed. Modeling in the component-oriented paradigm requires the construction of components including their behavior and the connection between the components. In this paradigm a component is considered as single unit of re-use. From that follows that component models them self should be as general as possible. In a certain system the behavior might be adapted and instantiated. However, the aim is to provide a library of component models that can be used in different systems. Beside re-use the component-oriented modeling paradigm also has the advantage that changes are also local. Either a component or a connection between two or more components has to be changed. Moreover, if necessary a component model that might belong to more component instances can be changed that causes automatically a change of all its instances. The question is now how the component-oriented modeling paradigm can be used in case of abductive reasoning? We now give an answer to this question and use again the constructed wetland example to illustrate the modeling approach.

The first step of modeling is to identify components and their connections. In case of the constructed wetland we have 5 components: Valves for controlling the inflow and the outflow, the basin where the water is kept, the plants, the weather, and the water itself. For simplicity we only consider one valve in our example. What we now have to do is to develop behavior models for our components, which should be as general as possible. For this model we use first order logic to represent the behavior of the components. Note that we use capitalized words to represent states of a component. For example the valve has two states, i.e., set appropriately or not set appropriately. We recommend Chang and Lee (1973) for an introduction into first order logic.

Valve: If there is water available at the input of a valve and the valve is set appropriately, then the amount of water at the output is within the expected boundary.

∀X: valve(X) → (SET_OK(X)∧ input(X,water) → output(X,water))

Basin: If there is no leak in the basin, the water level is stable.

∀X: basin(X) → (NO_LEAK(X) → water_level_stable(X))

Weather: If the air temperature is not too hot for several days, the water level is stable.

∀X: weather(X) → (NOT_HOT(X) → water_level_stable(X))

Wastewater: The wastewater should have a nitrate and phosphate concentration within given boundaries.

∀X: wastewater(X) → (OK(X) → ok_concentration(np,X))

Plants: If there is enough water available, and the nitrate and phosphate concentration is good enough the plants will grow. Growing plants have a green color.

∀X: plants(X) → (HEALTHY(X)∧ ok_concentration(np,X)∧ enough_water(X) → color(X, green))

The first order logic rules describe the correct behavior of each component. The correctness of a particular component is represented by a predicate written in upper case letters, which represent the state of a component. In order to compose a system from the components we first have to instantiate the components and to connect them. We do this by stating that a certain component with a particular name exists. For our example we use the following facts:

```
valve(in)∧ basin(cw)∧
weather(weather)∧ wastewater(ww)∧
plants(pl)
```

When combining these facts with the rules we finally obtain a set of grounded rules, i.e., rules without variables. Note that in this transformation process we eliminate those parts from the rules that are already known:

```
SET_OK(in)∧ input(in,water) →
output(in,water)
NO_LEAK(cw) → water_level_stable(cw)
NO_ HOT(weather) → water_level_
stable(weather)
OK(ww) → ok_concentration(np,ww)
HEALTHY (pl)∧ ok concentration(np,
pl)∧enough water(pl) → color(pl,
green)
```

What is still missing are the connections between the components. In our case we have to state that there are three causes for a stable water level and there is a connection between the nitrate and phosphate concentration of the water and the necessary concentration required by the plants.

```
output(in,water)∧ water_lev-
el_stable(cw)∧ water_level_
stable(weather) → enough_water(pl)
ok_concentration(np,ww) → ok_
concentration(np, pl)
```

Since all of the rules are grounded they can be directly used by the abductive diagnosis engine, which is based on propositional logic. However, in case of abductive reasoning we are interested in the faulty behavior of components because the faulty behavior is the cause of an observed undesired effect. But the logical model provided does not include the faulty behavior. We fix this problem and introduce the faulty behavior for each component. Moreover, we introduce rules stating inconsistencies like the color of a plant is yellow and green at the same time.

Valve: If the valve is not set appropriately, then the output is not within the nominal range.

∀X: valve(X) → (SET_FAULTY (X) → not_in_range(X,water))

∀X: valve(X) → (not_in_range(X,water)∧ output(X,water) → ⊥)

Basin: If there is a leak in the basin, the water level is decreasing.

∀X: basin(X) → (LEAK(X) → water_level_decreasing(X))

∀X: basin(X) → (water_level_stable(X)∧water_level_decreasing(X) → ⊥)

Weather: If the air temperature is too hot for several days, the water level will decrease.

∀X: weather(X) → (HOT(X) → water_level_decreasing(X))

∀X: weather(X) → (water_level_stable(X)∧ water_level_decreasing(X) → ⊥)

Wastewater: The wastewater should have a nitrate and phosphate concentration within given boundaries.

∀X: wastewater(X) → (WRONG_NP(X) → wrong_concentration(np,X))

∀X: wastewater(X) → (ok_concentration(np,X)∧wrong_concentration(np,X) → ⊥)

Plants: If there is either not enough water, or the nitrate and phosphate concentration is not within certain boundaries, the plants' color will be brown.

∀X: plants(X) → (DEAD(X)∧ wrong_concentration(np,X)∧no_enough_water(X) → color(X, brown))

∀X: plants(X) → (enough_water(X)∧ not_enough_water(X) → ⊥)

Again like for the correct behavior we have to instantiate the components and add rules stating the connections between the component instances:

```
not_in range(in,water) → not_enough_water(pl)
water_level_decreasing(cw) → not_enough_water(pl)
water_level_decreasing(weather) → not_enough_water(pl)
wrong_concentration(np,ww) → wrong_concentration(np, pl)
```

All grounded rules can be directly used as input to out abductive reasoning framework. The predicates written in upper case letters are the hypotheses.

We now summarize the modeling principles:

1. Develop component models. These models should capture the correct and the incorrect behavior of each component where the states or modes of a component are the hypotheses of the abduction problem. Do not forget to introduce rules stating impossibilities like that certain values cannot occur at the same time.
2. Introduce component instances and reduce the models using the component names.
3. State the interconnection between component instances using logical rules. The resulting model should be grounded and can be directly used by an abductive reasoning engine.

CONCLUSION

The purpose of the paper is two-fold. First, we discussed in detail why more traditional methodologies like decision trees and rule-based systems are less appropriate for environmental decision support systems. Using a small example, we illustrated several problems, e.g., the missing flexibility, the lack of re-use capabilities, and the need for larger changes of the knowledge base in case of changes in the underlying theory. All these problems are especially problematic in the environmental domain.

Second, we provided a methodology that solves the mentioned problems. In particular we presented abductive reasoning, which allows for computing a single explanation for a set of given observations. There are several advantages of abductive reasoning. The model is close to the models of systems described in textbooks where causes and their effects are described. Furthermore, the computed solution has no inherent restrictions like returning only one cause for a given problem.

In cases where there are more causes the system will find them all. Finally, beside writing a model and providing observations there is nothing to do else. The computation of solution is only based on the available knowledge.

We further introduced a probing process based on entropy. The result of probing is a test action that has to be carried out to get more observations in order to distinguish competing explanations. The presented algorithm guarantees to return a single solution in all cases where such a solution exists. We further discussed how to integrate abductive diagnosis and probing. Moreover, and most importantly we also tackle the modeling problem and propose a process for coming up with an abductive model. We illustrate the process on a small example from the environmental domain.

Future research will include an exhaustive study of the capabilities and computational requirements of abductive reasoning applied to practical examples. Moreover, we want to use the reasoning engine for teaching students logical modeling as part of their master program. A first trial with 5 teams of students was quite successful but also showed that the modeling issues is very difficult and that most of the effort has to go into teaching student how to use logic as means for presenting facts about the world. The last section before the conclusion is one part of this effort stating how modeling should be done ideally in order to support re-use and changes in the knowledge base. However, this methodology has to be supported with tools. Again providing design tools is part of future research.

ACKNOWLEDGMENT

The work described in this paper was partially funded by Austrian Academic Exchange Service (OeAD) under contract ES 19/2008.

REFERENCES

Bredeweg, B., Bouwer, A., Jellema, J., Bertels, D., Linnebank, F., & Liem, J. (2006). Garp3 – a new workbench for qualitative reasoning and modelling. In *Proceedings of the 20th International Workshop on Qualitative Reasoning (QR-06)*, Hanover, NH.

Chang, C. L., & Lee, R. C. T. (1973). *Symbolic Logic and Mechanical Theorem Proving*. New York: Academic Press.

Comas, J., Rodriguez-Roda, I., Sanchez-Marre, M., Cordes, U., Freixo, A., & Araez, J. (2003). A knowledge-based approach to the deflocculation problem: integrating on-line, off-line, and heuristic information. *Water Research, 37*, 2377–2387. doi:10.1016/S0043-1354(03)00018-6

Console, L., Theseider Dupré, D., & Torasso, P. (1991). On the relationship between abduction and deduction. *Journal of Logic and Computation, 1*(5), 661–690. doi:10.1093/logcom/1.5.661

Console, L., & Torasso, P. (1990). Integrating models of correct behavior into abductive diagnosis. In *Proceedings of the European Conference on Artificial Intelligence (ECAI)*, Stockholm (pp. 160–166). London: Pitman Publishing.

De Kleer, J. (1986a). An assumption-based TMS. *Artificial Intelligence, 28*, 127–162. doi:10.1016/0004-3702(86)90080-9

De Kleer, J. (1986b). Problem solving with the ATMS. *Artificial Intelligence, 28*, 197–224. doi:10.1016/0004-3702(86)90082-2

De Kleer, J. (1988). A general labeling algorithm for assumption-based truth maintenance. In *Proceedings of AAAI,* Saint Paul, MN (pp. 188-192). San Francisco: Morgan Kaufmann.

De Kleer, J., Mackworth, A. K., & Reiter, R. (1992). Characterizing diagnosis and systems. *Artificial Intelligence*, 56.

De Kleer, J., & Williams, B. C. (1986). Back to backtracking: Controlling the ATMS. In *Proceedings of the National Conference on Artificial Intelligence (AAAI)*, Philadelphia (pp. 910-917).

De Kleer, J., & Williams, B. C. (1987). Diagnosing multiple faults. *Artificial Intelligence*, *32*(1), 97–130. doi:10.1016/0004-3702(87)90063-4

(1988). Expert Systems in Engineering. InPham, D. T. (Ed.), *Artificial Intelligence in Industry*. Berlin: Springer-Verlag.

Forbus, K. D., & De Kleer, J. (1988). Focusing the ATMS. In *Proceedings of the National Conference on Artificial Intelligence (AAAI)*, Saint Paul, MN (pp. 193-198). San Francisco: Morgan Kaufmann.

Friedrich, G., Gottlob, G., & Nejdl, W. (1990). Hypothesis classification, abductive diagnosis and therapy. In *Proceedings of the International Workshop on Expert Systems in Engineering*, Vienna, Austria (LNAI 462).

Heller, U., & Struss, P. (1996). Transformation of Qualitative Dynamic Models – Application in Hydro-Ecology. In *Proceedings of the 10th International Workshop on Qualitative Reasoning* (pp. 83-92). AAAI Press.

Heller, U., & Struss, P. (1997). Conceptual Modeling in the Environmental Domain. In *Proceedings of the 15th IMACS World Congress on Scientific Computation, Modelling and Applied Mathematics,* Berlin (Vol. 6, pp. 147-152).

Heller, U., & Struss, P. (1998). Diagnosis and therapy recognition for ecosystems - usage of model-based diagnosis techniques. In *Proceedings of the 12th International Symposium Computer Science for Environmental Protection (UI-98)*, Bremen, Germany.

Heller, U., & Struss, P. (2002). Consistency-based problem solving for environmental decision support. *Computer-Aided Civil and Infrastructure Engineering*, *17*, 79–92. doi:10.1111/1467-8667.00257

Nilsson, N. J. (1980). *Principles of Artificial Intelligence*. Wellsboro, PA: Tioga Publishing Company.

Parsaye, K., & Chignell, M. (1988). *Expert Systems for Experts*. New York: John Wiley & Sons.

Partridge, D. (1987). The scope and limitations of first generation expert systems. *Future Generation Computer Systems*, *3*, 1–10. doi:10.1016/0167-739X(87)90038-0

Reiter, R. (1987). A theory of diagnosis from first principles. *Artificial Intelligence*, *32*(1), 57–95. doi:10.1016/0004-3702(87)90062-2

Salles, P., Bredeweg, B., & Araujo, S. (2003a). Qualitative models about stream ecosystem recovery: Exploratory studies. In *Proceedings of the 17th International Workshop on Qualitative Reasoning (QR)*, Brasilia, Brazil.

Salles, P., Bredeweg, B., & Bensusan, N. (2003b). The ants' garden: Qualitative models of complex interactions between populations. In *Proceedings of the 17th International Workshop on Qualitative Reasoning (QR)*, Brasilia, Brazil.

Struss, P. (1998). Artificial intelligence for nature - why knowledge representation and problem solving should play a key role in environmental decision support. In Herzog, O., & Günter, A. (Eds.), *KI-98: Advances in Artificial Intelligence (LNAI 1504)*. doi:10.1007/BFb0095425

This work was previously published in International Journal of Agricultural and Environmental Information Systems, Volume 2, Issue 1, edited by Petraq Papajorgji and François Pinet, pp. 63-82, copyright 2010 by IGI Publishing (an imprint of IGI Global).

Section 5
Survey on New Computer-Based Techniques Applied to Agriculture and Environment

Chapter 16
Modeling:
A Central Activity for Flexible Information Systems Development in Agriculture and Environment

P. Papajorgji
University of Florida, USA

F. Pinet
Cemagref, Clermont Ferrand, France

A. Miralles
Cemagref, Maison de la Télédétection, France

E. Jallas
ITK/CIRAD, France

P.M. Pardalos
University of Florida, USA

ABSTRACT

Enterprise information systems in agriculture and environment are becoming more complex and difficult to design and implement. This paper aims to show our vision on using model-based approaches to design complex and flexible agricultural and environmental information systems. At the center of this modeling approach is the Unified Modeling Language that facilitates expressing visually concepts of a problem domain and their relationships. UML has a core of notations that are generic and that can be used to model problems in any domain but can be extended to create profiles in order to take into consideration modeling concerns in a particular problem domain. UML profiles are created to use UML in designing spatial systems, ontologies, model driven architecture-based systems and Web-based systems and a recent profile makes it possible to use UML for business modeling purposes. UML is used to present design patterns; their use is crucial in designing complex and flexible information systems. Recently, UML is enriched with Object Constraint Language that is used to express constraints on modeling artifacts. The paper presents the state of the art in modeling agricultural and environmental systems and provides discussions for future directions.

DOI: 10.4018/978-1-4666-0333-2.ch016

INTRODUCTION

During the past decades agriculture has experienced substantial structural and technological changes. Currently, agricultural production is organized around large farms that can afford applying new technologies. At the same time, social consciousness about environmental issues is very strong. Land management is a complex process and to manage it successfully it requires that special attention should be paid to environmental issues while designing and developing agricultural and environmental information systems.

On the one hand, the advent of the Internet has opened vast opportunities for communication and business opportunities within the agricultural and environmental communities. On the other hand, it has substantially increased the amount of information farmers and territory managers need to analyze and synthesize to successfully administrate all the multi-facets of the agricultural production and land management. This new complex environment requires new and complex approaches to carry on farm production and land governance. Complex systems are not easy to design and develop unless a rigorous methodology and the right tools are used. Therefore, it is imperative that modern software engineering approaches be used in agriculture and environment-related areas.

Since the earliest days of software engineering, many methodologies have been designed and used to create better software. An important event that significantly influenced the software engineering world was the use of visual modeling tools. Modeling is a well-known engineering discipline as it helps one to understand reality. Models of complex systems are built because it is difficult to understand such a system in its entirety. Models are needed to express the structure and the behavior of complex systems. Using models makes it possible to visualize and better control system's architecture (Papajorgji & Shatar, 2004).

In the early 90s there were several modeling languages used by the software engineering community. The most well-known methodologies were Booch, Jacobson, (Object-Oriented Software Engineering) and Rumbaugh's (Object Modeling Technique). Other important methods were Fusion (Coleman et al., 1994), Shllaer-Mellor (Shlaer & Mellor, 1988), and Coad-Yourdon (Coad & Yourdon, 1991). All these methods had different strengths and weaknesses. An important event occurred in the mid 1990s when Booch, Jacobson, and Rumbaugh began adopting ideas from each other that led to the creation of the Unified Modeling Language or as it is known best, the UML (OMG, 2009b). Details about UML will be presented.

They define a software development process as the set of activities needed to transform a user's requirements into a software system (see Figure 1).

The various "ingredients" needed in the process of developing computer applications are presented in Figure 2. This model is called the 4 Ps model (Jacobson, Booch, & Rumbaugh, 1999). This model shows that the *Result* of a *Project* is a *Product* which requires different types of ingredients:

- *People* (actors) in order to describe the analyzed domain and *People* to develop and manage the *Project* (analysts, designers, programmers, etc.).

Figure 1. Software development process (Jacobson, Booch, & Rumbaugh, 1999).

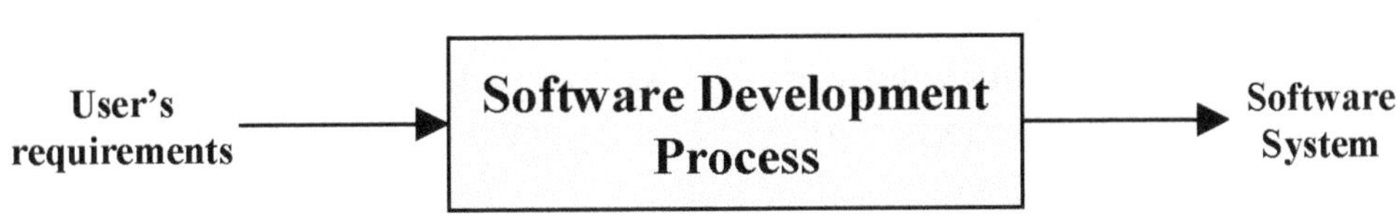

Figure 2. Computer application development (Jacobson, Booch, & Rumbaugh, 1999)

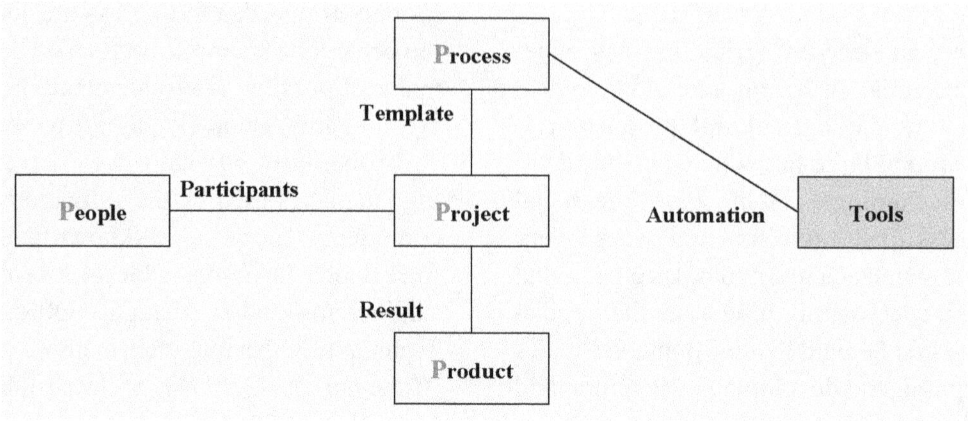

- A method (*Process*) to conduct the development following a framework (*Templates*) defining, organizing and explaining the successive steps of the development.
- *Tools* to facilitate the expression of the needs, the modeling, the project planning, the programming, and so forth.

Another important event that shaped the way software engineering was carried out is the design patterns movement. Design patterns are well-thought solutions for a large number of problems that have been built by experienced designers to be easily used by novice programmers (Papajorgji, 2009). A closer look into the design patterns will be presented.

UML diagrams often require additional information to precise the meaning of used modeling artifacts. The Object Constraint Language (OCL) is developed to add rigor to UML. We will describe, in detail, the OCL and its use as a modeling artifact in a later section.

Enterprise information systems must be developed to satisfy requirements of the business model. It is common that gaps are created between information systems and the business model. These gaps can have economic consequences and therefore must be eliminated. One way to approach this problem is to use the same formalism to model the business model and the corresponding information system. UML for business modeling is presented.

As agricultural and environmental processes occur in time and space, taking into account spatial considerations of the problem domain is imperative and modeling tools should provide the required support. A large number of modeling tools provide support for modeling spatially based problems. We will provide details about modeling spatially related problem domains in a later section.

One of the most important issues industry currently faces is the constant change caused by the change of underlying technologies (Frankel, 2003; Kleppe & Warmer, 2003). Often enterprises are forced to reproduce the same business model in a different and newer underlying technology. As this exercise is very costly, researchers are looking for other and efficient ways to insulate the intellectual investment from constant technological change. These efforts are the driving force for a new modeling approach referred to as the Model Driven Architecture (MDA). Details about MDA are presented.

Ontologies are used to better define the concepts and their relationships for a particular topic. Ontology development is a very recent phenomenon and a detailed view will be presented.

WHY UML?

Generalities

UML is a standard language for visualizing, specifying, constructing, and documenting object-oriented systems (Booch, Rumbaugh, & Jacobson, 1999). UML (OMG, 2009b) is based on a set of graphical symbols to abstract concepts and their relationships in a problem domain. Several types of diagrams are created to show different aspects of the problem (static, dynamic, organizational). Models created using UML are semantically richer than the ones expressed in any current object-oriented language and they can express syntax independently of any programming language. When a UML model is translated into a particular programming language, there is loss of information. A UML model is easy to read and interpret as it is expressed in plain English (or in any other language). Therefore, formal models raise the programming abstraction level above the 3GLs (Third Generation Language) in a profound way.

The event that confirmed UML as the modeling language in the industry is Microsoft's decision that Visual Studio 10 will feature UML support (Chairman Bill Gates told attendees at TechEd 2008 for Developers in Orlando, Florida).

Although UML is a general modeling language that can be used to model a large number of problem domains, as all modeling languages, its expressivity is limited as it is difficult to design a modeling language able to cover all problem domains. On the other hand the multitude of concepts used in some particular problem domains requires specific modeling artifacts. Thus, by design, UML has been empowered with an extensibility mechanism in order to adapt it for use in specific domains. The following section presents the UML extension mechanism.

Currently, there is a large number of vendors providing support for UML. Important is the number of open source UML tools as well. A number of vendors provide for free what is referred to as the "Community Version"; the users have a large selection of free and inexpensive tools to use. Therefore, the use of UML as a modeling language has considerably increased in the past 5 to 10 years.

Basic Extension Mechanisms of UML: Stereotypes, Tagged Values and Constraints

UML provides three modeling artifacts that can be used to extend the language: Stereotypes, Tagged Values and Constraints. Stereotypes are annotations associated to elements of diagrams; they are used to provide additional information to the annotated element. In general, stereotypes are used to introduce relevant categories or classifications of data of the annotated concept. Tagged values are also used to insert secondary categories of data. Often, tagged values are associated to stereotypes to refine the categories introduced by the stereotype. Constraints are seldom used to extend UML.

In UML, stereotypes are represented by an icon or text. Tagged value information can be completed by parameters. For example, in Objecteering (a UML tool, http://www.objecteering.com/) a tagged value can be used to indicate the persistence of a concept when this concept must be translated into a database table. This tagged value can be used by the designer in order to precise whether the concept is persistent or transient.

The use of stereotypes, tagged values and constraints increase the set of modeling business concepts available in UML and at the same time allows for expressing requirements from a particular problem domain.

Stereotypes, tagged values and constraints constitute business information that can be reused by introspection mechanism of the models to convert one model into another one. This introspection mechanism can be possible only in UML tools equipped with an internal programming language. This introspection mechanism is the key of Transformation notion of the MDA approach.

The Notion of Profile

The notion of profile was introduced for the first time in the version 1.3 ofUML, and finally became a standard in October 1999 (OMG, 1999). An important contribution to the definition of profile was given by a working group co-sponsored by the company Softeam (1999). Before October 1999, stereotypes, tagged values and constraints dedicated to a business domain were used as string-based extensions that could be attached to UML model elements in a flexible way (OMG, 2009a). To avoid the unorganized dissemination of the extension concepts into the UML metamodel, the profile notion was defined in 1999 to improve the structure and the definition of these concepts.

The first main service provided by a profile is to propose a predefined set of stereotypes, tagged values and constraints to create special UML modeling artifacts for a particular business domain. In other words, a profile is a tool that implements directly the UML extension mechanism to provide additional modeling services required on a particular domain. This tool can be considered like a library of the business functionalities used in a domain. This notion is very close to the plugging or add-on notion provided by several software.

Since the conception of the notion of profile, the OMG consortium has established a working program in order to define the profile specifications on certain strategic domains. The currently domains of which profiles have been standardized are the following (OMG, 2009a):

- Platform Independent Model (PIM) & Platform Specific Model (PSM) for Software Radio Components (also referred to as UML Profile for Software Radio)
- UML Profile for CORBA® and CORBA® Component Model (CCM) [This specification supersedes the separate profiles for CORBA® and the CORBA® Component Model.]
- UML Profile for Enterprise Application Integration (EAI)
- UML Profile for Enterprise Distributed Object Computing (EDOC)
- UML Profile for Modeling QoS and Fault Tolerance Characteristics and Mechanisms
- UML Profile for Schedulability, Performance and Time
- UML Profile for System on a Chip (SoC)
- UML Profile for Systems Engineering (SysML)
- UML Testing Profile

Note that the above list does not contain any profile for the domain of geomatics.

DESIGN PATTERNS

Generalities

Design patterns are one of the most important movement in the world of object-oriented design and development. The original idea of using a pattern language was introduced by an architect named Christopher Alexander (Alexander, 1977). Although in his book he describes the use of patterns in building architecture and urban planning, these ideas had a significant impact in the field of software engineering. According to Alexander, "a pattern describes a problem which occurs over and over again in our environment, and then describes the core of the solution to that problem, in such a way that you can use this solution a million of times over, without ever doing it the same way twice." The principal notion pointed by Christopher Alexander here is the repetitive nature of a problem which needs to apply the same solution all the time. In this context, this action becomes very quickly tedious and arduous.

Software patterns are reusable solutions to problems that occur often during the software development process. They are an elegant solution that makes it easier to reuse successful designs and architectures.

In mid 1990s, a group of four software engineers (Gamma, Helm, Johnson, & Vlissides, 1995) wrote the book titled *Design Patterns* that had a significant impact in the way software design was carried out. The book presents well-thought solutions for a large class of problems. The same way an architect uses prefabricated blocks for building complex constructions, a programmer will use patterns to develop complex software. The concept of design patterns allows novice programmers to use elegant solutions provided by experts. Using patterns makes the process of designing complex systems easier. Design patterns are expressed using UML class diagrams. UML provides all the necessary artifacts to express important information about the relationships between classes in a pattern. The use of design patterns allows a novice developer to create solutions to complex problems in relatively short period of time.

While design patterns are a software development norm in the industrial sector, their use in agricultural and environmental-related areas is fairly new. A serious effort was undertaken by (Papajorgji & Pardalos, 2006) to introduce complex examples of the use of design patterns in the crop simulation area. Although their book contains a good number of design patterns, more need to be done to present the use of recent developments in design patterns in agricultural and environmental-related areas. (Tambour et al., 2009) use design patterns to create flexible architectures of scientific plant models. For instance, the use of composite pattern offers a natural way of modeling plant topology at organ description level. In this architecture, each organ (attributes and behaviors) is individually described as a terminal class (e.g., leaf or fruit) or a composite class when the organ can bear other organs (e.g., branch or internode). This solution is suitable for all kind of plant topology without restriction (e.g., wheat, vine). The state pattern is also used to describe plant phonological stage. It allows to describe: 1) behavior which is noticeably different during the plant life cycle; 2) the conditions of transition between the phonological stages. In (Tambour et al., 2009), complex models are split into autonomous subcomponents or submodels (e.g., plant, disease, model, weather, or soil model). The use of interface pattern for each submodels allows to have a flexible architecture: it is possible to change a submodel by another submodel without changing the global architecture. It is useful when a scientist desires to change the use of a model by another one; for example, to change a coarse model of plant by a more detailed model. (Papajorgji, 2005) uses strategy design pattern to model a plug and play solution accessing database in a crop model simulation environment.

Miralles (Miralles, 2006; Miralles & Libourel, 2009b) have constructed spatial design patterns to create complex artifacts that facilitate modeling in the spatial-related domains. An example of such a spatial pattern is shown in Figure 3.

Conclusion

Design patterns are elegant solutions to complex problems and their use facilitates the construction of flexible software systems. The use of design patterns helps novice designers to use complex solutions prefabricated by experts. Their use not only facilitates the design process but also it considerably reduces the number of bugs in the system while transforming a model into code.

This conclusion is one of the results of a comparative study undertaken by (The Middleware Company, 2003). A team of analysts in charge of the comparing the development work of a traditional team and the development work of an MDA team, noted that the traditional team was obliged to correct bugs created during development while the MDA team did not have to deal with this issue because of the use of design patterns. Visibly, the correction of these bugs explains in part the 35% productivity benefit realized by the MDA team.

Figure 3. The 2D spatial design pattern

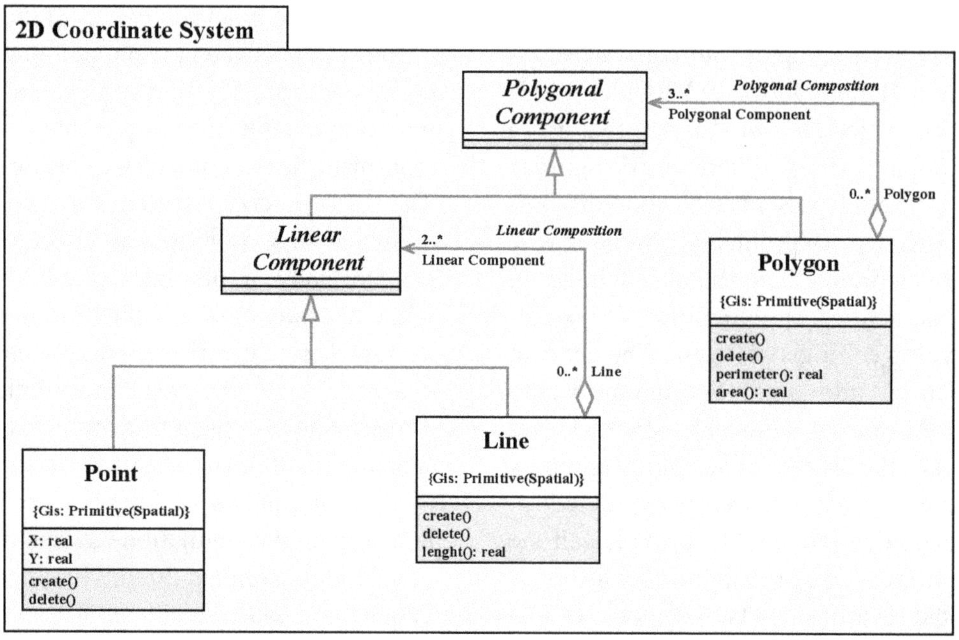

THE OBJECT CONSTRAINT LANGUAGE (OCL)

What is OCL?

According to the Object Management Group (OMG, 2007): "A UML diagram, such as a class diagram, is typically not refined enough to provide all the relevant aspects of a specification. There is, among other things, a need to describe additional constraints about the objects in the model. Such constraints are often described in natural language. Practice has shown that this will always result in ambiguities. To write unambiguous constraints, the so-called formal languages have been developed. The disadvantage of traditional formal languages is that they are designed to be used by people with a strong mathematical background, but are difficult to use by business or system modelers. OCL (Object Constraint Language) has been developed to fill this gap. It is a formal language and yet easy to read and write."

Currently, OCL is part of the UML2 standard (Schmid, Warmer, & Clark, 2002). OCL allows defining constraints on UML models. It is a textual language and integrates notations close to a spoken language to express constraints. Several concepts in OCL come from traditional object-oriented languages. OCL is used to specify invariants, that is, conditions that "must be true for all instances of a class at any time" (Schmid, Warmer, & Clark, 2002). OCL constraints can enrich the semantics of different UML diagrams; they improve the expressivity of the UML models. OCL is suitable for information system engineers. A growing number of system designers use this constraint language as a complement to their class diagram models.

OCL was first developed by a group of scientists at IBM around 1995 during a business modeling project. It was influenced by Syntropy, an object-oriented modeling language that relies to a large extent on mathematical concepts (Cook & Daniels, 1994). OCL is now part of the UML standard supported by the Object Management Group, and it plays an important role in the

Model Driven Architecture approach (Kleppe & Warmer, 2003).

OCL provides a platform-independent and generic method to model constraints. It can be interpreted by code engines/compilers to generate code automatically. Several tools are available in the market that allow producing integrity checking mechanisms in different languages (Java, C#, SQL, etc) from specifications of constraints expressed in OCL (Klasse Objecten, 2009). For instance, Dresden OCL Toolkit developed by the Technische Universität Dresden can generate SQL code from OCL constraints (Demuth, 2005; Demuth & Hussmann, 1999; Demuth, Hussmann, & Loecher, 2001; Demuth, Loecher, & Zschaler, 2004). The produced code can be used to check the integrity of a database before storing new data.

Examples of using OCL in Agriculture

While the use of OCL in the industrial sector is widely accepted (Klasse Objecten, 2009; TU Dresden University, 2009), its use in agriculture and environment is at its beginnings.

Papajorgji and Clark (2005) provide examples of using OCL constraints in a context of crop simulation systems (see Figure 4). The constraints are used to express the fact that a plant has a valid planting date and to calculate the phonological phase of the plant.

A spatial extension of OCL (referred to as Spatial OCL) has been proposed (Pinet, 2002; Pinet, Kang, & Vigier, 2005) to express spatial constraints in agricultural information systems. Spatial constraints model possible consistent spatial configurations of objects representing agricultural/environmental concepts. As an example, let us consider an information system that stores data in a database about forests and lakes. In this case, an example of spatial constraints could be: "a forest cannot be inside a lake." The representation of this constraint in Spatial OCL is:

Forest.allInstances() -> forAll (F | (1)

Lake.allInstances() -> forAll (L | (2)

(F.geometry) .disjoint (L.geometry))) (3)

Figure 4. Example of OCL constraints in crop simulation models

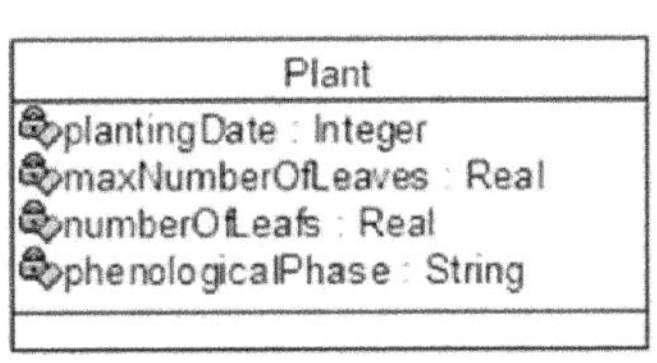


```
Plant
self.plantingDate>=0

Plant
phenologicalPhase =
(
if numberOfLeaves >
 maxNumberOfLeaves then
'reproductive'
)
```

The meaning of this constraint is:

For all instances F of the class Forest (1)

and for all instances L of the class Lake, (2)

the geometries of F and L should be disconnected (3)

Note that the shapes of forests and lakes are stored in attributes referred to as "geometry".

At the implementation level, the control of this constraint allows identifying inconsistent data or forbidding the insertion of data that do not verify it. For example, Dresden OCL Toolkit can produce automatically SQL queries from OCL constraints (Demuth, 2005; Demuth & Hussmann, 1999; Demuth, Hussmann, & Loecher, 2001; Demuth, Loecher, & Zschaler, 2004; Pinet et al., 2009). The generated queries can be used to check if the data to be inserted in the database verifie the constraints.

Spatial OCL has been employed to model the spatial constraints of a French agricultural information system for the traceability of fertilization activities (Bejaoui, Pinet, Schneider & Bedard, 2008; Pinet et al., 2009; Pinet, Duboisset, & Soulignac, 2007). Currently, this system records various plans for spreading organic matter on agricultural land in France. The specification and the control of constraints enabled check the quality of the data. Another application of Spatial OCL can be found in (Kang, Pinet, Schneider, Chanet, & Vigier, 2004). This work models communication constraints of an agricultural vehicles network.

The example of use proposed in Hasenohr and Pinet (2006) and Hasenohr, Pinet, Kay, Grenier, and Sevila (2005) concerns the European Common Agricultural Policy (CAP). CAP has been of very high importance to farmers for decades. The ongoing reform will have a major impact on the way European farmers must manage their farms. Until now, farmers received production based subsidies as a function of the sown area and the type of crop. From now on, a farmer will receive subsidies, not based on the amount produced, but for producing whilst respecting good agri-environmental conditions and using good farming practices. An example of good agri-environmental conditions cited in (Hasenohr & Pinet, 2006) is: "Finely tilled bare (unsown) seedbeds are not permitted over the winter." Furthermore European farmers will have the opportunity to subscribe to Farm Advisory Services (FAS) which will help them to comply with the applicable legislation and to manage their farm in an environmental friendly way while improving their farm sustainability. (Hasenohr & Pinet, 2006) presents how to design the farm advisory services systems with UML. A template (described by a class diagram) models different concepts of European CAP, for example, Farm, Crop, Soil Properties. The template can be adapted and extended to a national or regional case. It makes extensive use of OCL. Agri-environmental conditions are modeled by (Spatial) OCL constraints. In this case, OCL could ease the implementation of the modeled class diagram and it allows for a standardized representation of agri-environmental conditions.

Conclusion and Future Trends

OCL is currently used to model constraints. Several tools are now available (Klasse Objecten, 2009; Toval, Requena, & Fernández, 2003) and some of them allow to generate code. (Cabot & Teniente, 2006) present a recent comparison of tools that include code generators (for SQL and Java); this study describes the links between MDA and OCL.

Designers of agricultural/environmental systems mainly model spatial constraints with an extended OCL (see previous section). This shows that OCL can be easily adapted to a specific application domain (e.g., geographic information systems, Web services, mobile computing); it is possible to develop OCL extensions in order to create Domain Specific Languages (DSL). Unlike general-purpose languages (such as OCL or UML), DSL are languages designed for a specific kind of task (e.g., Spatial OCL or UML profiles) (Chiorean, Demuth, Gogolla, & Warmer, 2006). Syntax and semantic of OCL can be easily enriched in order to extend the language to address modeling issues in a specific domain.

OCL can be also employed to model queries on instances of UML class diagrams. This use should be generalized in the future. Several works study the use of OCL as query language (Akehurst & Bordbar, 2001; Mandel & Cengarle, 1999; Richardson, 2002). In this approach, the evaluation of an OCL expression on a dataset can return a set of instances. While the constraint presented in the previous section returns a Boolean value, its new version returns a set of instances:

Forest.allInstances() -> select (F | (1)

Lake.allInstances() -> forAll (L | (2)

(F.geometry) .disjoint (L.geometry))) (3)

The meaning of this query is:

Select all instances F of the class Forest (1)

such as for all instances L of the class Lake, (2)

the geometries of F and L is disconnected (3)

This approach provides the possibility to query directly class diagrams independently of the language of the implementation level (e.g., SQL, Java). Bold for Delphi enables users executing an OCL query over objects stored in main memory or in a database (Richardson, 2002). This tool supports several OCL operations.

Other researchers compare the expressivity of OCL with the ones of traditional database query language (Akehurst & Bordbar, 2001; Mandel & Cengarle, 1999).

UML FOR BUSINESS MODELING

Enterprise information systems should provide strong support to the enterprise business model and should be developed around the later. In the event there are discrepancies between the enterprise business model and the corresponding information systems it simply means that the information system cannot provide all the necessary functionalities required by the business model. If these discrepancies persist for long time, then the enterprise would suffer and the economic consequences may be severe. Therefore it is crucial for enterprise to have an information system that backs 100% the corresponding business model.

One way to assure a complete compliance between the enterprise business model and the corresponding information system is to use the same modeling formalism. The use of UML for modeling information systems is already a well-established fact and UML is the number one modeling language as presented previously. Using UML for business modeling is more recent phenomenon. Eriksson and Penker (Eriksson & Penker, 2000) present a UML extension for business modeling. They rightly argue that there are numerous advantages in using UML (or object-oriented concepts) for business modeling. A business can be described in terms of processes that need to achieve certain goals by collaborating with different types of resources. Sets of rules can define conditions and constraints as to how the processes and resources may relate to each other and how they may behave (Eriksson & Penker, 2000). Rules can be described using OCL. Furthermore, UML is a well-established language for visual modeling and it is extendible to adjust for particular notations in modeling particular areas of activity.

Business modeling could be created using other notations such as IDF0 (Abt, Vigier, & Schneider, 2009) but UML is a well-defined language and supported by a number of tools. Version 1.1 of UML for business modeling was prepared by a number of the most important players in the software industry such as Rational Software (http://ibm.com/rational), Microsoft (http://microsoft.com), Oracle (http://oracle.com), IBM (http://ibm.com) and Softeam (http://softeam.com) to name a few.

The use of UML for business modeling in agricultural/environmental related fields is very new. To our best knowledge only one publication (Ràbovà, 2005) discusses business modeling issues with UML. More should be done in this direction as we believe that using UML for developing information systems and for business modeling purposes will largely facilitate the implementation large enterprise systems in agriculture and environment.

UML FOR SPATIAL MODELING

In this section, we provide an example of a recent profile dedicated to the geomatics application domain. It is an interesting profile in the context of agricultural and environmental systems modeling; environmental events occur in time and space, and often data used by agricultural systems are geo-referenced.

Miralles (Miralles, 2006) has implemented into a UML profile the pictogramic language created at the Industrial Research Chair in Geospatial Databases for Decision Support. This language is widely described in (Bédard, 1999; Bédard, Larrivée, Proulx, & Nadeau, 2004; Miralles, 2006). The initial language was extended with the corresponding fuzzy concepts to indicate the spatial and temporal properties (Miralles, 2006). The SpatioTemporal Geographic Language Profile (STGL Profile) implements a variant of the pictogramic language.

The most important spatial and temporal concepts of SGTL that are used during the development of a Geographical Information System are Point, Line & Polygon and Time Point & Time Period. These five primitive concepts have been largely studied by researchers or standardization groups and are included in all of spatial and temporal languages currently in use (Filho, Sodré, Daltio, Rodrigues Júnior, & Vilela, 2004; Gayte, Libourel, Cheylan, & Lardon, 1997; ISO, 2002, 2003; Lbath & Pinet, 2000; OGC, 2005; Pantazis & Donnay, 1996; Parent, Spaccapietra, & Zimányi, 2006; Pinet, 2002; Tryfona & Jensen, 2000).

Like UML, STG language is fundamentally visual. The spatial and temporal primitives are implemented in the STGL Profile via UML stereotypes. Figure 5 shows, for each of the five primitives, the name of the stereotype and the associated pictogram.

In a spatial domain, there is important information that must be taken into account: it is the dimension of the space. A set of three Tagged values has been created in the STGL Profile to specify whether a business concept belongs to a space in which the coordinate system is 1D, 2D and/or 3D.

Figure 5. Table of the primitive concepts in STGL

	Primitive concept	Stereotype name	Associated pictogram
Spatial	Point	<<Point>>	
	Line	<<Line>>	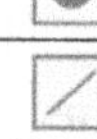
	Polygon	<<Polygon>>	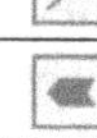
Temporal	Time Point	<<Time Point>>	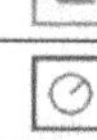
	Time Period	<<Time Period>>	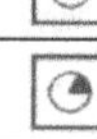

The names of these three Tagged values are: {In 1D Coordinate System}, {In 2D Coordinate System} and {In 3D Coordinate System}.

Using the STG language it is very easy to specify whether a business concept has spatial and/or temporal properties. In Figure 6, the business concept Parcel, often used in agricultural and in environmental field, has a polygonal geometry in a 2D space. Thus, a UML model enriched with pictograms is easy to read and allows identifying rapidly the business concepts.

The stereotypes and tagged values of the STGL Profile represent a source of information that can be used to during the model evolution. For example, an important evolution between analysis model and design model is the pictogram conversion into UML concepts. For this purpose, it is necessary to reify pictograms into UML concepts usable by the programming language generators. In the STGL Profile, two transformations are used to perform this evolution. The first one, referred to as the GIS Design Patterns transformation, enriches the design model with GIS design patterns (see section 0). This transformation is in charge of the reification of the primitive concepts into classes. The second one, referred to as the Pictogramic translation transformation, has the responsibility to convert the pictograms into relation between annotated business concepts and primitive classes. These two GIS transformation are described in detail by (Miralles, 2006; Miralles & Libourel, 2008, 2009a, 2009b).

THE MODEL DRIVEN ARCHITECTURE (MDA) APPROACH

The Model Driven Architecture (MDA) is a recent initiative of OMG (http://omg.org) aiming to make software development process independent from changes of the implementations technologies. MDA provides an open, vendor-neutral approach to the challenge of business of technology change (OMG 2009b). At the center of this approach are models; the software development process is driven by constructing models representing the software under development (Frankel, 2003; Kleppe & Warmer, 2003; Papajorgji, 2009). The MDA approach separates the business logic from the technicalities of the implementation technologies. It is a well-known fact that implementation technologies change often; almost every 7-10 years there is a new implementing technology and enterprises are often obliged to rewrite the same business model in a new underlying technology (Pastor et al., 2007).

MDA uses UML for to construct visual models. By its nature, UML allows for developing platform independent models (Papajorgji & Pardalos, 2006; Papajorgji & Shatar, 2004). These models present concepts from the application domain provided with data, behavior and relationships. Models expressed in UML are semantically richer than any other programming language and can be translated in any particular programming language and implementation environment.

Figure 6. Example of business concept annotated with Polygonal icon

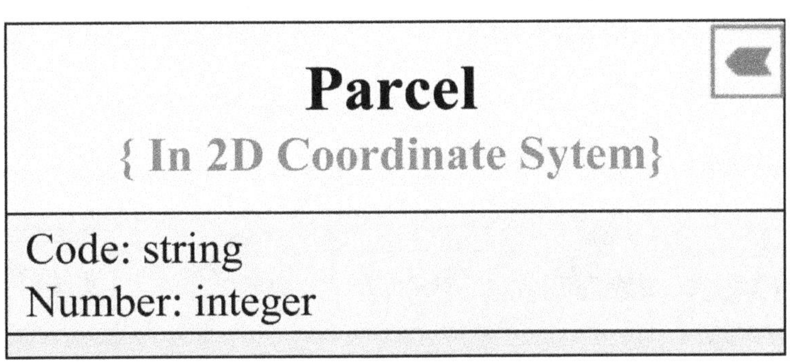

The MDA approach uses three levels of models as shown in Figure 7. The approach starts with a conceptual model expressing the depicted concepts from the problem domain and their relationships. This model is a Platform Independent Model (PIM). The advantage of a PIM is that allows for a larger participation of business specialists in model construction. During this phase, no technical details about the implementation environment and the underlying technologies are considered.

After all business issues related to the model construction are detailed and well-defined, then considerations about the implementation environment can be taken into account. The PIM enriched with details of the underlying technologies is referred to as a Platform Specific Model (PSM). Note that the same PIM could be transformed into several PSMs, considering the different implementation platforms that could be used. Thus, the same PIM could be implemented in .NET or Java environment.

Modeling Behavior

One of the most interesting features of the MDA approach is the ability to model behavior. The behavior of an object is provided by the set of methods defined in the object. The problem of generating code automatically is not new. Currently, many Integrated Development Environments (IDE) provide drag and drop capabilities (Papajorgji et. al., 2009). Although these capabilities have facilitated code generation, they have not solved the problem as only code for class skeleton is created; the user needs to manually add the code that implements the behavior. The issue of modeling behavior is approached by the researcher community mainly in two different ways.

The first approach in modeling behavior has been proposed by a group IBM engineers in Europe that proposed the Object Constraint Language (OCL) presented in section 0. This effort is largely inspired by the work of Meyer (Meyer, 1988), the creator of the "Design by Contract" approach. This approach is referred to as the elaborationist approach (McNeile, 2003). Although the amount of code that can be generated using OCL is significant, it has not totally addressed the issue of code generation (McNeile, 2003). Even the advocates of this approach (Kleppe & Warmer, 2003) do not pretend to generate the required code in all cases.

Another approach to modeling behavior is provided by Mellor (Mellor & Balcer, 2002; Starr, 2002) and it is referred to as the translationist approach (McNeile, 2003). According to this approach, the modeler defines the behavior of objects using a text-based action language (McNeile, 2003)(McNeile & Simmons, 2002). The main characteristics of this language are:

- The action language allows modelers to define behavioral specification at a higher level of abstraction.
- The language is independent of any specific implementation technology in the execution environment.

There are a number of vendors that have embraced the translationist approach such as Kennedy carter (http://www.kc.com), Accelerated Technology (http://acceleratedtechnology.com), Virtual Enterprise (http://intelliun.com) to name a few.

Figure 7. Process example of transformations applied to a model to obtain the next level model

MDA and Its Use in Agriculture and Environment

MDA has made considerable advances in the industrial sector while very timid steps in the agricultural and environmental-related areas. One of the initial efforts to use the MDA approach in crop simulation area is described in (Papajorgji et al., 2009). The authors use Oliva Nova (Pastor et al., 2007) to implement a full crop simulation model as shown in Figure 8. Another environmental example is presented in (Muzy et al., 2005). It concerns the use of MDA for a model of propagation processes applied to fire spread.

Transformation in Spatial Modeling

The spatial and temporal stereotype supporting the spatial and temporal primitive and the dimensional tagged value constitute relevant geomatics information that needs to be taken into consideration. For this purpose, the way adopted in STGL Profile (Miralles, 2006) is, in first step, to enrich the model with GIS Design patterns which reify the spatial and temporal primitives and, in a second step to convert the geomatics information annotating a business concept in a relation. The transformation creating the GIS Design patterns reifies the spatial and temporal primitives via the UML Class concept. The Pictogramic Translation Transformation is described as follows.

In the STGL Profile, the designer can configure the profile in order to convert geomatics information to association relationship or to generalization/specialization relationship (Miralles, 2006). Only the first conversion is depicted. In this case, when the designer annotates the business concept with the pictogram, two others tagged values are added by default to the concept: the first specifies the role of the primitive and second its cardinality. The default values of these tagged values can be subsequently modified by the designer.

Once the GIS Design patterns created, the business concept and the spatial or temporal primitives

Figure 8. Example of a MDA-based crop simulation model

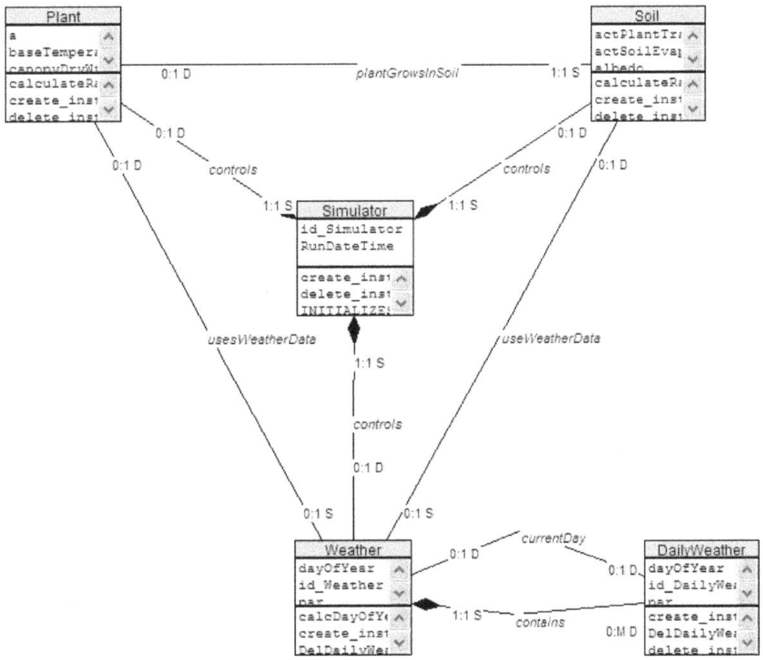

represented by the pictogram are totally disjoints (see Figure 9). In the example of this figure, the goal of the Pictogram Translation Transformation is to establish automatically a relationship between the business concept (Parcel) and the class reifying the primitive Polygon. This transformation creates an association, called Spatial Characteristic. The pictogram being a Polygon, the business concept is automatically annotated by default with the role tagged value {Spatial Role(Geometry)} and the cardinality tagged value {Spatial Cardinality(1)}. During the transformation, the role Geometry and cardinality 1 are assigned to the association end attached to the primitive Polygon. In this transformation, the name of the business concept is allocated to its role, Parcel in this example, and its cardinality's value is 0..1. Once the association has been created, the stereotype/pictogram and the three tagged values (dimension, role and cardinality) are deleted since this information becomes redundant.

Conclusion

One of the main benefits of the MDA paradigm is the productivity gain. It was the assertion of the Object Management Group to promote its new paradigm. To prove this assertion, the Compuware Corporation required to The Middleware Company to perform a productivity analysis of a software development (The Middleware Company, 2003). The same development was entrusted to two teams: a team working in a traditional way and another team applying the MDA paradigm and particularly the use of code generation tools. The result of this analysis of productivity was that MDA team developed its software more quickly than the traditional team with an advantage of 35% of devoted time. Only 330 hours were necessary to the MDA team to develop the software, time to compare with the 507 hours for the traditional team.

Figure 9. Rule for transforming the stereotype/pictogram couple into an association

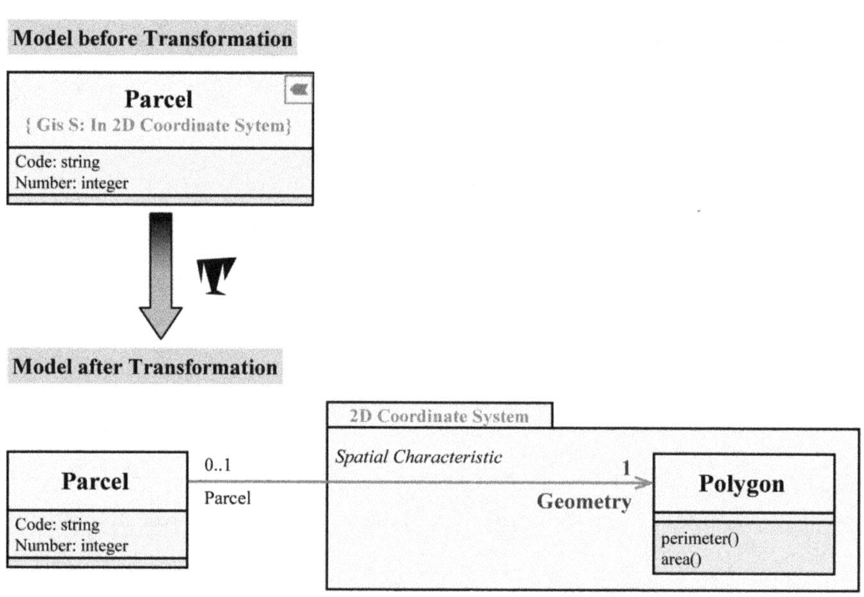

ONTOLOGY

Ontology: Definition

Ontology development is a very recent phenomenon. There are several definitions of the concept of ontology. An interesting definition of ontologies was proposed by Gruber as "ontology is a formal explicit specification of a shared conceptualization" (1993a). According to Gruber, conceptualization refers to an abstract model of phenomena in the world after identifying the relevant concepts of these phenomena. Explicit means that the type of concepts used, and the constraints on their use are explicitly defined. Formal refers to the fact that the ontology should be machine readable. Shared reflects the idea that ontology should capture consensual knowledge accepted by the communities (Gruber, 1993b).

Hendler defines ontology as "a set of knowledge terms, including the vocabulary, the semantic interconnections, and some simple rules of inference and logic for some particular topic" (2001). According to this definition an ontology specifies the meaning of relations between the concepts used and should enable some sort of reasoning.

An ontology is a model that represents a domain of knowledge. In practice, it can be used to formalize complex knowledge of Web resources, data standards, concepts dictionaries, and so on. Ontologies can take various forms (diagrams, formal models, thesauruses, etc.), and their use can be also diverse, but an ontology usually contain:

- A vocabulary of terms,
- A set of term definitions which identify concepts and fix the term interpretation (in the domain of knowledge considered),
- A representation of relationships between concepts,
- An agreement of a community of ontology users about term definitions and the ontology structure.

As a result of development of ontologies and Web-related technologies emerged the first ideas about the Semantic Web (Berners-Lee, Hendler, & Lassila, 2001). Researchers and practitioners of ontology development and application still face many challenges in their efforts toward the Semantic Web. One of the central problems that need to be addressed sooner rather than later is the exploration of new software engineering tools and methods for ontology development (Gasevic, Djuric, & Devedzic, 2006).

UML and Ontologies

The use of UML for ontology construction has only recently been present in specialized literature. As the focus of ontologies is to represent the meaning of relations between concepts used in a domain (Hendler, 2001), it is natural to use a language such as UML for ontology development. Cranefield (2001a) did the pioneering work to propose the use of UML for ontology construction. He analyzed standard concepts of UML and ontologies and presented analogies and differences. The following are the reasons for using UML for ontology development (Cranefield, 2001b):

- UML class diagrams provide a static modeling capability that is suitable for representing ontologies, that is, modeling the taxonomy of ontology's concepts and their relations.
- UML object diagrams can be interpreted as declarative representations of knowledge; therefore, they can be used for modeling instances of an ontology.
- OCL can be used for specifying constraints of an ontology.
- The same paradigm can be used for modeling both ontologies and knowledge.

After Cranefield, a number of researchers (e.g., Baclawski, Kokar, Kogut, Hart, Smith, Holmes, et al., 2002; Baclawski, Kokar, Kogut, Hart, Smith,

Letkowski, et al., 2002; Djuric, & Devedzic, 2005; Falkovych, Sabou, & Stuckenschmidt, 2003; Gasevic, Djuric, & Devedzic, 2006; Philippe, 2003; Schreiber, 2005) further explored the advantages and disadvantages of using UML as an ontology language and other software engineering initiatives such as MDA and its standards Meta-Object Facility (MOF) and XML Metadata Interchange (XMI). They acknowledged the benefits of using a standard modeling tool such as UML in ontology construction.

One main advantage of UML is that is widely used. Several tools are available for UML so designers can use them for describing their diagrams. UML is an open standard, and it is taught in many Universities. It allows expressing more complex ontology than traditional thesaurus. Users and developers are likely to be familiar with UML notations than formal Ontology-based languages such as Knowledge Interchange Format (KIF) (National Committee for Information Technology Standards, 1998), Description Logics (Donini, Lenzerini, Nardi, & Schaerf, 1996; Nardi & Brachman, 2002) or Web Ontology Language (OWL) (W3C, 2004).

There are several common features between UML and ontology-based languages (IBM, 2006). Several studies compare UML and formal languages for ontologies modeling (Atkinson, 2006; Cranefield & Purvis, 1999; IBM, 2006; Philippe, 2003; Schreiber, 2005). In an OMG recommendation (IBM, 2006), mappings among the meta-models of OWL and UML are proposed; OWL is usually considered as the flagship formal language for ontologies (Atkinson, 2006; Drummond & Shearer, 2006). Several common features between UML and OWL are shown. Table 1 presents the mapping between concepts of UML class diagrams and OWL.

The main drawback to UML is its lack of formal semantics. The formal semantics of traditional ontology-based language allows providing capabilities to develop a consistent set of concepts. These languages support some elements of reasoning and inference mechanisms to deduce new knowledge from a model. This is not the case with the standard UML.

Table 1. OWL and UML: the common features (see more details in IBM, 2006).

UML	OWL
class	class
subclass	subclass
instance	individual
attribute, association	property
mutliplicity	minCardinality, maxCardinality, cardinality

The official OMG documents defining the semantics of UML contain an informal description in English. Some researchers propose a mathematical model for UML (see Breu, Grosu, Huber, Rumpe, & Schwerin, 1997)). The work of Guizzardi, Falbo, and Pereira Filho (2002), Guizzardi, Wagner, Guarino, and Sinderen (2004) and (Djuric, & Devedzic, 2005; Gasevic, Djuric, & Devedzic, 2006) concerns the development of different methodological tools (UML profiles, design patterns, MDA) in order to build an ontology using UML formally and correctly. Cranefiel and Purvis develop an ontology using UML and Object Constraint Language (OCL); the concepts are described by UML classes and constraints of concepts are described in OCL (Cranefield & Purvis, 1999).

Because UML is widely used and supported by numerous tools, it could be viewed as a good candidate to develop ontologies. Nevertheless, UML is a semi-formal formalism and will not be sufficient to represent all the details required by complex reasoning processes (Cranefield, 2001b). OCL is sometimes considered as "a variant of [first order predicate logic] tuned for writing constraints on object structures" (Clark, Evans, Kent, Brodsky, & Cook, 2000), and might overcome this limitation in the future; in the OMG specification of OCL (OMG, 2007), an annex presents a first version of a formal semantic. Currently, this annex doesn't describe how to deduce new knowledge, or compute the logical correctness of OCL constraints.

As presented in this section, there is a wide consensus on advantages in using UML for ontology development. During our search in the literature, we found only a solitary voice (Beck et al., 2009) that states that "Whereas there are some notions in ontologies that are common to UML, mainly because they are both object design languages, it is a mistake to claim that UML can be used to model ontologies, or that ontologies can be used to design software in a way that is similar to UML. UML and ontologies are aimed at different purposes. UML does not support reasoning about concepts, and ontologies do not support software design." The paradox is that, this author seems to use a modified version of ArgoUML metamodel to develop his ontology editor (besides the visual similarities between the ontology editor and ArgoUML, error messages are the same as the ones delivered by ArgoUML).

Ontologies and their Application in Agriculture and Environment

While numerous ontologies have been proposed in different application domains, their current use in agriculture remains limited.

The well-known AGROVOC thesaurus is employed to develop the Agricultural Ontology Service (AOS) project (AGROVOC, 2007). AGROVOC was developed by the FAO and the Commission of the European Communities, in the early 1980s. It is a multilingual thesaurus concerning forestry, fisheries, food, environment and related domains. As presented in (AGROVOC, 2007), "it consists of words or expressions (terms), in different languages and organized in relationships (e.g., 'broader', 'narrower', and 'related'), used to identify or search resources". For example, the term "pollution" is associated with the term "pollutants" by the relationship "Related Terms" in the AGROVOC thesaurus. This means that the term "pollution" is related to the term "pollutants."

The French standard proposed by the project "Farm Information Management" (FIM - GIEA in French) (Pinet, Roussey, Brun, & Vigier, 2009; Pinet et al., 2006) has been formalized by an ontology. The proposed standard describes a large number of concepts related to farms. The final goal of the standard is to provide more complete data interchange formats in order to facilitate and to improve interoperability between information systems of the French Farms. The first step of the project was to carry out an inventory of the various previous standardization initiatives. Then, different terms, concepts and their relationships have been identified for different main fields of farm activities (land management, agricultural infrastructures and buildings, etc.) An important part of the FIM project consists in integrating and enhancing the definition of concepts and work on standardization already initiated by the various partners. All the members of the project can propose new concepts to the developed ontology. Data interchange formats are proposed on the basis of the vocabularies and the concepts of the ontology. The ontology is represented by UML class diagrams. Figure 10 summarizes the approach.

Perspectives: Toward Formal Ontologies with UML

An interesting perspective is to translate the ontology described in UML into a formal ontology (RDF, OWL, etc.) in order to offer new possibilities (e.g., produce reasoning). In this case, one solution consists in translating the UML diagrams into a formal ontology language, and then enriching the produced formal ontology. For example, it is possible to convert UML specifications into a formal ontology with Protégé and its UML Storage Backend Plug-In (Standford University, 2009). Then, additional specifications can be integrated into the ontology such as creation of individuals, modeling of inferences, and so on. Recently, a new product referred to as UML-OWL GEN (UML-OWL Gen, 2009), converts any Eclipse-based UML class diagram to OWL-DL ontology.

Figure 10. From standard to data interchange format

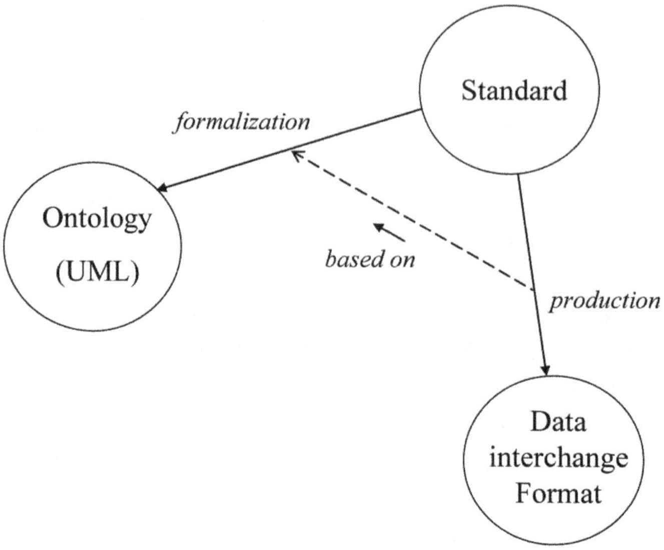

(Pinet et al., 2006) implement this approach to develop an ontology in agriculture. They propose to start modeling an ontology with a UML class diagram. Later, the UML specification is translated into OWL with Protégé and its UML Storage Backend Plug-In; see (Standford University, 2009). Then, inferences are defined with Description Logic in order to deduce new knowledge. A DIG reasoner is used to produce new information. This type of reasoners supports the standard interface defined by the Description logic Implementation Group (DIG - http://dl.kr.org/dig/). This standard provides a specification for connecting DL reasoners and different ontology modeling tools such as Protégé. A DIG reasoner provides an access interface which enables the reasoner to be accessed over HTTP.

CONCLUSION

As agricultural and environmental information systems are becoming more complex and difficult to design and develop, it is imperative to address this challenge with a multi-disciplinary team effort and with a rigorous methodology and the right tools. Model-based tools can be very useful to design, implement, and deploy complex and flexible information systems. Central to the presented model-based approach is the Unified Modeling Language (UML). Currently, UML is *the modeling language* as it is supported by all major players in the software industry. UML contains a core of notations that could be used for modeling any problem in different areas, but can be extended to take into consideration modeling specificities of a particular problem domain. Therefore, several profiles are created that make UML a general modeling tool to be used in distributed systems, spatial-related problems, and so forth. A UML profile for business modeling allows for using the same modeling formalism to design and develop enterprise information systems and enterprise business modeling; thus, creating a homogeneous modeling environment to reduce the discrepancies between enterprise information systems and enterprise business modeling.

Object Constraint Language (OCL), now part of UML, is a powerful tool to model constraints on attributes and behavior of objects. OCL added precision to UML diagrams and it is available to modelers to add the rigor to UML models that diagrams by themselves cannot achieve.

The advent of the Model Driven Architecture (MDA) approach increases the level of abstraction and makes modeling an activity that can be performed by specialist of the domain. MDA frees the software development process from the underlying technologies thus, creating a better environment for insulating intellectual investment from changes of the implementation technologies. The model-driven approach is a solid modeling methodology allowing for developing better, complex and flexible agricultural and environmental information systems.

As more complex methodologies and tools are available for use, the issue of preparing the researcher community to be able to efficiently use them, must be addressed sooner rather than later. One way to address this problem is to update the curriculum in Universities. Another way is to organize short courses tailored to specific needs. The Research Network "Design of Environmental and Agricultural Systems" (http://deas.research.free.fr) has taken the lead in organizing several short courses to large audiences of specialists in agriculture and environment.

REFERENCES

W3C. (2004). *OWL Web ontology language: Overview, W3C recommendation 2004-2-10.* Retrieved from http://www.w3.org/TR/owl-features/

Abt, V., Vigier, F., & Schneider, M. (2009). Enterprise business modelling languages applied to farm enterprise: A case study for IDEF0, GRAI Grid, and AMS languages. In P. Papajorgji & P. M. Pardalos (Eds.), *Advances in modeling agricultural systems* (pp. 167-192). New York: Springer.

AGROVOC. (2007). *A multilingual agricultural thesaurus.* Retrieved from ftp://ftp.fao.org/gi/gil/gilws/aims/references/flyers/agrovoc_en.pdf.

Akehurst, D., & Bordbar, B. (2001). On querying UML data models with OCL. In *«UML» 2001 — The Unified Modeling Language. Modeling Languages, Concepts, and Tools* (pp. 91-103).

Alexander, C. A. (1977). *Pattern language: Towns, buildings, constructions.* New York: Oxford University Press.

Atkinson, C. (2006, October 16). *Models versus Ontologies - What's the Difference and where does it Matter?* Paper prestented at VORTE2006, Regal Kowloon Hotel, Hong Kong.

Baclawski, K., Kokar, M., Kogut, P., Hart, L., Smith, J., Holmes, W., et al. (2002). *Extending UML to support ontology engineering for the Semantic Web.* Paper presented at the Fourth International Conference on the Unified Modeling Language.

Baclawski, K., Kokar, M., Kogut, P., Hart, L., Smith, J., & Letkowski, J. (2002). Extending the unified modeling language for ontology development. *Software and System Modeling, 1*(2), 142–156. doi:10.1007/s10270-002-0008-4

Beck, H., Morgan, K., Jung, Y., Wu, J., Grunwald, S., & Kwon, H.-y. (2009). Ontology-based simulation applied to soil, water, and nutrient management. In P. J. Papajorgji & P. M. Pardalos (Eds.), *Advances in modeling agricultural systems* (pp. 1-34). New York: Springer.

Bédard, Y. (1999). Visual modelling of spatial databases: Towards spatial PVL and UML. *Geomatica, 53*(2), 169–186.

Bédard, Y., Larrivée, S., Proulx, M.-J., & Nadeau, M. (2004, November). *Modeling Geospatial Databases with Plug-ins for Visual Languages: A Pragmatic Approach and the Impacts of 16 Years of Research and Experimentations on Perceptory.* Paper presented at the ER Workshops 2004 CoMoGIS, Shanghai, China.

Bejaoui, L., Pinet, F., Schneider, M., & Bedard, Y. (2008). An adverbial approach for the formal specification of topological constraints involving regions with broad boundaries. *Lecture Notes in Computer Science*, *5231*, 383–396. doi:10.1007/978-3-540-87877-3_28

Berners-Lee, T., Hendler, J., & Lassila, O. (2001). The Semantic Web. *Scientific American*, *284*(5), 34–43.

Booch, G., Rumbaugh, J., & Jacobson, I. (1999). *The unified modeling language user guide*. Upper Saddle River, NJ: Addison-Wesley.

Breu, R., Grosu, R., Huber, F., Rumpe, B., & Schwerin, W. (1997). *Towards a precise semantics for object-oriented modeling techniques. In Object-Oriented Technology, ECOOP '97 Workshop.*

Cabot, J., & Teniente, E. (2006). Constraint support in MDA tools: A survey. In D. Akehurst, R. Vogel, R. Paige (Eds.), *Model driven architecture – foundations and applications* (pp. 256-267). New York: Springer.

Chiorean, D., Demuth, B., Gogolla, M., & Warmer, J. (2006). In *Proceedings of the Sixth OCL Workshop OCL for (Meta-)Models in Multiple Application Domains (OCLApps 2006).*

Clark, T., Evans, A., Kent, S., Brodsky, S., & Cook, S. (2000). *A feasibility study in rearchitecting UML as a family of languages using a precise OO meta-modeling approach.* Retrieved from http://www.cs.york.ac.uk/puml/mmf/

Coad, P., & Yourdon, E. (1991). *Object-oriented analysis*. Englewood Cliffs, NJ: Prentice Hall.

Coleman, D., Arnold, P., Bodoff, S., Dollin, C., Gilchrist, H., Hayes, F., et al. (1994). *Object-oriented development - the fusion method.* Englewood Cliffs, NJ: Prentice-Hall.

Cook, S., & Daniels, J. (1994). *Designing object systems-object oriented modeling with Syntropy*. Englewood Cliffs, NJ: Prentice-Hall.

Cranefield, S. (2001a). Networked knowledge representation and exchange using UML and RDF. *Journal of Digital Information, 1*(8).

Cranefield, S. (2001b). *UML and the Semantic Web.* Paper presented at the International Semantic Web Working *Symposium*, Palo Alto, CA.

Cranefield, S., & Purvis, M. *(1999).* UML as an ontology modelling language. *In* Proceedings of the Workshop on Intelligent Information Integration,16th International Joint Conference on Artificial Intelligence (IJCAI-99).

Demuth, B. (2005). *The dresden OCL toolkit and the business rules approach.* Paper presented at the European Business Rules Conference (EBRC2005), Amsterdam, the Netherlands.

Demuth, B., & Hussmann, H. (1999). Using UML/OCL constraints for relational database design. In *«UML» '99 — The Unified Modeling Language* (pp. 751-751).

Demuth, B., Hussmann, H., & Loecher, S. (2001). OCL as a specification language for business rules in database applications. In *«UML» 2001 — The Unified Modeling Language. Modeling Languages, Concepts, and Tools* (pp. 104-117).

Demuth, B., Loecher, S., & Zschaler, S. (2004, September 15 - 17). *Structure of the Dresden OCL toolkit.* Paper presented at the 2nd International Fujaba Days "MDA with UML and Rule-based Object Manipulation", Darmstadt, Germany.

Djuric, D., & Devedzic, V. (2005). Ontology modeling and MDA. *Journal of Object Technology, 4*(1), 109–128.

Donini, F., Lenzerini, M., Nardi, D., & Schaerf, A. (1996). Reasoning in description logics. In *Principles of knowledge representation, studies in logic, language and information* (pp. 193-238). Stanford, CA: CSLI Publications.

Drummond, N., & Shearer, R. (2006). *The open world assumption or sometimes its nice to know what we don't know*. Manchester, UK: University of Manchester. Retrieved from http://www.cs.man.ac.uk/~drummond/presentations/OWA.pdf.

Eriksson, H. E., & Penker, M. (2000). *Business modeling with UML: Business patterns at eork*. New York: John Wiley & Sons.

Falkovych, K. I., Sabou, M., & Stuckenschmidt, H. (2003). UML for the Semantic Web: Transformation-based approaches. In *Knowledge Transformation for the Semantic Web, Frontiers in Artificial Intelligence and Applications, vol.95* (pp. 92-106). Amsterdam, the Netherlands: IOS.

Filho, J. L., Sodré, V. F., Daltio, J., Rodrigues Júnior, M. F., & Vilela, V. M. (2004). *A CASE tool for geographic database design supporting analysis patterns*. New York: Springer.

Frankel, D. (2003). *Model driven architecture: Applying Mda to enterprise computing*. New York: John Wiley & Sons.

Gamma, E., Helm, R., Johnson, R., & Vlissides, J. (1995). *Patterns elements of reusable object-oriented software*: Reading, MA: Addison-Wesley.

Gasevic, D., Djuric, D., & Devedzic, V. (2006). *Model driven architecture and ontology development*. New York: Springer.

Gayte, O., Libourel, T., Cheylan, J.-P., & Lardon, S. (1997). *Conception des systèmes d'information sur environnement*. Paris: Hermés.

Gruber, R. (1993a). *Towards principles for the design of ontologies used for knowledge sharing* (Tech. Rep. KSL-93-04). Paper presented at the International Workshop on Formal Ontology, Padova, Italy.

Gruber, R. (1993b). A translation approach to portable ontology specifications. *Knowledge Acquisition, 5*(2), 199–220. doi:10.1006/knac.1993.1008

Guizzardi, G., Falbo, R. A., & Pereira Filho, J. G. (2002). Using objects and patterns to implement domain ontologies. *Journal of Brazilian Computer Society, Special Issue on Software Engineering, 8*(1).

Guizzardi, G., Wagner, G., Guarino, N., & Sinderen, M. (2004). An ontologically well-founded profile for UML conceptual models. *Lecture Notes in Computer Science*, 3084.

Hasenohr, P., & Pinet, F. (2006). Modeling of a spatial DSS template in support to the Common agricultural policy. *Journal of Decision Systems, 15*(2), 181–196. doi:10.3166/jds.15.181-196

Hasenohr, P., Pinet, F., Kay, S., Grenier, G., & Sevila, F. (2005, June). *A case-study based methodology for modeling of spatial DSS using Spatial OCL*. Paper presented at the Joint Workshop on Decision Support Systems, Experimental Economics & e-Participation (GRAZ 2005), AUT.

Hendler, J. (2001). Agents and the semantic web. *IEEE Intelligent Systems and Their Applications, 16*(2), 30–37.

IBM. (2006, June). *Ontology Definition Metamodel.*

ISO. (2002). ISO 19108 - Geographic Information - Temporal schema.

ISO. (2003). ISO 19107 - Geographic Information - Spatial schema.

Jacobson, I., Booch, G., & Rumbaugh, J. (1999). *The unified software development process*. Reading, MA: Addison Wesley.

Kang, M. A., Pinet, F., Schneider, M., Chanet, J. P., & Vigier, F. (2004, April 29-May 1). *How to design geographic database? Specific UML profile and spatial OCL applied to wireless Ad Hoc networks*. Paper presented at the 7th Conference on Geographic Information Science (AGILE'2004), Heraklion, GRC.

Klasse Objecten. (2009). *OCL tools and services*. Retrieved from http://www.klasse.nl/ocl

Kleppe, A., & Warmer, J. (2003). *Object constraint language, the getting your models ready for MDA*. Reading, MA: Addison-Wesley.

Lbath, A., & Pinet, F. (2000). *The development and customization of GIS-based applications and Web-based GIS applications with the CASE tool AIGLE*. Paper presented at the 8th ACM Symposium on Advances in Geographic Information Systems, Washington DC.

Mandel, L., & Cengarle, M. (1999). On the expressive power of OCL. In *FM'99 — Formal Methods* (pp. 713-713).

McNeile, A. (2003). MDA The vision with the hole. Retrieved from http://www.metamaxim.com

McNeile, A., & Simons, N. (2004). Methods of behaviour modelling. *A commentary on behavior modelling techniques for MDA*.

Mellor, J. S., & Balcer, J. M. (2002). *Executable UML, a foundation for model-driven architecture*. Reading, MA: Addison Wesley.

Meyer, B. (1988). *Object-oriented software construction*. Englewood Cliffs, NJ: Prentice-Hall.

Miralles, A. (2006). *Ingénierie des modèles pour les applications environnementales*. Unpublished doctoral dissertation, Université Montpellier II, Montpellier.

Miralles, A., & Libourel, T. (2008). Modeling with enriched model driven architecture. In S. Shekhar & H. Xiong (Eds.), *Encyclopedia of geographical information sciences* (pp. 700-705). New York: Springer.

Miralles, A., & Libourel, T. (2009a). Application of a model transformation paradigm in agriculture: A simple environmental system case study. In P. Papajorgji & P. M. Pardalos (Eds.), *Advances in modeling agricultural systems,*(Vol. 25, pp. 37-54). New York: Springer.

Miralles, A., & Libourel, T. (2009b). A new methodology to automate the transformation of GIS models in an iterative development process. In P. Papajorgji & P. M. Pardalos (Eds.), *Advances in modeling agricultural systems* (Vol. 25, pp. 19-36). New York: Springer.

Muzy, A., Innocenti, E., Aïello, A., Santucci, J.-F., Santoni, P.-A., & Hill, D. R. C. (2005). Modelling and simulation of ecological propagation processes: application to fire spread. *Environmental Modelling & Software*, *20*(7), 827–842. doi:10.1016/j.envsoft.2004.04.011

Nardi, D., & Brachman, R. J. (2002). An introduction to description logics. In F. Baader, D. Calvanese, D. L. McGuinness, D. Nardi, & P. F. Patel-Schneide (Eds.), *Description logic handbook*. Cambridge University Press.

National Committee for Information Technology Standards. (1998). *Technical Committee T2 (Information Interchange and Interpretation): Knowledge Interchange Format, draft proposed American National Standard (dpANS)* (Rep. NCITS.T2/98-004). Retrieved from http://logic.stanford.edu/kif/dpans.html

OGC. (2005). Implementation specification for geographic information - simple feature access - part 1: Common architecture.

OMG. (1999). *Unified modeling language - specification - version 1.3*. Needham, MA: Author.

OMG. (2007). *Unified modelling language: OCL, version 2.0. OMG specification*. Needham, MA: Author.

OMG. (2009a, February). *OMG unified modeling language: Superstructure - version 2.2*. Retrieved June 2009, from http://www.omg.org/cgi-bin/apps/doc?formal/05-07-04.pdf

OMG. (2009b). *Unified modeling language (OMG UML), infrastructure version 2.2*. Needham, MA: Author.

Pantazis, D., & Donnay, J. P. (1996). *La conception SIG: méthode et formalisme*. Paris: Hermés.

Papajorgji, P. (2005). A plug and play approach for developing environmental models. *Environmental Modelling & Software, 20*(10), 1353–1357. doi:10.1016/j.envsoft.2004.11.016

Papajorgji, P. (2009). State of the art in modeling agricultural systems. In *Encyclopedia of optimization* (pp. 3693-3704). New York: Springer.

Papajorgji, P., Beck, H. W., & Braga, J. L. (2004). An architecture for developing service-oriented and component-based environmental models. *Ecological Modelling, 179*, 61–76. doi:10.1016/j.ecolmodel.2004.05.013

Papajorgji, P., & Clark, R. (2005). *A model driven approach to agricultural systems.* Paper presented at the EFITA 2005, Portugal.

Papajorgji, P., Clark, R., & Jallas, E. (2009). The model driven architecture approach: A framework for developing complex agricultural systems. In P. Papajorgji & P. M. Pardalos (Eds.), *Advances in modeling agricultural systems.* New York: Springer.

Papajorgji, P., Gargiulo, O., Jones, J. W., & Traore, S. (2009). A UML-based Plug&Play version of RothC. In P. Papajorgji & P. M. Pardalos (Eds.), *Advances in Modeling Agricultural Systems.* New York: Springer.

Papajorgji, P., & Pardalos, P. M. (2006). *Software engineering techniques applied to agricultural systems an object-oriented and UML approach.* New York: Springer.

Papajorgji, P., & Shatar, T. M. (2004). Using the unified modeling language to develop soil water-balance and irrigation-scheduling models. *Environmental Modelling & Software, 19*(5), 451–459. doi:10.1016/S1364-8152(03)00160-9

Parent, C., Spaccapietra, S., & Zimányi, E. (2006). *Conceptual modeling for traditional and spatio-temporal applications: The MADS approach.* New York: Springer-Verlag.

Pastor, O., & Molina, J. C. (2007). *Model-driven architecture in practice a software production environment based on conceptual codeling*. New York: Springer.

Philippe, M. (2003, May/June). *Translations between UML, OWL, KIF and the WebKB-2 Languages (For-Taxonomy, Frame-CG, Formalized English).*

Pinet, F. (2002). *Vers une Méthode de Conception pour la Télégéomatique: Intégration et Validation au sein d'un Atelier de Développement.* Unpublished doctoral dissertation, Institut National des Sciences Appliquées de Lyon, France.

Pinet, F., Duboisset, M., Demuth, B., Schneider, M., Soulignac, V., & Barnabe, F. (2009). Constraints modeling in Agricultural Databases. In *Advances in Modeling Agricultural Systems*. New York: Springer.

Pinet, F., Duboisset, M., & Soulignac, V. (2007). Using UML and OCL to maintain the consistency of spatial data in environmental information systems. *Environmental Modelling & Software, 22*(8), 1217–1220. doi:10.1016/j.envsoft.2006.10.003

Pinet, F., Kang, M. A., & Vigier, F. (2005). Spatial constraint modelling with a GIS extension of UML and OCL: application to agricultural information systems. *Lecture Notes in Computer Science, 3511*, 160–178.

Pinet, F., Roussey, C., Brun, T., & Vigier, F. (2009). The use of UML as a tool for the formalisation of standards and the design of ontologies in agriculture. In P. Papajorgji & P. M. Pardalos, (Eds.), *Advances in modeling agricultural systems*. New York: Springer.

Pinet, F., Ventadour, P., Brun, T., Papajorgji, P., Roussey, C., & Vigier, F. (2006, November). *Using UML for ontology construction: A case study in agriculture*. Paper presented at the 7th Agricultural Ontology Service (AOS) Workshop on Ontology-Based Knowledge Discovery: Using Metadata and Ontologies for Improving Access to Agricultural Information (in collaboration with AFITA 2006), Indian Institute of Science, Bangalore, India.

Ràbovà, I. (2005). The Formalisation of knowledge in agricultural industry business process. *Agricultural Economics*, *51*(7), 53–58.

Richardson, A. (2002, August). *Starting with Bold for Delphi / Bold for C++, Part 3: Object Constraint Language (OCL).*

Schmid, B., Warmer, J., & Clark, T. (2002). *Object modeling with the OCL: The relational behind the object constraint language*. New York: Springer.

Schreiber, G. (2005). *A UML Presentation Syntax for OWL Lite.*

Shlaer, S., & Mellor, S. J. (1988). *Object-oriented systems analysis - modeling the world in data.* Englewood Cliffs, NJ: Prentice-Hall.

Softeam. (1999). *UML profiles and the J language: Totally control your application development using UML*. Retrieved March 2002, from http://www.objecteering.com/pdf/whitepapers/us/uml_profiles.pdf

Standford University. (2009). *Protégé*. Retrieved from http://protege.stanford.edu

Starr, L. (2002). *Executable UML how to build class models*. Englewood Cliffs, NJ: Prentice Hall.

Tambour, L., Houlès, V., Cohen-Jonathan, L., Auffray, V., Escande, P., & Jallas, E. (2009). Design of a model-driven Web decision support system in agriculture: From scientific models to the final software. In P. Papajorgji & P. M. Pardalos (Eds.), *Advances in modeling agricultural systems* (pp. 67-102). New York: Springer.

The Middleware Company. (2003). *Model driven development for J2EE utilizing a model driven architecture (MDA) approach - productivity pnalysis* (case study).

Toval, A., Requena, V., & Fernández, J. L. (2003). Emerging OCL tools. *Software and Systems Modeling*, *2*(4), 248–261. doi:10.1007/s10270-003-0031-0

Tryfona, N., & Jensen, C. S. (2000, March 19-21). *Using abstractions for spatio-temporal conceptual modeling*. Paper presented at the 2000 ACM Symposium on Applied Computing, Villa Olmo (Italy).

TU Dresden University. (2009). *Dresden OCL toolkit - projects around the toolkit*. Retrieved from http://dresden-ocl.sourceforge.net/projects.html

This work was previously published in International Journal of Agricultural and Environmental Information Systems, Volume 1, Issue 1, edited by Petraq Papajorgji and François Pinet, pp. 1-25, copyright 2010 by IGI Publishing (an imprint of IGI Global).

Chapter 17
Data Mining Techniques in Agricultural and Environmental Sciences

Altannar Chinchuluun
University of Florida, USA

Petros Xanthopoulos
University of Florida, USA

Vera Tomaino
University of Florida, USA & University Magna Græcia of Catanzaro, Italy

P.M. Pardalos
University of Florida, USA

ABSTRACT

Data mining techniques are largely used in different sectors of the economy and they increasingly are playing an important role in agriculture and environment-related areas. This paper aims to show our vision on the importance of knowing and efficiently using data mining and machine learning-related techniques for knowledge discovery in the field of agriculture and environment. Efforts for searching hidden patterns in data are not a recent phenomenon. History shows that extensive observations on data have helped discover empirical laws in different fields of research. Therefore, it is important to provide researchers in agriculture and environmental-related areas with the most advanced knowledge discovery techniques. Data mining is the process of extracting important and useful information from large sets of data. This information can be converted into useful knowledge that could help to better understand the problem in study and to better predict future developments. The paper presents the state of the art in data mining and knowledge discovery techniques and provides discussions for future directions.

DOI: 10.4018/978-1-4666-0333-2.ch017

INTRODUCTION

The problem of searching for patterns in data is a fundamental one and has a long and successful history. There are many examples in different research areas that extensive observations of data has led to discovering empirical laws. As an example, the attentiv astronomical observations undertaken by several astronoms allowed Kepler to discover the laws of planetary motion (Bishop, 2006).

Over the years, several techniques have been developed to discover hidden patterns in data and these efforts led to the creation of a rigourous discipline known as data mining or knowledge discovery. Data mining is the process of finding useful patterns or correlations amongst data. These patterns, associations, or relationships between data can provide information about the problem in study and information can then be transformed into knowledge. The idea of using information hidden into relationships amongst data inspired researchers to apply these techniques for predicting future trends (Mucherino, Papajorgi, & Pardalos, 2009). Data mining techniques are developed from mainly three areas: statistics, artificial intelligence and machine learning. Although the roots of data mining may seem different, but essentially they aim the same target: discover a relationship that more or less maps measurements in one part of a data set to measurements in another, linked part of the data set (Pyle, 2003).

Regardless of the method used, the goal of data mining techniques is to split data in different categories, each of them representing some feature of interest the data may have (Mucherino et al., 2009). Thus, fundamental for the success of a data mining technique is the ability to group available data in disjoint categories, where each category contains data with similar properties. The similarity between data is usually measured using a distance function; similar data should belong to the same group or cluster. Therefore, the success of a data mining technique depends on the adequate definition of a suitable distance between data samples.

As the similarity between data samples is measured using a distance function, often it occurs that this distance needs to be optimal. Thus, many data mining techniques led to the formulation of a global optimization problem (Mucherino et al., 2009).

Data mining techniques can be grouped in three categories as shown in Figure 1.

Statistical Methods

Statistical methods such as Principal Component Analysis (PCA) and regression techniques are commonly used as simple methods for finding

Figure 1. A schematic representation of the classification of the data mining techniques

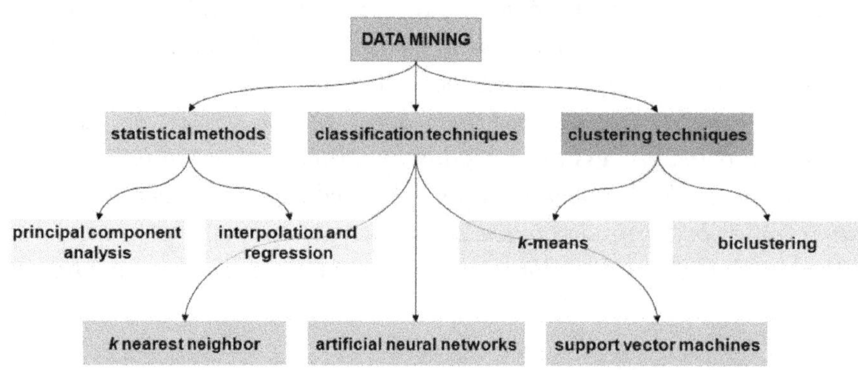

patterns in sets of data. PCA is a useful statistical technique that has found application in fields such as image compression, and is a common technique for finding patterns in big data sets. PCA helps identifying patterns in data, and expressing the data in such a way as to emphasize their similarities and differences. Since patterns in large data sets can be hard to find, because graphical representation of the data is not available, PCA is a powerful tool for analyzing data.

The main advantage of PCA is that once patterns in the data are identified data can be represented as components ordered by their relevance and it is possible then to discard components of low level of relevance without loss of important information and thus, reducing the complexity of the problem. In many cases, dimension reduction makes it possible to represent data graphically that enormously facilitates the understanding of discovered patterns.

Classification Techniques

Classification techniques use the well-known principle of Cicero *pares cum paribus facillime congregantur* (*birds of a feather flock together* or literally *equals with equals easily associate*). These techniques try to classify an unknown sample based on the known classification of its neighbors (Mucherino et al., 2009). Let us suppose that a set of samples with known classification is available, that is referred to as a training set. Each unknown sample should be classified considering its surrounding samples. Therefore, if the classification of a sample is unknown, then it could be predicted by considering the classification of its nearest neighbor samples. The main idea behind the classifications method is that given an unknown sample and a training set, all the distances between the unknown sample and all samples in the training set can be computed. The distance with the smallest value corresponds to the sample (or samples) in the training set closest to the unknown sample. Therefore, the unknown sample may be classified based on the classification of this nearest neighbor.

Clustering Techniques

Clustering techniques are used for partitioning a given set of data samples in clusters and there is no knowledge a priori about these data (Arulselvan, Baourakis, Boginski, Korchina, & Pardalos, 2008, Mucherino et al., 2009). The main idea behind the clustering algorithms is partitioning a set the data into a number *of* disjoint clusters by looking for inherent patterns in the set.

Clustering techniques are divided in *hierarchical* and *partitioning*. The hierarchical clustering approach builds a tree of clusters. The root of this tree can be a cluster containing all the data. Then, branch by branch, the initial big cluster is split in sub-clusters, until a partition having the desired number of clusters is reached. In this case, the hierarchical clustering is referred to as *divisive*. Furthermore, the root of the tree can also consist of a set of clusters, in which each cluster contains one and only one sample. Then, branch by branch, these clusters are merged together to form bigger clusters, until the desired number of clusters is obtained. In this case, the hierarchical clustering is referred to as *agglomerative*.

The paper is organized as follows. We review the main optimization based data mining techniques, including *k*-means clustering techniqe and support vector machine classifications, in the next section. The applications of these data mining techniques in agricultural engineering will be presented and future directions of this field are discussed.

Data Mining Algorithms

Data mining is the process of analyzing data using tools such as clustering, classification, feature selection and outlier's detection. Clustering techniques partition a given set of data into groups of similar samples according to some similarity criteria. Classification techniques determine classes of the test samples using known classification of a training data set. Feature selection techniques select a subset of features responsible for creating the condition corresponding to any class. Clustering is generally an initial step of data mining and it groups data into similar samples which can be used as a starting point of other techniques. Data clustering can be divided into two parts as hierarchical and partitional clustering (Jain, Murty, & Flynn, 1999). Single link and complete link are examples of hierarchical clustering while partitional clustering includes squared error algorithms (*k*-means), graph theoretic, mixture resolving (expectation maximization), mode seeking, and so forth. Bayesian classifier is a traditional statistical classification algorithm based on Bayes' theory. Bayesian classifiers use combination of conditional probability and posterior probabilities for classifying any information. Further details about Bayesian classifiers and applications can be found in Duda and Hart (1973), Marchant and Onyango (2003), Pernkopf (2005) and Yager (2006). Some other clustering techniques such as fuzzy clustering, artificial neural networks, nearest neighbor clustering and evolutionary approach based clustering are also becoming very popular tools for researchers. We discuss some of these clustering techniques in this section.

The *k*-means Algorithm

The *k*-means (MacQueen, 1967) method is one of the most popular unsupervised learning or clustering methods which have been applied in a variety of fields including pattern recognition, information retrieval, document extraction and microbiology analysis, and so forth. The method is called the *k*-means because it represents each of *k* number of clusters C_j (j= 1,2,…,k) by the mean (or the weighted average) of its points. The goal of this method is to classify a given data set through a certain number of clusters such that some metric relative to the centroids (or centers) of the clusters is minimized. We can define our problem mathematically as follows.

Suppose that we are given a set X of a finite number of points in d-dimensional Euclidean space R^r, that is, $X=(x^1,x^2,\dots x^n)$ where $x^i \in R^d$, $i=1,2,\dots,n$.

We aim at finding a partition C_j, $j=1,2,\dots,k$:

$$X = \bigcup_{j=1}^{k} C_j, \quad C_j \bigcap Cl = 0,$$

for all $j \neq l$, of X which minimizes the squared error function

$$f(C_1, C_2, \dots, C_k) = \sum_{j=1}^{k} \sum_{x^i \in C_j} \| c^j - x^i \|^2,$$

where ||.|| denotes the Euclidean norm, C^j is the center of the cluster C_j

$$c^j = \frac{1}{|C_j|} \sum_{x^i \in C_j} x^i, \quad j = 1, 2, \dots, k. \qquad (1)$$

Algorithm k-Means

Step 1: Initialize the centroids $C\frac{j}{0}$, $j=1,2,\dots,k$. Set q=0 (where q is the iteration counter).

Step 2: Assign each point x^i (i=1,2,…,n) to the cluster that has the closest centroid (or the center of the cluster) $c^j(j \in \{1,2,..k\}$, that is $j = ar\,gmin_{1 \leq l \leq k} \leq \| x^i - c_q^l \|^2$.

Step 3: When all points have been assigned, for j=1,2,…,k, calculate the new position c_{q+1}^j of the centroid j.

Step 4: If $c_q^j = c_{q+1}^j$ for all j=1,2,…,k, then stop, otherwise set $q=q+1$ and go to Step 2.

The k-means algorithms is easy to implement and its time complexity is order of n ($O(n)$), where n is number of patterns (Jain et al., 1999). However, it finds one of the many local solutions that depend on the initial starting points. To find a better solution, we can run the algorithm several times and choose the best one as the optimal solution. Unfortunately, repetition with different random selections (Duda & Hart, 1973) appears to be not a very efficient method. For this purpose, Bradley and Fayyad (1998) presented a procedure for computing a refined starting condition from a given initial one that is based on an efficient sampling technique for estimating the modes of a distribution, and their experiments presented that refined initial starting points indeed lead to improved solutions. Yager and Fillev (1994) developed the *mountain method* which is a simple and effective approach for approximate estimation of the cluster centers on the basis of the concept of a mountain function. It can be useful for obtaining the initial values of the clusters that are required by more complex cluster algorithms. Another drawback of the algorithm is that there are no efficient methods for defining the initial number of partitions. Many alternative methods to improve k-means were published in literature. Krishna and Murty (1999) proposed a new Genetic k-means algorithm (GKA) for global search and faster convergence. Zhang, Xiong, Mao, and Ou (2006) proposed parallel k-means algorithm for higher efficiency. Many algorithms similar to k-means have been appeared in the literature (Duda, Hart, & Stork, 2001; MacQueen, 1967).

Fuzzy *c*-means (FCM) Clustering

In the k-means algorithm, each sample can be assigned to only one cluster. Fuzzy clustering relaxes this restriction and an object can belong to several clusters at the same time but with certain degrees of memberships. The most known fuzzy clustering method is the fuzzy c-means method (FCM), introduced by Dunn (1974) and later generalized by Bezdek (1981). FCM partitions a data set $X=(x_1,x_2,...,x_n)\subset R^p$ of p features, into c fuzzy subsets where $u_{i,k}$ is the membership of x_k in class i (i=1,2,…,c). These classes are identified by their cluster centers v_i, (i=1,…,c). The objective of FCM is to find an optimal fuzzy c partition minimizing the objective function,

$$Jm(U,V:X)=\sum_{k=1}^{n}\sum_{i=1}^{c}u_{ik}^{m}\parallel x_k-v_i\parallel^2, \tag{2}$$

where the value of fuzzy partition matrix U is constrained in the range [0,1] such that

$$\sum_{i=1}^{c}u_{ij}=1, \quad k=1,2,...,n, \tag{3}$$

and

$$\sum_{k=1}^{n}u_{ik}\leq n, \quad i=1,2,...,c. \tag{4}$$

Here, $m\in[1,\infty)$ is an exponential weighting function that controls the fuzziness of the membership values, ||.|| is the Euclidean norm and $V=(v_1,v_2,...,v_c)$ is a matrix of unknown cluster centers $v_i\in R^p$ (i=1,…,c). Fuzzy c-means algorithm to minimize (2):

Step 1: Choose appropriate values for m and c, and a small positive number $\in$. Initialize randomly a fuzzy partition matrix U^0 and set iteration number t=0.

Step 2: For given membership values $u_{ik}^{(t)}$, calculate the cluster centers $v_i^{(t)}$ (i=1,2,…,c) as $v_i^{(t)}=\dfrac{\sum_{k=1}^{n}(u_{ik}^{(t)})^m x_k}{\sum_{k=1}^{n}(u_{ik}^{(t)})^m}$ (5)

Step 3: Given a new cluster center from *Step 2*, update membership values $u_{ik}^{(t+1)}$ using

$$u_{ik}^{(t+1)}=\left[\sum_{j=1}^{c}\left(\frac{\parallel x_k-v_i^{(t)}\parallel^2}{\parallel\sum_{k=1}^{n}\left(u_{ik}^{(t)}\right)^m\parallel}\right)\right]. \tag{6}$$

Step 4: Repeat *Step 2* and *3* until $|U-(t+1)-U^{(t)}|<\in$ or a pre-defined number of iterations is reached.

Methods discussed in previous sections are crisp/hard partitioning methods, which allow to partition data into a specified number of mutually exclusive datasets only, while fuzzy methods are soft partitioning methods where an object can belong to one or more data sets/partitions. Similarly to the crisp/hard partitioning methods, selection of initial matrix of centers plays an important role in convergence of FCM. Many times FCM does not guarantee the global optimal solutions due to randomized initialization of cluster centers and matrix U. Moreover, FCM solutions are also sensitive to noise and outliers. Hathway, Bezdek, and Hu (2000) have proposed a modified FCM using 1 norm distance to increase robustness against outliers. Hung and Yang's *psFCM* algorithm (Hung & Yang, 2001) finds the actual clusters' centers and refines initial value of FCM. This technique reduces the computational time by a large amount. Many other improved FCMs and their applications can also be found in (Hammah & Curran, 2001; Hathaway & Bezdek, 2001; Karayiannis, 1997; Zhang et al., 2006).

k-Nearest Neighbor Classification

The *k*-nearest neighborhood (Mitchell, 1997) method is widely adopted due to its efficiency. The key idea of the algorithm is to classify a new sample in the most frequent class of its closest neighbors in the training set. This is a majority voting formula on the class labels of its neighbors. A Euclidean distance measure is used to calculate how close each member of the training set is to the target data that is being examined. The *k*-nearest neighbor classification algorithm can be divided into two phases:

Training Phase

- Define a training set $S=\{(x_1,y_1),(x_2,y_2),\ldots,(x_n,y_n)$, where $x_i=(x_i^1,x_i^2,\ldots,x_i^d)$ is a *d*-dimensional feature vector of real numbers, for all $i=1,\ldots,n$.
- Define class labels y_i corresponding to each x_i for all i, $y_i \in C$ where $C=(1,2,\ldots N)$, N_c is the number of different classes.
- Task: determine y_{new} for C_{new}.

Testing Phase

- Find the closest point x_j to x_{new} with respect to Euclidean distance: $\sqrt{(x_j^1-x_{new}^1)^2+\cdots+(x_j^d-x_{new}^d)^d}$.
- Classify by $y_{new}=y_j$.

A serious drawback of this *k*-nearest neighbor technique is the computational complexity in searching the *k* nearest neighbors among those available training samples. Kuncheva (1997) claims to achieve better computational efficiency and higher classification accuracy by using genetic algorithms as editing techniques. Bermejo and Cabestany (2000) proposed a KNN algorithm with local Parzen window estimate to improve approximation quality. They also suggested an adaptive learning algorithm to allow fewer data points to be used in a training data set. Many other techniques have been proposed to reduce computational burden of *k*-nearest neighbor algorithms in Hwang and Wen (1998) and Pan, Qiao, and Sun (2004).

ARTIFICIAL NEURAL NETWORKS

Artificial neural networks (ANN) are efficient classification tools that have been applied to several applications including extracting regularities in data and classifying events in finance, marketing, Internet and biomedicine. The training process uses available examples to produce a model and to classify new events based on the extracted model. Neural networks are built from simple units, called neurons or cells by analogy with the way human brain works. Neurons are linked with each other by a set of weighted connections. The information to be analyzed is fed to the neurons of the input layer and then propagated to the neurons of the hidden layers (if any) for further processing. The result of this processing is then propagated to the next hidden layer and the process is continued until the output layer is reached. Each unit receives some information from other units and processes this information, which will be converted into the output of the unit. There are no specific methods of choosing the network parameters such as number of hidden layers and type of activation function. Generally one input and one output nodes are chosen for each data class.

The training set is known a priori and is used to fine-tune the network for future similar records. While in training phase, known data containing inputs and corresponding outputs are fed to the network, and the network learns to infer the relationship between these two.

The process of classification by ANN can be broadly defined as follows:

- Run a sample from the training set, by giving its attribute values as input.
- The summation of weights and activation functions are applied at each node of hidden and output layers, until an output is generated (feed-forward process).
- Compare output with the expected output from training set.
- If output does not match, go back layer to layer and modify arc weights and biases of nodes (back-propagation process).
- Run the next sample and process the same.
- Eventually the weights will converge and process stops.

Feed forward topology is widely used in multilayer perceptions networks. Feed forward network provides a general framework for representing non-linear functional mappings between a set of input variables and a set of output variables (Bishop, 1995). Below is a brief overview of feed forward network for each training sample *X* and for each hidden or output layer node *j*:

- Calculate input I_j to that node as $I_j = \sum_{i=1}^{d} w_{ji} x_i + w_{j0}$.
- Calculate output O_j from that node as $O_j = \dfrac{1}{1 + e^{-I_j}}$.

Back-propagation is widely used algorithm for the purpose of training the neural networks. Back-Propagation algorithm can be considered as a two-step process. In the first step, the derivatives of the error function with respect to the weights are evaluated. In second step, these derivatives are then used to compute the adjustments to be made to the weights by using gradient descent or any other optimization schemes. An overview of back-propagation can be given as below:

- For each node *j* in output layer, calculate the error as $Err_j = O_j(1 - O_j)(T - O_j)$.
- For each node *j* in hidden layer, calculate the error as $Err_j = O_j(1 - O_j)\left(\sum_k Err_k w_{jk}\right)$.
- For each weight w_{ij}, calculate weight increment as $\Delta w_{ij} = l \cdot Err_j \cdot O_i$.
- Now, update the previous weight as $w_{ij} = w_{ij} + \Delta w_{ij}$.
- For each bias θ calculate bias increment as $\Delta\theta_j = l \cdot Err_j$.
- Finally update bias with $\theta_j = \theta_j + \Delta\theta_j$.

Neural networks are widely used in classification but there are still many unsolved issues in applying neural networks such as scalability, misclassification, convergence, higher mean square errors, and so forth. Many researchers have tried to overcome these issues and proposed variety of neural networks with better performances. Jiang, Harvey, and Wah (2003) proposed a new approach of constructing and training neural networks to overcome the problems including local minima and the slow convergence of the learning process. In their approach, feed-forward network was constructed based on the data clusters generated based on locally trained clustering (LTC) and then further trained using standard algorithms operating on the global training set which converges rapidly due to its inherited knowledge with good generalization ability from the global training. Ji and Ma (1997) proposed a learning method based on combination of weak classifiers, which were found by a randomized algorithm, to achieve good generalization and fast training time on both the test problems and the real applications. They showed that if the weakness factor was chosen according to the critical value given by their theory, the combinations of weak classifiers could achieve a good generalization performance with polynomial space- and time-complexity. Yu, Chen, and Cheng (1995) proposed dynamic learning using derivative information instead of fixed learning rate to optimize back propagation. The information gathered from the forward and backward propagation was used for dynamic learning; with this technique they achieved higher convergence rate and significant reduction in learning process. The probability of misclassification of any random sample can be termed as the generalization error of a classifier. Many researchers have used ensemble methods to reduce misclassification or generalization errors (Hansen & Salamon, 1990; Hashem & Schmeiser, 1995). Further information on classification errors, learning and generalization, and some of the recent developments in neural networks can be found in Kulkarni, Lugosi, and Venkatesh (1998), Solazzia and Uncinib (2004), and Zhang (2000). Recent studies in data mining techniques (Wu et al., 2008) do not consider ANN as a top 10 data mining technique.

SUPPORT VECTOR MACHINES

Support Vector Machine (SVM) is a state of the art machine learning algorithm (Cortes & Vapnik, 1995; Vapnik, 1995). The main idea of SVM is to separate the input space in two half spaces using the hyperplane $x^T w - y = 0$ which maximizes the margin between the two classes $\{A,B\}$. In other words, given n points $x_i \in R^d$, $i=1,2,\ldots,n$, and the corresponding labels $y^i = \begin{cases} 1 \; if \; x^i \in A \\ -1 \; if \; x^i \in B \end{cases}$, we need to find two parallel hyperplanes $x^T w - y = \pm 1$ which separates the points between two classes. The margin between two classes is represented by the distance between the two hyperplanes, $\frac{2}{\| w \|}$. Thus the optimization problem is to minimize the norm of w, with constraints to correctly classified points of both classes:

$$\min \frac{1}{2} \| w \|^2$$
$$s.t. \; y^i (x^{i^T} w - y) \geq 1.$$
$$i = 1, 2, \ldots, n$$

Here we note that minimizing $||w||$ is the same as minizing $||w||^2$. This method works when there is a perfect linear separation between the two classes. Therefore it is called the hard margin classifier. When the two classes are not linearly separable, the soft margin classifier is used. This classifier finds a hyperplane that allows a few points to violate the separation. The number of points that are not correctly classified should be minimized in this case. Then the optimization problem is reduced to the following:

$$\min \frac{1}{2} \| w \|^2 + \frac{C}{2} \sum_{i=1}^{n} \xi^{i^2}$$
$$s.t. \, y^i (x^{i^T} w - y) \geq 1 - \xi^i .$$
$$i = 1, 2, \dots, n$$

These problems belong to the quadratic convex programming problems and can be solved using standard quadratic programming techniques (Bertsekas, 1995). For nonlinear classification, the SVM is used with kernel functions (Mangasarian & Wild, 2006) and the basic solution technique is through linear programming.

GENERALIZED EIGENVALUE CLASSIFICATION

Generalized eigenvalue classification methods proposed by Mangasarian and Wild (2004) (Guarracino, Cifarelli, Seref, & Pardalos, 2007) have a classification accuracy comparable with those obtained by SVM at a lower execution time. This is also a support vector machine classification where each plane of two non-parallel planes is generated such that it is closest to one of the two data sets and as far as possible from the other data set. This problem leads to two simple generalized eigenvalue problems. The main idea is to classify two sets of points, *A* and *B*, using two hyperplanes, each closest to one set of points, and furthest from the other. Let $x^T w\text{-}y=0$ be a hyperplane in R^d. To satisfy the previous condition for the points in *A*, the hyperplanes can be obtained by solving the following optimization problem:

$$\min_{w,y \neq 0} \frac{\| Aw - ey \|^2}{\| Bw - ey \|^2} . \tag{8}$$

The hyperplane for the *B* can be obtained by minimizing the inverse of the objective function in (8). If we use the notations $G=[A,-e]^T[A,-e], H=[B,-e]^T[B,-e], z=[w^T,y]^T$, then Equation (8) becomes:

$$\min_{z \in R^d} \frac{z^T G z}{z^T H z} \tag{9}$$

The expression in (9) is the Raleigh quotient of the generalized eigenvalue problem $G_x=\lambda Hx$. The stationary points are obtained only at the eigenvectors of (9), where the value of the objective function is given by the eigenvalues. When H is positive definite, the Raleigh quotient is bounded and it ranges over the interval determined by minimum and maximum eigenvalues (Parlett, 1998). H is positive definite under the assumption that the columns of [B, - *e]* are linearly independent. The inverse of the objective function in (4) has the same eigenvectors and reciprocal eigenvalues. Let $zm_{in=}[w1\,y1_{,}]$ and $zm_{ax=}[w2\,y2_{,}]$ be the eigenvectors related to the eigenvalues of smallest and largest modulo, respectively. Then $xT^w1\_y1\_0$ is the closest hyperplane to the set of points in A and the furthest from those in B and $xT^w2\_y2\_0$ is the closest hyperplane to the set of points in B, and the furthest from those in A. In order to regularize the problem, we can solve:

$$\underset{w,y \neq 0}{min} \frac{\| Aw - ey \|^2 + \delta \| \widetilde{B} \|^2}{\| Bw - ey \|^2 + \delta \| \widetilde{A} \|^2} .$$

where the $\widetilde{A}$ and $\widetilde{B}$ are the diagonals of matrices [*Aw−ey*] and [*Bw−ey*], respectively.

By choosing the eigenvectors related to the new minimum and maximum eigenvalue, we obtain solutions that are close to the ones of the original problem. A point is classified according to the closest hyperplane or the class.

More recently, Cifarelli, Guarracino, Seref, Cuciniello, and Pardalos (2007) introduced IReGEC, an incremental technique capable of reducing the training set in the learning phase of the supervised classification, with the advantage of lower overfitting of data and improved classification accuracy. This method together with some other classification methods has been applied to different biomedical data sets (Felici, Bertolazzi, Guarracino, Chinchuluun, & Pardalos, 2009; Guarracino et al., 2007). Incremental subset selection permits to construct a small set of points that retains the information of the entire training set and provides comparable accuracy results. A kernel built from a smaller subset is computationally more efficient in predicting new elements, compared to the one that uses the entire training set. Furthermore, a smaller set of points reduces the probability of overfitting the problem. Finally, as new points are available, the cost to retrain the algorithm decreases if the influence of those new points on classification is only evaluated with respect to that subset, rather than the whole training set.

APPLICATIONS

Historically one of the first papers dealing with data discrimination into different classes was directly applied to an agricultural problem. This is Fisher's Linear Discriminant Analysis (LDA) algorithm published in the annals of Eugenics (later renamed Annals of Human Genetics) and had been applied in order to discriminate 150 flowers into three classes of flowers based on four quantitative features namely: sepal width, sepal length, pedal width and pedal length (Fisher, 1936). Today this famous dataset (best known in data mining community as IRIS dataset) can easily be found in any open dataset repositories (e.g., UCI, http://archive.ics.uci.edu/ml/) and serves as an example in many undergraduate data mining courses around the world.

Although the first use data mining techniques was in agriculture, mathematical tools were not extensively used in this field for many years. The change came especially due to recent technological advances that made it possible to store and process large amount of data in home computers. Accumulation of data generated the challenge and the need for processing and analysis generating several well defined mathematical problems. Today many data miners identify and attack problems coming from agricultural science.

Clustering for example has been used (and more especially the famous *k-means* method) to address problems that arise during the fermentation process of wine. In Urtubia, Perez-Correa, Meurens, and Agosin (2004), the authors used the k-means to predict how good the fermentation process will be. This was achieved by recording a number of features related to sugars, alchohols organic acids and nitrogen sources. Initially, PCA was used to reduce the dimensionality of the problem and then the k-means clustering method was applied to determine the different clusters.

The k-means algorithm also has been used for image segmentation in the area of machine vision. As part of it, Leemans and Destain used k-means method for grading apples and identifying visually defected products before shipping them to the end customers (2004). They describe a conveyer based system able to analyze four apples in 1 second. The same authors extended their machine vision-based data mining using other methods such as the Linear Discriminant Analysis (Leemans, Magein, & Destain, 2002) and Bayesian (Leemans, Magein, & Destain, 1999) to solve the same problem.

On the other side Artificial Neural Networks (ANN) that are popular tools for especially among computer scientists for supervised classification problems. ANN's have demonstrated good results in practice, especially if the testing dataset is wisely chosen and the structure of the network is such that it is robust against overfitting. In Aerts, Jans, Halloy, Gustin, and Berckmans (2004), the authors utilize ANNs to detect abnormal coughing sounds in a herd. Abnormal coughing sound usually is associated with some disease so it is very crucial for the farmer to distinguish if there is any potentially diseased animal in the herd. Thus, data mining can be used to prevent and control the spread of dangerous contagious diseases.

Another application of ANN's in agriculture was presented by Shahin, Tollner, and McClendon (2001) where they are used as intelligent classifiers on Magnetic Resonance Imaging (MRI) scans of apples in order to detect internal defects in apples

(Shahin et al., 2001). From this perspective, neural network classification is used as a quality control tool to decrease the number of defected products. By utilizing X-ray images Schatzki, Haff, Young, Can, Le, and Toyofuku were able to detect about 90% of the defected apples (1997).

Liu, Lyon, Windham, Lyon, and Savage (2004) use PCA to analize chicken breast quality. In order to analyze the meat quality and the deboning time, a set of 36 chicken carcasses has been considered and randomly divided into four subgroups, each one containing 9 carcasses. These subgroups are designed for different deboning times. Chickens in the different groups have been deboned after 2, 4, 6 and 24 hours after the death. After deboning, the breasts have been cut in two parts, and each part has been subject to a different set of analysis. After using the PCA to decrease the dimensionality of the problem, results show that the first seven principal components are able to represent about the 70% percent of the total variations on the data. Moreover, the first four principal components represent about 50% of the total variations. In particular, the first principal component takes 23.3% of the variations, the second one 13.6%, the third one 8.8% and finally the fourth one 6.9%. An analysis on the data showed that the first component was mainly defined by the sheer force and by the attributes decided by the group of panelists supervising this process. Therefore, these attributes are the most important variables for the evaluation of the chicken breast quality (Mucherino et al., 2009).

In Jagtap, Jones, LaRow, Ajayan, and O'Brien (2006) a *k*-NN algorithm is used for the recalibration of the precipitation outputs from the FSU-GSM (Florida State University *Global* Spectral Model) and FSU-RSM (Florida State University *Regional* Spectral Model) climate models. These climate models may not produce sufficiently accurate daily weather variable outputs to use in crop models. The objective is to find *k* neighboring years which have the forecasts closest to those of a target year. It is therefore assumed that the climate during a target year is a replication of the weather recorded in the past. The *k*-NN method resulted able to improve the accuracy of the monthly precipitation forecasts across all sites used in the study.

FUTURE DIRECTIONS

Data mining and knowledge discovery techniques are relatively new to agricultural and environmental fields. Their use is associated and conditioned with the use of research operations sets of tools. There are a number of research papers that show that agricultural and environmental sciences can really benefit from the use of mathematical tools and modern technology. It is important to note that the a number of published papers are purely research and have not yet been applied to be part of the set of tools farmers or practitioners use everyday. As an example, the study that use a Artificial Neural Network (Aerts et al., 2004) describes the process of how a pig goes through the system designed to record pig coughs to discover whether the animal has health problems but it does not address the issue of scaling the proposed system to be applied to the entire herd. The proposed system is complex and can examine only one animal at a time. It will be difficult to see the proposed system be applied in a herd of hundreds of pigs where each animal must be examined individually. The difficulties would be operational and finacial.

Often occurs that producers and practitioners in agriculture and environment have more trust on traditional decision making methods and they might see suspiciously automated decision making tools. This phenomenon has in general happened in all areas where data mining and knowledge discovery techniques have been introduced.

The answer that data miners provide is that classification, clustering and generally any other mathematical tool, doesn't aim to replace the human expert. On the contrary, they serve as expert assisting tools that help humans make sounder and better decisions. Data mining and Knowledge discovery are totally human activities. Algorithms would provide the results that need to be interpreted by data miners in order to reach a final conclusion about the problem under study.

Many methods such as classical k-means clustering, ANNs, support vector machines, k-NNR classification have been applied in a variety of problems. But still there are many state of the art methods whose potential applications in agriculture have not been explored in full yet (e.g., biclustering). We envision more and more data mining application related papers with different problems and different algorithms being published in journals of interest to agricultural/environmental related scientists and practitioners.

We think that it is the time to introduce data mining and knowledge discovery techniques in the curriculum of agricultural and environmental departments, thus students can become familiar with these promising techniques. The recently published book titled "Data Mining in Agriculture" (Mucherino et al., 2009) specifically tailored for students of agricultural and environmental fields will be a useful source of knowledge as it provides an exhaustive inventory of data mining and knowledge discovery techniques applied in agricultural and environmental sciences.

A number of academic institutions are organizing "Summer Schools" where in a week or two they provide a good introduction in data mining and knowledge discovery techniques. More of similar efforts will certainly help in making data mining and knowledge discovery techniques familiar to students, researchers and practitioners in the field of agriculture and environment.

REFERENCES

Aerts, J.-M., Jans, P., Halloy, D., Gustin, P., & Berckmans, D. (2004). Labeling of cough data from pigs for on-line disease monitoring by sound analysis. *American Society of Agricultural and Biological Engineers*, *48*(1), 351–354.

Arulselvan, A., Baourakis, G., Boginski, V., Korchina, E., & Pardalos, P. M. (2008). Analysis of food industry market using network approaches. *British Food Journal*, *110*(9), 916–928. doi:10.1108/00070700810900611

Bermejo, S., & Cabestany, J. (2000). Adaptive soft k-nearest-neighbour classifiers. *Pattern Recognition*, *33*, 1999–2005. doi:10.1016/S0031-3203(99)00186-7

Bertsekas, D. P. (1995). *Nonlinear programming*. Belmont, MA: Athena Scientific.

Bezdek, J. (1981). *Pattern recognition with fuzzy objective function algorithms*. New York: Plenum Press.

Bishop, C. M. (1995). *Neural networks for pattern recognition*. Oxford, UK: Oxford University Press.

Bishop, M. C. (2006). *Pattern recognition and machine learning*. New York: Information Sciences and Statistics, Springer.

Bradley, S., & Fayyad, M. (1998, July 24-27). Refining initial points for k-means clustering. In J. Shavlik (Ed.), *Proceedings of the 15th International Conference on Machine Learning (ICML98)*, Madison, WI (pp. 91-99). San Francisco: Morgan Kaufmann.

Chinchuluun, R., Won Suk, L., Bhorania, J., & Pardalos, P. M. (2009). Clustering and classification algorithms in food and agricultural applications: A survey. In P. Papajorgji and P. M. Pardalos (Eds.), *Advances in modeling agricultural systems* (pp. 1-22). New York: Springer.

Cifarelli, C., Guarracino, M. R., Seref, O., Cuciniello, S., & Pardalos, P. M. (2007). Incremental classification with generalized eigenvalues. *Journal of Classification*, *24*(2), 205–219. doi:10.1007/s00357-007-0012-z

Cortes, C., & Vapnik, V. (1995). Support vector machines. *Machine Learning*, *20*, 273–279.

Duda, R. O., & Hart, P. E. (1973). *Pattern classification and scene analysis*. New York: John Wiley & Sons.

Duda, R. O., Hart, P. E., & Stork, D. G. (2001). *Pattern classification* (2nd ed.). New York: John Wiley & Sons.

Dunn, J. (1974). A fuzzy relative of the ISODATA process and its use in detecting compact well separated clusters. *Journal of Cybernetics*, *3*(3), 32–57. doi:10.1080/01969727308546046

Felici, G., Bertolazzi, P., Guarracino, M. R., Chinchuluun, A., & Pardalos, P. M. (2008. November). Logic formulas based knowledge discovery and its application to the classification of biological data. In R. P. Mondaini (Ed.), 8th International Symposium on Mathematical and Computational Biology (*BIOMAT 2008),* Sao Paulo, Brazil (pp. 223-234). World Scientific.

Fisher, R. A. (1936). The use of multiple measurements in taxonomic problems. *Annals of Eugenics*, *7*, 179–188.

Guarracino, M. R., Chinchuluun, A., & Pardalos, P. M. (in press). Decision rules for efficient classification of biological data. *Optimization Letters*.

Guarracino, M. R., Cifarelli, C., Seref, O., & Pardalos, P. M. (2007). A classification algorithm based on generalized eigenvalue problems. *Optimization Methods and Software*, *22*(1), 73–81. doi:10.1080/10556780600883874

Hammah, R. E., & Curran, J. H. (2000). Validity measures for the fuzzy cluster analysis of orientations. *IEEE Transactions on Pattern Analysis and Machine Intelligence*, *22*(12), 1467–1472. doi:10.1109/34.895981

Hansen, L. K., & Salamon, P. (1990). Neural network ensembles. *IEEE Transactions on Pattern Analysis and Machine Intelligence*, *12*(10), 993–1001. doi:10.1109/34.58871

Hashem, S., & Schmeiser, B. (1995). Improving model accuracy using optimal linear combinations of trained neural networks. *IEEE Transactions on Neural Networks*, *6*(3), 792–794. doi:10.1109/72.377990

Hathaway, R., & Bezdek, J. (2001). Fuzzy *c*-means clustering of incomplete data. *IEEE Transactions on Systems, Man, and Cybernetics - Part B, 31*(5), 735-744.

Hathaway, R., Bezdek, J., & Hu, Y. (2000). Generalized fuzzy c-means clustering strategies using L norm distances. *IEEE transactions on Fuzzy Systems*, *8*(5), 576–582. doi:10.1109/91.873580

Hung, M., & Yang, D. (2001, November 29-December 1). An efficient fuzzy c-means clustering algorithm. In *Proceedings of the IEEE International Conference on Data Mining,* San Jose, CA (pp. 225-232).

Hwang, W. J., & Wen, K. W. (1998). Fast k classification algorithm based on partial distance search. *Electronics Letters*, *34*(21), 2062–2063. doi:10.1049/el:19981427

Jagtap, S. S., Jones, J. W., LaRow, T., Ajayan, A., & O'Brien, J. J. (2006). (Manuscript submitted for publication). Statistical recalibration of precipitation outputs from coupled climate models. *Journal of Applied Meteorology*.

Jain, A. K., Murty, M. N., & Flynn, P. J. (1999). Data clustering: A review. *ACM Computing Surveys*, *31*(3), 264–323. doi:10.1145/331499.331504

Ji, C., & Ma, S. (1997). Combinations of weak classifiers. *IEEE Transactions on Neural Networks*, *8*(1), 32–42. doi:10.1109/72.554189

Jiang, X., Harvey, A., & Wah, K. S. (2003). Constructing and training feed-forward neural networks for pattern classification. *Patt. Recog.*, *36*, 853–867. doi:10.1016/S0031-3203(02)00087-0

Karayiannis, N. B. (1997). A methodology for constructing fuzzy algorithms for learning vector quantization. *IEEE Transactions on Neural Networks*, *8*(3), 505–518. doi:10.1109/72.572091

Krishna, K., & Murty, M. (1999). Genetic k-means algorithm. *IEEE Transactions on Systems, Man, and Cybernetics - Part B, 29*(3), 433-439.

Kulkarni, S. R., Lugosi, G., & Venkatesh, S. S. (1998). Learning pattern classification - a survey. *IEEE Transactions on Information Theory*, *44*(6), 2178–2206. doi:10.1109/18.720536

Kuncheva, L. I. (1997). Fitness functions in editing *k*-NN reference set by genetic algorithms. *Patt. Recog.*, *30*(6), 1041–1049. doi:10.1016/S0031-3203(96)00134-3

Leemans, V., & Destain, M. F. (2004). A real time grading method of apples based on features extracted from defects. *Journal of Food Engineering*, *61*, 83–89. doi:10.1016/S0260-8774(03)00189-4

Leemans, V., Magein, H., & Destain, M.-F. (1999). Defect segmentation on 'Jonagold' apples using colour vision and Bayesian method. *Computers and Electronics in Agriculture*, *23*, 43–53. doi:10.1016/S0168-1699(99)00006-X

Leemans, V., Magein, H., & Destain, M.-F. (2002). On-line apple grading according to European standards using machine vision. *Biosystems Engineering*, *83*(4), 397–404. doi:10.1006/bioe.2002.0131

Liu, Y., Lyon, B. G., Windham, W. R., Lyon, C. E., & Savage, E. M. (2004). Principal component analysis of physical, color, and sensory characteristics of chicken breasts deboned at two, four, six, and twenty-four hours postmortem. *Poultry Science*, *83*, 101–108.

MacQueen, J. B. (1967). Some methods for classification and analysis of Multivariate Observations. In *Proceedings of 5th Berkeley Symposium on Mathematical Statistics and Probability: Vol. 1* (pp. 281-297). Berkeley, CA: University of California Press.

Mangasarian, O. L., & Wild, E. W. (2004). *Multisurface proximal support vector classification via generalized eigenvalues* (Techn. Rep. 04-03). Madison, WI: Data Mining Institute, Computer Science Department, University of Wisconsin.

Mangasarian, O. L., & Wild, E. W. (2006). *Nonlinear knowledge-based classification* (Tech. Rep. 06-04). Madison, WI: Data Mining Institute, Computer Science Department, University of Wisconsin.

Marchant, J. A., & Onyango, C. M. (2003). Comparison of a Bayesian classifier with a multilayer feed-forward neural network using the example of plant/weed/soil discrimination. *Computers and Electronics in Agriculture*, *39*, 3–22. doi:10.1016/S0168-1699(02)00223-5

Mitchell, T. M. (1997). *Machine learning*. New York: McGraw-Hill.

Mucherino, A., Papajorgji, P., & Pardalos, P. M. (2009). Survey of Data Mining Techniques Applied to Agriculture. *Operational Research -. International Journal (Toronto, Ont.)*, *9*, 121–140.

Mucherino, A., Papajorgji, P., & Pardalos, P. M. (2009). *Data mining in agriculture*. New York: Springer.

Pan, J. S., Qiao, Y. L., & Sun, S. H. (2004). A fast k nearest neighbors classification algorithm. *IEICE Transactions on Fundamentals of Electronics, Communications and Computer Sciences. E (Norwalk, Conn.)*, *87-A*(4), 961–963.

Parlett, B. N. (1998). The symmetric eigenvalue problem. *SIAM, 20,* 357.

Pernkopf, F. (2005). Bayesian network classifiers versus selective *k*-NN classifier. *Pattern Recognition*, *38*(1), 1–10. doi:10.1016/j.patcog.2004.05.012

Pyle, D. (2003). *Business modeling and data mining*. San Francisco, CA: Morgan Kaufmann Publishers.

Schatzki, T. F., Haff, R. P., Young, R., Can, I., Le, L.-C., & Toyofuku, N. (1997). Defect detection in apples by means of x-ray imaging. *Transactions of the American Society of Agricultural Engineers*, *40*(5), 1407–1415.

Shahin, M. A., Tollner, E. W., & McClendon, R. W. (2001). Artificial intelligence classifiers for sorting apples based on watercore. *Journal of Agricultural Engineering Research*, *79*(3), 265–274. doi:10.1006/jaer.2001.0705

Solazzia, M., & Uncinib, A. (2004). Regularizing neural networks using flexible multivariate activation function. *Neural Networks*, *17*(2), 247–260. doi:10.1016/S0893-6080(03)00189-8

Urtubia, A., Perez-Correa, J. R., Meurens, M., & Agosin, E. (2004). Monitoring large scale wine fermentations with infrared spectroscopy. *Talanta*, *64*(3), 778–784. doi:10.1016/j.talanta.2004.04.005

Vapnik, V. (1995). *The nature of statistical learning theory*. New York: Springer-Verlag.

Wu, X., Kumar, V., Quinlan, J. R., Ghosh, J., Yang, Q., & Motoda, H. (2008). Top 10 algorithms in data mining. *Knowledge and Information Systems*, *6*(14), 1–37. doi:10.1007/s10115-007-0114-2

Yager, R. R. (2006). An extension of the naive Bayesian classifier. *Information Science*, *176*(5), 577–588. doi:10.1016/j.ins.2004.12.006

Yager, R. R., & Filev, D. P. (1994). Approximate clustering via the mountain method. *IEEE Transactions on Systems, Man, and Cybernetics*, *24*(8), 1279–1284. doi:10.1109/21.299710

Yu, X., Chen, G., & Cheng, S. (1995). Dynamic learning rate optimization of the backpropagation algorithm. *IEEE Transactions on Neural Networks*, *6*(3), 669–677. doi:10.1109/72.377972

Zhang, G. P. (2000). Neural networks for classification: a survey. *IEEE Transactions on Systems, Man, and Cybernetics*, *30*(4), 451–462. doi:10.1109/5326.897072

Zhang, Y., Xiong, Z., Mao, J., & Ou, L. (2006, June). The study of parallel k-means algorithm. In *Proceedings of the 6th World Congress on Intelligent Control and Automation: Vol.2,* Dalian, China (pp. 5868-5871). IEEE.

This work was previously published in International Journal of Agricultural and Environmental Information Systems, Volume 1, Issue 1, edited by Petraq Papajorgji and François Pinet, pp. 26-40, copyright 2010 by IGI Publishing (an imprint of IGI Global).

Chapter 18
The Role of Crop Systems Simulation in Agriculture and Environment

K. J. Boote
University of Florida, USA

J. W. Jones
University of Florida, USA

G. Hoogenboom
University of Georgia, USA

J. W. White
USDA-ARS, Arizona, USA

ABSTRACT

Simulation of crop systems has evolved from a neophyte science into a robust and increasingly accepted discipline. Our vision is that crop systems simulation can serve important roles in agriculture and environment. Important roles and uses of crop systems simulation are in five primary areas: (1) basic research synthesis and integration, where simulation is used to synthesize our understanding of physiology, genetics, soil characteristics, management, and weather effects, (2) strategic tools for planning and policy to evaluate strategies and consequences of genetic improvement or resource management, (3) applications for management purposes, where crop systems simulations are used to evaluate impacts of weather and management on production, water use, nutrient use, nutrient leaching, and economics, (4) real time decision support to assist in management decisions (irrigation, fertilization, sowing date, harvest, yield forecast, pest management), and (5) education in class rooms and farms, to explain how crop systems function and are managed.

INTRODUCTION

Simulation of crop systems has advanced greatly over the past 30 to 40 years. From a neophyte science with inadequate computing power, the field has evolved into a robust and increasingly accepted science supported by improved software, languages, development tools, and computer capabilities, but the foundation continues to be scientific insights from plant physiology, soil

DOI: 10.4018/978-1-4666-0333-2.ch018

science, agroclimatology, and related fields. Crop system simulators contain mathematical equations describing basic flow and conversion processes of carbon, water, and nitrogen balance that are integrated daily or hourly by the computer program to predict the time course of crop growth, nutrient uptake, and water use, as well as to predict final yield and other plant traits and outputs. The goal of this paper is to give our vision on how crop systems simulation can serve important future roles in agriculture and environment and suggests how to prioritize research to better support these roles. The paper leads off with an historical overview describing how crop system models began, then discusses five primary roles and uses of crop systems simulation in agriculture and environment, concludes with a challenge for potential linkage of crop models with molecular biology-genetics, and suggests the need for continued improvement of the science in crop system models.

AN HISTORICAL OVERVIEW

The use of crop system models and simulation had its start in crop physiology, soil physics, and soil-crop-water processes. Early models focused mainly on the crop carbon (C) balance under optimum conditions, where only solar radiation and temperature were the driving variables. Simulation of crop canopy photosynthesis from leaf-level parameters was a primary focus (DeWit, 1965; Duncan, 1971), along with predicting crop development as described through their growth stages and examining strategies for increasing reproductive yield. These patriarchs of crop modeling soon advanced to developing simple whole crop models. Concurrently, the early agricultural engineers and soil physicists were developing soil-plant-water balance models that predicted daily crop evapotranspiration, crop water uptake, and water flow processes in soils (Whisler et al., 1986). See Whisler et al. (1986) for an overview and history of crop simulation models up to the mid-1980s, including typical processes considered, data required, model testing, and applications. The crop aspects of many of the early soil-water-balance models were often fairly simple, estimating daily growth from light-interception and radiation-use-efficiency. The soil water balance models vary from tipping bucket one-dimensional water balance (Ritchie, 1985, 1998) to more complex Darcy-driven water flow with two dimensional flow such as 2-DSOIL (Ahuja, Ma, & Timlin, 2006) and RZWQM (Ma et al., 2003). The next improvement in crop system models came with the simulation of soil nitrogen (N) balance with a simple tipping bucket plug-flow of nitrate N to allow simulation of N leaching, but success was limited until improvement in two major components had occurred: first, the crop C balance routines needed to estimate crop N demand accurately and second, accurate routines to estimate soil organic matter mineralization are needed to estimate the supply of soil mineral N beyond that coming from applied fertilizer N. There are many published soil organic matter models (e.g., see Smith et al., 1997, who compared nine different soil organic matter models). The most frequently cited organic matter models are CENTURY (Parton, Stewart, & Cole, 1988) and ROTHC (Jenkinson & Rayner, 1977), and these models often serve as reference models for many studies (Traore, Bostick, Jones, Koo, Goita, & Bado, 2008). Each of these models has shortcomings, and there are many difficulties correctly simulating soil organic matter dynamics, even after 20-30 years of progress, because soils are so variable and soil organic matter is complex.

Over the past 10 to 20 years, crop system model developers have succeeded in linking good crop C balance (N demand) with good soil water balance and good soil-crop N balance. The DSSAT V3.5 models (Hoogenboom, Wilkens, Porter, Batchelor, & Hunt, 1999; Jones et al., 1998) were among the early models to succeed in this full linkage, but APSIM (Keating et al., 2003; McCown, Hammer, Hargreaves, Holzworth, & Freebairn, 1996) and other models are also at this stage of development.

The DSSAT-CSM V4.0 model (Hoogenboom et al., 2004; Jones et al., 2003) was a further improvement, in its use of a land-unit module, which is the interface of crop-soil-weather, where the soil organic matter module used can be the CENTURY model (Gijsman, Hoogenboom, Parton, & Kerridge, 2002) or the older Godwin soil organic matter model (Godwin & Jones, 1991; Godwin & Singh, 1998). Table 1 lists five current crop system models, APSIM, CROPSYST, CSM-CERES, CSM-CROPGRO, and EPIC, all of which have capabilities of simulating cropping sequences.

More modular systems of crop model development evolved over the 1990s and 2000s that allowed components of the crop models to separate along scientific disciplinary lines, and allowed smaller groups of modelers to improve those modules or to replace them with new modules or to add new modules (Jones, Keating, & Porter, 20001; Keating et al. 2003; Van Kraalingen, Rappoldt, & van Laar, 2003). This modular development was a natural evolution of the models, but progress had to build upon the first comprehensive models because one needed the rest of the model to properly predict most of the important driving variables of the soil-plant-atmosphere system, even while improving any particular module. APSIM (Keating et al., 2003; McCown et al., 1996) was one of the first modeling groups to develop modular crop models, with modules that could "plug and play" into the main simulation engine. Crop modules could be pulled out and replaced by alternate crop modules. The DSSAT-CSM (V4.0) was another of the early models to emphasize this modularity (Jones et al., 2001; Jones et al., 2003), with a single soil water module, two soil organic matter modules possible, a weather module, a management inputs module, and a crop template module with common source code, but one that allowed simulating different crops by defining species input files. The other modules operated around a soil-plant-atmosphere module that handled the competition for water, light, among soil, plant, and atmosphere. This modular approach built on early developments in the 1990s with the CROPGRO model, which showed the value of using a common source code for many crops, but with all species, ecotype, and cultivar coefficients pulled out of the code, and placed in read-in files (Boote, Jones, & Hoogenboom, 1998; Boote, Jones, Hoogenboom, & Pickering, 1998; Hoogenboom et al., 1994; Hoogenboom et al., 1999). This modular approach with one common source code facilitated develop-

Table 1. Examples of crop systems simulators, simple description, and background references on selected crop system models

Name	Crops	Description	Reference Describing the Model
APSIM	Maize, sorghum, wheat, barley, and many others	Radiation use efficiency. Daily time step. Allows cropping sequences.	Keating et al., 2003; McCown et al., 1996.
CropSyst	Barley, maize, sorghum, soybean, wheat, and others	Radiation use efficiency. Daily time step. Allows cropping sequences.	Stockle, Donatelli, & Nelson, 2003; Stockle, Martin, & Campbell, 1994.
CSM-CERES	Barley, maize, millet, sorghum, and wheat	Radiation use efficiency. Daily time step. Allows cropping sequences.	Jones et al., 2003; Hoogenboom et al., 2004.
CSM-CROPGRO	Common bean, faba bean, peanut, soybean,, cotton, and others	Farquhar-type photosynthesis on an hourly basis with hedge-row light interception. Considers growth and maintenance respiration. Allows cropping sequences.	Boote, Jones, & Hoogenboom, 1998, Boote, Jones, Hoogenboom, & Pickering, 1998; Hoogenboom et al., 2004; Jones et al., 2003.
EPIC	Maize, millet, rice, sorghum, soybean, wheat, and others	Radiation use efficiency. Allows cropping sequences and can model effects of tillage and soil erosion.	Williams, Jones, Kiniry, & Spanel, 1989.

ment of additional models for other crop species such as faba bean (Boote, Mínguez, & Sau, 2002).

Model applications are a logical outcome of desires of scientists to develop crop simulation models for a broad range of uses, depending on individual scientist interest and interpreted need of farmers and policy specialists. Whisler et al. (1986) reviewed a range of applications attempted up to the mid-1980s. Boote, Jones, and Pickering (1996) in their paper reviewed potential uses and limitations of crop models, discussing a wide range of applications in a number of areas. Now, about 10 years later, we in the current paper, review important roles and uses of crop systems simulation models in the following five areas, which surprisingly, differ little from those visualized previously by Whisler et al. (1986) and Boote et al. (1996).

RESEARCH SYNTHESIS: MODELS ENCOURAGE INTERDISCIPLINARY COLLABORATION AMONG RESEARCHERS

Much of the first 30 years of crop systems simulation emphasized improving basic research synthesis and integration of our understanding in physiology, genetics, soil fertility, soil-water-traits, and ecology. Systems simulation provided an important tool, by which the research was done to understand the system and how the components interacted. Then this understanding was integrated or written, however imperfectly, into the code of models, to allow predicting the behavior of a system for given conditions. If the behavior was sufficiently well predicted over many conditions, then the optimistic goal was for the model predictions to be used to manage or control the system. Along the historical path of crop system model development, the crop modeling community learned some important lessons. We began to appreciate how crop system model development required a holistic effort involving scientists in many different disciplines, working together, to share their knowledge and data, in a continuing process whereby code was written and simulations tested, to help understand the relationships, to help synthesize and integrate the relationships and knowledge of crop physiology, genetics, soil physics, soil nutrients, and weather impacts. Crop simulation models have the potential to integrate the processes of your favorite discipline into crop system models, to allow you to see how the whole system works, and to see just how important a given process, enzyme, or genetic trait may be. Systems simulation provides a good focal point for engaging scientists in different disciplines as they work together on a given weather-crop-soil-water-N situation: the scientists get to see how their component processes fit, and ideally those scientists can provide help putting their processes into the models the way they think they should. One lesson is that crop systems simulation, for a given actual field situation, cannot just ignore certain aspects, but it must consider all processes and forcing factors that influence the crop C balance, soil-crop N balance, and soil-crop water balance. This encourages researchers to consider the broader context of cropping systems, and consider driving processes they might otherwise ignore. This process often forces one to consider other disciplines. So far, our discussion has focused on the biophysical; however as the models are improved and are increasingly used as decision support tools in industry and government, it has become evident that there is a need to include economists, social scientists, and political scientists in this interdisciplinary discussion.

The value of synthesis and integration should not to be underestimated, in an era where researchers have gone progressively deeper and more detailed in their individual laboratory investigations, to the point where researchers may lose sight of the importance of their individual contribution. Too often, and with understandable pride, these researchers may point to major potential production improvements possible with their particular findings, but where these findings are not sufficiently tested in field situations. This is particularly true with plant genomics scientists or plant physiologists who do not get into the field, and those who study processes in test tubes, or study plants such as *Arabidopsis* grown in artificial soil media in un-bordered pots in low-light growth chambers. There is a need to test the effects of given genetics or fertility or management treatments of crops in whole stands on problem soils under normal weather variability. In many cases, the phenotype is more determined by environment than by genotype.

We will give several examples of how optimistic promises from plant physiology, were scaled back in reality, when considered at the whole crop level and considering the whole system of inputs and outputs. After the advent of good infrared gas analyzers and the discovery of the C-4 vs. C-3 photosynthetic pathways, there was a major push during the 1970s and 1980s to select for higher single leaf photosynthesis. Researchers such as Dornhoff and Shibles (1970) found variation for increased single leaf light-saturated photosynthesis (A_{max}) that was associated with increased specific leaf weight (SLW) among soybean cultivars. This genetic variation was as much as 48% on a leaf area basis, but only 17% on a leaf mass basis. However, the yield of cultivars with higher leaf rate is not increased in proportion to the increase in single leaf rate. When the higher leaf photosynthesis is scaled up in model simulation of leaf-to-canopy assimilation and then to crop yield, the response is much less. With no coupling of A_{max} to SLW, a 10% higher leaf rate gives a 4 to 5% increase in canopy assimilation (but only for high light times of day). Therefore, the degree of benefit is scaled back considerably. The more realistic case is that the increased A_{max} is actually coupled or linked to increased SLW (this can be considered as a pleiotropic effect that causes the greater leaf rate). In this case, the simulated increase in canopy assimilation is much less, about 2 to 3%, because growing thicker leaves results in lower LAI (from the same leaf mass) and gives lower light interception (Boote & Tollenaar, 1994; Boote, Jones, Batchelor, Nafziger & Myers, 2003). Thus, the promised large benefit at leaf photosynthesis level is diminished greatly going to whole crop stands and on to final grain yield. There are other ways, via systems analysis of the whole crop, to envision a way to capitalize on the higher leaf A_{max}, but that requires a higher sowing density and narrow row spacing to push up the LAI to re-gain the light capture given up by thick leaves. Not surprisingly over the past 30 to 40 years, soybean producers have evolved toward using narrow row spacing and higher density sowing, such that some of the benefit is partially realized in current soybean cultivars (Boote et al., 2003). This example shows the need to synthesize and integrate an individual discipline finding into a whole crop systems simulation to see what the yield and economic value might be. This will be particularly important as molecular biologists propose improvements resulting from single genes that affect yield processes, in contrast to the current efforts on genes for herbicide tolerance, pest resistance, or crop composition.

Another example of the benefit of simulating crop C balance and energy cost, is that of genetically altering the composition of harvested products. A key concern is that there are differential energetic costs for producing different compounds. Simulations by Boote and Tollenaar (1994) showed that soybean yield was reduced almost 1% for each 1% increase in protein or 1% increase in oil, because proteins and oils require more energy to synthesize than starchy compounds. These simulations dispel the dream of plant breeders to just genetically "break" the observed negative linkage of high protein with lower yield potential. The problem here is one of limiting energy (or limited C or limited reduced N) that can be used to synthesize tissue, but it is not a genetics problem. The so-called genetic linkage is actually a physiological C:N:energy balance linkage. Indeed, if you were to give soybean a maize-seed composition, you could increase soybean yield by 32 to 41% (Boote & Tollenaar, 1994).

There have been major promises made for genes related to drought tolerance and water-use-efficiency. Crop systems simulations that consider energy balance, must trade (exchange) CO_2 gain for water vapor loss, because the pathway is the same. If stomata are partly closed to slow water loss rates, then the plant must decrease its CO_2 gain, and also endure higher foliage temperature. The physics of gradients for CO_2 in, and water vapor out, depend on relative humidity, leaf temperature, and light intensity (to drive CO_2-sink strength of leaf). The ratio of leaf internal CO_2 to external CO_2, appears to be highly regulated (Pearcy & Bjorkman, 1983) in crop plants (C-3 at 0.7 and C-4 at 0.4). There are only minor ways to modify the system. Thicker leaves and higher leaf protein concentration can decrease the intercellular CO_2 (C_i) of these species, thus providing a very minor improvement in WUE. Otherwise, there is little opportunity for improvement, despite promises in the literature. The concept of crop tolerance to desiccation, which is promoted by molecular genetics scientists, is almost irrelevant to annual seed-producing crops, because you do not want plants to enter a dormant state and stop producing economic yield, or you have already lost too much yield potential. However, survival under extended drought may have application to sustain stands of perennials used for grazing or forage, to avoid the necessity to re-plant the crop after extended drought.

Our future vision is that using models to integrate and synthesize research across disciplines will become more important into the future. Current funding agencies are also moving in this direction, by requiring participation of multiple disciplines to be represented in the development of grant proposals. There are now fairly accurate and well-tested mechanistic crop system simulation models that broadly consider all the external driving forces that can be found in field conditions, including the soil and weather environment and crop management inputs. These models can be used as platforms to allow the detail specialists to improve the crop models for depth and mechanism in their discipline area and then to test at the levels of what they understand to be intermediate test measurement outcomes, such as tissue nutrient concentrations or canopy transpiration, for example. But importantly, final growth and yield outcomes of those detail processes will also be available.

MODELS AS STRATEGIC TOOLS FOR RESEARCH PLANNING AND POLICY

There are many possible applications of crop system models that relate to making policy and strategy decisions at national, state, and local levels, where the use is not necessarily for real-time decisions. Government policy makers, agri-genetics firms (plant breeders), agrichemical companies, agricultural equipment companies, and research organizations can use crop system simulations as tools to evaluate strategies to plan for investments in research toward crop genetic improvement, crop management, and product development. This tool use differs from using models for integrating and synthesizing research, although it will require intelligent users and economic or political motivation. For example, a plant breeding company or a university plant breeder could use the model as a tool to evaluate aspects of genetic improvement versus germplasm that they hope to improve. A company selling fertilizer, fungicide, or herbicides could do the same to determine efficacy of a product.

Impacts of potential climate change on crop production, water use, nutrient leaching, and regional or national economics can be projected or hypothesized for future climate change scenarios, using crop system simulators along with other spatial and economics tools. This requires that the crop models be robust and correct for their response to elevated CO_2, temperature, rainfall, and solar radiation, and respond correctly to sowing dates, N fertility, and cultivar variation. Such tool use also requires the availability of reliable soils information, weather data, and knowledge of cropping practices (sowing date, cultivar, and fertilization) for the regions of interest. Various government agencies, universities, World Bank, and international centers such as International Food Policy Research Institute are interested in such applications. Potential uses include: to predict crop production/food security under a range of projected climate scenarios, to subsequently compute the economic consequences of the altered production, to compute the water use (irrigation required) and fertilizer use involved, to hypothesize possible adaptation/mitigation strategies under climate change such as shifted sowing dates, alternate crops/cultivars, genetic improvement for climatic stress (Boote, Jones, & Hoogenboom, 2008).

At regional and state levels with the same models, soils, and weather data availability over spatial regions, the crop system simulators could be used: 1) to project the consequences of N fertilization on production and N leaching over small to large watersheds for a given season or over long term weather, 2) to project crop water use (irrigation demand) for given watersheds for a given year or over long term weather, and 3) to hypothesize consequences of land use change to produce alternative crops, such as a biofuel crop, with the goal of understanding the water and fertilization requirements as well as the environmental and economic impact of shifting from one crop or land use to another. This could be particularly relevant in the current discussion of whether to produce biofuel crops rather than food crops, as it would allow evaluating overall economic, water, and resource use requirements. In order to take a good look at the options, one must have available good crop system models for forests, sugarcane, sweet sorghum, switchgrass, corn, and other food crop commodities.

For the state of Florida, Jones and Boote adapted and tested the DSSAT crop models with the goal of developing a best management practice (BMP) tool to maximize N use efficiency and to minimize N leaching. This required testing the models for correct uptake of N and correct tissue N concentrations, with the strategy that this would assist in predicting potential N leaching. Generally data on N leaching are scarce which limits model testing. The interest of the state of Florida, however, is to scale up and extend predictions of N use, N uptake, and N leaching over large-scale watersheds. In this particular case, the crop model is planned to be placed as one component within a watershed model.

Crop models have potential uses to evaluate possible "yield gap" causing reduced growth and yield, especially in under-developed country situations. For this purpose, the models must be developed and calibrated for potential growth in fields that do not have limiting factors of water, nutrients, or pests. In other words, the model response is primarily to temperature, solar radiation, and day length, along with genetic traits and crop management. If the same model over-predicts the slope of dry matter gain and/or final yield in a new field with the same basic weather, then one can suspect limiting factors such as soil fertility or pest limitations. This "yield gap" simulation approach was followed by Naab, Singh, Boote, Jones, & Marfo (2004) who determined that peanut yield in the rainy savannah of Ghana, was only partially limited by rainfall and soil water. Foliar diseases proved to be the main limitation and soil infertility was a minor additional contributing factor. The CROPGRO model in the DSSAT V3.5 and later DSSAT V4.0 is programmed to accept scouting inputs of pest damage (such as percent defoliation and percent necrosis), which are coupled to the model to create simulated reduction in leaf area index and simulated reduction in photosynthesis and eventually reduced growth and yield (Batchelor, Jones, Boote, & Pinnschmidt, 1993; Boote, Batchelor, Jones, & Pinnschmidt, 1993).

REAL-TIME DECISION SUPPORT TOOLS FOR MANAGEMENT DECISIONS

For many years, agricultural engineers and extension specialists have worked toward and anticipated the use of crop system simulation to assist with management decisions, both strategic pre-season as well as in-season, for a given field. The management decisions could include irrigation, N fertilization, sowing date, projected harvest, yield forecast, and pest management. The realization of this dream has been slow and sporadic in its development, for several reasons. The first requirement is for crop models with sufficient ability to respond to these desired management practices as well as to the soils and weather experienced. Good spatial soils data are a second requirement because soils vary across a region and even within a given field. The third requirement is availability of long-term weather records (which allow a strategic pre-season evaluation of practices). A fourth requirement is for continuous access to current weather data to allow updating the simulations and projecting the crop growth and yield to the current date. Most of the limitations of these types of applications relate to access to accurate and local weather data (Hoogenboom, 2000). Typically, the models are additionally configured to run the remainder of the season with multiple "run-out" seasons of weather, to evaluate the range and probabilities of possible outcomes, depending on whether one irrigated or fertilized the crop today (or not), or otherwise manipulated the crop. Linkage to concurrently running pest simulators or input of scouting crop damage is another aspect that was attempted or envisioned. Prediction of growth stage for application of herbicides or growth regulators is a further use of in-season simulation.

The GOSSYM-COMAX system for cotton was one of the first simulators to attempt real-time field decision making, but a number of difficulties limited sustainable use of this system (McKinion & Baker, 1989). We attempted this use with the CROPGRO-Soybean model in a United Soybean Board (USB) project. The USB asked that the technology be transferred to a commercial firm as soon as possible, and the tool called PC-Yield (Welch, Jones, Brennan, Reeder, & Jacobson, 2002) was available for several years from a company named EMERGE which also sold weather services. However, the company went out of business in 2001, and the tool no longer is supported. One important lesson from the project, was the need to ensure ready access to data bases on cultivar coefficients, soils profiles, typical production practices, and the weather data. Lastly, a convenient simple interface with farmer/producers was very important. In order to develop this interface and what decision to offer a producer, we needed to have a good understanding of how the producers made their actual decisions.

Current versions of real-time decision support are being developed to provide producers with current weather data, a future 3-6 month expected climate projection (based on ENSO signal), and predicted crop performance. This approach has been successful with the Southeastern Climate Consortium (SCC) where the delivered product is primarily the past climate, along with simple computations such as chilling days, growing degree days, and pest favorability ratings (Fraisse et al., 2006; Paz et al., 2007). The SCC and other weather delivery service agencies are now examining how to provide available crop model simulations that run with the weather data for given locations, but this effort is proceeding slowly, so as not to overload the users. The agrometeorological weather service in India is actively moving in this direction (Boote et al., 2008). Issues to be resolved include learning what information the farmers and stakeholders really want, and how to develop their confidence in the information being delivered (Crane, Roncoli, Paz, Breuer, Broad, & Hoogenboom, 2009; Roncoli, Paz, Breuer, Ingram, Hoogenboom, & Broad, 2006). Again, we see the valuable role of socio-economic scientists in understanding how farmers make their decisions.

CROP SYSTEM SIMULATORS AS EDUCATIONAL TOOLS

Crop simulation models are valuable tools for the class room, to teach students and scientists the importance of considering and integrating all the driving variables that affect crop growth and yield in a given production situation. The first crop modeling courses were primarily to teach students about how to use the crop models, but increasingly the crop models are now being used in more general undergraduate/graduate classes to teach students about the integrational aspects of agriculture. Crop modeling courses (on how to model) were first offered by the Dutch modeling group in 1981 (K. Boote was an attendee). The next group to offer a crop modeling course, were Jones and Boote in 1984 at the University of Florida. About that time (1983-1984), the International Benchmark Sites Network for Agrotechnology Transfer (IBSNAT) group was formed, and these scientists developed the Decision Support System for Agrotechnology Transfer (DSSAT) models, and began to conduct crop modeling workshops (Tsuji, Hoogenboom, & Thorton, 1998). This DSSAT modeling group of mostly the same scientists, over the next 20 years conducted crop modeling course every 1 to 2 years to present, for U.S. and international scientists. The DSSAT scientists now operate under a new umbrella, International Consortium for Agricultural Systems Applications (ICASA). The number of courses offered and the number of scientists taking the courses has increased over time, until in 2008, the DSSAT course was offered to 55 persons at one time. Most of the attendees have been scientists with PhDs and MSs with backgrounds in agronomy, agricultural engineering, agricultural economics, entomology, pathology, and information systems. Maybe a third of the attendees were actually graduate students at the time of taking the course, so most attendees were actually already in academia, government, and industry positions. G. Hoogenboom estimates that a total of 500 persons have taken the DSSAT course training and over 2,500 scientists have obtained a copy of the software through the International Consortium for Agricultural Systems Applications (ICASA, http://www.ICASA.net). J. W. Jones and K. J. Boote have offered essentially the same course as a University of Florida course annually or biannually since 2000. The crop modeling courses have been quite valuable to train individuals how to use the DSSAT software, but more importantly, to train the scientists to think about the whole systems function, to learn about the particular processes involved and how they are linked to the crop, and to learn how the systems models synthesize and link all the processes.

A FUTURE CHALLENGE: MODELING AND GENETICS/ MOLECULAR BIOLOGY

Advances in plant molecular biology are providing opportunities for crop modelers to improve the characterization of cultivar effects in models and to refine descriptions of plant processes. Although gene-based characterization of cultivars was first used in the 1990s (Messina, Jones, Boote, & Vallejos, 2006; White & Hoogenboom, 1996), broader application has been limited by the number of physiological gene loci described for any crop and the difficulty of obtaining genotype data for individual cultivars. Molecular biology is both increasing the list of major loci and providing low cost services for genotyping. Where information on specific loci is lacking, quantitative trait loci (QTLs) have been used. Again with decreasing costs of molecular characterizations, one option is to parameterize cultivar differences using QTLs. Even without QTL information, Mavromatis, Boote, Jones, Irmak, Shinde, and Hoogenboom (2001) and Mavromatis, Boote, Jones, Wilkerson, and Hoogenboom (2002) demonstrated that reliable genetic coefficients for simulating different cultivars could be derived from field performance trials on soybean with sufficient precision for crop model prediction under field conditions. With such genetic coefficients, crop models have already been used to evaluate genetic yield potential (Boote, Kropff, & Bindraban, 2001; Boote et al., 2003), especially over multi-environment test sites and yield testing procedures relative to genetic traits of crops such as peanut (Anothai, Pathanothai, Pannangpetch, Jogloy, Boote, & Hoogenboom, 2009; Putto, Pathanothai, Jogloy, Pannangpetch, Boote, & Hoogenboom, 2009).

Plant biology has also recognized the value of simulation modeling for synthesizing and integrating information, a process that can be viewed as a renewal of the interests of physiologists that fueled early simulation modeling (Loomis, Rabbinge, & Ng, 1979). Thus, simulation models are being developed for specific processes such as floral induction (Welch, Roe, Das, Dong, He, & Kirkham, 2005) and photosynthesis (Zhu, Portis, & Long, 2004). The iPlant Collaborative www.iplantcollaborative.org), a major project supported by the U.S. National Science Foundation, has recognized the need to develop an integrated cyberinfrastructure to facilitate modeling from the gene network level up to whole plant or crop system levels.

FUTURE NEEDS: MORE SCIENCE IN THE MODELS AND MODEL IMPROVEMENT

An important question is whether the crop models and their applications are data-limited or whether they are science-limited. Certainly, many of us have seen applications moving rather quickly in certain cases, and possibly ahead of the science. This is inevitable, considering the desire for users to promote novel applications (usually driven by grant funding), even before the crop modeling tools are full developed and tested. Funding is generally for applications, whereas funding for tool (crop model) development is rarer. In some cases, the science is already substantially developed (i.e., has the answers), but the crop model developers have too much inertia or are too busy, and basically delay making these changes because the changes may be too much work.

In answer to the above question, we believe that models are science-limited and are not being improved fast enough from the scientific knowledge that is already known, in part because the past emphases has been on software interfaces and other topics. Thus, we suggest several areas where we feel there is a need for more discipline-science-focused contribution to crop system model improvement: 1) description of root growth and nutrient uptake, especially for P and K, 2) coupling of foliar disease models (rusts or leafspots) to crop models to predict damage, 3) full coupling of energy balance to improve simulation of water-use-efficiency and responses to climate change, 4) full integration of crop models with global circulation models and regional circulation models, which may require the same energy balance as above, 5) addition of nutrient balance in models for nutrients beyond N and P, and 6) incorporation of information from genetics and molecular biology.

REFERENCES

Ahuja, L. R., Ma, L., & Timlin, D. J. (2006). Trans-disciplinary soil physics research critical to synthesis and modeling of agricultural systems. *Soil Science Society of America Journal*, *70*, 311–326. doi:10.2136/sssaj2005.0207

Anothai, J., Patanothai, A., Pannangpetch, K., Jogloy, S., Boote, K. J., & Hoogenboom, G. (2009). Multi-environment evaluation of peanut lines by model simulation with the cultivar coefficients derived from a reduced set of observed field data. *Field Crops Research*, *110*, 111–121. doi:10.1016/j.fcr.2008.07.009

Batchelor, W. D., Jones, J. W., Boote, K. J., & Pinnschmidt, H. O. (1993). Extending the use of crop models to study pest damage. *Transactions of the ASAE. American Society of Agricultural Engineers, 36*, 551–558.

Boote, K. J., Batchelor, W. D., Jones, J. W., Pinnschmidt, H., & Bourgeois, G. (1993). Pest damage relations at the field level. In F. W. T. Penning de Vries et al. (Ed.), *Systems Approaches for Agricultural Development* (pp. 277-296). Dordrecht, The Netherlands: Kluwer Academic Publishers.

Boote, K. J., Jones, J. W., Batchelor, W. D., Nafziger, E. D., & Myers, O. (2003). Genetic coefficients in the CROPGRO-soybean model: Links to field performance and genomics. *Agronomy Journal, 95*, 32–51.

Boote, K. J., Jones, J. W., & Hoogenboom, G. (1998). Simulation of crop growth: CROPGRO Model. In R. M. Peart & R. B. Curry (Eds.), *Agricultural systems modeling and simulation* (pp. 651-692). New York: Marcel Dekker.

Boote, K. J., Jones, J. W., & Hoogenboom, G. (2008). Crop simulation models as tools for agro-advisories for weather and disease effects on production. *Journal of Agrometeorology, 10*, 9–17.

Boote, K. J., Jones, J. W., Hoogenboom, G., & Pickering, N. B. (1998). The CROPGRO Model for grain legumes. In G. Y Tsuji, G. Hoogenboom, & P. K. Thornton (Eds.), *Understanding options for agricultural production* (pp. 99-128). Dordrecht, The Netherlands: Kluwer Academic Publishers.

Boote, K. J., Jones, J. W., & Pickering, N. B. (1996). Potential uses and limitations of crop models. *Agronomy Journal, 88*, 704–716.

Boote, K. J., Kropff, M. J., & Bindraban, P. S. (2001). Physiology and modelling of traits in crop plants: implications for genetic improvement. *Agricultural Systems, 70*, 395–420. doi:10.1016/S0308-521X(01)00053-1

Boote, K. J., Mínguez, M. I., & Sau, F. (2002). Adapting the CROPGRO legume model to simulate growth of faba bean. *Agronomy Journal, 94*, 743–756.

Boote, K. J., & Tollenaar, M. (1994). Modeling genetic yield potential. In K. J. Boote, J. M. Bennett, T. R. Sinclair, & G. M. Paulsen (Eds.), *Physiology and Determination of Crop Yield* (pp. 533-565). Madison, WI: ASA-CSSA-SSSA.

Crane, T. A., Roncoli, C., Paz, J. O., Breuer, N., Broad, K., Hoogenboom, G., & Ingram, K. T. (2009). Seasonal climate forecasts and risk management among Georgia farmers. *Weather, Climate and Society* (Submitted for publication).

DeWit, C. T. (1965). *Photosynthesis of leaf canopies* (Agr. Res. Rep. 663). Pudoc, Wageningen.

Dornhoff, G. M., & Shibles, R. M. (1971). Varietal differences in net photosynthesis of soybean leaves. *Crop Science, 10*, 42–45.

Duncan, W. G. (1971). Leaf angles, leaf area, and canopy photosynthesis. *Crop Science, 11*, 482–485.

Fraisse, C. W., Breuer, N. E., Zierden, D., Bellow, J. G., Paz, J., & Cabrera, V. E. (2006). AgClimate: A climate forecast information system for agricultural risk management in the southeastern USA. *Computers and Electronics in Agriculture, 53*(1), 13–27. doi:10.1016/j.compag.2006.03.002

Gijsman, A. J., Hoogenboom, G., Parton, W. J., & Kerridge, P. C. (2002). Modifying DSSAT crop models for low-input agricultural systems using a soil organic matter-residue module from CENTURY. *Agronomy Journal, 94*, 462–474.

Godwin, D. C., & Jones, D. C. (1991). Nitrogen dynamics in soil-plant systems. In J. Hanks & J. T. Ritchie (Eds.), *Modeling plant and soil systems. Agronomy Monographs No. 31* (pp 287-321). Madison, WI: ASA, CSSA, and SSSA.

Godwin, D. C., & Singh, U. (1998). Nitrogen balance and crop response to nitrogen in upland and lowland cropping systems. In G. Y. Tsuji, G. Hoogenboom, & P. K. Thornton (Eds.), *Understanding options for agricultural production. System approaches for sustainable agricultural development* (pp. 55-77). Dordrecht, The Netherlands: Kluwer Academic Publishers.

Hoogenboom, G. (2000). Contribution of agrometeorology to the simulation of crop production and its applications. *Agricultural and Forest Meteorology, 103*(1-2), 137–157. doi:10.1016/S0168-1923(00)00108-8

Hoogenboom, G., Jones, J. W., Wilkens, P. W., Batchelor, W. D., Bowen, W. T., Hunt, L. A., et al. (1994). Crop models. In G. Y. Tsuji, G. Uehara, & S. Balas (Eds.), *DSSAT Version 3, Volume 2* (pp. 95-244). Honolulu, HI: University of Hawaii.

Hoogenboom, G., Jones, J. W., Wilkens, P. W., Porter, C. H., Batchelor, W. D., Hunt, L. A., et al. (2004). *Decision support system for agrotechnology transfer version 4.0* Honolulu, HI: University of Hawaii.

Hoogenboom, G., Wilkens, P. W., Thornton, P. K., Jones, J. W., Hunt, L. A., & Imamura, D. T. (1999). Decision support system for agrotechnology transfer V3.5. In G. Hoogenboom, P. W. Wilkens, & G. Y. Tsuji (Eds.), *DSSAT Version 3, Volume 4* (pp. 1-36). Honolulu, HI: University of Hawaii.

Jenkinson, D. S., & Rayner, J. H. (1977). Turnover of soil organic-matter in some of Rothamsted classical experiments. *Soil Science, 123*, 298–305. doi:10.1097/00010694-197705000-00005

Jones, J. W., Hoogenboom, G., Porter, C. H., Boote, K. J., Batchelor, W. D., & Hunt, L. A. (2003). DSSAT cropping system model. *European Journal of Agronomy, 18*, 235–265. doi:10.1016/S1161-0301(02)00107-7

Jones, J. W., Keating, B. A., & Porter, C. (2001). Approaches for modular model development. *Agricultural Systems, 70*(2), 421–443. doi:10.1016/S0308-521X(01)00054-3

Jones, J. W., Tsuji, G. Y., Hoogenboom, G., Hunt, L. A., Thornton, P. K., Wilkens, P., et al. (1998). Decision support system for agrotechnology transfer: DSSAT v3. In G. Y. Tsuji, G. Hoogenboom, & P. K. Thornton (Eds.), *Understanding options for agricultural production* (pp. 157-177). Boston: Kluwer Academic Press.

Keating, B. A., Carberry, P. S., Hammer, G. L., Probert, M. E., Robertson, M. J., & Holzworth, D. (2003). An overview of APSIM, a model designed for farming systems simulation. *European Journal of Agronomy, 18*, 267–288. doi:10.1016/S1161-0301(02)00108-9

Kraalingen, D., Van, W. G., Rappoldt, C., & van Laar, H. H. (2003). The Fortran simulation translator, a simulation language. *European Journal of Agronomy, 18*, 359–361. doi:10.1016/S1161-0301(02)00131-4

Loomis, R. S., Rabbinge, R., & Ng, E. (1979). Explanatory models in crop physiology. *Annual Review of Plant Physiology, 30*, 339–367. doi:10.1146/annurev.pp.30.060179.002011

Ma, L., Nielsen, D. C., Ahuja, L. R., Malone, R. W., Saseendran, S. A., & Rojas, K. W. (2003). Evaluation of RZWQM under varying irrigation levels in Eastern Colorado. *Transactions of the ASAE. American Society of Agricultural Engineers, 46*, 39–49.

Mavromatis, T., Boote, K. J., Jones, J. W., Irmak, A., Shinde, D., & Hoogenboom, G. (2001). Developing genetic coefficients for crop simulation models with data from crop performance trials. *Crop Science, 41*, 40–51.

Mavromatis, T., Boote, K. J., Jones, J. W., Wilkerson, G. G., & Hoogenboom, G. (2002). Repeatability of model genetic coefficients derived from soybean performance trials across different states. *Crop Science*, *42*, 76–89.

McCown, R. L., Hammer, G. L., Hargreaves, J. N. G., Holzworth, D. P., & Freebairn, D. M. (1996). APSIM: A novel software system for model development, model testing and simulation in agricultural systems research. *Agricultural Systems*, *50*, 255–271. doi:10.1016/0308-521X(94)00055-V

McKinion, J. M., & Baker, D. N. (1989). Application of the GOSSYM/COMAX system to cotton crop management. *Agricultural Systems*, *31*, 55–65. doi:10.1016/0308-521X(89)90012-7

Messina, C. D., Jones, J. W., Boote, K. J., & Vallejos, C. E. (2006). A gene-based model to simulate soybean development and yield responses to environment. *Crop Science*, *46*, 456–466.

Naab, J. B., Singh, P., Boote, K. J., Jones, J. W., & Marfo, K. O. (2004). Using the CROPGRO-peanut model to quantify yield gaps of peanut in the Guinean savanna zone of Ghana. *Agronomy Journal*, *96*, 1231–1242.

Parton, W. J., Stewart, J. W. B., & Cole, C. V. (1988). Dynamics of C, N, P and S in grassland soils: A model. *Biogeochemistry*, *5*, 109–131. doi:10.1007/BF02180320

Paz, J. O., Fraisse, C. W., Hatch, L. U., Garcia y Garcia, A., Guerra, L. C., & Uryasev, O. (2007). Development of an ENSO-based irrigation decision support tool for peanut production in the southeastern US. *Computers and Electronics in Agriculture*, *55*(1), 28–35. doi:10.1016/j.compag.2006.11.003

Pearcy, R. W., & Bjorkman, O. (1983). Physiological effects. In E. R. Lemon (Ed.), *CO_2 and plants: The response of plants to rising levels of atmospheric carbon dioxide. AAAS Selected Symp. 84* (pp. 65-105). Boulder, CO: Westview Press.

Putto, C., Pathanothai, A., Jogloy, S., Pannangpetch, K., Boote, K. J., & Hoogenboom, G. (2009). Determination of efficient test sites for evaluation of peanut breeding lines using the CSM-CROPGRO-peanut model. *Field Crops Research*, *110*, 272–281. doi:10.1016/j.fcr.2008.09.007

Ritchie, J. T. (1985). A user-oriented model of the soil water balance in wheat. In E. Fry & T. K. Atkin (Eds.), *Wheat growth and modeling* (pp. 293-305). New York: Plenum Press.

Ritchie, J. T. (1998). Soil water balance and plant water stress. In G. Y. Tsuji, G. Hoogenboom, & P. K. Thornton (Eds), *Understanding options for agricultural production* (pp. 41-54). Dordrecht, The Netherlands: Kluwer Academic Publishers.

Roncoli, C., Paz, J., Breuer, N., Ingram, K., Hoogenboom, G., & Broad, K. (2006). *Understanding farming decisions and potential applications of climate forecasts in South Georgia* (Tech. Rep. Series SECC-06-006). Gainesville, FL: Southeast Climate Consortium.

Smith, P., Smith, J. U., Powlson, D. S., McGill, W. B., Arah, J. R. M., & Chertov, O. G. (1997). A comparison of the performance of nine soil organic matter models using datasets from seven long-term experiments. *Geoderma*, *81*, 153–225. doi:10.1016/S0016-7061(97)00087-6

Stockle, C. O., Donatelli, M., & Nelson, R. (2003). CropSyst, a cropping systems simulation model. *European Journal of Agronomy*, *18*, 239–307. doi:10.1016/S1161-0301(02)00109-0

Stockle, C. O., Martin, S. A., & Campbell, G. S. (1994). CropSyst, a cropping systems simulation model: Water/nitrogen budgets and crop yield. *Agricultural Systems*, *46*, 335–359. doi:10.1016/0308-521X(94)90006-2

Traore, P. S. C., Bostick, W. M., Jones, J. W., Koo, J., Goita, K., & Bado, B. V. (2008). A simple soil organic matter model for biomass data assimilation in community-level carbon contracts. *Ecological Applications*, *18*(3), 624–636. doi:10.1890/07-1133.1

Tsuji, G. Y., Hoogenboom, G., & Thornton, P. K. (1998). *Understanding options for agricultural production. Systems approaches for sustainable agricultural development.* Dordrecht, The Netherlands: Kluwer Academic Publishers. ISBN 07923-4833-8.

Welch, S. M., Jones, J. W., Brennan, M. W., Reeder, G., & Jacobson, B. M. (2002). PCYield: Model-based decision support for soybean production. *Agricultural Systems*, *74*, 79–98. doi:10.1016/S0308-521X(02)00022-7

Welch, S. M., Roe, J. L., Das, S., Dong, Z., He, R., & Kirkham, M. B. (2005). Merging genomic control networks and soil-plant-atmosphere-continuum models. *Agricultural Systems*, *86*, 243–274. doi:10.1016/j.agsy.2004.07.019

Whisler, F. D., Acock, B., Baker, D. N., Fye, R. E., Hodges, H. F., & Lambert, J. R. (1986). Crop simulation models in agronomic systems. *Advances in Agronomy*, *40*, 141–208. doi:10.1016/S0065-2113(08)60282-5

White, J. W., & Hoogenboom, G. (1996). Simulating effects of genes for physiological traits in a process-oriented crop model. *Agronomy Journal*, *88*, 416–422.

Williams, J. R., Jones, C. A., Kiniry, J. R., & Spanel, D. A. (1989). The EPIC crop growth model. *ASAE*, *32*, 497-511.

Zhu, X. G., Portis, A. R. Jr, & Long, S. P. (2004). Would transformation of C3 crop plants with foreign Rubisco increase productivity? A computational analysis extrapolating from kinetic properties to canopy photosynthesis. *Plant, Cell & Environment*, *27*, 155–165. doi:10.1046/j.1365-3040.2004.01142.x

This work was previously published in International Journal of Agricultural and Environmental Information Systems, Volume 1, Issue 1, edited by Petraq Papajorgji and François Pinet, pp. 41-54, copyright 2010 by IGI Publishing (an imprint of IGI Global).

Compilation of References

(1988). Expert Systems in Engineering. InPham, D. T. (Ed.), *Artificial Intelligence in Industry*. Berlin: Springer-Verlag.

Abal, E. G., Greenfield, P. F., Bunn, S. E., & Tarte, D. M. (2006, January 16-18). Healthy Waterways: Healthy Catchments –An Integrated Research/Management Program to Understand and Reduce Impacts of Sediments and Nutrients on Waterways in Queensland, Australia. In *Frontiers of WWW Research and Development – Proceedings of APWeb 2006: 8th Asia-Pacific Web Conference*, Harbin, China (pp. 1126-1135).

Abelló, A. (2002). YAM2 (Yet Another Multidimensional Model): An Extension of UML. In *Proceedings of the International Database Engineering & Applications Symposium (IDEAS 02)*, Edmonton, Canada (pp. 172-181). Washington, DC: IEEE Computer Society.

Abt, V., Vigier, F., & Schneider, M. (2009). Enterprise business modelling languages applied to farm enterprise: A case study for IDEF0, GRAI Grid, and AMS languages. In P. Papajorgji & P. M. Pardalos (Eds.), *Advances in modeling agricultural systems* (pp. 167-192). New York: Springer.

Aerts, J.-M., Jans, P., Halloy, D., Gustin, P., & Berckmans, D. (2004). Labeling of cough data from pigs for on-line disease monitoring by sound analysis. *American Society of Agricultural and Biological Engineers*, *48*(1), 351–354.

Agostinho, S. (2008). Learning design representations to document, model, and share teaching practice. In Lockyer, L., Bennet, S., Agostinho, S., & Harper, B. (Eds.), *Handbook of research on learning design and learning objects: Issues, applications and technologies* (pp. 1–19). Hersey, PA: IGI Global.

Agrawal, R., & Srikant, R. (1994). Fast algorithms for mining association rules in large databases. In *Proceedings of the 20th International Conference on Very Large Data Bases* (pp. 487-499).

AGROVOC. (2007). *A multilingual agricultural thesaurus.* Retrieved from ftp://ftp.fao.org/gi/gil/gilws/aims/references/flyers/agrovoc_en.pdf.

Agumya, A., & Hunter, G. J. (2002). Responding to the consequences of uncertainty in geographical data. *International Journal of Geographical Information Science*, *16*(5), 405–417. doi:10.1080/13658810210137031

Ahlberg, M., Aanisma, P., & Dillon, P. (2005). Education for Sustainable Living: Integrating theory, practice, design, and development. *Scandinavian Journal of Educational Research*, *49*(2), 167–186. doi:10.1080/00313830500048923

Ahlberg, M., & Ahoranta, V. (2002). Two improved educational theory based tools to monitor and promote quality of geographical education and learning. *International Research in Geographical and Environmental Education*, *11*(2), 119–137. doi:10.1080/10382040208667475

Ahlqvist, O., & Ban, H. (2007). Categorical measurement semantics: A new second space for geography. *Geography Compass*, *1*(3), 536–555. doi:10.1111/j.1749-8198.2007.00026.x

Ahuja, L. R., Ma, L., & Timlin, D. J. (2006). Transdisciplinary soil physics research critical to synthesis and modeling of agricultural systems. *Soil Science Society of America Journal*, *70*, 311–326. doi:10.2136/sssaj2005.0207

Akehurst, D., & Bordbar, B. (2001). On querying UML data models with OCL. In *«UML» 2001 — The Unified Modeling Language. Modeling Languages, Concepts, and Tools* (pp. 91-103).

Al Haj Khaled, R., Duru, M., Decruyenaere, V., Jouany, C., & Cruz, P. (2006). Using leaf traits to rank native grasses according to their nutritive value. *Rangeland Ecology and Management, 59*, 648–654. doi:10.2111/05-031R2.1

Alberti, M., Solera, G., & Tsetsi, V. (1994). *La città sostenibile*. Milano, Italy: FrancoAngeli.

Alexander, C. A. (1977). *Pattern language: Towns, buildings, constructions*. New York: Oxford University Press.

Altieri, M. A. (1999). The ecological role of biodiversity in agroecosystems. *Agriculture Ecosystems & Environment, 74*, 19–31. doi:10.1016/S0167-8809(99)00028-6

Amorim, R. R., Lama, M., Sánchez, E., Riera, A., & Vila, X. A. (2006). A Learning design ontology based on the IMS specification. *Journal of Educational Technology & Society, 9*(1), 38–57. Retrieved from http://www.ifets.info/journals/9_1/5.pdf.

Anderson, L. W., & Krathwohl, D. R. (Eds.). (2001). *A taxonomy for learning, teaching and assessing: A revision of Bloom's Taxonomy of educational outcomes, Complete edition*. New York: Longman.

Andrienko, G. L., & Andrienko, N. V. (1999). Knowledge-based visualization to support spatial data mining. In D. J. Hand, J. N. Kok, & M. R. Berthold (Eds.), *Proceedings of the Third International Symposium on Advances in Intelligent Data Analysis* (LNCS 1642, pp. 149-160).

Andrienko, N., Andrienko, G., & Gatalsky, P. (2003). Exploratory spatio-temporal visualization: an analytical review. *Journal of Visual Languages and Computing, 14*(6), 503–541. doi:10.1016/S1045-926X(03)00046-6

Andrieu, N., Poix, C., Josien, E., & Duru, M. (2007). Simulation of forage management strategies considering farm-level land diversity: Example of dairy farms in the Auvergne. *Computers and Electronics in Agriculture, 55*, 36–48. doi:10.1016/j.compag.2006.11.004

Anothai, J., Patanothai, A., Pannangpetch, K., Jogloy, S., Boote, K. J., & Hoogenboom, G. (2009). Multi-environment evaluation of peanut lines by model simulation with the cultivar coefficients derived from a reduced set of observed field data. *Field Crops Research, 110*, 111–121. doi:10.1016/j.fcr.2008.07.009

Anselin, L. (1995). Local indicators of spatial association – LISA. *Geographical Analysis, 27*, 93–115. doi:10.1111/j.1538-4632.1995.tb00338.x

ANSI Z535.6. (2006). *American national standard for product safety signs and labels*.

Ansquer, P., Al Haj Khaled, R., Cruz, P., Theau, J. P., Therond, O., & Duru, M. (2009). Characterizing and predicting plant phenology in species-rich grasslands. *Grass and Forage Science, 64*, 57–70. doi:10.1111/j.1365-2494.2008.00670.x

Ansquer, P., Duru, M., Theau, J. P., & Cruz, P. (2008). Functional traits as indicators of fodder provision over a short time scale in species-rich grasslands. *Annals of Botany, 103*, 117–126. doi:10.1093/aob/mcn215

Antle, J. M., Capalbo, S. M., Elliott, E. T., Hunt, H. W., Mooney, S., & Paustian, K. H. (2001). Research needs for understanding and predicting the behavior of managed ecosystems: Lessons from the study of agroecosystems. *Ecosystems (New York, N.Y.), 4*, 723–735. doi:10.1007/s10021-001-0041-0

Appice, A., Ceci, M., Lanza, A., Lisi, F. A., & Malerba, D. (2003). Discovery of spatial association rules in georeferenced census data: A relational mining approach. *Intelligent Data Analalysis, 7*(6).

Arigon, A. M., Miquel, M., & Tchounikine, A. (2007). Multimedia Data Warehouses: a Multiversion Model and a Medical Application. *Journal of Multimedia Tools and Applications, 35*(1), 91–108. doi:10.1007/s11042-007-0118-7

Aronne, G., & De Micco, V. (2008). The use of multivariate analysis for honey characterisation. In G. Aronne, M. Buonanno, & V. De Micco (Eds.), *Improving honey production and marketing: new approaches to botanical and geographical characterization* (pp. 24-31). ISBN 88-901941-3-8

Aronne, G., Buonanno, M., & De Micco, V. (2008). Identification and characterisation of produceable honeys in a geographical area. In G. Aronne, G., Buonanno, M., & De Micco, V. (Eds.). (2008). Improving honey production and marketing: new approaches to botanical and geographical characterization (pp. 32-46). ISBN 88-901941-3-8

Aronne, G., De Micco, V., & Buonanno, M. (2008). Analysis of honey variability in different years. In G. Aronne, M. Buonanno, & V. De Micco (Eds.), *Improving honey production and marketing: new approaches to botanical and geographical characterization* (pp. 47-52). ISBN 88-901941-3-8

Aronne, G., & De Micco, V. (2010). (in press). Traditional melissopalynology integrated by multivariate analysis and sampling methods to improve botanical and geographical characterization of honeys. *Plant Biosystems*.

Arulselvan, A., Baourakis, G., Boginski, V., Korchina, E., & Pardalos, P. M. (2008). Analysis of food industry market using network approaches. *British Food Journal*, *110*(9), 916–928. doi:10.1108/00070700810900611

Asconit Consultants. (2003). *État initial du site et évaluation environnementale du SCoT du pays de Montbéliard*. Paris, France: Direction Générale de l'Environnement de Franche-Comté.

Asconit Consultants. (2005a). *IVB, Utilisation d'un SIG pour l'expression des enjeux de l'Etat dans le cadre d'un SCoT, application au territoire du SCoT Sud-Loire – guide méthodologique*. Paris, France: DIREN Rhône-Alpes.

Asconit Consultants. (2005b). *IVB, Utilisation d'un système d'information géographique pour l'expression des enjeux de l'Etat, application au territoire du SCoT Sud Loire – guide technique*. Paris, France: DIREN Rhône-Alpes.

Atkinson, C. (2006, October 16). *Models versus Ontologies - What's the Difference and where does it Matter?* Paper prestented at VORTE2006, Regal Kowloon Hotel, Hong Kong.

Auger, P., Kooi, B., Bravo de la Parra, R., & Poggiale, J. C. (2006). Bifurcation analysis of a predator-prey model with predators using hawk and dove tactics. *Journal of Theoretical Biology*, *238*, 597–607. PubMed doi:10.1016/j.jtbi.2005.06.012

Baclawski, K., Kokar, M., Kogut, P., Hart, L., Smith, J., Holmes, W., et al. (2002). *Extending UML to support ontology engineering for the Semantic Web*. Paper presented at the Fourth International Conference on the Unified Modeling Language.

Baclawski, K., Kokar, M., Kogut, P., Hart, L., Smith, J., & Letkowski, J. (2002). Extending the unified modeling language for ontology development. *Software and System Modeling*, *1*(2), 142–156. doi:10.1007/s10270-002-0008-4

Bailey, C., Zalfan, M. T., Davis, H. C., Fill, K., & Conole, G. (2006). Panning for gold: Designing pedagogically inspired learning nuggets. *Journal of Educational Technology & Society*, *9*(1), 113–122. Retrieved from http://eprints.soton.ac.uk/19642/.

Bailey, T. C., & Gatrell, A. C. (1995). *Interactive spatial data analysis*. Harlow, UK: Prentice Hall.

Bammer, G. (2005). Integration and Implementation Sciences: Building a New Specialization. *Ecology and Society*, *10*(2), 6.

Batchelor, W. D., Jones, J. W., Boote, K. J., & Pinnschmidt, H. O. (1993). Extending the use of crop models to study pest damage. *Transactions of the ASAE. American Society of Agricultural Engineers*, *36*, 551–558.

Battesti, M. J., & Goeury, C. (1992). Efficacité de l'analyse melissopalynologie quantitative pour la certification des origines géographiques et botaniques des miels: le modèle des miels corses. *Review of Palaeobotany and Palynology*, *75*, 77–102. doi:10.1016/0034-6667(92)90151-6

Batton-Hubert, M., Bonnevialle, M., Joliveau, T., & Paran, F. (2007a). *IVB dans un Scot: Ateliers*. Paper presented at the Rencontres internationales Géomatique et Territoire (Paper presented at the Sageo, Saint-Etienne, France).

Batton-Hubert, M., Bonnevialle, M., Joliveau, T., & Paran, F. (2007b). *Les méthodes d'identification des Infrastructures Vertes et Bleues (IVB) au service des projets d'aménagement et de développement durable des SCoT*. Paper presented at the Semaine régionale du développement durable, Saint-Etienne, France.

Batton-Hubert, M., Bonnevialle, M., Joliveau, T., & Paran, F. (2008). *Intégrer une dimension écologique et paysagère dans la planification territoriale – Méthode et questionnement à propos de la démarche Infrastructures Vertes et Bleues (IVB) dans les SCoT de la Loire (France).* Paper presented at the Les outils pour décider ensemble, Université de Laval, Québec, Canada.

Batty, M., & Jiang, B. (2000). Multi-agent simulation: Computational dynamics within GIS. In Atkinson, P., & Martin, D. (Eds.), *Innovation in GIS VII: GIS and geo-computation* (pp. 55–71). London, UK: Taylor & Francis.

Bayarri, M. J. (2007). A framework for validation of computer models. *Technometrics, 49*(2), 138–154. doi:10.1198/004017007000000092

Beck, H., Morgan, K., Jung, Y., Wu, J., Grunwald, S., & Kwon, H.-y. (2009). Ontology-based simulation applied to soil, water, and nutrient management. In P. J. Papajorgji & P. M. Pardalos (Eds.), *Advances in modeling agricultural systems* (pp. 1-34). New York: Springer.

Becu, N., Bousquet, F., Barreteau, O., Perez, P., & Walker, A. (2003). A methodology for eliciting and modelling stakeholders' representations with agent based modelling. In D. Hales, B. Edmonds, E. Norling, & J. Rouchier (Eds.), *Proceedings of the 4th International Workshop on Multi-Agent Based Simulation III* (LNCS 2927, pp. 131-149).

Bédard, Y. (1998). *Site Perceptory*. Retrieved from http:// sirs.scg.ulaval.ca/ perceptory/

Bédard, Y. (1999a, décembre 15). *PERCEPTORY: Nouvel Atelier de Génie Logiciel pour la modélisation des bases de données spatiales et spatio-temporelles.* Paper presented at the Journée scientifique du Cemagref-Engref, Montpellier, France.

Bédard, Y., & Paquette, F. (1989, 2-7 April). *Extending entity/relationship formalism for spatial information systems, AUTO-CARTO 9.* Paper presented at the Ninth International Symposium on Computer-Assisted Cartography, Baltimore, MD.

Bédard, Y., Larrivée, S., Proulx, M.-J., & Nadeau, M. (2004, November). *Modeling Geospatial Databases with Plug-ins for Visual Languages: A Pragmatic Approach and the Impacts of 16 Years of Research and Experimentations on Perceptory.* Paper presented at the ER Workshops 2004 CoMoGIS, Shanghai, China.

Bédard, Y., Pageau, J., & Caron, C. (1992, 1-14 August). *Spatial Data Modeling: The Modul-R Formalism and CASE Technology.* Paper presented at the ISPRS Symposium, Washington, DC.

Bédard, Y. (1986). A Study of data using a communication based conceptual framework of land information systems. *Le Géomètre Canadien, 40*(4), 449–460.

Bédard, Y. (1999b). Visual modelling of spatial databases: towards spatial PVL and UML. *Geomatica, 53*(2), 169–186.

Bédard, Y., & Larrivée, S. (2008). Modeling with Pictogrammic Languages. In Shekhar, S., & Xiong, H. (Eds.), *Encyclopedia of Geographical Information Sciences* (pp. 716–725). New York: Springer.

Bédard, Y., & Larrivée, S. (2008). Spatial database modeling with pictrogrammic languages. In Shekhar, S., & Xiong, H. (Eds.), *Encyclopedia of GIS* (pp. 716–725). New York: Springer Verlag. doi:10.1007/978-0-387-35973-1_812

Bédard, Y., Merrett, T., & Han, J. (2001). Fundamentals of spatial data warehousing for geographic knowledge discovery. In Miller, H. J., & Han, J. (Eds.), *Geographic data mining and knowledge discovery.* doi:10.4324/9780203468029_chapter_3

Bédard, Y., Proulx, M., Bernier, E., & Gosselin, P. (2008). *How the new analytical geomatics technologies can help environmental health professionals and decision-makers to make further use of mapping than what is offered traditionally by geographic information systems (GIS) and web mapping.* Vancouver, Canada: The National Collaborating Centre for Environmental Health.

Bédard, Y., Proulx, M., Rivest, S., & Badard, T. (2006). Merging Hypermedia GIS with Spatial On-Line Analytical Processing: Towards Hypermedia SOLAP. In Stefanakis, E., Peterson, M., & Delis, V. (Eds.), *Geographic Hypermedia: Concepts and Systems* (pp. 167–185). Berlin: Springer. doi:10.1007/978-3-540-34238-0_10

Bédard, Y., Rivest, S., & Proulx, M. J. (2005). Spatial on-line analytical processing (SOLAP): concepts, architectures and solutions from a geomatics engineering perspective. In Koncillia, W. R. (Ed.), *Data warehouses and OLAP: Concepts, architectures and solutions.*

Bejaoui, L., Pinet, F., Schneider, M., & Bedard, Y. (2008). An adverbial approach for the formal specification of topological constraints involving regions with broad boundaries. *Lecture Notes in Computer Science, 5231*, 383–396. doi:10.1007/978-3-540-87877-3_28

Ben Messaoud, R., Boussaid, O., & Loudcher Rabaséda, S. (2006). Efficient Multidimensional Data Representation Based on Multiple Correspondence Analysis. In *Proceedings of the 2th ACM SIGKDD International Conference on Knowledge Discovery and Data Mining (KDD'2006)*, Philadelphia (pp. 662-667). New York: ACM Press.

Bentayeb, F., Boussaid, O., Darmont, J., Harbi, N., & Loudcher, S. (Eds.). (2010). Warehousing and Mining Complex Data: Applications to Biology, Medicine, Behavior, Health and Environment. *International Journal of Biomedical Engineering and Technology, 3*(1-2).

Bermejo, S., & Cabestany, J. (2000). Adaptive soft k-nearest-neighbour classifiers. *Pattern Recognition, 33*, 1999–2005. doi:10.1016/S0031-3203(99)00186-7

Berners-Lee, T., Hendler, J., & Lassila, O. (2001). The Semantic Web. *Scientific American, 284*(5), 34–43.

Berthoud, G. (2001). *Les Corridors biologiques en Isère: Projet de Réseau écologique Départemental de l'Isère (REDI)*. Paris, France: Conseil Général de l'Isère.

Berthoud, G., Lebeau, R., & Righetti, A. (2004). *Le Réseau Ecologique National (REN) Suisse Cahiers de l'environnement, Office Fédéral de l'Environnement des Forêts et du Paysage* (p. 373). Berne, Suisse: OFEFP.

Bertin, J., & Bonin, S. (1992). *La graphique et le traitement graphique de l'information*. Paris: Flammarion.

Bertsekas, D. P. (1995). *Nonlinear programming*. Belmont, MA: Athena Scientific.

Bezdek, J. (1981). *Pattern recognition with fuzzy objective function algorithms*. New York: Plenum Press.

Bimonte, S. (2007). *Vers l'intégration de l'informtion géographique dans les entrepôts de données et l'analyse en ligne: de la modélisation à la visualisation*. Unpublished doctoral dissertation, INSA Lyon, France.

Bimonte, S., Tchounikine, A., & Miquel, M. (2005). Towards a Spatial Multidimensional Model. In *Proceedings of the 8th ACM International Workshop on Data warehousing and OLAP (DOLAP 05)* (pp. 39-46). New York: ACM Press.

Bimonte, S., Tchounikine, A., & Miquel, M. (2006). GeoCube, a Multidimensional Model and Navigation Operators Handling Complex Measures: Application in Spatial OLAP. In T. Yaknho, M. Tatyana, & E. Neuhold (Eds.), *Proceedings of International Conference in Advances in Information Systems* (pp. 100-109). Berlin: Springer.

Bimonte, S., Wehrle, P., Tchounikine, A., & Miquel, M. (2006). GeWOlap: A Web Based Spatial OLAP Proposal. In R. Meersman, Z. Tari, & P. Herrero (Eds.), *Proceedings of Workshop on Semantic-Based Geographical Information Systems* (pp. 1596-1605). Berlin: Springer.

Bimonte, S., Tchounikine, A., Miquel, M., & Laurini, R. (2007). *Proposition pour l'intégration de l'analyse spatiale et de l'analyse multidimensionnelle*. Revue des Nouvelles Technologies de l'Information, Numéro Spécial. Cépaduès-Éditions.

Bishop, C. M. (1995). *Neural networks for pattern recognition*. Oxford, UK: Oxford University Press.

Bishop, M. C. (2006). *Pattern recognition and machine learning*. New York: Information Sciences and Statistics, Springer.

Bishr, Y. (1998). Overcoming the semantic and other barriers to GIS interoperability. *International Journal of Geographical Information Science, 12*, 299–314. doi:10.1080/136588198241806

Bodon, F. (2004). Surprising results of trie-based fim algorithms. In *Proceedings of the Workshop on Frequent Itemset Mining Implementations* (Vol. 126).

Bogorny, V., Valiati, J., Camargo, S., Engel, P., Kuijpers, B., & Alvares, L. O. (2006). Mining maximal generalized frequent geographic patterns with knowledge constraints. In *Proceedings of the IEEE International Conference on Data Mining* (pp. 813-817). Washington, DC: IEEE Computer Society.

Boin, A. T. (2008). *Exposing Uncertainty: Communicating spatial data quality via the Internet* (p. 197). Unpublished doctoral dissertation, University of Melbourne, Melbourne, Australia.

Boltanski, L., & Thévenot, L. (2006). *On justification: The economies of worth*. Princeton, NJ: Princeton University Press.

Booch, G., Rumbaugh, J., & Jacobson, I. (1999). *The unified modeling language user guide*. Upper Saddle River, NJ: Addison-Wesley.

Boote, K. J., & Tollenaar, M. (1994). Modeling genetic yield potential. In K. J. Boote, J. M. Bennett, T. R. Sinclair, & G. M. Paulsen (Eds.), *Physiology and Determination of Crop Yield* (pp. 533-565). Madison, WI: ASA-CSSA-SSSA.

Boote, K. J., Batchelor, W. D., Jones, J. W., Pinnschmidt, H., & Bourgeois, G. (1993). Pest damage relations at the field level. In F. W. T. Penning de Vries et al. (Ed.), *Systems Approaches for Agricultural Development* (pp. 277-296). Dordrecht, The Netherlands: Kluwer Academic Publishers.

Boote, K. J., Jones, J. W., & Hoogenboom, G. (1998). Simulation of crop growth: CROPGRO Model. In R. M. Peart & R. B. Curry (Eds.), *Agricultural systems modeling and simulation* (pp. 651-692). New York: Marcel Dekker.

Boote, K. J., Jones, J. W., Hoogenboom, G., & Pickering, N. B. (1998). The CROPGRO Model for grain legumes. In G. Y Tsuji, G. Hoogenboom, & P. K. Thornton (Eds.), *Understanding options for agricultural production* (pp. 99-128). Dordrecht, The Netherlands: Kluwer Academic Publishers.

Boote, K. J., Jones, J. W., Batchelor, W. D., Nafziger, E. D., & Myers, O. (2003). Genetic coefficients in the CROPGRO-soybean model: Links to field performance and genomics. *Agronomy Journal*, *95*, 32–51.

Boote, K. J., Jones, J. W., & Hoogenboom, G. (2008). Crop simulation models as tools for agro-advisories for weather and disease effects on production. *Journal of Agrometeorology*, *10*, 9–17.

Boote, K. J., Jones, J. W., & Pickering, N. B. (1996). Potential uses and limitations of crop models. *Agronomy Journal*, *88*, 704–716.

Boote, K. J., Kropff, M. J., & Bindraban, P. S. (2001). Physiology and modelling of traits in crop plants: implications for genetic improvement. *Agricultural Systems*, *70*, 395–420. doi:10.1016/S0308-521X(01)00053-1

Boote, K. J., Mínguez, M. I., & Sau, F. (2002). Adapting the CROPGRO legume model to simulate growth of faba bean. *Agronomy Journal*, *94*, 743–756.

Borges, K. A. V., Davis, C. A. Jr, & Laender, A. H. F. (2001). OMT-G: An Object-Oriented Data Model for Geographic Applications. *GeoInformatica*, *5*(3), 221–260. doi:10.1023/A:1011482030093

Borruso, G., & Schoier, G. (2004). Density analysis on large geographical databases. Search for an index of centrality of services at urban scale. In A. Laganá, M. L. Gavrilova, V. Kumar, Y. Mun, C. J. K. Tan, & O. Gervasi (Eds.), *Proceedings of the International Conference on Computational Science and its Applications* (LNCS 3044, pp. 1009-1015). doi: 10.1007/978-3-540-24709-8_106

Boussaid, O., Darmont, J., Bentayeb, F., & Loudcher, S. (2008). Warehousing complex data from the Web. *International Journal of Web Engineering and Technology*, *4*(4), 408–433. doi:10.1504/IJWET.2008.019942

Bradley, S., & Fayyad, M. (1998, July 24-27). Refining initial points for k-means clustering. In J. Shavlik (Ed.), *Proceedings of the 15th International Conference on Machine Learning (ICML98),* Madison, WI (pp. 91-99). San Francisco: Morgan Kaufmann.

Bredeweg, B., Bouwer, A., Jellema, J., Bertels, D., Linnebank, F., & Liem, J. (2006). Garp3 – a new workbench for qualitative reasoning and modelling. In *Proceedings of the 20th International Workshop on Qualitative Reasoning (QR-06)*, Hanover, NH.

Breiman, L., Meisel, W., & Purcell, E. (1977). Variable kernel estimates of multivariate densities. *Technometrics*, *19*, 135–144. doi:10.2307/1268623

Breu, R., Grosu, R., Huber, F., Rumpe, B., & Schwerin, W. (1997). *Towards a precise semantics for object-oriented modeling techniques. In Object-Oriented Technology, ECOOP '97 Workshop.*

Brodeur, J. (n.d.). *Interopérabilité des données géospatiales: élaboration du concept de proximité géosémantique* (p. 247). Unpublished doctoral dissertation, Université Laval, Mondreal, Canada.

Brodeur, J., & Badard, T. (2008). Modeling with ISO 191xx Standards. In Shekhar, S., & Xiong, H. (Eds.), *Encyclopedia of Geographical Information Sciences* (pp. 705–716). New York: Springer.

Brouns, F., Koper, R., Manderveld, J., Van Bruggen, J., Sloep, P., & Van Rosmalen, P. (2005). A first exploration of an inductive analysis approach for detecting learning design patterns. *Journal of Interactive Media in Education, 3*, 1–10. Retrieved from http://jime.open.ac.uk/2005/03.

Brunk, C., Kelly, J., & Kohavi, R. (1997). An integrated system for data mining. In *Proceedings of Knowledge Discovery in Databases* (pp. 135–138). Mineset.

Buckheit, J., & Donoho, D. L. (1995). *Wavelab and reproducible research.* http://www-stat.stanford.edu/~donoho/Reports/1995/

Bunn, S., Abal, E., Greenfield, P., & Tarte, D. (2007). Making the connection between healthy waterways and healthy catchments: South east Queensland, Australia. *Water Science and Technology: Water Supply, 7*(2), 93–100. doi:10.2166/ws.2007.044

Burel, F., & Baudry, J. (1999). *Écologie du paysage: Concepts, méthodes et applications*. Paris, France: Tech. & Doc.

Burt, J. E., & Barber, G. M. (1996). *Elementary statistics for geographers* (2nd ed.). New York, NY: Guilford Press.

Cabot, J., & Teniente, E. (2006). Constraint support in MDA tools: A survey. In D. Akehurst, R. Vogel, R. Paige (Eds.), *Model driven architecture – foundations and applications* (pp. 256-267). New York: Springer.

Camagni, R., Capello, R., & Nijkamp, P. (1998). Towards sustainable city policy: An economy-environment technology nexus. *Ecological Economics, 24*(1), 103–118. doi:10.1016/S0921-8009(97)00032-3

Camagni, R., Gibelli, M. C., & Rigamonti, P. (2002). Urban mobility and urban form: The social and environmental costs of different patterns of urban expansion. *Ecological Economics, 40*(2), 199–216. doi:10.1016/S0921-8009(01)00254-3

Camossi, E., Bertolotto, M., & Bertino, E. (2008). Querying Multigranular Spatio-temporal Objects. In S. Bhowmick, J. Küng, & R. Wagner (Eds.), *Proceedings of International Conference on Database and Expert Systems Applications* (pp. 390-403). Berlin: Springer.

Cao, H., Mamoulis, N., & Cheung, D. W. (2005). Mining frequent spatio-temporal sequential patterns. In *Proceedings of the 5th International Conference on Data Mining* (pp. 82-89). Washington, DC: IEEE Computer Society.

Cao, L. (2008). Domain driven data mining (d3m). In *Proceedings of the International Conference on Data Mining* (pp. 74-76). Washington, DC: IEEE Computer Society.

Capuano, N., De Santo, M., Marsella, M., Molinara, M., & Salerno, S. (2000, August). A multi-agent architecture for intelligent tutoring. In *Proceedings of the International Conference on Advances in Infrastructure for Electronic Business, Science, and Education on the Internet,* L'Aquila, Italy. Retrieved from http://www.capuano.biz/Papers/SSGRR_2000_P1.pdf

Caron, C., Bédard, Y., & Gagnon, P. (1993). MODUL-R: un formalisme individuel adapté pour les SIRS. *Revue de géomatique, 3*(3), 283-306.

Cavailhès, J., Peeters, D., Sekeris, E., & Thisse, J. (2004). The Periurban city: Why to live between the suburbs and the countryside. *Regional Science and Urban Economics, 34*(6), 681–703. doi:10.1016/j.regsciurbeco.2003.08.003

Ceci, M., Appice, A., & Malerba, D. (2007). Discovering emerging patterns in spatial databases: A multi-relational approach. In J. N. Kok, J. Koronacki, R. Lopez de Mantaras, S. Matwin, D. Mladenic, & A. Skowron (Eds.), *Proceedings of the 11th European Conference on Principles and Practice of Kowledge Discovery in Databases* (LNCS 4702, pp. 390-397).

Celik, M., Kang, J. M., & Shekhar, S. (2007). Zonal co-location pattern discovery with dynamic parameters. In *Proceedings of the IEEE International Conference on Data Mining* (pp. 433-438). Washington, DC: IEEE Computer Society.

Chakhar, S. (2006). *Cartographie décisionnelle multicritère: formalisation et implémentation informatique.* Unpublished doctoral dissertation, Université Paris Dauphine, Paris, France.

Chakhar, S., & Mousseau, V. (2008). GIS-based multicriteria spatial modeling generic framework. *International Journal of Geographical Information Science*, *22*(11), 1159–1196. doi:10.1080/13658810801949827

Chang, C. C., & Lin, C. J. (2001). *LIBSVM: a library for support vector machines*. Retrieved November 23, 2009, from http://www.csie.ntu.edu.tw/~cjlin/libsvm

Chang, C. L., & Lee, R. C. T. (1973). *Symbolic Logic and Mechanical Theorem Proving*. New York: Academic Press.

Charrier, J. B. (1994). *Geografia dei rapporti città-campagna*. Milano, Italy: FrancoAngeli.

Chau, K. W. (2007). An ontology-based knowledge management system for flow and water quality modelling. *Advances in Engineering Software*, *38*(3), 172–181. doi:10.1016/j.advengsoft.2006.07.003

Chen, V. C. P. (2003). A review of design and modeling in computer experiments. In Khattree, R., & Rao, C. R. (Eds.), *Handbook of Statistics* (*Vol. 22*). Elsevier.

Cherfi, S. S., Akoka, J., & Comyn-Wattiau, I. (2007). Perceived vs. measured quality of conceptual schemas: an experimental comparison. In *Proceedings of ACM Int.* (Vol. 334, pp. 185-190).

Chinchuluun, R., Won Suk, L., Bhorania, J., & Pardalos, P. M. (2009). Clustering and classification algorithms in food and agricultural applications: A survey. In P. Papajorgji and P. M. Pardalos (Eds.), *Advances in modeling agricultural systems* (pp. 1-22). New York: Springer.

Chiorean, D., Demuth, B., Gogolla, M., & Warmer, J. (2006). In *Proceedings of the Sixth OCL Workshop OCL for (Meta-)Models in Multiple Application Domains (OCLApps 2006)*.

Chuffart, F., Faure, T., Dumoulin, N., Reuillon, R., & Deffuant, G. (2008). Design and Analysis of Computer Experiments with SimExplorer. In *Proceedings of the 2008 European Simulation and Modelling Conference (ESM 08)*, Le Havre, France (pp. 575-577).

Chuffart, F., Dumoulin, N., Faure, T., & Deffuant, G. (2010). SimExplorer: Programming Experimental Design on Models and Managing Quality of Modelling Process. *International Journal of Agricultural and Environmental Information Systems*, *1*(1), 55–68.

Cifarelli, C., Guarracino, M. R., Seref, O., Cuciniello, S., & Pardalos, P. M. (2007). Incremental classification with generalized eigenvalues. *Journal of Classification*, *24*(2), 205–219. doi:10.1007/s00357-007-0012-z

Claramunt, C., Coulondre, S., & Libourel, T. (1997). Autour des méthodes orientées objet pour la conception des SIG. *Revue internationale de géomatique, 7*(3-4), 233-257.

Clark, T., Evans, A., Kent, S., Brodsky, S., & Cook, S. (2000). *A feasibility study in rearchitecting UML as a family of languages using a precise OO meta-modeling approach*. Retrieved from http://www.cs.york.ac.uk/puml/mmf/

Clegg, S. R., Hardy, C., & Nord, W. R. (1996). *Handbook of organization studies*. London, UK: Sage.

Coad, P., & Yourdon, E. (1991). *Object-oriented analysis*. Englewood Cliffs, NJ: Prentice Hall.

Codd, E., Codd, S., & Salley, C. (1994). *Providing OLAP (On-line Analytical Processing) to User-Analysts: An IT Mandate*. E. F. Codd Associates.

Coleman, D., Arnold, P., Bodoff, S., Dollin, C., Gilchrist, H., Hayes, F., et al. (1994). *Object-oriented development - the fusion method*. Englewood Cliffs, NJ: Prentice-Hall.

Colonnese, G., Manhães, R., Montenegro, S., Carvalho, R., & Tanaka, A. (2005). PostGeoOlap: an Open-Source Tool for Decision Support. In *Proceedings of the 2nd Simpósio Brasileiro de Sistemas de Informação*.

Comas, J., Rodriguez-Roda, I., Sanchez-Marre, M., Cordes, U., Freixo, A., & Araez, J. (2003). A knowledge-based approach to the deflocculation problem: integrating on-line, off-line, and heuristic information. *Water Research*, *37*, 2377–2387. doi:10.1016/S0043-1354(03)00018-6

Conole, G. (2007). Describing learning activities: Tools and resources to guide practice. In Beetham, H., & Sharpe, R. (Eds.), *Rethinking pedagogy for a digital age* (pp. 81–91). Oxford, UK: Routledge Falmer.

Conole, G. (2008). Capturing practice, the role of mediating artefacts in learning design. In Lockyer, L., Bennet, S., Agostinho, S., & Harper, B. (Eds.), *Handbook of research on learning design and learning objects: Issues, applications and technologies* (pp. 187–207). Hershey, PA: IGI Global.

Console, L., & Torasso, P. (1990). Integrating models of correct behavior into abductive diagnosis. In *Proceedings of the European Conference on Artificial Intelligence (ECAI)*, Stockholm (pp. 160–166). London: Pitman Publishing.

Console, L., Theseider Dupré, D., & Torasso, P. (1991). On the relationship between abduction and deduction. *Journal of Logic and Computation*, *1*(5), 661–690. doi:10.1093/logcom/1.5.661

Cook, S., & Daniels, J. (1994). *Designing object systems-object oriented modeling with Syntropy*. Englewood Cliffs, NJ: Prentice-Hall.

Coquillard, P., & Hill, D. R. C. (1997). *Modélisation et simulation d'écosystèmes: Des modèles déterministes aux simulations à évènements discrets*. Paris, France: Masson.

Cortes, C., & Vapnik, V. (1995). Support vector machines. *Machine Learning*, *20*, 273–279.

Crane, T. A., Roncoli, C., Paz, J. O., Breuer, N., Broad, K., Hoogenboom, G., & Ingram, K. T. (2009). Seasonal climate forecasts and risk management among Georgia farmers. *Weather, Climate and Society* (Submitted for publication).

Cranefield, S. (2001b). *UML and the Semantic Web.* Paper presented at the International Semantic Web Working *Symposium*, Palo Alto, CA.

Cranefield, S., & Purvis, M. *(1999).* UML as an ontology modelling language. *In* Proceedings of the Workshop on Intelligent Information Integration,16th International Joint Conference on Artificial Intelligence (IJCAI-99).

Cranefield, S. (2001a). Networked knowledge representation and exchange using UML and RDF. *Journal of Digital Information*, *1*(8).

Crozier, M. (1964). *The bureaucratic phenomenon*. Chicago, IL: University of Chicago Press.

Crozier, M., & Friedberg, E. (1977). *L'acteur et le système: Les contraintes de l'action collective*. Paris, France: Seuil.

CUAHSI. (2009). *Hydrological Information System (HIS)*. Retrieved March 8, 2010, from http://his.cuahsi.org/

Damiani, M., & Spaccapietra, S. (2006). Spatial Data Warehouse Modeling. In *Processing and Managing Complex Data for Decision Support* (pp. 1–27). Hershey, PA: IGI Global. doi:10.4018/978-1-59140-655-6.ch001

Danese, M., Lazzari, M., & Murgante, B. (2008). Kernel density estimation methods for a geostatistical approach in seismic risk analysis: The case study of Potenza Hilltop Town (southern Italy). In O. Gervasi, B. Murgante, A. Laganà, D. Taniar, Y. Mun, & M. L. Gavrilova (Eds.), *Proceedings of the International Conference on Computational Science and its Applications* (LNCS 5072, pp. 415-427). doi: 10.1007/978-3-540-69839-5_31

Danese, M., Lazzari, M., & Murgante, B. (2009). Geostatistics in historical macroseismic data analysis. *Transactions on Computational Sciences*, *6*(5730), 324–341. doi: 10.1007/978-3-642-10649-1_19

Darmont, J., Boussaid, O., Ralaivao, J. C., & Aouiche, K. (2005). An Architecture Framework for Complex Data Warehouses. In *Proceedings of the 7th International Conference on Enterprise Information Systems (ICEIS 05)*, Miami, FL (pp. 370-373).

Das Neves, F., Gordillo, S., Mostaccio, C., & Levato, A. (1997, April 16-18). *Toward a foundation for Object Oriented GIS Design.* Paper presented at the Joint European Conference on Geographical Information, Vienna, Austria.

De Coligny, F. (2006, November 13-17). Efficient building of forestry modelling software with the capsis methodology. In *Proceedings of the Second International Symposium on Plant Growth Modeling, Simulation, Visualization and Applications*, Beijing, China.

De Coligny, F. (2007). Efficient building of forestry modelling software with the capsis methodology. In Fourcaud, T., & Zhang, X. P. (Eds.), *Proceedings of PMA06: Plant Growth Modeling and Applications* (pp. 216–222). Los Alamitos, CA: IEEE Computer Society.

De Kleer, J. (1988). A general labeling algorithm for assumption-based truth maintenance. In *Proceedings of AAAI*, Saint Paul, MN (pp. 188-192). San Francisco: Morgan Kaufmann.

De Kleer, J., & Williams, B. C. (1986). Back to backtracking: Controlling the ATMS. In *Proceedings of the National Conference on Artificial Intelligence (AAAI)*, Philadelphia (pp. 910-917).

De Kleer, J. (1986a). An assumption-based TMS. *Artificial Intelligence*, *28*, 127–162. doi:10.1016/0004-3702(86)90080-9

De Kleer, J. (1986b). Problem solving with the ATMS. *Artificial Intelligence*, *28*, 197–224. doi:10.1016/0004-3702(86)90082-2

De Kleer, J., Mackworth, A. K., & Reiter, R. (1992). Characterizing diagnosis and systems. *Artificial Intelligence*, 56.

De Kleer, J., & Williams, B. C. (1987). Diagnosing multiple faults. *Artificial Intelligence*, *32*(1), 97–130. doi:10.1016/0004-3702(87)90063-4

De Marchi, F., Flouvat, F., & Petit, J.-M. (2005). Adaptive strategies for mining the positive border of interesting patterns: Application to inclusion dependencies in databases. In J.-F. Boulicaut, L. De Raedt, & H. Mannila (Eds.), *Proceedings of the Conference* on *Constraint-Based Mining and Inductive Databases* (LNCS 3848, pp. 81-101).

Demuth, B. (2005). *The dresden OCL toolkit and the business rules approach.* Paper presented at the European Business Rules Conference (EBRC2005), Amsterdam, the Netherlands.

Demuth, B., & Hussmann, H. (1999). Using UML/OCL constraints for relational database design. In *«UML» '99 — The Unified Modeling Language* (pp. 751-751).

Demuth, B., Hussmann, H., & Loecher, S. (2001). OCL as a specification language for business rules in database applications. In *«UML» 2001 — The Unified Modeling Language. Modeling Languages, Concepts, and Tools* (pp. 104-117).

Demuth, B., Loecher, S., & Zschaler, S. (2004, September 15 - 17). *Structure of the Dresden OCL toolkit.* Paper presented at the 2nd International Fujaba Days "MDA with UML and Rule-based Object Manipulation", Darmstadt, Germany.

Devillers, R., Bédard, Y., Jeansoulin, R., & Moulin, B. (2007). Towards spatial data quality information analysis tools for experts assessing the fitness for use of spatial data. *International Journal of Geographical Information Science*, *21*(3), 261–282. doi:10.1080/13658810600911879

DeWit, C. T. (1965). *Photosynthesis of leaf canopies* (Agr. Res. Rep. 663). Pudoc, Wageningen.

DialogPlus Project. (2008). *Taxonomy of learning activities*. Retrieved from http://www.jisc.ac.uk/uploaded_documents/DialogPlus%20taxonomy.doc

Diaz, S., & Cabido, M. (2001). Vive la difference: Plant functional diversity matters to ecosystem processes. *Trends in Ecology & Evolution*, *16*, 646–655. doi:10.1016/S0169-5347(01)02283-2

Dillon, J. L. (1979). An evaluation of the state of affairs in Farm Management. *South African Journal of Agricultural Economics*, *1*, 7–13.

Dillon, P., & Ahlberg, M. (2006). Integrativism as a theoretical and organisational framework for e-learning and practitioner research. *Technology, Pedagogy and Education*, *15*(1), 7–30. doi:10.1080/14759390500435747

DiMaggio, P., & Powell, W. (1991). *The new institutionalism in organizational analysis*. Chicago, IL: University of Chicago Press.

DIMENC/SGNC & BRGM. (2005). *Carte géologique de la Nouvelle-Caldonie au 1/50000.* Retreived from http://dimenc.gouv.nc/portal/page/portal/dimenc/telechargements/tele_geologie/temp_3.pdf

Djuric, D., & Devedzic, V. (2005). Ontology modeling and MDA. *Journal of Object Technology*, *4*(1), 109–128.

Donini, F., Lenzerini, M., Nardi, D., & Schaerf, A. (1996). Reasoning in description logics. In *Principles of knowledge representation, studies in logic, language and information* (pp. 193-238). Stanford, CA: CSLI Publications.

Dornhoff, G. M., & Shibles, R. M. (1971). Varietal differences in net photosynthesis of soybean leaves. *Crop Science*, *10*, 42–45.

Drummond, N., & Shearer, R. (2006). *The open world assumption or sometimes its nice to know what we don't know*. Manchester, UK: University of Manchester. Retrieved from http://www.cs.man.ac.uk/~drummond/presentations/OWA.pdf.

DTSI/SGT. (2008). *Cartographie de l'occupation du sol de la Nouvelle-Caldonie au 1/50000.*

Dubos-Paillard, E., Guermond, Y., & Langlois, P. (2003). *Analyse de l'évolution urbaine par automate cellulaire, UMR 6063 IDEES.* Paris, France: Laboratoire MTG Rouen.

Duda, R. O., & Hart, P. E. (1973). *Pattern classification and scene analysis*. New York: John Wiley & Sons.

Duda, R. O., Hart, P. E., & Stork, D. G. (2001). *Pattern classification* (2nd ed.). New York: John Wiley & Sons.

Duncan, W. G. (1971). Leaf angles, leaf area, and canopy photosynthesis. *Crop Science*, *11*, 482–485.

Dunn, J. (1974). A fuzzy relative of the ISODATA process and its use in detecting compact well separated clusters. *Journal of Cybernetics*, *3*(3), 32–57. doi:10.1080/01969727308546046

Duru, M., Adam, M., Cruz, P., Martin, G., Ansquer, P., & Ducourtieux, C. (2009). Modelling above-ground herbage mass for a wide range of grassland community types. *Ecological Modelling*, *220*, 209–225. doi:10.1016/j.ecolmodel.2008.09.015

Duru, M., Cruz, P., & Theau, J. P. (2008). Un modèle générique de digestibilité des graminées des prairies semées et permanentes pour raisonner les pratiques agricoles. *Fourrages*, *193*, 79–102.

Duru, M., & Hubert, B. (2003). Management of grazing systems: from decision and biophysical models to principles for action. *Agronomie*, *23*, 689–703. doi:10.1051/agro:2003051

Dutfoy, A., Dutka-Malen, I., Pasanisi, A., Lebrun, R., Mangeant, F., Sen Gupta, J., et al. (2009). OpenTURNS, an open source initiative to treat uncertainties, risks'N statistics in a structured industrial approach. In *41èmes Journées de Statistique*. Bordeaux: SFdS.

Ecosystem Health Monitoring Program (EHMP). (2009). *Annual Report Cards*. Retrieved March 8, 2010, from http://www.healthywaterways.org/EcosystemHealthMonitoringProgram/ProductsandPublications/AnnualReportCards.aspx

Ecosystem Health Monitoring Program (EHMP). (2010). *Home*. Retrieved March 8, 2010, from http://www.ehmp.org/

Eder, J., Koncilia, C., & Morzy, T. (2002). The COMET Metamodel for Temporal Data Warehouses. In *Proceedings of the 14th International conference on Advanced Information Systems Engineering (CAiSE 02)*, Toronto, Canada (LNCS 2348, pp. 83-89). New York: Springer.

Eldred, M. S., Bichon, B. J., Adams, B. M., & Mahadevan, S. (2008). Overview of reliability analysis and design capabilities in DAKOTA with application to shape optimization of MEMS. In Tsompanakis, Lagaros & Papadrakakis (Ed.), *Structural Design Optimization Considering Uncertainties. Structures and Infrastructures series, 1*, 401-432.

El-Gemayel, J., Sibertin-Blanc, C., & Chapron, P. (2009). Impact of tenacity upon the behaviors of social actors. In *Proceedings of the 1st International Workshop on Applied agent based simulator engineering for complex system study*, Nagoya, Japan.

Ellerkamp, P. (2001). *Etude de l'extension urbaine autour d'Avignon et de Montpellier de 1936 à 1999 par la simulation de dynamiques spatiales à l'aide de modèles cellulaires d'interaction locale*. Unpublished doctoral dissertation, University of Avignon, Avignon, France.

Elzbieta, M., & Zimányi, E. (2008). *Advanced Data Warehouse Design*. New York: Springer.

Epanechnikov, V. A. (1969). Nonparametric estimation of a multivariate probability density. *Theory of Probability and Its Applications*, *14*, 153–158. doi:10.1137/1114019

Eriksson, H. E., & Penker, M. (2000). *Business modeling with UML: Business patterns at eork*. New York: John Wiley & Sons.

Escribano, A., Gómez, L., Kuijpers, B., & Vaisman, A. (2007). Piet: a GIS-OLAP implementation. In I. Song & T. Pedersen (Eds.), *Proceedings of the International Conference on DataWarehousing and Knowledge Discovery* (pp. 73-80). New York: ACM

European Commission Directorate General for Energy & Transport. (2006). *Education on energy: Teaching tomorrow's energy consumers*. Brussels, Belgium: European Communities. Retrieved from http://ec.europa.eu/energy/action_plan_energy_efficiency/doc/education_en.pdf

European Environment Agency (EEA). (1999). *Environmental indicators: Typology and overview* (Tech. Rep. No. 25). Copenhagen, Denmark: Author.

European Environment Agency (EEA). (2002). *Energy and environment in the European Union (Environmental Issue Rep. No. 31)*. Copenhagen, Denmark: Author.

European Environment Agency (EEA). (2008). *Indicators*. Retrieved from http://www.eea.europa.eu/themes/energy/indicators

Falconer, I., Beetham, H., Oliver, R., Lockyer, L., & Littlejohn, A. (2007). *Mod4L final report: Representing learning designs*. Retrieved from http://www.academy.gcal.ac.uk/mod4l/

Falkovych, K. I., Sabou, M., & Stuckenschmidt, H. (2003). UML for the Semantic Web: Transformation-based approaches. In *Knowledge Transformation for the Semantic Web, Frontiers in Artificial Intelligence and Applications, vol. 95* (pp. 92-106). Amsterdam, the Netherlands: IOS.

Faure, T., & Deffuant, G. (2006). SimExplorer: A software tool for programming and executing designs on complex models. In *SCS International Conference on Modeling and Simulation - Methodology, Tools, Software Applications*. Calgary, Canada.

Faure, T., & Deffuant, G. (2006). *SimExplorer: A Software Tool for Programming and Executing Designs on Complex Models*. Paper presented at the SCS Conference on Modeling and Simulation Methodology, Tools, Software Applications, Calgary, Canada.

Felici, G., Bertolazzi, P., Guarracino, M. R., Chinchuluun, A., & Pardalos, P. M. (2008. November). Logic formulas based knowledge discovery and its application to the classification of biological data. In R. P. Mondaini (Ed.), 8th International Symposium on Mathematical and Computational Biology (*BIOMAT 2008),* Sao Paulo, Brazil (pp. 223-234). World Scientific.

Ferber, J. (1999). *Multi-agent systems, an introduction to distributed artificial intelligence*. Reading, MA: Addison-Wesley.

Few, S. (2006). *Information dashboard design: The effective visual communication of data*. Sebastopol, CA: O'Reilly Media.

Figueira, J., Greco, S., & Ehrgott, M. (2005). *Multiple criteria decision analysis: State of the art surveys* (*Vol. 57*). London, UK: Springer.

Filho, J. L. (1998, November 16-19). *GeoFrame Applying Analysis Patterns in GIS Domain.* Paper presented at the 10th Colloquium of Spatial Information Research Center, Otago, New Zealand.

Filho, J. L., & Iochpe, C. (1999). Specifying analysis patterns for geographic databases on the basis of a conceptual framework. In *Proceedings of the 7th ACM international symposium on Advances in geographic information systems*, Kansas City, MO (pp. 7-13). New York: ACM Press.

Filho, J. L., Sodré, V. F., Daltio, J., Rodrigues Júnior, M. F., & Vilela, V. M. (2004). *A CASE tool for geographic database design supporting analysis patterns*. New York: Springer.

Filho, J. L., & Iochpe, C. (2008). Modeling with a UML Profile. In Shekhar, S., & Xiong, H. (Eds.), *Encyclopedia of Geographical Information Sciences* (pp. 691–700). New York: Springer.

Fisher, R. A. (1936). The use of multiple measurements in taxonomic problems. *Annals of Eugenics*, *7*, 179–188.

Fitzgerald, A., & Long, A. (2008). *OAK Law Project Report - A Review and Analysis of Academic Publishing Agreements and Open Access Policies.* Retrieved March 8, 2010, from http://www.oaklist.qut.edu.au/files/OAK_Law_Review_and_Analysis_v1_20080212.pdf

Flouvat, F., De Marchi, F., & Petit, J.-M. (2009). The iZi project: Easy prototyping of interesting pattern mining algorithms. In *Proceedings of the 13th Pacific-Asia International Conference on Knowledge Discovery and Data Mining: New Frontiers in Applied Data Mining* (pp. 1-15).

Forbus, K. D., & De Kleer, J. (1988). Focusing the ATMS. In *Proceedings of the National Conference on Artificial Intelligence (AAAI)*, Saint Paul, MN (pp. 193-198). San Francisco: Morgan Kaufmann.

Forman, R. T. T., & Godron, M. (1986). *Landscape ecology*. New York, NY: John Wiley & Sons.

Fraisse, C. W., Breuer, N. E., Zierden, D., Bellow, J. G., Paz, J., & Cabrera, V. E. (2006). AgClimate: A climate forecast information system for agricultural risk management in the southeastern USA. *Computers and Electronics in Agriculture*, *53*(1), 13–27. doi:10.1016/j.compag.2006.03.002

Frank, A. U. (2007). Data quality ontology: An ontology for imperfect knowledge. In *Proceedings of COSIT 2007* (LNCS 4736, pp. 406-420).

Frankel, D. (2003). *Model driven architecture: Applying Mda to enterprise computing*. New York: John Wiley & Sons.

Friedberg, E. (1993). *Le pouvoir et la règle: Dynamiques de l'action organisée*. Paris, France: Seuil.

Friedrich, G., Gottlob, G., & Nejdl, W. (1990). Hypothesis classification, abductive diagnosis and therapy. In *Proceedings of the International Workshop on Expert Systems in Engineering*, Vienna, Austria (LNAI 462).

Gamma, E., Helm, R., Johnson, R., & Vlissides, J. (1995). *Patterns elements of reusable object-oriented software*: Reading, MA: Addison-Wesley.

Gasevic, D., Djuric, D., & Devedzic, V. (2006). *Model driven architecture and ontology development*. New York: Springer.

Gatrell, A. C., Bailey, T. C., Diggle, P. J., & Rowlingson, B. S. (1996). Spatial point pattern analysis and its application in geographical epidemiology. *Transactions of the Institute of British Geographers*, *21*(1), 256–274. doi:10.2307/622936

Gay, D., Rouet, I., Mangeas, M., Dumas, P., & Selmaoui, N. (2007). Assessment of classification methods for soils erosion risks. In *Proceedings of the International Congress on Modelling and Simulation*, Christchurch, New Zealand (pp. 2659-2665).

Gayte, O., Libourel, T., Cheylan, J.-P., & Lardon, S. (1997). *Conception des systèmes d'information sur environnement*. Paris: Hermés.

Genero, M., Manso, E., Visaggio, A., Canfora, G., & Piattini, M. (2007). Building measure-based prediction models for UML class diagram maintainability. *Empirical Software Engineering*, *12*, 517–549. doi:10.1007/s10664-007-9038-4

Getis, A., & Ord, J. K. (1992). The analysis of spatial association by use of distance statistics. *Geographical Analysis*, *24*, 189–206. doi:10.1111/j.1538-4632.1992.tb00261.x

Gijsman, A. J., Hoogenboom, G., Parton, W. J., & Kerridge, P. C. (2002). Modifying DSSAT crop models for low-input agricultural systems using a soil organic matter-residue module from CENTURY. *Agronomy Journal*, *94*, 462–474.

Glorio, O., & Trujillo, J. (2008). An MDA Approach for the Development of Spatial Data Warehouses. In I. Song, J. Eder, & T. Nguyen (Eds.), *Proceedings of International Conference of DataWarehousing and Knowledge Discovery* (pp. 23-32). New York: ACM.

Godwin, D. C., & Jones, D. C. (1991). Nitrogen dynamics in soil-plant systems. In J. Hanks & J. T. Ritchie (Eds.), *Modeling plant and soil systems. Agronomy Monographs No. 31* (pp 287-321). Madison, WI: ASA, CSSA, and SSSA.

Godwin, D. C., & Singh, U. (1998). Nitrogen balance and crop response to nitrogen in upland and lowland cropping systems. In G. Y. Tsuji, G. Hoogenboom, & P. K. Thornton (Eds.), *Understanding options for agricultural production. System approaches for sustainable agricultural development* (pp. 55-77). Dordrecht, The Netherlands: Kluwer Academic Publishers.

Griffiths, D. (2005). *From primitives to patterns: a discussion paper, UNFOLD Project*. Retrieved from http://www.unfoldproject.net/ providers_folder/papers/ terms/primitives-patterns.pdf

Griffiths, D., & Blat, J. (2005). The role of teachers in editing and authoring units of learning using IMS Learning Design. *Advanced Technology for Learning*, *2*(4), 1–9.

Grime, J. P. (1973). Competition and Diversity in Herbaceous Vegetation. *Nature*, *244*, 310–311. doi:10.1038/244311a0

Grimm, V. (1999). Ten years of individual-based modelling in ecology: What have we learn and what could we learn in the future? *Ecological Modelling*, *115*, 129–148. doi:10.1016/S0304-3800(98)00188-4

Gruber, R. (1993a). *Towards principles for the design of ontologies used for knowledge sharing* (Tech. Rep. KSL-93-04). Paper presented at the International Workshop on Formal Ontology, Padova, Italy.

Gruber, R. (1993b). A translation approach to portable ontology specifications. *Knowledge Acquisition*, *5*(2), 199–220. doi:10.1006/knac.1993.1008

Grüber, T. R. (1995). Towards principles for the design of ontologies used for knowledge sharing. *International Journal of Human-Computer Studies*, *43*(5-6), 907–928. doi:10.1006/ijhc.1995.1081

Guarracino, M. R., Cuciniello, S., Feminiano, D., Toraldo, G., & Pardalos, P. M. (2008). Current Classification Algorithms for Biomedical Applications. In *Proceedings of the Centre de Recherches Mathématiques CRM*.

Guarracino, M. R., Chinchuluun, A., & Pardalos, P. M. (in press). Decision rules for efficient classification of biological data. *Optimization Letters*.

Guarracino, M. R., Cifarelli, C., Seref, O., & Pardalos, P. M. (2007). A classification algorithm based on generalized eigenvalue problems. *Optimization Methods and Software*, *22*(1), 73–81. doi:10.1080/10556780600883874

Gudmundsson, J., Laube, P., & Wolle, T. (2008). Movement patterns in spatio-temporal data. In Shekhar, S., & Xiong, H. (Eds.), *Encyclopedia of GIS* (pp. 726–732). New York, NY: Springer. doi:10.1007/978-0-387-35973-1_823

Guerois, M., & Pumain, D. (2001). *Urban sprawl in France (1950-2000)*. Milano, Italy: FrancoAngeli.

Guizzardi, G., Falbo, R. A., & Pereira Filho, J. G. (2002). Using objects and patterns to implement domain ontologies. *Journal of Brazilian Computer Society, Special Issue on Software Engineering, 8*(1).

Guizzardi, G., Wagner, G., Guarino, N., & Sinderen, M. (2004). An ontologically well-founded profile for UML conceptual models. *Lecture Notes in Computer Science*, 3084.

Hadzilacos, T., & Tryfona, N. (1997). An Extended Entity-Relationship Model for Geographic Applications. *SIGMOD Record*, *26*(3), 24–29. doi:10.1145/262762.262766

Hammah, R. E., & Curran, J. H. (2000). Validity measures for the fuzzy cluster analysis of orientations. *IEEE Transactions on Pattern Analysis and Machine Intelligence*, *22*(12), 1467–1472. doi:10.1109/34.895981

Han, J., & Kamber, M. (2006). *Data mining, second edition: Concepts and techniques (the Morgan Kauffman series in data management systems)*. San Francisco, CA: Morgan Kauffman.

Han, J., Pei, J., Yin, Y., & Mao, R. (2004). Mining frequent patterns without candidate generation: A frequent-pattern tree approach. *Data Mining and Knowledge Discovery*, *8*(1), 53–87. doi:10.1023/B:DAMI.0000005258.31418.83

Hansen, L. K., & Salamon, P. (1990). Neural network ensembles. *IEEE Transactions on Pattern Analysis and Machine Intelligence*, *12*(10), 993–1001. doi:10.1109/34.58871

Hanski, I., & Gilpin, M. E. (1997). *Metapopulation biology: Ecology, genetics, and evolution*. New York, NY: Academic Press.

Hasenohr, P., Pinet, F., Kay, S., Grenier, G., & Sevila, F. (2005, June). *A case-study based methodology for modeling of spatial DSS using Spatial OCL*. Paper presented at the Joint Workshop on Decision Support Systems, Experimental Economics & e-Participation (GRAZ 2005), AUT.

Hasenohr, P., & Pinet, F. (2006). Modeling of a spatial DSS template in support to the Common agricultural policy. *Journal of Decision Systems*, *15*(2), 181–196. doi:10.3166/jds.15.181-196

Hashem, S., & Schmeiser, B. (1995). Improving model accuracy using optimal linear combinations of trained neural networks. *IEEE Transactions on Neural Networks*, *6*(3), 792–794. doi:10.1109/72.377990

Hathaway, R., & Bezdek, J. (2001). Fuzzy *c*-means clustering of incomplete data. *IEEE Transactions on Systems, Man, and Cybernetics - Part B, 31*(5), 735-744.

Hathaway, R., Bezdek, J., & Hu, Y. (2000). Generalized fuzzy c-means clustering strategies using L norm distances. *IEEE transactions on Fuzzy Systems*, *8*(5), 576–582. doi:10.1109/91.873580

Heller, U., & Struss, P. (1996). Transformation of Qualitative Dynamic Models – Application in Hydro-Ecology. In *Proceedings of the 10th International Workshop on Qualitative Reasoning* (pp. 83-92). AAAI Press.

Heller, U., & Struss, P. (1997). Conceptual Modeling in the Environmental Domain. In *Proceedings of the 15th IMACS World Congress on Scientific Computation, Modelling and Applied Mathematics,* Berlin (Vol. 6, pp. 147-152).

Heller, U., & Struss, P. (1998). Diagnosis and therapy recognition for ecosystems - usage of model-based diagnosis techniques. In *Proceedings of the 12th International Symposium Computer Science for Environmental Protection (UI-98)*, Bremen, Germany.

Heller, U., & Struss, P. (2002). Consistency-based problem solving for environmental decision support. *Computer-Aided Civil and Infrastructure Engineering*, *17*, 79–92. doi:10.1111/1467-8667.00257

Helton, J. C., Johnson, J. D., Salaberry, C. J., & Storlie, C. B. (2006). Survey of sampling based methods for uncertainty and sensitivity analysis. *Reliability Engineering & System Safety*, *91*, 1175–1209. doi:10.1016/j.ress.2005.11.017

Hendler, J. (2001). Agents and the semantic web. *IEEE Intelligent Systems and Their Applications*, *16*(2), 30–37.

Hernandez, V., Voss, A., Göhring, W., & Hopmann, C. (2005). Sustainable decision support by the use of multi-level and multi-criteria spatial analysis on the Nicaragua Development Gateway. In *Proceedings of the 8th International Conference on the Global Spatial Data Infrastructure*.

Herrero, B., & Valencia-Barrera, R. M. (2001). Palynological analysis of honeys from Palencia Province (Spain). *Acta Botanica Croatica*, *60*(1), 11–24.

Hidding, M., Needham, B., & Wisserhof, J. (2000). Discourses of town and country. *Landscape and Urban Planning*, *48*(3-4), 121–130. doi:10.1016/S0169-2046(00)00036-0

Hoogenboom, G., Jones, J. W., Wilkens, P. W., Batchelor, W. D., Bowen, W. T., Hunt, L. A., et al. (1994). Crop models. In G. Y. Tsuji, G. Uehara, & S. Balas (Eds.), *DSSAT Version 3, Volume 2* (pp. 95-244). Honolulu, HI: University of Hawaii.

Hoogenboom, G., Jones, J. W., Wilkens, P. W., Porter, C. H., Batchelor, W. D., Hunt, L. A., et al. (2004). *Decision support system for agrotechnology transfer version 4.0* Honolulu, HI: University of Hawaii.

Hoogenboom, G., Wilkens, P. W., Thornton, P. K., Jones, J. W., Hunt, L. A., & Imamura, D. T. (1999). Decision support system for agrotechnology transfer V3.5. In G. Hoogenboom, P. W. Wilkens, & G. Y. Tsuji (Eds.), *DSSAT Version 3, Volume 4* (pp. 1-36). Honolulu, HI: University of Hawaii.

Hoogenboom, G. (2000). Contribution of agrometeorology to the simulation of crop production and its applications. *Agricultural and Forest Meteorology*, *103*(1-2), 137–157. doi:10.1016/S0168-1923(00)00108-8

Horsburgh, J. S., Tarboton, D. G., Maidment, D. R., & Zaslavsky, I. (2008). A Relational Model for Environmental and Water Resources Data. *Water Resources Research*, 44.

Huang, Y., Shekhar, S., & Xiong, H. (2004). Discovering colocation patterns from spatial data sets: A general approach. *IEEE Transactions on Knowledge and Data Engineering*, *16*(12), 1472–1485. doi:10.1109/TKDE.2004.90

Hung, M., & Yang, D. (2001, November 29-December 1). An efficient fuzzy c-means clustering algorithm. In *Proceedings of the IEEE International Conference on Data Mining*, San Jose, CA (pp. 225-232).

Hunter, J., & Mauger, B. (2008). *Health-e-Waterways Database Survey Report*. Retrieved March 8, 2010, from http://www.health-e-waterways.org/reports.php

Hwang, W. J., & Wen, K. W. (1998). Fast k classification algorithm based on partial distance search. *Electronics Letters*, *34*(21), 2062–2063. doi:10.1049/el:19981427

IBM. (2006, June). *Ontology Definition Metamodel.*

Imieliński, T., Khachiyan, L., & Abdulghani, A. (2002). Cubegrades: Generalizing Association Rules. *Data Mining and Knowledge Discovery*, *6*(3), 219–257. doi:10.1023/A:1015417610840

IMS Global Learning Consortium. (2003). *IMS Learning Design specification ver. 1.0*. Retrieved from http://www.imsglobal.org/learningdesign/index.cfm

Inmon, W. (2005). *Building the Data Warehouse* (4th ed.). New York: John Wiley & Sons.

Inmon, W. H. (1996). *Building the Data Warehouse*. New York: John Wiley & Sons.

INRA. (2007). *Alimentation des bovins, ovins et caprins. Besoins des animaux – Valeur des aliments*. Paris: Quae Editions.

Institute of Electrical and Electronics Engineers (IEEE). (2000). Standard for modeling and simulation (MetS), high level architecture (HLA) *(Tech. Rep. No 1516-2000)*. Washington, DC: IEEE Computer Society.

Institute of Electrical and Electronics Engineers Learning Technology Standards Committee. (2002). *IEEE standard for Learning Object Metadata*. Retrieved from http://ieeeltsc.org/wg12LOM/lomDescription

International Atomic Energy Agency (IAEA) & International Energy Agency. (IEA). (2001). *Indicators for sustainable energy development*. New York: United Nations.

International Atomic Energy Agency (IAEA). United Nations Department of Economic and Social Affairs, International Energy Agency (IEA), Eurostat, & European Environment Agency (EEA). (2005). *Energy indicators for sustainable development: Guidelines and methodologies*. Vienna, Austria: IAEA.

ISO. (2002). ISO 19108 - Geographic Information - Temporal schema.

ISO. (2003). ISO 19107 - Geographic Information - Spatial schema.

Jackson, M. O. (2008). *Social and economic networks*. Princeton, NJ: Princeton University Press.

Jacobson, I., Booch, G., & Rumbaugh, J. (1999). *The unified software development process*. Reading, MA: Addison Wesley.

Jagtap, S. S., Jones, J. W., LaRow, T., Ajayan, A., & O'Brien, J. J. (2006). (Manuscript submitted for publication). Statistical recalibration of precipitation outputs from coupled climate models. *Journal of Applied Meteorology*.

Jain, A. K., Murty, M. N., & Flynn, P. J. (1999). Data clustering: A review. *ACM Computing Surveys*, *31*(3), 264–323. doi:10.1145/331499.331504

Jappiot, M., Blanchi, R., & Alexandrian, D. (2000, October 25-27). *Cartographie du risque d'incendie de forêt: besoins, méthodes et données, essai de normalisation*. Paper presented at the Journées GIS Incendie de Forêt, Hyères, France.

Jeltsch, F., Milton, S. J., Dean, W. R. J., & van Rooyen, N. (1996). Tree spacing and coexistence in semiarid savannas. *Journal of Ecology*, *84*, 583–595. doi:10.2307/2261480

Jeltsch, F., Milton, S. J., Dean, W. R. J., van Rooyen, N., & Moloney, K. A. (1998). Modelling the impact of small-scale heterogeneities on tree-grass coexistence in semi-arid savannas. *Journal of Ecology*, *86*, 780–794. doi:10.1046/j.1365-2745.1998.8650780.x

Jeltsch, F., Moloney, K. A., & Milton, S. J. (1999). Detecting process from snap-shot pattern: lessons from tree spacing in the southern Kalahari. *Oikos*, *85*, 451–467. doi:10.2307/3546695

Jenkinson, D. S., & Rayner, J. H. (1977). Turnover of soil organic-matter in some of Rothamsted classical experiments. *Soil Science*, *123*, 298–305. doi:10.1097/00010694-197705000-00005

Jiang, X., Harvey, A., & Wah, K. S. (2003). Constructing and training feed-forward neural networks for pattern classification. *Patt. Recog.*, *36*, 853–867. doi:10.1016/S0031-3203(02)00087-0

Ji, C., & Ma, S. (1997). Combinations of weak classifiers. *IEEE Transactions on Neural Networks*, *8*(1), 32–42. doi:10.1109/72.554189

Jiggins, J., & Roling, N. (2000). Adaptive management: potential and limitations for ecological governance. *International Journal of Agricultural Resources. Governance and Ecology*, *1*, 28–42.

Jones, J. W., Tsuji, G. Y., Hoogenboom, G., Hunt, L. A., Thornton, P. K., Wilkens, P., et al. (1998). Decision support system for agrotechnology transfer: DSSAT v3. In G. Y. Tsuji, G. Hoogenboom, & P. K. Thornton (Eds.), *Understanding options for agricultural production* (pp. 157-177). Boston: Kluwer Academic Press.

Jones, J. W., Hoogenboom, G., Porter, C. H., Boote, K. J., Batchelor, W. D., & Hunt, L. A. (2003). DSSAT cropping system model. *European Journal of Agronomy*, *18*, 235–265. doi:10.1016/S1161-0301(02)00107-7

Jones, J. W., Keating, B. A., & Porter, C. (2001). Approaches for modular model development. *Agricultural Systems*, *70*(2), 421–443. doi:10.1016/S0308-521X(01)00054-3

Jones, M. C., Marron, J. S., & Sheather, S. J. (1996). A brief survey of band-width selection for density estimation. *Journal of the American Statistical Association*, *91*, 401–407. doi:10.2307/2291420

Jongman, R. H., & Pungetti, G. (2004). *Ecological networks and greenways: Concept, design, implementation*. Cambridge, UK: Cambridge University Press. doi:10.1017/CBO9780511606762

Jouven, M., & Baumont, R. (2008). Simulating grassland utilization in beef suckler systems to investigate the trade-offs between production and floristic diversity. *Agricultural Systems*, *96*, 260–272. doi:10.1016/j.agsy.2007.10.001

JPIVOT. (n.d.). Retrieved October 10, 2009, from http://jpivot.sourceforge.net

Kang, M. A., Pinet, F., Schneider, M., Chanet, J. P., & Vigier, F. (2004, April 29-May 1). *How to design geographic database? Specific UML profile and spatial OCL applied to wireless Ad Hoc networks.* Paper presented at the 7th Conference on Geographic Information Science (AGILE'2004), Heraklion, GRC.

Karayiannis, N. B. (1997). A methodology for constructing fuzzy algorithms for learning vector quantization. *IEEE Transactions on Neural Networks*, *8*(3), 505–518. doi:10.1109/72.572091

Keating, B. A., Carberry, P. S., Hammer, G. L., Probert, M. E., Robertson, M. J., & Holzworth, D. (2003). An overview of APSIM, a model designed for farming systems simulation. *European Journal of Agronomy*, *18*, 267–288. doi:10.1016/S1161-0301(02)00108-9

Keating, B. A., & McCown, R. L. (2001). Advances in farming systems analysis and intervention. *Agricultural Systems*, *70*, 555–579. doi:10.1016/S0308-521X(01)00059-2

Keeney, R. L., & Raiffa, H. (1976). *Decisions with multiple objectives: Preferences and values trade-offs*. New York, NY: John Wiley & Sons.

Keim, D. A. (2002). Information visualization and visual data mining. *IEEE Transactions on Visualization and Computer Graphics*, *8*(1), 1–8. doi:10.1109/2945.981847

Kelley, J. F. (1984). An iterative design methodology for user-friendly natural language information applications. *ACM Transactions on Office InformationSystems*, *2*(1), 26–41. doi:10.1145/357417.357420

Kemp, D. R., & Michalk, D. L. (2007). Towards sustainable grassland and livestock management. *The Journal of Agricultural Science*, *145*, 543–564. doi:10.1017/S0021859607007253

Kimball, R. (1996). *The Data Warehouse Toolkit: Practical Techniques for Building Dimensional Data Warehouses*. New York: John Wiley & Sons.

Kimball, R., & Caserta, J. (2004). *The Data Warehouse ETL Toolkit: Practical Techniques for Extracting, Cleaning, Conforming, and Delivering Data*. New York: John Wiley & Sons.

Kimball, R., & Ross, M. (2002). *The Data Warehouse Toolkit* (2nd ed.). New York: John Wiley & Sons.

Klasse Objecten. (2009). *OCL tools and services*. Retrieved from http://www.klasse.nl/ocl

Kleijnen, J. P. C., Sanchez, S. M., Lucas, T. W., & Cioppa, T. M. (2005). State-of-the-art review: a user's guide to the brave new world of designing simulation experiments. *INFORMS Journal on Computing*, *17*(3), 263–289. doi:10.1287/ijoc.1050.0136

Kleppe, A., & Warmer, J. (2003). *Object constraint language, the getting your models ready for MDA*. Reading, MA: Addison-Wesley.

Knight, C., Gašević, D., & Richards, G. (2006). An ontology-based framework for bridging learning design and learning content. *Journal of Educational Technology & Society*, *9*(1), 23–37. Retrieved from http://www.ifets.info/journals/9_1/4.pdf.

Koperski, K., & Han, J. (1995). Discovery of spatial association rules in geographic information databases. In M. J. Egenhofer & J. R. Herring (Eds.), *Proceedings of the 4th International Symposium on Advances in Spatial Databases* (LNCS 951, pp. 47-66).

Kösters, G., & Pagel, B.-U. (1996, November 15-16). *The GeoOOA-Tool and Its Interface to Open Software Development Environments for GIS.* Paper presented at the 4th Workshop on Advances in Geographic Information Systems, Rockville, MD.

Kösters, G., Pagel, B.-U., & Six, H.-W. (1995, December 1-2). *Object-Oriented Requirements Engineering for GIS Applications.* Paper presented at the 3rd International ACM Workshop on Advances in Geographic Information Systems, Baltimore, MD.

Kösters, G., Pagel, B.-U., & Six, H.-W. (1997). GIS-application development with GeoOOA. *International journal of GIS, 11*(4), 307-355.

Kotis, K., & Vouros, G. (2006). Human-centered ontology engineering: The HCOME methodology. *International Journal of Knowledge and Information Systems, 10*(1), 109–131. doi:10.1007/s10115-005-0227-4

Kraalingen, D., Van, W. G., Rappoldt, C., & van Laar, H. H. (2003). The Fortran simulation translator, a simulation language. *European Journal of Agronomy, 18*, 359–361. doi:10.1016/S1161-0301(02)00131-4

Krishna, K., & Murty, M. (1999). Genetic k-means algorithm. *IEEE Transactions on Systems, Man, and Cybernetics - Part B, 29*(3), 433-439.

Kuhn, W. (2005). Geospatial semantics: Why, of what, and how. *Journal on Data Semantics, III*, 1–24. doi:10.1007/11496168_1

Kulkarni, S. R., Lugosi, G., & Venkatesh, S. S. (1998). Learning pattern classification - a survey. *IEEE Transactions on Information Theory, 44*(6), 2178–2206. doi:10.1109/18.720536

Kumazawa, T., Saito, O., Kozaki, K., Matsui, T., & Mizoguchi, R. (2009, March). *Knowledge structuring tool for sustainability science based on ontology engineering.* Paper presented at the International Conference on Complex, Intelligent and Software Intensive Systems, Fukuoka, Japan.

Kuncheva, L. I. (1997). Fitness functions in editing *k*-NN reference set by genetic algorithms. *Patt. Recog., 30*(6), 1041–1049. doi:10.1016/S0031-3203(96)00134-3

Lampin, C., Jappiot, M., Long, M., Borgniet, L., & Dumas, E. (2005, June 20-23). *Outils de caractérisation et cartographie des interfaces habitat-forêt.* Paper presented at the International Colloquium of Spatial Analysis and Geomatics, Avignon, France.

Langley, P., Goodchild, M., Maguire, D., & Rhind, D. (Eds.). (2001). *Geographic Information Systems and Science* (p. 517). New York: John Wiley & Sons.

Langlois, A., & Phipps, M. (1997). *Automates cellulaires: Application à la simulation urbaine*. Paris, France: Hermes.

Larrivée, S., Bédard, Y., & Pouliot, J. (2005). *How to Enrich the Semantics of Geospatial Databases by Properly Expressing 3D Objects in a Conceptual Schema.* Paper presented at the OTM Workshops 2005.

Larrivée, S., Bédard, Y., & Pouliot, J. (2006). Fondement de la modélisation conceptuelle des bases de données géospatiales 3D. *Revue internationale de géomatique, 16*(1), 9-27.

Larsen, G. (2007). *Modelling hydrodynamic processes within Pumicestone Passage, Northern Moreton Bay, Queensland.* Retrieved March 8, 2010, from http://eprints.qut.edu.au/16634/1/Genevieve_Larsen_Thesis.pdf

Larsen, R. J., & Marx, M. L. (2005). *An Introduction to Mathematical Statistics and Its Applications.* Upper Saddle River, NJ: Perason.

Laurini, R., & Thompson, D. (2001). *Fundamentals of Spatial Information Systems*. New York: Academic Press.

Lavorel, S., & Garnier, E. (2002). Predicting changes in community composition and ecosystem functioning from plant traits: revisiting the Holy Grail. *Functional Ecology, 16*, 545–556. doi:10.1046/j.1365-2435.2002.00664.x

Lbath, A., & Aufaure-Portier, M.-A. (1995). AIGLE: Un atelier visuel de génération d'applications géographiques. *Revue Internationale de géomatique, 5*(2), 179-195.

Lbath, A., & Pinet, F. (2000). *The development and customization of GIS-based applications and Web-based GIS applications with the CASE tool AIGLE.* Paper presented at the 8th ACM Symposium on Advances in Geographic Information Systems, Washington DC.

Leclerc, D., & Poumay, M. (2005). *The 8 learning events model*. Liege, Belgium: University of Liege, LabSET. Retrieved from http://www.labset.net/media/prod/8LEM.pdf

Lee, J. H. W., Huang, Y., Dickman, M., & Jayawardena, A. W. (2003). Neural Network modelling of coastal algal blooms. *Ecological Modelling, 159*(2-3), 179–201. doi:10.1016/S0304-3800(02)00281-8

Leemans, V., & Destain, M. F. (2004). A real time grading method of apples based on features extracted from defects. *Journal of Food Engineering, 61*, 83–89. doi:10.1016/S0260-8774(03)00189-4

Leemans, V., Magein, H., & Destain, M.-F. (1999). Defect segmentation on 'Jonagold' apples using colour vision and Bayesian method. *Computers and Electronics in Agriculture, 23*, 43–53. doi:10.1016/S0168-1699(99)00006-X

Leemans, V., Magein, H., & Destain, M.-F. (2002). On-line apple grading according to European standards using machine vision. *Biosystems Engineering, 83*(4), 397–404. doi:10.1006/bioe.2002.0131

Lemaire, G., & Gastal, F. (1997). N uptake and distribution in plant canopies. In Lemaire, G. (Ed.), *Diagnosis of the Nitrogen Status in Crops* (pp. 3–44). Berlin: Springer Verlag. doi:10.1007/978-3-642-60684-7_1

Leung, C. K.-S., Irani, P., & Carmichael, C. L. (2008). Wifisviz: Effective visualization of frequent itemsets. In *Proceedings of the International Conference on Data Mining* (pp. 875-880). Washington, DC: IEEE Computer Society.

Levesque, M.-A., Bédard, Y., Gervais, M., & Devillers, R. (2007). Towards managing the risks of data misuse for geospatial datacubes de données. In *Proceedings of ISSDQ 2007*.

Lisi, F. A., & Malerba, D. (2004). Inducing multi-level association rules from multiple relations. *Machine Learning, 55*(2), 175–210. doi:10.1023/B:MACH.0000023151.65011.a3

List, B., Bruckner, R. M., Machaczek, K., & Schiefer, J. (2002). A Comparison of Data Warehouse Development Methodologies Case Study of the Process Warehouse. In *Proceedings of the 13th International Conference on Database and Expert Systems Applications (DEXA 02)*, Aix-en-Provence, France (LNCS 2453, pp. 203-215). New York: Springer.

Liu, Y., Hill, D., Rodrigues, A., Marini, L., Kooper, R., Myers, J., et al. (2009, May 18-22). *A New Framework for On-Demand Virtualization, Repurposing and Fusion of Heterogeneous Sensors.* Paper presented at the Sensor Web Enablement Workshop 2009, 2009 International Symposium on Collaborative Technologies and Systems, Baltimore, MD.

Liu, Y., Lyon, B. G., Windham, W. R., Lyon, C. E., & Savage, E. M. (2004). Principal component analysis of physical, color, and sensory characteristics of chicken breasts deboned at two, four, six, and twenty-four hours postmortem. *Poultry Science, 83*, 101–108.

Longley, P., Goodchild, M., Maguire, D., & Rhind, D. (Eds.). (2001). *Geographic Information Systems and Science*. New York: John Wiley & Sons.

Loomis, R. S., Rabbinge, R., & Ng, E. (1979). Explanatory models in crop physiology. *Annual Review of Plant Physiology, 30*, 339–367. doi:10.1146/annurev.pp.30.060179.002011

Lopes De Oliveira, J., Pires, F., & Bauzer Medeiros, C. (1997). An Environment for Modeling and Design of Geographic Applications. *GeoInformatica, 1*(1), 29–58. doi:10.1023/A:1009704100446

Louveaux, J., Maurizio, A., & Vorwohl, G. (1978). Methods of Melissopalynology. *Bee World, 59*, 139–157.

Luján-Mora, S., Trujillo, J., & Song, I. Y. (2006). A UML Profile for Multidimensional Modeling in Data Warehouses. *Data & Knowledge Engineering, 59*(3), 725–769. doi:10.1016/j.datak.2005.11.004

MacEachren, A., & Kraak, M. (1997). Exploratory cartographic visualization: advancing the agenda. *Computers & Geosciences, 23*(4), 335–343. doi:10.1016/S0098-3004(97)00018-6

MacQueen, J. B. (1967). Some methods for classification and analysis of Multivariate Observations. In *Proceedings of 5th Berkeley Symposium on Mathematical Statistics and Probability: Vol. 1* (pp. 281-297). Berkeley, CA: University of California Press.

Madin, J., Bowers, S., Schildhauer, M., Krivov, S., Pennington, D., & Villa, F. (2007). An Ontology for Describing and Synthesizing Ecological Observation Data. *Ecological Informatics*, *2*(3), 279–296. doi:10.1016/j.ecoinf.2007.05.004

Mahboubi, H., Ralaivao, J. C., Loudcher, S., Boussaid, O., Bentayeb, F., & Darmont, J. (2009). X-WACoDa: An XML-based approach for Warehousing and Analyzing Complex Data. Advances in Data Warehousing and Mining. In Bellatreche, L. (Ed.), *Data Warehousing Design and Advanced Engineering Applications: Methods for Complex Construction* (pp. 38–54). Hershey, PA: IGI Global.

Mahévas, S., & Pelletier, D. (2004). ISIS-Fish, a generic and spatially-explicit simulation tool for evaluating the impact of management measures on fisheries dynamics. *Ecological Modelling*, *171*, 65–84. doi:10.1016/j.ecolmodel.2003.04.001

Maillard, M. (2008). *MultiAgents formalisation of the sociology of the organized action.* Unpublished doctoral dissertation, University of Toulouse, Toulouse, France.

Maillé, E. (2008). *Intégration conceptuelle et opérationnelle de modèles spatiaux-dynamique*: *Application à la dynamique du risque d'incendie de forêt.* Unpublished doctoral dissertation, Paul Cezanne University, Marseille, France.

Maillé, E., & Espinasse, B. (2007, June 18-19). *Un cadre de modélisation pour la simulation de dynamiques spatiales complexes.* Paper presented at the International Colloquium of Spatial Analysis and Geomatics, Clermont-Ferrand, France.

Maillé, E., & Bouillon, C. (2005). *Consommation des espaces agricoles et naturels par le processus d'urbanisation discontinue. Aix-en Provence*. France: Cemagref.

Maillé, E., & Espinasse, B. (2009). Spatial agents for scale integration of spatial dynamics models. *European Journal of Geomatics and Spatial Analysis*, *19*(4), 523–549.

Ma, L., Nielsen, D. C., Ahuja, L. R., Malone, R. W., Saseendran, S. A., & Rojas, K. W. (2003). Evaluation of RZWQM under varying irrigation levels in Eastern Colorado. *Transactions of the ASAE. American Society of Agricultural Engineers*, *46*, 39–49.

Malerba, D. (2008). A relational perspective on spatial data mining. *International Journal of Data Mining. Modelling and Management*, *1*(1), 103–118.

Malinowski, E., & Zimányi, E. (2004). Representing spatiality in a conceptual multidimensional model. In D. Pfoser, I. Cruz, & M. Ronthaler (Eds.), *Proceedings of ACM International Workshop on Geographic Information Systems* (pp. 12-22). New York: ACM Press.

Malinowski, E., & Zimányi, E. (2007). Logical Representation of a Conceptual Model for Spatial Data Warehouses. *GeoInformatica*, *11*(4), 431–457. doi:10.1007/s10707-007-0022-3

Mandel, L., & Cengarle, M. (1999). On the expressive power of OCL. In *FM'99 — Formal Methods* (pp. 713-713).

Mangasarian, O. L., & Wild, E. W. (2004). *Multisurface proximal support vector classification via generalized eigenvalues* (Techn. Rep. 04-03). Madison, WI: Data Mining Institute, Computer Science Department, University of Wisconsin.

Mangasarian, O. L., & Wild, E. W. (2006). *Nonlinear knowledge-based classification* (Tech. Rep. 06-04). Madison, WI: Data Mining Institute, Computer Science Department, University of Wisconsin.

Mannila, H., & Toivonen, H. (1997). Levelwise search and borders of theories in knowledge discovery. *Data Mining and Knowledge Discovery*, *1*(3), 241–258. doi:10.1023/A:1009796218281

MAPINFO. *MapXtreme Java* (n.d.). Retrieved October 10, 2009, from http://extranet.mapinfo.com/products/Overview.cfm?productid=1162

Marchand, P., Brisebois, A., Bédard, Y., & Edwards, G. (2003). Implementation and evaluation of a hypercube-based method for spatio-temporal exploration and analysis. *Journal of the International Society of Photogrammetry and Remote Sensing*, *59*(1), 6–20. doi:10.1016/j.isprsjprs.2003.12.002

Marchant, J. A., & Onyango, C. M. (2003). Comparison of a Bayesian classifier with a multilayer feed-forward neural network using the example of plant/weed/soil discrimination. *Computers and Electronics in Agriculture*, *39*, 3–22. doi:10.1016/S0168-1699(02)00223-5

Martin, G., Martin-Clouaire, R., Rellier, J. P., & Duru, M. (2009b). A simulation model to design grassland-based livestock systems. *Environmental Modelling & Software*.

Martin-Clouaire, R., & Rellier, J. P. (2009). Modelling and simulating work practices in agriculture. *International Journal of Metadata. Semantics and Ontologies*, *4*(1-2), 42–53. doi:10.1504/IJMSO.2009.026253

Martin, G., Hossard, L., Theau, J. P., Therond, O., Josien, E., & Cruz, P. (2009a). Characterizing potential flexibility in grassland use - An application to the French Aubrac region. *Agronomy for Sustainable Development*, *29*, 381–389. doi:10.1051/agro:2008063

Martin, G., Theau, J. P., Therond, O., Carre, J., Cruz, P., & Jouany, C. (2009c). *Bases et premier exemple d'application d'une démarche articulant diagnostic et simulation de systèmes fourragers pour évaluer et améliorer l'efficience d'utilisation de l'herbe*. Fourrages.

Matias, R., & Moura-Pires, J. (2007). Revisiting the OLAP Interaction to Cope with Spatial Data and Spatial Analysis. In J. Cardoso, J. Cordeiro, & J. Filipe (Eds.), *Proceedings of 9th International Conference on Enterprise Information Systems*. Portugal: INSTICC.

Mavromatis, T., Boote, K. J., Jones, J. W., Irmak, A., Shinde, D., & Hoogenboom, G. (2001). Developing genetic coefficients for crop simulation models with data from crop performance trials. *Crop Science*, *41*, 40–51.

Mavromatis, T., Boote, K. J., Jones, J. W., Wilkerson, G. G., & Hoogenboom, G. (2002). Repeatability of model genetic coefficients derived from soybean performance trials across different states. *Crop Science*, *42*, 76–89.

McAndrew, P., & Goodyear, P. (2007). Representing practitioner experiences through learning design and patterns. In Beetham, H., & Sharpe, R. (Eds.), *Rethinking pedagogy for a digital age* (pp. 92–102). Oxon, UK: Routledge.

McCown, R. L. (2002). Changing systems for supporting farmers' decisions: problems, paradigms, and prospects. *Agricultural Systems*, *74*, 179–220. doi:10.1016/S0308-521X(02)00026-4

McCown, R. L., Brennan, L. E., & Parton, K. A. (2006). Learning from the historical failure of farm management models to aid management practice. Part 1. The rise and demise of theoretical models of farm economics. *Australian Journal of Agricultural Research*, *57*, 143–156. doi:10.1071/AR05051

McCown, R. L., Hammer, G. L., Hargreaves, J. N. G., Holzworth, D. P., & Freebairn, D. M. (1996). APSIM: A novel software system for model development, model testing and simulation in agricultural systems research. *Agricultural Systems*, *50*, 255–271. doi:10.1016/0308-521X(94)00055-V

McKeown, R. (2002). *Education for sustainable development toolkit*. Knoxville, TN: University of Tennessee, Energy, Environment and Resources Center.

McKinion, J. M., & Baker, D. N. (1989). Application of the GOSSYM/COMAX system to cotton crop management. *Agricultural Systems*, *31*, 55–65. doi:10.1016/0308-521X(89)90012-7

McNeile, A. (2003). MDA The vision with the hole. Retrieved from http://www.metamaxim.com

McNeile, A., & Simons, N. (2004). Methods of behaviour modelling. *A commentary on behavior modelling techniques for MDA*.

Mellor, J. S., & Balcer, J. M. (2002). *Executable UML, a foundation for model-driven architecture*. Reading, MA: Addison Wesley.

Mennis, J. L., Peuquet, D. J., & Qian, L. (2000). A conceptual framework for incorporating cognitive principles into geographical database representation. *International Journal of Geographical Information Science*, *14*(6), 501–520. doi:10.1080/136588100415710

Meot, A., Hubert, B., & Lasseur, J. (2003). Organisation of the pastoral territory and grazing management: joint modelling of grazing management practices and plant cover dynamics. *Agricultural Systems*, *76*, 115–139. doi:10.1016/S0308-521X(02)00105-1

Merot, A., Bergez, J. E., Wallach, D., & Duru, M. (2008). Adaptation of a functional model of grassland to simulate the behaviour of irrigated grasslands under a Mediterranean climate: The Crau case. *European Journal of Agronomy*, *29*, 163–174. doi:10.1016/j.eja.2008.05.006

Messina, C. D., Jones, J. W., Boote, K. J., & Vallejos, C. E. (2006). A gene-based model to simulate soybean development and yield responses to environment. *Crop Science*, *46*, 456–466.

Meyer, B. (1988). *Object-oriented software construction*. Englewood Cliffs, NJ: Prentice-Hall.

Miller, H. J., & Han, J. (2009). *Geographic data mining and knowledge discovery*. Boca Raton, FL: Taylor & Francis.

Millington, J. D. A. (2007). *Modelling land-use/cover change and wildfire regimes in a Mediterranean landscape*. Unpublished doctoral dissertation, London King's College, London, UK.

Miralles, A. (2006). *Ingénierie des modèles pour les applications environnementales*. Unpublished doctoral dissertation, Université Montpellier II, Montpellier.

Miralles, A., & Libourel, T. (2008). Modeling with enriched model driven architecture. In S. Shekhar & H. Xiong (Eds.), *Encyclopedia of geographical information sciences* (pp. 700-705). New York: Springer.

Miralles, A., & Libourel, T. (2009a). Application of a model transformation paradigm in agriculture: A simple environmental system case study. In P. Papajorgji & P. M. Pardalos (Eds.), *Advances in modeling agricultural systems*, (Vol. 25, pp. 37-54). New York: Springer.

Miralles, A., & Libourel, T. (2009b). A new methodology to automate the transformation of GIS models in an iterative development process. In P. Papajorgji & P. M. Pardalos (Eds.), *Advances in modeling agricultural systems* (Vol. 25, pp. 19-36). New York: Springer.

Mitchell, T. M. (1997). *Machine learning*. New York: McGraw-Hill.

Mitchell, S. (2005). The ESRI Guide to GIS Analysis: *Vol. 2. Spatial Measurements and Statistics*. Redlands, CA: ESRI Press.

Mohan, P., & Brooks, C. (2003). Learning objects on the semantic web. In *Proceedings of the 3rd IEEE international conference of Advanced Learning Technologies* (pp. 195-199). Washington, DC: IEEE Computer Society.

Mohan, P., Greer, J., & McCalla, G. (2003). Instructional planning with learning objects. In P. Baumgartner, P.A. Cairns, M. Kohlhase, & E. Melis (Eds.), *Proceedings of the IJCAI 2003 Workshop on Knowledge Representation and Automated Reasoning for E-learning Systems* (pp. 52-58). Koblenz-Landau, Germany: University of Koblenz-Landau.

Molines, N. (2006). *Méthode et outils pour la planification des grandes infrastructures linéaires et leur évaluation environnementale*. Unpublished doctoral dissertation, Université Jean Monnet, Saint-Etienne, Université de Laval, Québec, Canada.

Montgomery, J. L., Harmon, T., Kaiser, W., Sanderson, A., Haas, C. N., & Hooper, R. (2007). The WATERS Network: An integrated environmental observatory network for water research. *Environmental Science & Technology*, *41*, 6642–6647. doi:10.1021/es072618f

Moody, D. L. (2005). Theoretical and practical issues in evaluating the quality of conceptual models: Current state and future directions. *Data & Knowledge Engineering*, *15*(3), 243–276. doi:10.1016/j.datak.2004.12.005

Moran, P. (1948). The interpretation of statistical maps. *Journal of the Royal Statistical Society. Series A (General)*, *10*, 243–251.

Mouillot, F., Rambal, S., & Lavorel, S. (2001). A generic process-based simulator for Mediterranean landscapes (SIERRA): Design and validation exercises. *Forest Ecology and Management*, *147*, 75–97. doi:10.1016/S0378-1127(00)00432-1

Mucherino, A., Papajorgji, P., & Pardalos, P. M. (2009). *Data mining in agriculture*. New York: Springer.

Mucherino, A., Papajorgji, P., & Pardalos, P. M. (2009). Survey of Data Mining Techniques Applied to Agriculture. *Operational Research -. International Journal (Toronto, Ont.)*, *9*, 121–140.

Mulino. (2004). *MULtisectoral INtegrated and Operational decision support system, or mDSS*. Retrieved from http://www.netsymod.eu/mdss/

Murgante, B., & Las Casas, G. (2004). G.I.S. and fuzzy sets for the land suitability analysis. In A. Laganá, M. L. Gavrilova, V. Kumar, Y. Mun, C. J. K. Tan, & O. Gervasi (Eds.), *Proceedings of the International Conference on Computational Science and its Applications* (LNCS 3044, pp. 1036-1045). doi: 10.1007/978-3-540-24709-8_109

Murgante, B., Las Casas, G., & Danese, M. (2008). Where are the slums? New approaches to urban regeneration. In Liu, H., Salerno, J., & Young, M. (Eds.), *Social computing, behavioral modeling and prediction* (pp. 176–186). Berlin, Germany: Springer-Verlag. doi:10.1007/978-0-387-77672-9_20

Murgante, B., Las Casas, G., & Sansone, A. (2008). A spatial rough set for extracting periurban fringe. *Revue des Nouvelles Technologies de l'Information, 857*, 101–125.

Muttil, N., & Chau, K. W. (2006). Neural network and genetic programming for modelling coastal algal blooms. *International Journal of Environment and Pollution, 28*(3-4), 223–238. doi:10.1504/IJEP.2006.011208

Muttil, N., & Chau, K. W. (2007). Machine learning paradigms for selecting ecologically significant input variables. *Engineering Applications of Artificial Intelligence, 20*(6), 735–744. doi:10.1016/j.engappai.2006.11.016

Muzy, A., Innocenti, E., Aïello, A., Santucci, J.-F., Santoni, P.-A., & Hill, D. R. C. (2005). Modelling and simulation of ecological propagation processes: application to fire spread. *Environmental Modelling & Software, 20*(7), 827–842. doi:10.1016/j.envsoft.2004.04.011

Naab, J. B., Singh, P., Boote, K. J., Jones, J. W., & Marfo, K. O. (2004). Using the CROPGRO-peanut model to quantify yield gaps of peanut in the Guinean savanna zone of Ghana. *Agronomy Journal, 96*, 1231–1242.

Napoleone, C. (2005). *Prix fonciers et immobiliers, et localisation des ménages au sein d'une agglomération, urbaine*. Unpublished doctoral dissertation, Aix-Marseille University, Marseille, France.

Nardi, D., & Brachman, R. J. (2002). An introduction to description logics. In F. Baader, D. Calvanese, D. L. McGuinness, D. Nardi, & P. F. Patel-Schneide (Eds.), *Description logic handbook.* Cambridge University Press.

Nassauer, J. I., & Opdam, P. (2008). Design in science: extending the landscape ecology paradigm. *Landscape Ecology, 23*, 633–644. doi:10.1007/s10980-008-9226-7

National Committee for Information Technology Standards. (1998). *Technical Committee T2 (Information Interchange and Interpretation): Knowledge Interchange Format, draft proposed American National Standard (dpANS)* (Rep. NCITS.T2/98-004). Retrieved from http://logic.stanford.edu/kif/dpans.html

Nejdl, W., Wolf, B., Qu, C., Decker, S., Sintek, M., Naeve, A., et al. (2002, May). *EDUTELLA: a P2P networking infrastructure based on RDF*. Paper presented at the 11th International Conference on World Wide Web, Honolulu, HI.

Nickul, D., Reitman, L., Ward, J., & Wilber, J. (2005). *Service oriented architecture (SOA) and specialized messaging patterns.* Retrieved from http://www.adobe.com/ government/pdfs/ SOA-technical-whitepaper.pdf

Niedrite, L., Treimanis, M., Solodovnikova, D., & Grundmane, L. (2009). Development of Data Warehouse Conceptual Models - Method Engineering Approach. In Taniar, D. (Ed.), *Progressive Methods in Data Warehousing and Business Intelligence: Concepts and Competitive Analytics* (pp. 1–23). Hershey, PA: IGI Global.

Nilsson, N. J. (1980). *Principles of Artificial Intelligence*. Wellsboro, PA: Tioga Publishing Company.

North, M. J., Tatara, E., Collier, N. T., & Ozik, J. (2007). Visual agent-based model development with repast simphony. In *Proceedings of the Agent 2007 Conference on Complex Interaction and Social Emergence*. Argonne, USA: Argonne National Laboratory.

O'Hagan, R., Atkinson, R., Cox, S., Fitch, P., Lemon, D., & Walker, G. (2007). *A Reference Model for a Water Resources Observation Network.* Paper presented at MODSIM07.

Oden, J. T. (2006). *Revolutionizing engineering science through simulation. Blue Ribbon Panel on Simulation-Based Engineering Science. National Science Foundation*. NSF.

OGC. (2005). Implementation Specification for Geographic information - Simple feature access - Part 1: Common architecture.

Ohe, W. V. D., Oddo, L. P., Piana, M. L., Morlot, M., & Martin, P. (2004). Harmonized methods of melissopalynology. *Apidologie*, *35*, 18–25. doi:10.1051/apido:2004050

OMG. (1999). *Unified modeling language - specification - version 1.3.* Needham, MA: Author.

OMG. (2003, March). *Unified Modeling Language - Specification - Version 1.5*. Retrieved from http://www.omg.org/cgi-bin/apps/doc?formal/03-03-01.pdf

OMG. (2007). *Unified modelling language: OCL, version 2.0. OMG specification.* Needham, MA: Author.

OMG. (2009a, February). *OMG unified modeling language: Superstructure - version 2.2*. Retrieved June 2009, from http://www.omg.org/cgi-bin/apps/doc?formal/05-07-04.pdf

OMG. (2009b). *Unified modeling language (OMG UML), infrastructure version 2.2*. Needham, MA: Author.

OOAD. (1997). *Object Oriented Analysis and Design*. Retrieved from http://www.ooad.org/

Open Geospatial Consortium (OGC). (2009). *Observations and Measurements*. Retrieved March 8, 2010, from http://www.opengeospatial.org/standards/om

ORACLE. (n.d.). *Oracle9i® Database*. Retrieved October 10, 2009, from http://www.oracle.com/technology/products/oracle9i/index.html

O'Sullivan, D., & Unwin, D. (2002). *Geographic information analysis*. New York, NY: John Wiley & Sons.

O'Sullivan, D., & Wong, D. W. S. (2007). A surface-based approach to measuring spatial segregation. *Geographical Analysis*, *39*, 147–168. doi:10.1111/j.1538-4632.2007.00699.x

Pan, J. S., Qiao, Y. L., & Sun, S. H. (2004). A fast k nearest neighbors classification algorithm. *IEICE Transactions on Fundamentals of Electronics, Communications and Computer Sciences. E (Norwalk, Conn.)*, *87-A*(4), 961–963.

Pantazis, D., & Donnay, J. P. (1996). *La conception SIG: méthode et formalisme*. Paris: Hermés.

Pantazis, D. (1994). *Analyse Méthodologique des Phases de Conception et de Développement d'un SIG*. Belgium: Université de Liège.

Pantazis, D., & Donnay, J. P. (1996). *La conception SIG: méthode et formalisme*. Paris: Hermés.

Papadimitriou, V. (2004). Prospective primary teachers' understanding of climate change, greenhouse effect, and ozone layer depletion. *Journal of Science Education and Technology*, *13*(2), 299–307. doi:10.1023/B:JOST.0000031268.72848.6d

Papajorgji, P. (2009). State of the art in modeling agricultural systems. In *Encyclopedia of optimization* (pp. 3693-3704). New York: Springer.

Papajorgji, P., & Clark, R. (2005). *A model driven approach to agricultural systems.* Paper presented at the EFITA 2005, Portugal.

Papajorgji, P., & Pardalos, P. M. (2006). *Software engineering techniques applied to agricultural systems an object-oriented and UML approach*. New York: Springer.

Papajorgji, P., Clark, R., & Jallas, E. (2009). The model driven architecture approach: A framework for developing complex agricultural systems. In P. Papajorgji & P. M. Pardalos (Eds.), *Advances in modeling agricultural systems.* New York: Springer.

Papajorgji, P., Gargiulo, O., Jones, J. W., & Traore, S. (2009). A UML-based Plug&Play version of RothC. In P. Papajorgji & P. M. Pardalos (Eds.), *Advances in Modeling Agricultural Systems.* New York: Springer.

Papajorgji, P. (2005). A plug and play approach for developing environmental models. *Environmental Modelling & Software*, *20*(10), 1353–1357. doi:10.1016/j.envsoft.2004.11.016

Papajorgji, P., Beck, H. W., & Braga, J. L. (2004). An architecture for developing service-oriented and component-based environmental models. *Ecological Modelling*, *179*, 61–76. doi:10.1016/j.ecolmodel.2004.05.013

Papajorgji, P., & Shatar, T. M. (2004). Using the unified modeling language to develop soil water-balance and irrigation-scheduling models. *Environmental Modelling & Software*, *19*(5), 451–459. doi:10.1016/S1364-8152(03)00160-9

Paquette, G. (2007). An ontology and a software framework for competency modelling and management. *Educational Technology and Society: Special Issue on Advanced Technologies for Life-long Learning, 10*(3), 1-21. Retrieved from www.ifets.info/journals/10_3/ets_10_3.pdf

Paran, F. (2005). *Représentations territoriales pour la gestion équilibrée d'un patrimoine écologique et anthropique dans le domaine de l'eau.* Unpublished doctoral dissertation, Ecole Nationale Supérieure des Mines, Saint-Etienne, France.

Parent, C., Spaccapietra, S., & Zimányi, E. (2006). *Conceptual modeling for traditional and spatio-temporal applications: The MADS approach.* New York: Springer-Verlag.

Parent, C., Spaccapietra, S., & Zimányi, E. (2006). *Conceptual Modeling for traditional and Spatio-Temporal Applications: The MADS Approach.* New York: Springer Verlag.

Parent, C., Spaccapietra, S., & Zimányi, E. (2008). Modeling and Multiple Perceptions. In Shekhar, S., & Xiong, H. (Eds.), *Encyclopedia of Geographical Information Sciences* (pp. 682–690). New York: Springer.

Parent, C., Spaccapietra, S., Zimányi, E., Donini, P., Plazanet, C., & Vangenot, C. (1997). MADS, modèle conceptuel spatio-temporel. *Revue Internationale de Géomatique, 7*(3-4), 317–352.

Parlett, B. N. (1998). The symmetric eigenvalue problem. *SIAM, 20,* 357.

Parsaye, K., & Chignell, M. (1988). *Expert Systems for Experts.* New York: John Wiley & Sons.

Parsons, A. J. (1988). The effect of season and management on the grass growth of grass sward. In Jones, M. B., & Lazenby, A. (Eds.), *The grass crop* (pp. 129–178). London: Chapmann and Hall. doi:10.1007/978-94-009-1187-1_4

Parton, W. J., Stewart, J. W. B., & Cole, C. V. (1988). Dynamics of C, N, P and S in grassland soils: A model. *Biogeochemistry, 5,* 109–131. doi:10.1007/BF02180320

Partridge, D. (1987). The scope and limitations of first generation expert systems. *Future Generation Computer Systems, 3,* 1–10. doi:10.1016/0167-739X(87)90038-0

Pasin, M. (2005). *Semantic learning narratives* (Tech. Rep. KMI-05-07). London: The Open University, Knowledge Media Institute.

Pastor, O., & Molina, J. C. (2007). *Model-driven architecture in practice a software production environment based on conceptual codeling.* New York: Springer.

Pawlowski, J. M., & Bick, M. (2006). Managing & reusing didactical expertise: The didactical object model. *Journal of Educational Technology & Society, 9*(1), 84–96.

Paz, J. O., Fraisse, C. W., Hatch, L. U., Garcia y Garcia, A., Guerra, L. C., & Uryasev, O. (2007). Development of an ENSO-based irrigation decision support tool for peanut production in the southeastern US. *Computers and Electronics in Agriculture, 55*(1), 28–35. doi:10.1016/j.compag.2006.11.003

Pearcy, R. W., & Bjorkman, O. (1983). Physiological effects. In E. R. Lemon (Ed.), *CO_2 and plants: The response of plants to rising levels of atmospheric carbon dioxide. AAAS Selected Symp. 84* (pp. 65-105). Boulder, CO: Westview Press.

PENTAHO. (n.d.). Retrieved October 10, 2009, from http://mondrian.pentaho.org

Pernin, J. P., & Lejeune, A. (2006, June). *Models for the re-use of learning scenarios.* Paper presented at the IFIP conference, Ålesund, Norway. Retrieved from http://dspace.learningnetworks.org/bitstream/1820/580/1/Models.pdf

Pernkopf, F. (2005). Bayesian network classifiers versus selective *k*-NN classifier. *Pattern Recognition, 38*(1), 1–10. doi:10.1016/j.patcog.2004.05.012

Persano Oddo, L., & Bogdanov, S. (2004). Determination of botanical origin: problems and issues. *Apidologie, 35*(1), S2–S3. doi:10.1051/apido:2004044

Persano-Oddo, L., Piana, M. L., & Ricciardelli D'Albore, G. (2007). *I mieli regionali italiani. Caratterizzazione melissopalinologica* (p. 139). Agnesotti, VT: Ministero delle Politiche Agricole Alimentari e Forestali e CRA.

Peuquet, D. J. (1994). It's about time: A conceptual framework for the representation of temporal dynamics in geographic information system. *Annals of the Association of American Geographers. Association of American Geographers, 84*(3), 441–461. doi:10.1111/j.1467-8306.1994.tb01869.x

Phan, D., & Varenne, F. (2009). Agent-based models and simulations in economics and social sciences: From conceptual exploration to distinct ways of experimenting. *Journal of Artificial Societies and Social Simulation*, 10.

Philippe, M. (2003, May/June). *Translations between UML, OWL, KIF and the WebKB-2 Languages (For-Taxonomy, Frame-CG, Formalized English).*

Pinet, F. (2002). *Vers une Méthode de Conception pour la Télégéomatique: Intégration et Validation au sein d'un Atelier de Développement.* Unpublished Thèse de docteur en Informatique, Institut National des Sciences Appliquées de Lyon, Lyon, France.

Pinet, F., & Lbath, A. (2003, April 24-26). *Object-Oriented Modelling of Communicating GIS-Based Applications.* Paper presented at the 6th AGILE European Conference on Geographic Information Science, Lyon, France.

Pinet, F., & Schneider, M. (2009). A unified object constraint model for designing and implementing multidimensional systems. *Journal of Data Semantics, 13.*

Pinet, F., Duboisset, M., Demuth, B., Schneider, M., Soulignac, V., & Barnabe, F. (2009). Constraints modeling in Agricultural Databases. In *Advances in Modeling Agricultural Systems*. New York: Springer.

Pinet, F., Roussey, C., Brun, T., & Vigier, F. (2009). The use of UML as a tool for the formalisation of standards and the design of ontologies in agriculture. In P. Papajorgji & P. M. Pardalos, (Eds.), *Advances in modeling agricultural systems.* New York: Springer.

Pinet, F., Ventadour, P., Brun, T., Papajorgji, P., Roussey, C., & Vigier, F. (2006, November). *Using UML for ontology construction: A case study in agriculture.* Paper presented at the 7th Agricultural Ontology Service (AOS) Workshop on Ontology-Based Knowledge Discovery: Using Metadata and Ontologies for Improving Access to Agricultural Information (in collaboration with AFITA 2006), Indian Institute of Science, Bangalore, India.

Pinet, F., Duboisset, M., & Soulignac, V. (2007). Using UML and OCL to maintain the consistency of spatial data in environmental information systems. *Environmental Modelling & Software, 22*(8), 1217–1220. doi:10.1016/j.envsoft.2006.10.003

Pinet, F., Kang, M. A., & Vigier, F. (2005). Spatial constraint modelling with a GIS extension of UML and OCL: application to agricultural information systems. *Lecture Notes in Computer Science, 3511*, 160–178.

Poe, V. (1996). *Building a Data Warehouse for Decision Support*. Upper Saddle River, NJ: Prentice-Hall.

Preusch, M. (2009, July 29). *Acoustic tags help track movement of salmon through dams*. Retrieved February 7, 2010, from http://www.oregonlive.com/ environment/index.ssf/ 2009/07/acoustic_tags_help_track_movem.html

Prevosto, B., Hill, D., & Coquillard, P. (2003). Individual-based modelling of pinus sylvestris invasion after grazing abandonment in the French Massif Central. *Plant Ecology, 166*(1), 121–137. doi:10.1023/A:1024404214782

Price, R., Ramamohanarao, K., & Srinivasan, B. (1999). *Spatio-temporal unified modeling language.* Paper presented at the Workshop on Spatio- Models and Languages, IEEE DEXA'99, Florence, Italy.

Proulx, M.-J., Larrivée, S., & Bédard, Y. (2002). Représentation multiple et généralisation avec UML et l'outil Perceptory. In *Généralisation et représentation multiple* (pp. 113–130). Paris: Hermès Sciences Publications.

Putto, C., Pathanothai, A., Jogloy, S., Pannangpetch, K., Boote, K. J., & Hoogenboom, G. (2009). Determination of efficient test sites for evaluation of peanut breeding lines using the CSM-CROPGRO-peanut model. *Field Crops Research, 110*, 272–281. doi:10.1016/j.fcr.2008.09.007

Pyle, D. (2003). *Business modeling and data mining*. San Francisco, CA: Morgan Kaufmann Publishers.

Quang, P. T., & Chartier-Kastler, C. (1991). *Merise in Practice*. MacMillan Education.

Quesnel, G., Duboz, R., & Ramat, E. (2009). The virtual laboratory environment -- an operational framework for multi-modelling, simulation and analysis of complex dynamical systems. *Simulation Modelling Practice and Theory, 17*, 641–653. doi:10.1016/j.simpat.2008.11.003

Ràbovà, I. (2005). The Formalisation of knowledge in agricultural industry business process. *Agricultural Economics*, *51*(7), 53–58.

Rafanelli, M. (2003). *Multidimensional databases: Problems and solutions*. Hershey, PA: IGI Global.

Rafanelli, M. (2003). Operators for Multidimensional Aggregate Data. In Rafanelli, M. (Ed.), *Multidimensional databases: problems and solutions* (pp. 116–165). Hershey, PA: IGI Global. doi:10.4018/978-1-59140-053-0.ch005

Ramat, E. (2004). *Contribution à la modélisation et à la simulation des systèmes complexes, habilitation à diriger les recherches*. Unpublished doctoral dissertation, Littoral Cote d'Opale University, Brest, France.

Rao, F., Zhang, L., Yu, X., Li, Y., & Chen, Y. (2003). Spatial Hierarchy and OLAP-Favored Search in Spatial Data Warehouse. In S. Rizzi & Y. Song (Eds.), *Proceedings ACM International Workshop on Data Warehousing and OLAP* (pp. 48-55). New York: ACM Press.

Ravat, F., Teste, O., Tournier, R., & Zurfluh, G. (2007). Integrating Complex Data into a Data Warehouse. In *Proceedings of the 9th International Conference on Software Engineering Knowledge Engineering (SEKE 07)*, Boston, MA (p. 483).

Reijers, H. A. (2006, June 26-28). Workflow flexibility: The forlorn promise. In *Proceedings of the 15th IEEE International Workshops on Enabling Technologies: Infrastructures for Collaborative Enterprises,* Manchester, UK (pp. 271-272).

Reiter, R. (1987). A theory of diagnosis from first principles. *Artificial Intelligence*, *32*(1), 57–95. doi:10.1016/0004-3702(87)90062-2

Rellier, J. P. (2005). *DIESE: un outil de modélisation et de simulation de systèmes d'intérêt agronomique. Internal report UBIA-INRA, Toulouse-Auzeville.* Retrieved from http://carlit.toulouse.inra.fr/diese/docs/ri_diese.pdf

Renn, O. (1998). Three decades of risk research: accomplishments and new challenges. *Journal of Risk Research*, *1*(1), 49–71. doi:10.1080/136698798377321

Richardson, A. (2002, August). *Starting with Bold for Delphi / Bold for C++, Part 3: Object Constraint Language (OCL).*

Ritchie, J. T. (1985). A user-oriented model of the soil water balance in wheat. In E. Fry & T. K. Atkin (Eds.), *Wheat growth and modeling* (pp. 293-305). New York: Plenum Press.

Ritchie, J. T. (1998). Soil water balance and plant water stress. In G. Y. Tsuji, G. Hoogenboom, & P. K. Thornton (Eds), *Understanding options for agricultural production* (pp. 41-54). Dordrecht, The Netherlands: Kluwer Academic Publishers.

Rivest, S., Bédard, Y., Proulx, M., Nadeaum, M., Hubert, F., & Pastor, J. (2005). SOLAP: Merging Business Intelligence with Geospatial Technology for Interactive Spatio-Temporal Exploration and Analysis of Data. *Journal of International Society for Photogrammetry and Remote Sensing*, *60*(1), 17–33. doi:10.1016/j.isprsjprs.2005.10.002

Robinson, J. (2004). Squaring the circle? Some thoughts on the idea of sustainable development. *Ecological Economics*, *48*, 369–384. doi:10.1016/j.ecolecon.2003.10.017

Robinson, S. (2004). *Simulation: The Practice of Model Development and Use*. Chichester, UK: Wiley.

Rocha, L. V. d., Edelweiss, N., & Iochpe, C. (2001). *GeoFrame-T: A Temporal Conceptual Framework for Data modeling.* Paper presented at the 9th ACM international symposium on Advances in geographic information systems, Atlanta, GA.

Rocha, L. V. d. (2001). *Framework Conceitual Temporal para Aplicações de Sistemas de Informações Geográficas*. Porto Alegre, Brazil: Universidade Federal do Rio Grande Do Sul.

Rodriguez, A., & Raper, J. (1999). Defining spatial agents. In Camara, A. S., & Raper, J. (Eds.), *Spatial multimedia and virtual reality research, monograph* (1st ed., pp. 111–129). New York, NY: Taylor & Francis.

Rogner, H.-H., & Popescu, A. (2000). An introduction to energy. In *World Energy Assessment: Energy and the Challenge of Sustainability* (pp. 31-37). New York: UNDP.

Romera, A. J., Morris, S. T., Hodgson, J., Stirling, W. D., & Woodward, S. J. R. (2004). A model for simulating rule-based management of cow-calf systems. *Computers and Electronics in Agriculture*, *42*, 67–86. doi:10.1016/S0168-1699(03)00118-2

Roncoli, C., Paz, J., Breuer, N., Ingram, K., Hoogenboom, G., & Broad, K. (2006). *Understanding farming decisions and potential applications of climate forecasts in South Georgia* (Tech. Rep. Series SECC-06-006). Gainesville, FL: Southeast Climate Consortium.

Rosch, E. (1978). Principles of categorization. In Rosch, E., & Lloyd, B. (Eds.), *Cognition and Categorization* (pp. 27–77).

Rouet, I., Gay, D., Allenbach, M., Selmaoui, N., Ausseil, A.-G., Mangeas, M., et al. (2009). Tools for soil erosion mapping and hazard assessment: Application to New Caledonia, SW Pacific. In *Proceedings of the International Congress on Modelling and Simulation*, Cairns, Australia (pp. 1986-1992).

Rumbaugh, J., Blaha, M., Lorensen, W., Eddy, F., & Premerlani, W. (1990). *Object-Oriented Modeling and Design*. Upper Saddle River, NJ: Prentice Hall.

Saaty, T. L. (1982). *Decision making for leaders: The analytical hierarchy process for decisions in a complex world*. Belmont, CA: Life-Time Learning Publications.

Saaty, T. L., & Shih, H.-S. (2009). Structures in decision making: On the subjective geometry of hierarchies and networks. *European Journal of Operational Research, 199*(3), 867–872. doi:10.1016/j.ejor.2009.01.064

Sacks, J., Welch, W. J., Mitchell, T. J., & Wynn, H. P. (1989). Design and analysis of computer experiments. *Statistical Science, 4*, 409–435. doi:10.1214/ss/1177012413

Salles, P., Bredeweg, B., & Araujo, S. (2003a). Qualitative models about stream ecosystem recovery: Exploratory studies. In *Proceedings of the 17th International Workshop on Qualitative Reasoning (QR)*, Brasilia, Brazil.

Salles, P., Bredeweg, B., & Bensusan, N. (2003b). The ants' garden: Qualitative models of complex interactions between populations. In *Proceedings of the 17th International Workshop on Qualitative Reasoning (QR)*, Brasilia, Brazil.

Salsano, E. (1998). *Fondamenti di urbanistica*. Bari, Italy: Laterza.

Saltelli, A., Ratto, M., Andres, T., Campolongo, F., Cariboni, J., & Gatelli, D. (2008). *Global Sensitivity Analysis*. New York: John Wiley & Sons.

Saltelli, A., Tarantola, S., Campolongo, S., & Ratto, M. (2004). *Sensivity analysis in practice - A guide to assessing scientific models*. Willey.

Sampaio, M., Sousa, A., & Baptista, C. (2006). Towards a logical multidimensional model for spatial data warehousing and OLAP. In Y. Song & P. Vassiliadis (Eds.), *Proceedings of ACM International Workshop on Data Warehousing and OLAP* (pp. 83-90). New York: ACM Press.

Santner, T. J., Williams, B. J., & Notz, W. I. (2003). *Design and Analysis of Computer Experiments*. New York: Springer Verlag.

Sawhney, R. (2006). Interplay between uncertainty and flexibility across the value-chain: towards a transformation model of manufacturing flexibility. *Journal of Operations Management, 42*, 476–493. doi:10.1016/j.jom.2005.11.008

Sboui, T., Bédard, Y., Brodeur, J., & Badard, T. (2007). A conceptual framework to support semantic interoperability of geospatial datacubes. In *Proceedings of ER/2007 SeCoGIS workshop* (LNCS 4802, pp. 378-387). New York: Springer.

Scala, M., Aronne, G., & De Micco, V. (2004). Variability in pollen spectra of chestnut honeys from different sited and years. *Polen, 14*, 403.

Scala, M., De Micco, V., & Aronne, G. (2004). Multivariate analysis of the accompanying pollen for the determination of the geographical origin of honey. *Polen, 14*, 109.

Schatzki, T. F., Haff, R. P., Young, R., Can, I., Le, L.-C., & Toyofuku, N. (1997). Defect detection in apples by means of x-ray imaging. *Transactions of the American Society of Agricultural Engineers, 40*(5), 1407–1415.

Scheller, M. R., & Domingo, J. B. (2006). *LANDIS-II core model description, Landis-II models v 5.1 user guide*. Retrieved from http://www.landis-ii.org/exts/age-only-succession/LANDIS-II%20Age-Only%20Succession%20v2.0%20User%20Guide.pdf

Schmid, B., Warmer, J., & Clark, T. (2002). *Object modeling with the OCL: The relational behind the object constraint language*. New York: Springer.

Schneeweis, C. (1995). Hierarchical structure in organisation: a conceptual framework. *European Journal of Operational Research, 81*, 4–31. doi:10.1016/0377-2217(95)00058-X

Scholefield, D., Lockyer, D. R., Whitehead, D. C., & Tyson, K. C. (1991). A model to predict transformations and losses of nitrogen in UK pastures grazed by beef cattle. *Plant and Soil, 132*, 165–177.

Schreiber, G. (2005). *A UML Presentation Syntax for OWL Lite.*

Scotch, M., & Parmanto, B. (2006). Development of SOVAT: a numerical-spatial decision support system for community health assessment research. *International Journal of Medical Informatics, 34*(10), 771–784. PubMed doi:10.1016/j.ijmedinf.2005.10.008

SEEK. (2006). *SEEK KR Ontologies.* Retrieved March 8, 2010, from http://seek.ecoinformatics.org/Wiki.jsp?page=KROntologies

Serment, J. (2007). *Une infrastructure d'intégration à base d'agents logiciels pour l'élaboration de systèmes d'aide à la décision environnementale: Application à la gestion hydraulique camarguaise*. Unpublished doctoral dissertation, Paul Cezanne University, Marseille, France.

Shahin, M. A., Tollner, E. W., & McClendon, R. W. (2001). Artificial intelligence classifiers for sorting apples based on watercore. *Journal of Agricultural Engineering Research, 79*(3), 265–274. doi:10.1006/jaer.2001.0705

Sharp, H., Manns, M. L., & Eckstein, J. (2000). The Pedagogical Patterns Project. In *Addendum to the Proceedings of the Conference on Object-oriented Programming, Systems, Languages, and Applications* (pp. 139-140).

Shekhar, S., & Huang, Y. (2001). Discovering spatial colocation patterns: A summary of results. In C. S. Jensen, M. Schneider, B. Seeger, & V. J. Tsotras (Eds.), *Proceedings of the 7th International Symopisum on Advances in Spatial and Temporal Databases* (LNCS 2121, pp. 236-256).

Shekhar, S., Coyle, M., Goyal, B., Liu, D.-R., & Sarkar, S. (1997). *Data Models in Geographic Information Systems.*

Shekhar, S., Lu, C., Tan, X., Schawla, S., & Vatsavai, R. (2001). Map Cube: A Visualization Tool for Spatial Data Warehouses. In Miller, H., & Han, J. (Eds.), *Geographic Data Mining and Knowledge Discovery* (pp. 74–109). London: Taylor & Francis. doi:10.4324/9780203468029_chapter_4

Shekhar, S., & Xiong, H. (2008). *Encyclopedia of Gis* (1st ed.). New York: Springer.

Shlaer, S., & Mellor, S. J. (1988). *Object-oriented systems analysis - modeling the world in data.* Englewood Cliffs, NJ: Prentice-Hall.

Shneiderman, B., & Plaisant, C. (2004). *Designing the User Interface: Strategies for Effective Human-Computer Interaction.* Reading, MA: Addison-Wesley.

Sibertin-Blanc, C., Amblard, F., & Mailliard, M. (2006). A coordination framework based on the sociology of organized action. In O. Boissier, J. Padget, V. Dignum, & G. Lindemann (Eds.), *Proceedings of the Conference on Coordination, Organizations, Institutions and Norms in Multi-Agent Systems* (LNCS 3913, pp. 3-17).

Sillverman, B. W., & Jones, M. C. (1989). E. Fix and J. L. Hodges (1951): An important contribution to nonparametric discriminant analysis and density estimation. *International Statistical Review, 57*(3), 233–247. doi:10.2307/1403796

Silva, J., Times, V., Salgado, A., Souza, C., Fidalgo, R., & Oliveira, A. (2008). A set of aggregation functions for spatial measures. In I. Song & A. Abelló (Eds.), *Proceedings of the ACM International Workshop on Data Warehousing and OLAP* (pp. 25-32). New York: ACM Press.

Simon, H. A. (1982). *Models of bounded rationality: Behavioral economics and business organization* (*Vol. 1-2*). Cambridge, MA: MIT Press.

Skanavis, C., Petreniti, V., & Giannopoulou, K. (2004, August). Educators and environmental education in Greece. In *Proceedings of the International Conference for the Protection and Restoration of the Environment VII*, Mykonos, Greece. Retrieved from http://www.srcosmos.gr/srcosmos/showpub.aspx?aa=4353

Smith, M., & Grice, A. (2005). Ecosystem Health Monitoring Program. In Abal, E. G., Bunn, S. E., & Dennison, W. C. (Eds.), *Healthy Waterways, Healthy Catchments: Making the connection in south east Queensland* (pp. 149–182). Brisbane, QLD, Australia: Moreton Bay and Catchments Partnership.

Smith, M., & Storey, A. W. (2001). *Design and Implementation of Baseline Monitoring (DIBM3): Developing an Ecosystem Health Monitoring Program for Rivers and Streams in Southeast Queensland*. Brisbane, QLD, Australia: South East Queensland Regional Water Quality Management Strategy.

Smith, P., Smith, J. U., Powlson, D. S., McGill, W. B., Arah, J. R. M., & Chertov, O. G. (1997). A comparison of the performance of nine soil organic matter models using datasets from seven long-term experiments. *Geoderma*, *81*, 153–225. doi:10.1016/S0016-7061(97)00087-6

Softeam. (1999). *UML profiles and the J language: Totally control your application development using UML*. Retrieved March 2002, from http://www.objecteering.com/pdf/whitepapers/us/uml_profiles.pdf

Solazzia, M., & Uncinib, A. (2004). Regularizing neural networks using flexible multivariate activation function. *Neural Networks*, *17*(2), 247–260. doi:10.1016/S0893-6080(03)00189-8

South East Queensland Healthy Waterways Partnership (SEQ HWP). (2009). *Healthy Waterways.* Retrieved March 8, 2010, from http://www.healthywaterways.org/

Spaccapietra, S., Al-Jadir, L., & Yu, S. (2004, April 4). Somebody, sometime, somewhere, something. In *Proceedings of the International Workshop on Ubiquitous Data*, Tokyo, Japan (pp. 6-13).

SPIRE. (2009). *Spire Research Group, Spire Ontology Tool (SPOTTER).* Retrieved March 8, 2010, from http://spire.umbc.edu/spotter/

Spiropoulou, D., Antonakaki, T., Kontaxaki, S., & Bouras, S. (2007). Primary teachers' literacy and attitudes on education for sustainable development. *Journal of Science Education and Technology*, *16*, 443–450. doi:10.1007/s10956-007-9061-7

Standford University. (2009). *Protégé*. Retrieved from http://protege.stanford.edu

Starr, L. (2002). *Executable UML how to build class models.* Englewood Cliffs, NJ: Prentice Hall.

Staub, P., Gnagi, H. R., & Morf, A. (2008). Semantic interoperability through the definition of conceptual model transformations. *Transactions in GIS*, *12*(2), 193–207. doi:10.1111/j.1467-9671.2008.01095.x

Stefanovic, N., Han, J., & Koperski, K. (2000). Object-Based Selective Materialization for Efficient Implementation of Spatial Data Cubes. *IEEE Transactions on Knowledge and Data Engineering*, *12*(6), 938–958. doi:10.1109/69.895803

Stockle, C. O., Martin, S. A., & Campbell, G. S. (1994). CropSyst, a cropping systems simulation model: Water/nitrogen budgets and crop yield. *Agricultural Systems*, *46*, 335–359. doi:10.1016/0308-521X(94)90006-2

Stolte, C., Tang, D., & Hanrahan, P. (2003). Multiscale Visualization Using Data Cubes. *IEEE Transactions on Visualization and Computer Graphics*, *9*(2), 176–187. doi:10.1109/TVCG.2003.1196005

Struss, P. (1998). Artificial intelligence for nature - why knowledge representation and problem solving should play a key role in environmental decision support. In Herzog, O., & Günter, A. (Eds.), *KI-98: Advances in Artificial Intelligence (LNAI 1504)*. doi:10.1007/BFb0095425

Sutton, R. S., & Barto, A. G. (1998). *Reinforcement learning: An introduction*. Cambridge, MA: MIT Press.

SWEET. (2009). *Semantic Web for Earth and Environmental Terminology.* Retrieved March 8, 2010, from http://sweet.jpl.nasa.gov/

Tambour, L., Houlès, V., Cohen-Jonathan, L., Auffray, V., Escande, P., & Jallas, E. (2009). Design of a model-driven Web decision support system in agriculture: From scientific models to the final software. In P. Papajorgji & P. M. Pardalos (Eds.), *Advances in modeling agricultural systems* (pp. 67-102). New York: Springer.

Tang, D., Stolte, C., & Bosch, R. (2003). Design Choices when Architecting Visualizations. In *Proceedings of IEEE Symposium on Information Visualization* (pp. 65-79). Washington, DC: IEEE Computer Society.

Tan, P.-N., Steinbach, M., & Kumar, V. (2005). *Introduction to data mining*. Reading, MA: Addison-Wesley.

Tardieu, H., Rochfeld, A., & Colletti, R. (1983). La méthode Merise: *Vol. 1. Principes et outils*. Paris: Éditions d'Organisation.

Tavares-Pereira, F., Figueira, J., Mousseau, V., & Roy, B. (2007). Multiple criteria districting problems. *Annals of Operations Research, 154*(1), 69–92. doi:10.1007/s10479-007-0181-5

TenCompetence Project. (2006). *M6.1- Templates of the pedagogical models to be used in authoring environment: Model that combines classical and new forms of assessment*. Retrieved from http://dspace.learningnetworks.org/bitstream/1820/876/4/M6-1.pdf

The Middleware Company. (2003). *Model driven development for J2EE utilizing a model driven architecture (MDA) approach - productivity pnalysis* (case study).

Tjioe, H. C., & Taniar, D. (2005). Mining Association Rules in Data Dare Warehouses. *International Journal of Data Warehousing and Mining, 1*(3), 28–62.

Toval, A., Requena, V., & Fernández, J. L. (2003). Emerging OCL tools. *Software and Systems Modeling, 2*(4), 248–261. doi:10.1007/s10270-003-0031-0

Traore, P. S. C., Bostick, W. M., Jones, J. W., Koo, J., Goita, K., & Bado, B. V. (2008). A simple soil organic matter model for biomass data assimilation in community-level carbon contracts. *Ecological Applications, 18*(3), 624–636. doi:10.1890/07-1133.1

Trujillo, J., Palomar, M., Gomez, J., & Song, I. S. (2001). Designing Data Warehouses with OO Conceptual Models. *Computer, 34*(12), 66–75. doi:10.1109/2.970579

Tryfona, N., & Jensen, C. S. (1998). *CHOROCHRONOS: A Network for Spatiotemporal Database Systems - Conceptual Data Modeling for Spatiotemporal Applications* (Tech. Rep. No. CH-98-08). European Commission DG XII Science, Research and Development.

Tryfona, N., & Jensen, C. S. (2000, March 19-21). *Using abstractions for spatio-temporal conceptual modeling.* Paper presented at the 2000 ACM Symposium on Applied Computing, Villa Olmo (Italy).

Tryfona, N., & Jensen, C. S. (2000, March 19-21). *Using Abstractions for Spatio-Temporal Conceptual Modeling.* Paper presented at the 2000 ACM Symposium on Applied Computing, Villa Olmo, Italy.

Tsuji, G. Y., Hoogenboom, G., & Thornton, P. K. (1998). *Understanding options for agricultural production. Systems approaches for sustainable agricultural development.* Dordrecht, The Netherlands: Kluwer Academic Publishers. ISBN 07923-4833-8.

TU Dresden University. (2009). *Dresden OCL toolkit - projects around the toolkit*. Retrieved from http://dresden-ocl.sourceforge.net/projects.html

Ullrich, C. (2004, November). Description of an instructional ontology and its application in web services for education. In *Proceedings of the Workshop on Applications of Semantic Web Technologies for E-learning, SW-EL'04,* Hiroshima, Japan. Retrieved from http://www.activemath.org/pubs/Ullrich-InstructionalOntology-SWEL-2004.pdf

Uno, T., Asai, T., Uchida, Y., & Arimura, H. (2003). An efficient algorithm for enumerating frequent closed item sets. In *Proceedings of Frequent Itemset Mining Implementations* (*Vol. 90*). LCM.

Urtubia, A., Perez-Correa, J. R., Meurens, M., & Agosin, E. (2004). Monitoring large scale wine fermentations with infrared spectroscopy. *Talanta, 64*(3), 778–784. doi:10.1016/j.talanta.2004.04.005

Vaisman, A. A., & Zimànyi, E. (2009). A multidimensional model representing continuous fields in spatial data warehouses. In *Proceedings of the 17th ACM SIGSPATIAL international Symposium on Advances in Geographic Information Systems (ACM SIG 09)*, Seattle, WA (pp. 168-177). New York: ACM Press.

Van Den Berg, L., & Wintjes, A. (2000). New rural lifestyle estates in The Netherlands. *Landscape and Urban Planning, 48*(3-4), 169–176. doi:10.1016/S0169-2046(00)00040-2

van der Ploeg, J. D. (1994). Styles of Farming: an Introductory Note on Concepts and Methodology. In van der Ploeg, J. D., & Long, A. (Eds.), *Born from within. Practice and perspectives of endogenous rural development* (pp. 7–30). Assen, The Netherlands: Van Gorcum.

van Ittersum, M. K., & Rabbinge, R. (1997). Concepts in production ecology for analysis and quantification of agricultural input-output combinations. *Field Crops Research, 52*, 197–208. doi:10.1016/S0378-4290(97)00037-3

Vangenot, C. (2001). *Multi-représentation dans les bases de données géographiques.* Unpublished Thèse de docteur es sciences, Université de Bourgogne, France.

Vangenot, C. (2004). Les ontologies spatiales. *European Journal of Geomatics and Spatial Analysis, 14*(2).

Vapnik, V. (1995). *The nature of statistical learning theory.* New York: Springer-Verlag.

Vapnik, V. N. (1998). *Statistical learning theory.* New York: John Wiley & Sons.

Vautier, C., Roggero, P., Adreit, F., & Sibertin-Blanc, C. (2009). Evaluation by simulation of the social acceptability of agricultural policies for water quality. In *Proceedings of the International ICSC Symposium on Information Technologies on Environmental Engineering*, Thessaloniki, Greece (pp. 478-490).

Vega-Gorgojo, G., Bote-Lorenzo, M. L., Gómez-Sánchez, E., Asensio-Pérez, J. I., Dimitriadis, Y., & Jorrín-Abellán, I. M. (2008). Ontoolcole: Supporting educators in the semantic search of CSCL tools. *Journal of Universal Computer Science, 14*(1), 27–58. Retrieved from http://www.jucs.org/jucs_14_1/ontoolcole_supporting_educators_in/jucs_14_1_0027_0058_vega.pdf.

Veldkamp, A., & Fresco, L. O. (1996). CLUE: A conceptual model to study the conversion of land use and its effect. *Ecological Modelling, 85*, 253–270. doi:10.1016/0304-3800(94)00151-0

W3C. (2004). *OWL Web ontology language: Overview, W3C recommendation 2004-2-10.* Retrieved from http://www.w3.org/TR/owl-features/

Weibel, R., & Dutton, G. (2001). Generalizing Spatial Data and Dealing with Multiple Representations. In Longley, P., Goodchild, M., Maguire, D., & Rhind, D. (Eds.), *Geographic Information Systems and Science* (pp. 125–15). New York: John Wiley & Sons.

Weinberg, G. M. (1975). *An introduction to general systems thinking.* New York, NY: John Wiley & Sons.

Welch, S. M., Jones, J. W., Brennan, M. W., Reeder, G., & Jacobson, B. M. (2002). PCYield: Model-based decision support for soybean production. *Agricultural Systems, 74*, 79–98. doi:10.1016/S0308-521X(02)00022-7

Welch, S. M., Roe, J. L., Das, S., Dong, Z., He, R., & Kirkham, M. B. (2005). Merging genomic control networks and soil-plant-atmosphere-continuum models. *Agricultural Systems, 86*, 243–274. doi:10.1016/j.agsy.2004.07.019

Whisler, F. D., Acock, B., Baker, D. N., Fye, R. E., Hodges, H. F., & Lambert, J. R. (1986). Crop simulation models in agronomic systems. *Advances in Agronomy, 40*, 141–208. doi:10.1016/S0065-2113(08)60282-5

White, J. W., & Hoogenboom, G. (1996). Simulating effects of genes for physiological traits in a process-oriented crop model. *Agronomy Journal, 88*, 416–422.

White, T. A., Barker, D. J., & Moore, K. J. (2004). Vegetation diversity, growth, quality and decomposition in managed grasslands. *Agriculture Ecosystems & Environment, 101*, 73–84. doi:10.1016/S0167-8809(03)00169-5

Williams, J. R., Jones, C. A., Kiniry, J. R., & Spanel, D. A. (1989). The EPIC crop growth model. *ASAE, 32*, 497-511.

Wooldridge, M. J. (2002). *An introduction to multiagent systems.* Chichester, UK: John Wiley & Sons.

World Economic Forum. Global Leaders for Tomorrow Environment Task Force, Yale Center for Environmental Law and Policy, Yale University, Center for International Earth Science Information Network, & Columbia University. (2001). *2001 Environmental Sustainability Index.* New Haven, CT: Yale Center.

Wu, C. L., & Chau, K. W. (2006). Mathematical model of water quality rehabilitation with rainwater utilization — a case study at Haigang. *International Journal of Environment and Pollution, 28*(3-4), 534–545. doi:10.1504/IJEP.2006.011227

Wu, X., Kumar, V., Quinlan, J. R., Ghosh, J., Yang, Q., & Motoda, H. (2008). Top 10 algorithms in data mining. *Knowledge and Information Systems, 6*(14), 1–37. doi:10.1007/s10115-007-0114-2

Yager, R. R. (2006). An extension of the naive Bayesian classifier. *Information Science, 176*(5), 577–588. doi:10.1016/j.ins.2004.12.006

Yager, R. R., & Filev, D. P. (1994). Approximate clustering via the mountain method. *IEEE Transactions on Systems, Man, and Cybernetics, 24*(8), 1279–1284. doi:10.1109/21.299710

Yoo, J. S., & Shekhar, S. (2006). A joinless approach for mining spatial colocation patterns. *IEEE Transactions on Knowledge and Data Engineering*, *18*(10), 1323–1337. doi:10.1109/TKDE.2006.150

Yougworth, P. (1995). OLAP spells success for users and developers. *Data Based Advisor,* 38-49.

Yu, X., Chen, G., & Cheng, S. (1995). Dynamic learning rate optimization of the backpropagation algorithm. *IEEE Transactions on Neural Networks*, *6*(3), 669–677. doi:10.1109/72.377972

Zhang, Y., Xiong, Z., Mao, J., & Ou, L. (2006, June). The study of parallel k-means algorithm. In *Proceedings of the 6th World Congress on Intelligent Control and Automation: Vol.2,* Dalian, China (pp. 5868-5871). IEEE.

Zhang, G. P. (2000). Neural networks for classification: a survey. *IEEE Transactions on Systems, Man, and Cybernetics*, *30*(4), 451–462. doi:10.1109/5326.897072

Zhang, J., & Fujiwara, A. (2007). *Development of the DPSIR+C framework for measuring the social capacity of environmental management.* Retrieved from.http://home.hiroshima-u.ac.jp/~hicec/coe/products/DP2007/DP2007-4.pdf

Zhao, M. Y., Cheng, C. T., Chau, K. W., & Li, G. (2006). Multiple criteria data envelopment analysis for full ranking units associated to environment impact assessment. *International Journal of Environment and Pollution*, *28*(3-4), 448–464. doi:10.1504/IJEP.2006.011222

Zhu, X. G., Portis, A. R. Jr, & Long, S. P. (2004). Would transformation of C3 crop plants with foreign Rubisco increase productivity? A computational analysis extrapolating from kinetic properties to canopy photosynthesis. *Plant, Cell & Environment*, *27*, 155–165. doi:10.1046/j.1365-3040.2004.01142.x

About the Contributors

Petraq Papajorgji is a research scientist at the center for Applied Optimization (CAO) at the Industrial and Systems Engineering, University of Florida, Gainesvillle, Fl, USA. His research interests are formal modeling of information systems, UML and the Model Driven Architecture (MDA). Dr. Papajorgji has coauthored and edited a number of books on software engineering techniques applied to Agriculture and Environment. Dr. Papajorgji is member of the group of researchers honored with the highest award of Albania "The Prize of the Republic" for the study "The conditions of olive trees and measures for its further improvement". Dr. Papajorgji teaches several international courses on modeling agricultural and environmental systems using UML and MDA. He is member of several international organizations and member of organizing committees of international conferences. Dr. Papajorgji teaches two graduate courses at the University of Florida.

François Pinet received his M.Sc in Computer Science in 1997 (ENS and University of Lyon). He received his PhD in Computer Science in 2002 (INSA Lyon, France). He is currently a first class researcher at the French Institute for Agricultural and Environmental Engineering. His field of research is in environmental information systems and geomatics. He belongs to several scientific committees of different conferences and journals in these fields. Dr. Pinet has coauthored over 50 scientific papers and is involved in numerous national and international IT projects. He has been a co-organizer for several workshops on environmental systems. In France, he also teaches several graduate courses (M.Sc and PhD degrees) on IT techniques applied to environmental sciences.

* * *

Eva Abal's (Associate Professor) scientific expertise and research interests include coordination of multidisciplinary projects, ecophysiology of marine communities, with emphasis on biological indicators, effective science communication, monitoring and evaluation, and strategic research planning and facilitating linkages between scientists and managers/stakeholders. Eva oversees the management of scientific and ecosystem health monitoring activities of the South East Queensland Healthy Waterways Partnership in Australia, including strategic research plan development and dissemination of information to stakeholders. In addition to her role in the SEQHWP, Eva also works as part-time Chief Scientific Officer (CSO) of the Great Barrier Reef Foundation (GBRF) and part-time chief scientist of the International Water Centre based in Brisbane, Australia. Eva's passion is in the synthesis and effective communication of scientific information, making science relevant and useful to managers.

Françoise Adreit is an associate professor in Computer Science at Université Toulouse 2 – le Mirail and a researcher at IRIT (Institut de Recherche en Informatique de Toulouse) in the Cooperative Multi-Agent Systems (SMAC) team; her field of research is the design of complex systems.

Abdulmonem Alabri is a PhD student at the eResearch lab at The University of Queensland. Abdulmonem is a key researcher on the Health-e-Waterways project which is developing cyber-infrastructure to address data management and decision-making challenges concerning Queensland's water quality. His PhD topic is "Enhancing the Quality of Citizen Science Data" and involves employing new techniques to improve the quality and reliability of data captured by citizen scientists. Abdulmonem has also been a researcher on the BioMANTA (Modeling and Analysis of Biological Network Activity) project and Australian Research Enabling Environment (ARCHER).

Giovanna Aronne is associate professor of environmental and applied botany Department of Fruitculture, Botany and Plant Pathology University of Naples Federico II.

Mirellie Batton-Hubert is an Associate Professor in Environmental Information Systems and administrator of the IDEE (Information, Decision and Environmental Evaluation Department), Ecole Nationale Supérieure des Mines de Saint-Etienne. Ph.Dr.es Sciences of the University Jean Monnet thesis (2006). She is specialized in determinist and cognitive modelling, data mining, GIS, applied mathematics, multicriteria decision analysis and optimization.

Peter Becker has a degree in computer science from the Philipps University in Marburg, Germany. He started his academic career as a researcher in Formal Concept Analysis and Conceptual Knowledge Processing for the Distributed Systems Technology Centre in Australia. After working as a consultant for various commercial enterprises, he began his current position at the eResearch Lab at the University of Queensland, Brisbane, Australia. His roles include project management and technical lead for the health-e-waterways project.

Yvan Bédard is professor of GIS and Spatial Databases at Laval University, Quebec City, Canada. He holds a Canada NSERC Industrial Research Chair in Geospatial Database. He is an active member of the Centre for Research in Geomatics where he acted as Director for 7 years, and of Canada's GEOIDE network of centers of excellence. Dr Bédard has a multi-million dollars record in both fundamental and applied research. He has contributed to over 100 full-refereed papers and 300 non-refereed papers and conferences. His research interest focuses on geospatial databases modeling, Spatial OLAP and data quality. He co-founded Intelli3, a private company merging GIS and Business Intelligence solutions.

SandroBimonte is researcher at Cemagref (TSCF). Born in 1978, He speaks fluently French, English, and Italian. He obtained his PhD at INSA Lyon, France (2004-2007). From 2007-2008, he carried out researches at IMAG, France. He is managing editor of Journal of Decision Systems, editorial board member of International Journal of Decision Support System Technology and International Journal of Data Mining, Modelling and Management. He is member of the Commission on GeoVisualization of the International Cartographic Association. His research activities concern spatial data warehouses and spatial OLAP, visual languages, geographic information systems, spatiotemporal databases and geovisualization.

Marie Bonnevialle is the project manager at the Natural Parks of Massif central organization, since August 2009. Working on "identification of Massif central ecological networks" Project.

Kenneth J. Boote is a Professor in the Agronomy Department at the University of Florida. He earned his Ph.D. degree from Purdue University. His research focuses on measuring and modeling crop growth and yield in response to climate, management, soils, and genetic factors. He is a co-developer of crop simulation models for grain legumes and perennial forages. He teaches crop physiology and crop simulation courses. Dr. Boote has authored or co-authored 165 refereed journal articles, 62 non-refereed articles, 36 book chapters, and 320 abstracts. He has served as Associate Editor for Agronomy Journal and Crop Science, as well as Division Chair for A-1, C-2, and Software Scene in the American Society of Agronomy and the Crop Science Society of America. He is Fellow of those two Societies and in 2008 was awarded the Crop Science Research Award and the Carl Sprengel Agronomic Research Award. He is active in the Biological Systems Simulation Group. Because of his expertise on climate impacts and in crop growth modeling, Dr. Boote is in high demand as an invited speaker at many national and international conferences, as well as international training courses. He was a major contributor to the recent Synthesis Assessment Product 4.3 report by USDA on climate change impacts on crop production in USA.

Jean-Pierre Chanet received his Electronic Engineer degree in 1990 (Polytech Clermont-Fd). He received his M. Sc in Robotic in 1992 and his PhD in Computer Science in 2007 from the university of Clermont-Fd. Since 1991 he works as researcher at Cemagref, the French Institute for Agricultural and Environmental Engineering. He is the head of a team working on Information System for agriculture and environment. His field of research is in wireless sensor networks and data integration in information systems. He belongs to several scientific committees of different conferences and journals in these fields. He is the leader of several national and international IT projects in environmental sciences.

Altannar Chinchuluun received a B.Sc. in Mathematics in 2002 from the National University of Mongolia and a B.Sc. in Business Administration in 2002 from the Mongolian University of Science and Technology. He received his M.Sc and a PhD in Operations Research at the University of Florida in December. Altannar has received a number of prestigious awards such as Outstanding International Student Award, College of Engineering, University of Florida, 2007, Graduate Student Award for Excellence in Research, Department of Industrial and Systems Engineering, University of Florida, 2006, Outstanding International Student Award, College of Engineering, University of Florida, 2006, INFORMS 2006 Doctoral Colloquium participant, Department of Industrial and Systems Engineering, University if Florida, 2006, etc. Altannar has published a number of papers in the most well-known International Journals and edited several books. He is Associate Editor, Journal of Global Optimization, Springer, Guest Editor, Special issues on "Optimization and Optimal Control", Optimization, Taylor & Francis, 2009. Currently he is a post doctoral associate at the Imperial College London, UK.

Florent Chuffart has been post-doctorate researcher in Cemagref from 2008 to 2009. He received a PhD in sofware engineering in 2007 from the University of Caen (France). His main contributions to SimExplorer development are generic integration for numerical model exploration, use case definition and tutorial design.

Maria Danese is Postdoctoral Researcher at the Archaeological and monumental heritage institute, National Research Council, Italy. She is interested in Spatial Statistics applied to planning, landscape archaeology and damage scenarios, publishing several scientific papers in this research field. She has also conducted research at National Centre for Geocomputation (NCG) of National University of Ireland, Maynooth, directed by Professor Stewart Fotheringham.

Veronica De Micco is researcher in Environmental and Applied Botany at the University of Naples Federico II where she graduated in Agricultural Sciences and Technologies in 2000 and was appointed Research Doctor (PhD in Woody Cultures) in 2004. Since 1998, she had significant scientific experiences at research centers, including Washington State University, University of Leiden, Swiss Federal Research Institute, ESTEC - European Space Agency, University of Rome "La Sapienza" and CNRS-Grenoble, where she met scientists she is still collaborating with. Her research experience mainly regards species auto-ecology and morpho-functional adaptive strategies of plants in variable environmental conditions. Major interests are in the study of the vascular system in higher plants, focusing on water transport, cell wall formation, and the development of new methodologies for anatomical analysis. Researches are conducted with field work and laboratory analyses including microscopy and digital image analysis. Author of more than 50 refereed journal articles, book chapters and 2 books. Winner of 3 international and 2 national awards. Member of editorial advisory board and referee for several international journals and projects. Funded by ESA (FOTON-M2 Flight campaign), EU and French Research Council. Board member of the Working Group of "Plant Biorythms and Phenology" of the Italian Society of Botany.

Guillaume Deffuant is research director in Cemagref and responsible for the "Laboratoire d'Ingénierie des Systèmes Complexes" (LISC). He received a PhD in cognitive sciences from University of Paris 6 (France). His research interests are related to modeling complex systems. Dr. Deffuant has coauthored over 60 scientific papers. He coordinates the European project PATRES (Pattern resilience) which investigates systematic ways to determine robust regularities in complex dynamics, and develops new methods for managing the viability and resilience of such dynamics.

Guillaume Deffuant is research director in Cemagref and responsible for the "Laboratoire d'Ingénierie des Systèmes Complexes" (LISC). He received a PhD in cognitive sciences from University of Paris 6 (France). His research interests are related to modeling complex systems. Dr. Deffuant has coauthored over 60 scientific papers. He coordinates the European project PATRES (Pattern resilience) which investigates systematic ways to determine robust regularities in complex dynamics, and develops new methods for managing the viability and resilience of such dynamics.

Nicolas Dumoulin is a computer scientist in Cemagref. He received a Master's degree in 2003 from University of Rennes (France). His main works are related to the development of tools for modelling ecosystems, especially using individual-based models. He is one of the main designers and developers of SimExplorer.

Michel Duru is senior researcher at INRA Toulouse and head of the laboratory AGIR (Agrosystems and territorial development). His main research area concerns grassland management at field and farm levels through building sward indicators and decision support systems based on biophysical and decisional models.

Bernard Espinasse received an engineer diploma from the Ecole Nationale Supérieure d'Arts et Métiers (1977) and a Doctorate in Information Systems from Aix-Marseille University (1981). He is currently Full Professor in Computer Sciences at Aix-Marseille University and project leader at LSIS UMR CNRS research laboratory. He has been Associate Professor at the Laval University of Québec, Canada (1983-1987). His main research topic is Software Agents for Information Retrieval and Decision Support Systems.

Thierry Faure is a researcher in Cemagref LISC. He got a PhD in physics in 2000, from University of Clermont-Ferrand (France). His researches focus on computer methods for efficiently exploring complex models. He is one of the main designers and developers of SimExplorer.

Frédéric Flouvat is an associate professor at the University of New Caledonia (Noumea, New Caledonia), where he is teaching algorithmic and databases. He is also a member of the matter and environment multidisciplinary research center. This laboratory brings together geologists, physicists and computer scientists to address both fundamental and applied questions relating to the concepts of risk and sustainable development. His research interest is spatio-temporal data mining and application to environmental sciences.

Dominique Gay is currently a post-doctoral researcher at Orange Labs, France Telecom Research and Development (Lannion, France) in 2010. He received a PhD degree in Computer Science from the "Université de la Nouvelle-Calédonie" at Nouméa (New-Caledonia) and the "Institut National des Sciences Appliquées" at Lyon (France) in 2009. His research interest is about data mining with a focus on pattern mining and its use in supervised classification.

Mario Guarracino is researcher at High Performance Computing and Networking Institute of the Italian National Research Council. He received a PhD in Mathematics defending a thesis on projection techniques for parallel sparse linear algebra and an Ms in Applied Mathematics. His postdoctoral training from National Research Council focused on low cost high performance architectures for scientific computing. He has been collaborating with Center for Applied Optimization at University of Florida since 2005. He has taught various undergraduate courses in both computer science and mathematics. His research interests include machine learning methods for computational biology and planning and development of high performance computational components for parallel and distributed problem solving environments. He is author of over several publications in the field of high performance scientific computing, computational biology and machine learning.

Dias Haralambopoulos is Professor, Deputy Department Chairman, and Director of Energy Management Lab in the Dept of Environment at the University of Aegean. He graduated from the Dept of Mechanical Engineering at National Technical University of Athens, Greece, in 1978. He received an MSc in Industrial Management, and a PhD in Mechanical Engineering, from the University of Birmingham, UK, in 1980 and 1983 respectively. He is responsible TRS member for the courses: Energy Conversion and the Environment, Renewable Energy Sources, Energy Analysis and Planning. His current research interests include Energy Conversion and Environmental Impacts, Energy Planning, Energy Policy and Modelling, Renewable Energy Sources, Energy Utilisation in Industry and Buildings.

Gerrit Hoogenboom is a Professor in Crop Modeling and Agrometeorology and Coordinator for Research, Extension, and Instruction of the Department of Biological and Agricultural Engineering at the University of Georgia on the Griffin Campus in Griffin, Georgia, USA. Dr. Hoogenboom was born in the Netherlands, where his family owned 1 ha of greenhouses for vegetable production. After completing his M.Sc. at Wageningen University with a dual major in Horticulture and Theoretical Production Ecology, he moved to the USA, where he received his Ph.D. in Agronomy and Soils from Auburn University, Alabama. He spent four years at the University of Florida as a Postdoctoral Research Associate and joined the University of Georgia in 1989 as an Assistant Professor. Since the start of his career, Dr. Hoogenboom has been interested in combining crop physiology and mathematics. His current research emphasis is on the development and application of crop simulation models and decision support systems for yield prediction and resource management. This ranges from gene-based models to the application of crop models for climate change, climate variability, biofuels, and food security. He coordinates the development of the Decision Support System for Agrotechnology Transfer (DSSAT), a software system that includes over 25 different crop models, data utilities, and application programs. He has conducted training workshops on crop modeling in many countries. He has published > 160 scientific papers in refereed journals as well as numerous book chapters and proceedings. He is an Editor for Agricultural Systems, Journal of Agricultural Science, Climate Research, and the Brazilian Journal of Agrometeorology.

Jane Hunter is Professor of eResearch at the School of Information Technology and Electrical engineering at the University of Queensland. She leads the eResearch Lab, specializing in the development and application of innovative semantic web technologies to the analysis and management of research data collections. She is currently a member of the Academy of Sciences Committee for Data in Science and on the Editorial Board of IEEE Multimedia, the Elsevier Journal of Web Semantics and the International Journal of Digital Curation. She has published over 80 peer-reviewed papers on the semantic web, digital libraries, data management and preservation. She is currently a CI on projects associated with water information management, marine sciences, protein crystallography, cultural heritage preservation and Australian Literature.

Eric Jallas is the founder, President & CEO of the ITK, Montpelier, France. ITK is a company specialized in developing decision support systems with focus in agriculture and environment. He received his Ph.D. degree from Mississippi State University, USA. As student at Mississippi State University Dr. Jallas was honored with the Outstanding International Student Award, College of Engineering, MSU, 1998. His research focuses on crop modeling and decision support System for agriculture. Dr. Jallas is the main developer of the COTONSimbad DSS, the first DSS integrating a mechanistical crop simulation model with a plant architecture model, and a 3D visualization tool and artificial intelligence-based tools (Genetic Algorithm and Rule based expert systems). Dr. Jallas has authored or co-authored more than 70 refereed journal articles, non-refereed articles and book chapters. In 2002 and 2003, he was recipient of the prestigious "Concours National d'Aide à la Creation d'Entreprise de Technologie Innovante" award from the French Research Secretary and the Europe Union Council. In 2003 his innovative entrepreneurial efforts were honored with the "Figaro Entreprises" award and the "Concours Tremplin Entreprise"award from the French Senat and the Ecole Superieure de Sciences Economiques et Commerciales. Dr Jallas has been for 15 years Board Member of Cirad (Centre International de Recherche Aide pour le Development).

Thierry Joliveau is Professor of Geography and Geographical Information Systems at the University of Saint-Etienne, France and Director of MAGIS, the French National Research Network in Geomatics. His research interest include Geographical Information theory and applications for environmental and landscape management.

James W. Jones is a Distinguished Professor of Agricultural and Biological Engineering, Institute of Food and Agricultural Sciences, University of Florida. Dr. Jones research focuses on developing and applying models of cropping systems aimed at understanding interactions between climate, crops, soil, and management. His research also focuses on application of those models for improving management of cropping systems, on use of climate forecasts to understand and manage crop production risks, and on assessing impacts of climate change on agricultural production. Dr. Jones is co-developer of the Southeastern Climate Consortium, which is developing methods for applying climate forecasts to reduce risks to agricultural and natural systems and developing decision support tools to help farmers and natural resource managers reduce those risks. He is author of more than 200 scientific journal articles. He is a Fellow of the American Society of Agricultural and Biological Engineers, Fellow of the American Society of Agronomy, Fellow of the Soil Science Society of America, and serves on several international science advisory boards. Dr. Jones also teaches a graduate course on techniques for developing and applying biological and agricultural models that respond to climate variability, soil characteristics, and management.

Sotirios Karetsos is a PhD candidate in the Dept of Environment at the University of Aegean. He graduated from the Electrical and Computer Engineering Dept of the Polytechnic School at the Aristotle University of Thessaloniki, Greece, in 1987. He received an MSc in Environmental and Ecological Engineering from the Dept of Environment at the University of Aegean, Greece, in 2005. He is member of the Energy Management Lab research group, Dept. of Environment, University of the Aegean. His current research interests include Sustainable Energy Education, Learning design, Learning Models, Environmental Knowledge engineering, Sustainable Energy ontologies, Semantic annotation and retrieval of learning resources.

Konstantinos Kotis was born in 1973 in Samos, Greece. He graduated from the Computation Department at the University of Manchester (UMIST), UK in 1995. He has obtained his PhD in Knowledge Management and Representation from the Department of Information and Communication Systems Eng. at the University of Aegean. He has worked as post-doc researcher at Ai-Lab, Dept. of Information and Communication Systems Eng. University of the Aegean (2005-2006), where he works as a researcher since 2006. He is a member of the Hellenic Artificial Intelligence Society and reviewer in several international journals and conferences. He has authored more than 40 publications related to Semantic Web technologies, Ontology engineering, Semantic Web Services and Grid.

Hadj Mahboubi was born in 1981 in Tlemcen, Algeria. Since December 2008, Hadj Mahboubi holds a PhD in computer science from the University of Lyon 2, France. He received his master degree from INSA Lyon in 2005 and his engineering degree in computer science in 2003 from the University of Tlemcen, Algeria. He was a research assistant in the University of Lyon 2, France. Now, he is a post-doctoral researcher at Cemagref (Clermont-Ferrand). His research interests lie in the field of classical and data warehouse design and performance optimization.

Eric, Maillé is a researcher at the Environmental Science and Technology Research Institute (Cemagref Aix-en-Provence, France). He received an agronomist engineer diploma from the Agronomic High School of Toulouse and a Doctorate in Computer Sciences from the Aix-Marseille University. His current field of research is in spatial and environmental systems dynamic modelling for land management decision support systems. His main publications concern land cover change models integration for forest fire risk dynamic representation.

Guillaume Martin holds a Master degree in Plant Science (2006) from Wageningen University (Netherlands) and a Ph. D. (2009) in Agronomy from Toulouse University (France). He is currently working as a postdoctoral researcher at INRA Toulouse. His research interests concern grassland ecology and grassland management as well as the development of methods (mainly model-based) for design and analysis of integrated farming systems.

R. Martin-Clouaire holds a Masters degree in Biomedical Engineering (1982) from Saskatchewan Univ. (Canada) and a Ph. D. (1986) in Artificial Intelligence from Toulouse University (France). He joined INRA (Institut National de la Recherche Agronomique) in 1987 as a research scientist. His main research area concerns the modelling and simulation of agricultural production systems and, in particular, decision-making processes involved in production management. He is currently director of the laboratory "Unité de Biométrie et Intelligence Artificielle".

Pierre-Olivier Mazagol holds a PhD in Geography, GIS Engineer at Saint-Étienne University, he is specialized in spatial analysis applied to environmental problems (landscape ecology, wetlands characterization).

André Miralles received his Mechanical Engineer degree in 1976 (INSA Toulouse). In 1978, he was employed as researcher at the French Institute for Agricultural and Environmental Engineering where he works since. He leads during 5 years hydraulic investigations on three-dimensional stream generating into the fishway devices equipping the dams to facilitate the upstream migration of the anadromous fishes (salmon, etc.). Then, he is in charge of the laboratory doing on one hand standardized tests of the sprayers applying pesticides on the cereals or on the orchards and, on the other hand, the researches to improve the spraying techniques and the application techniques of the products. It is involved in different European projects and leads some of them. He assumes also the charge of President of the CIETAP, French committee where the manufacturers of sprayers, the industrialists of chemistry, the representatives of the government and the scientists exchange ideas and organize conferences or the exposures. In 2000, he moves to a position of computing researcher and received his PhD in Computing Science in 2006 (University of Montpellier). Currently, he works in the Joint Research Unit named Territories, Environment, Remote Sensing & Spatial Information. His main interest is focused on the design of new methodologies for software development allowing a better flexibility and a better interactivity with the users. For that, he carries out researches on the Model Driven Architecture and on business design patterns of the agricultural and environmental domains. He applies his currently research in data processing in an important project involving many research teams working on pesticides in order to improve the data capitalisation but also the diffusion and the sharing of the knowledge. He is Associate Editor of the

International Journal of Agricultural and Environmental Information Systems and Guest Editor of the Special issue on Environmental and agricultural data processing for water and territory management.

Beniamino Murgante is Assistant Professor of Urban and Regional Planning at the University of Basilicata (Southern Italy). His main research interests are urban and regional planning with a focus on the use of spatial information for developing decision support systems based on multi-criteria evaluation, fuzzy sets, rough sets and geo-statistical methods. He has published papers and books in the field of spatial analysis, modelling, geocomputation and planning. He has been member of scientific committees of several international conferences and he has been Chair/Co-chair of several international workshops.

Frédéric Paran is a PhD doctor of environmental sciences, Research engineer at the IDEE department (Information, Decision and Environmental Evaluation Department), Ecole Nationale Supérieure des Mines de Saint-Etienne. He is specialized in water resources and territorial management.

Panos M. Pardalos is a Distinguished Professor of Industrial and Systems Engineering department at the University of Florida and director of the Center for Applied Optimization (CAO). Professor Pardalos is editor in chief of five internationally well-known journals and member of the editorial board of more than 15 international journals in the field of applied mathematics. He is member of Academy of Sciences of Ukraine, Spain, Russia and Lithuania and Mongolia. Professor Pardalos is winner of The William Pierskalla best paper award for research excellence in health care management science. He has a long list of publications and books.

J.-P. Rellier joined INRA in 1976. In his early career, he worked as a statistical analyst in the area of cropping systems. In the mid-eighties, he became a member of the national group on agricultural expert system development. Since 1988, he has been a software engineer in the "Unité de Biométrie et Intelligence Artificielle" where his main areas of interest concern the methodological aspects of complex system modelling and simulation.

Pascal Roggero is a professor in Sociology and the co-director of LEREPS (Laboratoire d'Etudes et de Recherches sur l'Economie de la Production) at Université Toulouse1-Capitole; his field of research is the introduction of the complex thinking and systems in Sociology.

Isabelle Rouet is currently a post-doctoral researcher at the matter and environment multidisciplinary research center of the University of New Caledonia. She received a PhD degree in Geology from the University of New Caledonia. Her research interest is soil erosion and geomatics.

Mehrdad Salehi holds a Ph.D. of Geomatics Sciences from Laval University Canada, specializing in Geographical Information Systems (GIS) and spatial databases. He also holds a Master of Science and Bachelor of Science of Surveying and Geomatics Engineering from the University of Tehran, Iran. Mr. Salehi's research interests include spatiotemporal databases, spatial datacubes, spatial OLAP, and spatial data quality. His professional background includes extensive GIS software development, spatial database design and development, and GIS analysis experience.

Tarek Sboui holds a M.Sc. in Computer Sciences from Laval University. Currently, he is a Ph.D. candidate in Geomatics Sciences at Laval University. He is member of the NSERC Industrial research chair in geospatial databases for decision-support. His professional interests include semantic interoperability, artificial intelligence, human computer interactions, geospatial databases modeling, multidimensional modeling, geospatial data warehouse, Spatial OLAP (SOLAP), and metadata management.

Nazha Selmaoui-Folcher is associate Professor at the University of New Caledonia since 1998. She is member of the matter and environment multidisciplinary research center since 2008. She is teaching Computer Sciences and Mathematics. She received her PhD degree at the "Institut National des Sciences Appliquées" at Lyon (France) in 1992 on Image Analysis. Her research interest is actually on Spatio-Temporal Data Mining and application to environmental sciences. In 1987, she received a Laurea Degree in Agriculture from the University of Naples, Italy. During the period 1989-1994, she moved at the University of Aberdeen, UK where she completed a Ph.D. in Plant Science. In 1994, she returned to Italy where she worked as a university researcher in general Botany at the University of Naples Federico II and started up the activities of the Laboratory of Botany and Plant Reproductive Ecology. In 2002, winning a national competition, she became associate professor of Environmental and Applied Botany at the Faculty of Agriculture of the University of Naples Federico II. She has been consulting for the Italian Ministry of Environment and other government authorities. She serves on the scientific board of the Inter-university Centre of Biodiversity (Rome) and on the managing board of the National Park of Vesuvius. She is member of several scientific affiliations and she has been principal investigator and coordinator of numerous research projects. She is author of 4 monographs and over 160 scientific publications and reports.

Christophe Sibertin-Blanc is a professor in Computer Science at Université Toulouse 1 - Capitole and a researcher at IRIT-SMAC. His current field of research is the formalization, analysis and simulation of social organization theories.

Vera Tomaino received the M.S. Degree in Industrial Engineering from the University of Calabria, Italy, in 2007. She is currently pursuing her Ph.D. in the Department of Biomedical Engineering at the University Magna Græcia, Catanzaro. Since November 2008, she is working as a visiting scholar at the Center for Applied Optimization, University of Florida. Her current research interests include Data Mining in Biomedical Applications and Optimization. She is currently working on data mining applications in the field of biomedicine particularly applied to cancer research data. Vera Tomaino has published her work in well-known journals and in international conferences. She closely works with the medical industry and research labs dealing with cancer research data applies data mining techniques and Optimization.

Catharine van Ingen, from the Microsoft eResearch Lab in San Francisco, is Chief Architect in Microsoft Research's eScience Group and is a CI on the Berkeley Water Centre's (BWC) Russian River and Ameriflux projects. Catharine has a wealth of experience in hardware, including work with the Alpha machine and MIPS processor teams, and in industrial-strength software for algorithms used to manage water flows, logging data from particle accelerator detectors.

Claude Vautier has a PhD in Sociology and is an associate researcher at LEREPS; his field of research is the design of complex systems and its application to public organizations.

Jeffrey W.White is a plant physiologist (PhD, UC Berkeley, 1981) who emphasizes use of ecophysiological models and geospatial tools to understand how crops respond to environment. His early research on common bean, conducted at CIAT near Cali, Colombia, sought to understand the physiological and genetic basis of cultivar differences in yield potential, adaptation to water deficits, and responses to photoperiod and temperature. Moving to CIMMYT in 1995, he established the Center's first GIS and crop modeling laboratory. His group revised CIMMYT's mega-environment classifications for maize and wheat research, and they successfully promoted use of geospatial tools at national and regional scales such as for efforts to develop drought tolerant maize in eastern and southern Africa. In 2003, Dr. White joined USDA-ARS in Arizona, where he currently works on predicting crop response to global change, continuing his work with models and geospatial tools. His work on incorporating genetic information in models suggests new avenues for collaboration among plant sciences, ecophysiology and global change biology. He has also long advocated improved data management, most recently through the International Crop Information System (ICIS), the International Consortium for Agricultural Systems Analysis (ICASA) and the iPlant Collaboration. He has coauthored over 60 journal papers and supervised over 15 MSc and PhD theses.

Franz Wotawa received a M.Sc. in Computer Science (1994) and a PhD in 1996 both from the Vienna University of Technology. He is currently professor of software engineering at the Graz University of Technology and Dean of the Computer Science Faculty. Since the founding of the Institute for Software Technology in 2003 to 2009 FranzWotawa had been the head of the institute. His research interests include model-based and qualitative reasoning, theorem proving, mobile robots, verification and validation, and software testing. Currently, Franz Wotawa works on applyingmodel-based diagnosis to software debugging and on test-case generation and repair. Themain focus of his work has always been on providing algorithms and models for debugging and testing in order to make them applicable in practice. During his career Franz Wotawa has written more than 200 papers for journals, books, conferences, and workshops. He supervised 40 master and 13 PhD students. Franz Wotawa has been member of the a various number of program committees and organized several workshops and special issues of journals. He is a member of IEEE Computer Society, ACM, AAAI, the Austrian Computer Society (OCG), and the Austrian Society for Artificial Intelligence.

Petros Xanthopoulos received the Dipl. Eng. degree in electronics and computer engineering from the Electronics and Computer Engineering Department, Technical University of Crete, Chania, Greece, in 2005 and a MSc from Industrial and Systems Engineering, University of Florida. He is currently working toward the Ph.D. degree in the Industrial and Systems Engineering Department, University of Florida, Gainesville. He has published his work in journals of IEEE, Elsevier and Wiley and he has edited a special issue for the journal of Combinatorial Optimization entitled "Data Mining in Biomedicine". Currently he is editing a book on Computational Neuroscience that will be published from Springer. He has also co-organized (with Dr. Panos M. Pardalos) many special sessions in international meeting and two conferences related to applications of data mining in biomedicine. Petro Xanthopoulos has a patent with well-known researchers titled "*Time Frequency Transformation Analysis for Detection and Quantification of Epileptiform Activity Load in Generalized Epilepsies*" UF-731P. He is currently a Research Assistant at the Center for Applied Optimization, University of Florida.

Index

E

F

G

H

I

J

K

L

M

N

O

P

Q

R

S

T

U

V

W

Y

www.ingramcontent.com/pod-product-compliance
Lightning Source LLC
LaVergne TN
LVHW062140200926
848484LV00007B/1838